# NANOSTRUCTURES: NOVEL ARCHITECTURE

# NANOSTRUCTURES: NOVEL ARCHITECTURE

**MIRCEA V. DIUDEA**
**EDITOR**

**Nova Science Publishers, Inc.**
*New York*

**LIBRARY OF CONGRESS CATALOGING-IN-PUBLICATION DATA**
Nanostructures : novel architecture / Mircea V. Diudea (editor).
p. cm.
Includes bibliographical references and index.
ISBN 1-59454-499-9
1. Nanostructures. I. Diudea, Mircea V.
QC176.8.N35N373 2005
620'.5--dc22 2005011969

*Published by Nova Science Publishers, Inc. ✣ New York*

# CONTENTS

# PREFACE

Novel carbon allotropes, such as spherical fullerenes and nanotubes, have been added, in the last three decades, to the traditionally recognized diamond and graphite. Although fullerene C60 has been speculated about for a long time. A fullerene is, according to a classical definition, an all-carbon molecule consisting entirely of pentagons (exactly 12) and hexagons (n/2-10). Non-classical fullerene extensions to include rings of other sizes have been considered. Fullerenes are commonly synthesized by arc-discharge or laser ablation methods. Spherical fullerenes became nowadays parts of real chemistry: they can be functionalized or inserted in supramolecular assemblies.

The authors of Chapter 1 discuss about nanoscience, a new science of materials with revolutionary electrical, optical, magnetic, etc., properties and its coming of age is seeing in connection with nanotechnology and its impact on everyday life. It is monitored according to the database of the Institute of Scientific Information, the Science Citation Index, by a scientometric evaluation of the growth and distribution of journal papers during the period 1986-2002.

The discovery of new "hybrid" ($0D$+$1D$) nanostructures, *viz.* fullerenes incapsulated inside carbon nanotubes (the so-called fullerene peapods $C_{60}$@NT) has triggered in the last time intensive experimental and theoretical studies aimed at synthesis, investigation of properties, and simulation of the electronic structure of these unique nanosystems. The basic research findings along these lines have been analyzed in Chapter 2.

In Chapter 3, the authors discuss about the encapsulation of fullerene cages inside single-walled nanotubes SWNT can be performed very efficiently in both vapor phase and in solution. In the inner space of SWNT, zero dimensional molecules, like $C_{60}$, are ordered in one-dimensional arrays. Such structures are called nanopeapods. Annealing peapods over 800°C, results in double-walled carbon nanotubes DWNT. The growing of the inner tube is a multi-step process, within which the inner tube increases its diameter up to the van der Waals distance to the external parent tube. The coalescence of $C_{60}$ molecules is described by a sequence of Stone-Wales SW bond rotations. The process passes through the peanut-shaped intermediates and terminates with a capped (polyhex) nanotube, *i.e.*, tubulene. The strain relief from the spherical fullerene to the tubulene provides the driving force of the transformation. Tubulenes can further be transformed, by SW, in other tubular forms. The finally resulted DWNT could be the precursor of a distinct-walled torus DWT, with the apex zones decorated by non-hexagonal faces. Semiempirical calculations suggest that such tori are far more stable than the classical polyhex ones, at least at small and moderate size. Topology of fullerenes, tubulenes and related objects can be described by using the spiral code and/or highly discriminating topological indices operating on layer matrices, either on the parent associated graphs or their transforms by medial and dual operations on maps.

Tubulenes are capped nanotubes built up from two caps and a distancing nanotube/neck of diverse decoration. Ways of "in silico" construction and isomerization of several finite nanostructures, temptingly classified by their neck and/or caps, are presented in Chapter 4. Periodicity is considered both as topology (constitutive typing enumeration) and $\pi$-electronic structure, in the Hückel model. Spectral regularities appearing in families of tubulenes are summarized in some rules for their $\pi$-electronic structure. The older and newer rules are (re)written in terms of the classical $C_{N(k6^k(56)^k-A[2k,n])}$ cylinders.

The theoretical basis for remodeling a graphene sheet into the third dimension is briefly presented. Nanocones have been first observed in the 1960s, and discussed since the 1970s. They have been called by various names: graphite (graphene, graphitic) cones or molecular cones, but most authors now prefer *nanocones*. Chapter 5 discusses predominantly single-wall nanocones (SWNCs) made from carbon atoms. However, SWNCs made from BN, Si, or from polymers are also known. Nanotubes may be considered to be a particular case of nanocones with a cap of $m$ = six 60° sectors of the graphene lattice. The other five possible graphitic SWNCs with positive curvature are discussed in detail. They can be obtained from $m$ = one through five 60°-sectors of the graphene lattice by linking the dangling bonds in various ways. A few additional cone-helix structures are also possible. Multi-wall graphitic nanocones may have much sharper tips than the SWNC with $m = 1$. The related *nanohorns* with less regular shape are also discussed. Various properties and possible applications of nanocones are summarily reviewed, and a comprehensive bibliography is attached.

Construction of polyhex (6,3) tori from square tiled (4,4) tori is given in detail in Chapter 6. Topological description by $\alpha$-spiral code was adapted for polyhex toroidal nets. Strain energy calculation, in terms of the POAV1 theory, revealed a dependence by $D^{-2}$ (or $h_n^{-2}$) at constant tube cross-section, with $D$ and $h_n$ being the large diameter and the number of hexes on the $n$-dimension (counting the cross-sections around the large hollow) of the torus, respectively. Varying the tube section showed dependence by $h_c^{-1}$, where $h_c$ is the number of hexes on the $c$-dimension, counting the number of atoms around the tube. The $\pi$-electronic structure is evaluated at the simple Hückel level of theory. Criteria for metallic character in (6,3) tori are given in terms of the $[c,n]$ dimensions of the net. These criteria superimpose over those for perfect Clar structure and support the newly proposed perfect Corannulenic structure, in a valence bond representation of total resonant sextet molecules. Resonance energy calculations support the idea of further increasing stability/aromaticity by highly organized corannulenic polyhex systems.

Counting of all distances in a molecular graph can be expressed as a single number, known as the Wiener index. This number equals the first derivative of the Hosoya polynomial (in $x = 1$), defined as a distance-based increasing power sequence. Analytical formulas and/or recursions for calculating Wiener index and Hosoya polynomial, in several classes of tubular and toroidal nets, are given in Chapter 7.

An algorithm for calculating the number of Kekulé structures of a polyhex torus (in polynomial time), based on a *conjecture*, is presented. The algorithm, called B, is given in the very short in Chapter 8, therefore the interesting reader find more information (the algorithm A) in ref. 2.

Fully-resonant-azulenoid networks - those that in chemical terms can be described as a set of essentially disconnected azulene nuclei - are defined a little more carefully than in earlier work, and several examples are illustrated with a view to the generation of tubes and tori. A method of generating certain members of this set by conversion of toroidal benzenoids is described, and its relevance for a partial characterization pointed out. The speculative possibility of novel properties is briefly discussed in Chapter 9.

The topological coordinate method for 3-connected planar graphs is extended. The two-dimensional planar representation of these graphs was obtained only relying on the topological information contained in the connectivity structure. The results were presented in Chapter 10, by drawing Schlegel diagrams for fullerenes.

Tiling of a tubular nanostructure can be modified, from the tetragonal (4,4) to hexagonal (6,3) and other coverings of interest, by following various routes: (i) cutting procedures; (ii) Stone-Wales SW edge rotations and (iii) operations on maps. Cutting procedures, as implemented in the Torus software program, provide tubular/toroidal objects of various size, embedding and covering. Stone-Wales edge rotations, by CageVersatile program, enable covering changes in both spherical and tubular objects. Several new coverings and routes of possible isomerization reactions were identified by this procedure. The most versatile and deeply involved in the electronic structure of their transforms are some geometric operations on maps, including the newly proposed Capra operation. Examples are given for all the three classes of transformations.

A corannulenic Cor covering is proposed, by analogy to the well-known perfect Clar structure, to account for some supra-organized fully-resonant $\pi$-electronic aromatic systems of fullerenes. The constructive demands for such DCor coverings are given.

Supramolecular graphitoids, as infinite periodic surfaces of negative curvature and genus higher than zero, is shown to be constructible by self-assembling of various repeat units, some of them appearing as 3D junctions of nanotubes. In this respect, new representations of the Dyck graph are proposed in Chapter 11.

Covering operations in nanostructure modeling are presented in Chapter 12, as algorithms implemented in the original software program CageVersatile 1.1. The basic operations described are: Dual, Medial, Stellation, Truncation, Leapfrog, Quadruple and Capra. Output file objects are illustrated.

Structures for low-density zeolite-like possible carbon and boron nitride allotropes can be constructed from networks of trigonal ($sp^2$) carbon atoms containing hexagons, heptagons, and/or octagons. Chapter 13 surveys the geometry, topology, and symmetry of these structures. The most symmetrical such structures are obtained by leapfrog transformations of either the Dyck tessellation of 12 octagons or the Klein tessellation of 24 heptagons. Such tessellations can be embedded into genus 3 infinite periodic minimal surfaces of negative curvature. The automorphism group, ${}^4O$, of the Dyck tessellation is a solvable group of order 96, the structure of which relates to the dual of the Dyck tessellation as the regular tripartite graph $K_{4,4,4}$. However, the automorphism group, ${}^7O$, of the Klein tessellation is a simple group of order 168, which can be generated from the prime number 7 in a way analogous to the generation of the icosahedral rotation group $I$ ($\approx$ the alternating group $A_5$) from the prime number 5.

The full genealogical tree of the group of automorphisms of the Dyck graph is derived. Three different subgroups of index 2 are identified, and their topological relevance is demonstrated. Two of these correspond to orientation-preserving subgroups for regular maps on a triple torus and a torus that form a pair of Petrie duals. Zeolite-type and nanotube-type molecular realizations of the Dyck graph are discussed in Chapter 14.

Spongy carbon nano-structures, also called schwarzites, have been recently synthesized. They consist of highly connected $sp^2$ carbon covalent networks, periodic in the three dimensions of the Euclidean space. The intimate structure of schwarzites show the topology of triply periodic minimal surfaces TPMSs. These negatively curved surfaces can be variously tessellated by using some map operations, of crucial importance being two twin septupling-open operations that introduce heptagons within a polyhedral lattice. These new operations were theorized and exemplified in Chapter 15. Several schwarzites, with the associate graph embedded in D-, P- and FRD-type surfaces are illustrated and the topology of their dual lattice is rationalized in terms of map operations filiation.

Amphiphiles made of long alkyl chains attached to a cubanoid core have been designed with the aid of both computer graphics and molecular mechanics calculations. These novel molecules are expected to self-assemble in water solution in a way analogous to that of lipid bilayers. The resulting superstructures investigated in Chapter 16, display interesting architectures which can be stabilized by further polymerization reactions. The encapsulation of electro-active molecules like, for example, fullerenes paves the way towards the formation of nanostructured functional materials that have the ability to mimic complex biological processes such as the transduction of signals through cell membranes.

The authors of Chapter 17 discuss about the hydridic-to-protonic interaction, or dihydrogen bonding, between negatively charged hydrogen atoms of M-H (M=Al,B,Ga,Ir,Mo,Mn,Os, Re,Ru,W) bonds and conventional proton donors, can influence reactivity and stereoselectivity in solution and in solid state directs crystal packing, thus finding utility in supramolecular synthesis. These bonds were used in construction of novel one-, two-, and three dimensional extended structures with various topolgies. Some of these systems can react topochemically in solid state, releasing hydrogen and exchanging the weak O-H···H-M interactions for strong O-M covalent bond to form ordered, self-assembled nanostructured materials.

Metal-oxygen clusters are the most important representatives of nanosized inorganic molecular clusters. They are also called polyoxometalate/metal oxide-based clusters. Over the past ten years, a new generation of very large polyoxometalate/metal oxide-based clusters with supramolecular properties has emerged and exerted a significant impact on our overall perception on nanosized systems. The first part of Chapter 18 is dedicated to the basic structures of the new polyoxometalate clusters. Three basic types and their subtypes are discussed and exemplified by representative clusters: (1) the giant ellipsoid-type (with 63 metal centers); (2) the giant sphere-type, which includes complete giant sphere (with 132 metal centers) and down-sized giant sphere (with 102 metal centers) and (3) the giant wheel-type, which includes common giant wheel or the Bielefeld structure (with 154 metal centers) and enlarged giant wheel (with 176 metal centers). Certain illustrious molecules, such as the $[Mo_{132}]$ giant sphere and the $[Mo_{154}]$ giant wheel, can be considered as true prototypes of the inorganic metal-oxygen clusters. The close relationship between giant spheres and giant wheels, with topological implications, is also discussed.

Metal-oxygen clusters, also called polyoxometalate/metal oxide-based clusters, are the main representatives of nanosized inorganic molecular clusters. The second part of Chapter 19 is dedicated to the special structures of the new generation of polyoxometalate/metal oxide-based clusters with supramolecular properties. The special structures are divided into mutant-class and hybrid-class, as they are related to one basic structure or they share elements specific to at least two basic structures. Five special structures, named after the corresponding cluster, are described and explained: [$Mo_{75}V_{20}$] (with 95 metal centers), [$Mo_{120}Pr_6$] (with 126 metal centers), [$Mo_{248}$] (with 248 metal centers), [$Mo_{256}Eu_8$] (with 264 metal centers) and [$Mo_{368}$] or the Müller structure (with 368 metal centers). The [$Mo_{368}$] giant hedgehog cluster, with Müller structure, is the largest synthesized and structurally characterized molecule.

In: Nanostructures: Novel Architecture
Editor: Mircea V. Diudea, pp. 1-7

ISBN: 1-59454-499-9


*Chapter 1*

# A SIMPLISTIC APPROACH OF REVEALING THE COMING OF AGE OF NANOSCIENCE

***Tibor Braun*[*a,b] and Ildikó Dióspatonyi**[b]

[a]Institute of Inorganic and Analytical Chemistry
L. Eötvös University, POB 123, 1443 Budapest, Hungary
[b]Information Science and Scientometrics Research Unit (ISSRU)
Institute for Science Organization, Hungarian Academy
of Sciences, Budapest, Hungary

## ABSTRACT

Nanoscience - a new science of materials with revolutionary electrical, optical, magnetic, etc., properties and its coming of age as seen in connection with nanotechnology and its impact on everyday life. It is monitored according to the database of the Institute of Scientific Information, the Science Citation Index, by a scientometric evaluation of the growth and distribution of journal papers during the period 1986-2002.

Nanoscience and nanotechnology deal with materials and structures having dimensions that measure up to a billionth of a meter (nanometer). Multitudes of basic properties of materials are defined by length scales in the range of 10 to 200 Ångstroms. When a solid is produced in a way that at least one of its dimensions is smaller than the length scale of some property, that property gets "confined" and will depend on the size and shape of what is called a "quantum crystal". At nanoscale one enters a world where physics and chemistry meet and novel properties of matter develop.[1],[2] Based on these properties, a universe emerges, the characteristics of which can be applied to a full range of other disciplines including technologies that can, *e.g.*, manipulate individual atoms with tiny probe elements.

Visionaries[3] are seeing, in addition to the birth of a new nanoscience of materials with revolutionary electrical, optical, magnetic, *etc.*, properties, the coming of age of technologies making molecular engines and robots which like cells could populate nanofactories or could make copies of themselves.

* E-mail address: braun@mail.iif.hu

One way tomonitor the emergence of new fields is to peruse the many papers and reviews published on the topic. A simpler and perhaps more enlightening approach to the growth of the field could be offered by the bibliometric (scientometric) quantification of a related specific term in title of journal papers during a determined period of time. In this sense, we have built a database on the frequency of usage of the prefix nano- in the title of meanstream science and technology journal papers during the period 1986–2002, based on the *Science Citation Index* database of the Institute of Scientific Information, Philadelphia, USA. We have also counted the citations to some path breaking publications on the topic during the same period in the same Index.

There were hundreds of different terms in journal article titles during 1986–2002 including the prefix nano-, ranging from "nanoamorphous" to "nanowhishker" From these about a hundred were excluded for being irrelevant to the topic (*e.g*, nanoampère, nanogram, nanoleakage, etc.). As visible on Figure 1 the total number of papers having the nano-prefix in their title during 1986–2002 amounted to many thousands; the exponential growth is clearly visible and the doubling time[4],[5] is 1.6 years, indicating a very fast growth rate.

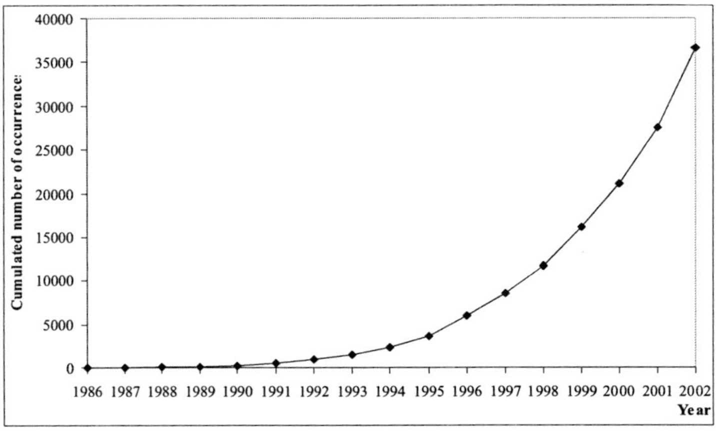


(a) Liniar

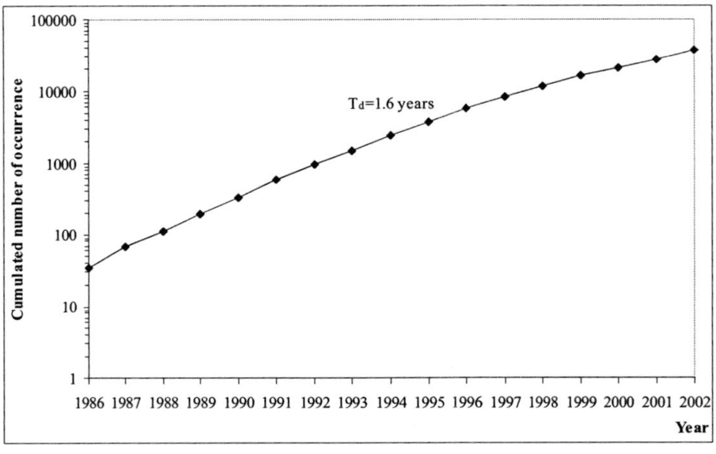


(b) Logarithmic; Doubling time ($T_a$) 1.6 years

Figure 1: The growth rate of terms with the prefix nano- in the title of journal papers, 1986-2002.

The distribution of the most frequently used nano-terms and their growth can be seen on Figures 2, a–h. The figures also allow to distinguish the main orientations by which the general nano-field progresses.

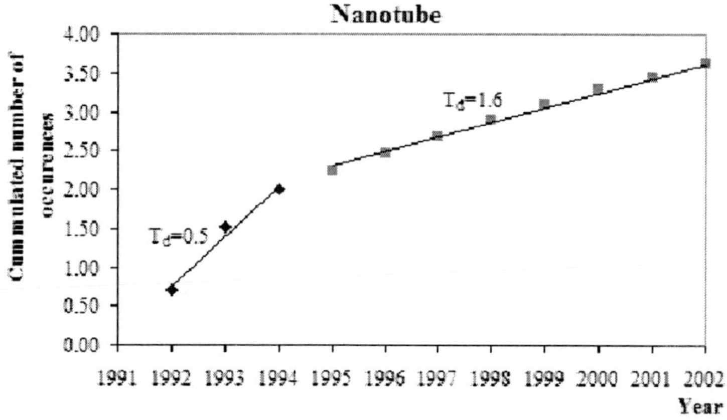


(2, a)

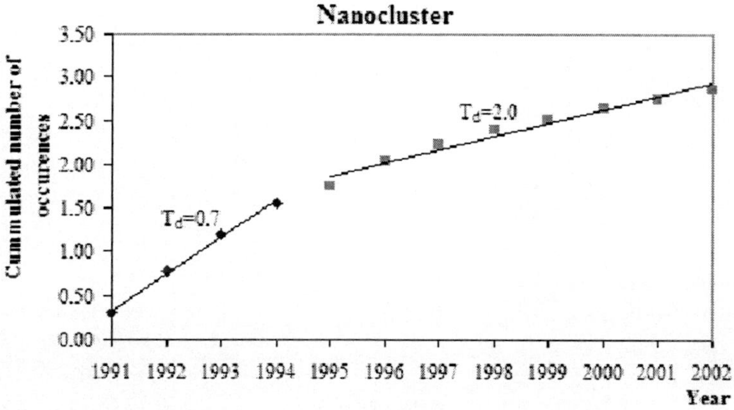


(2, b)

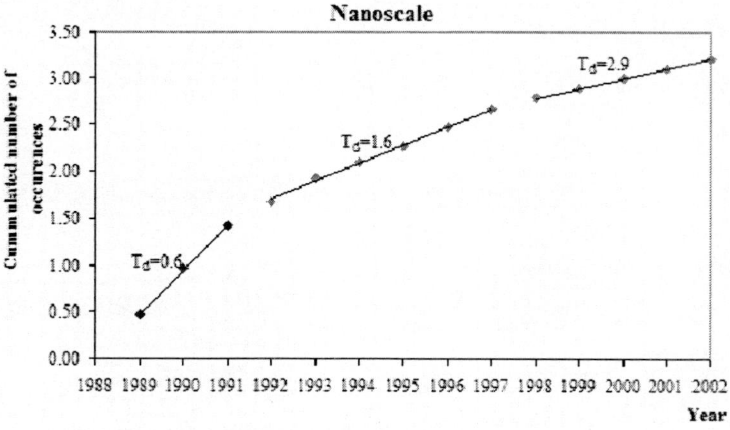


(2, c)

Figure 2 continued on next page.

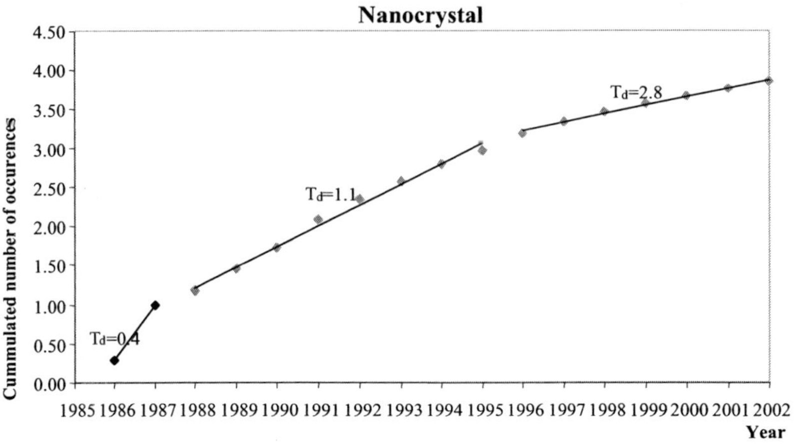


(2, d)

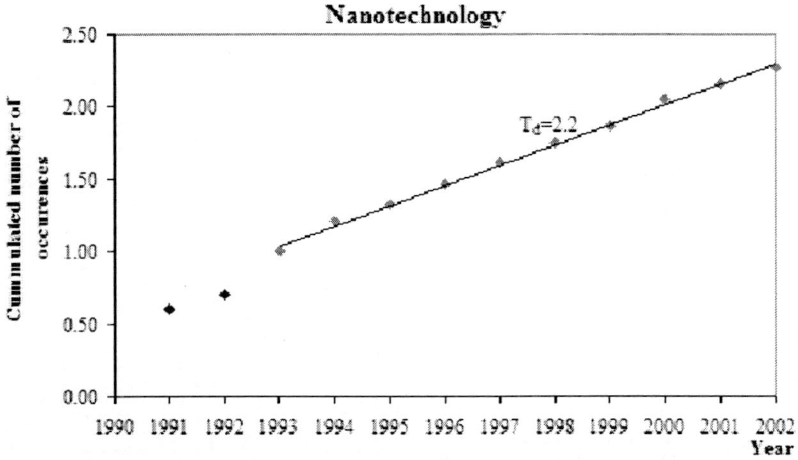


(2, e)

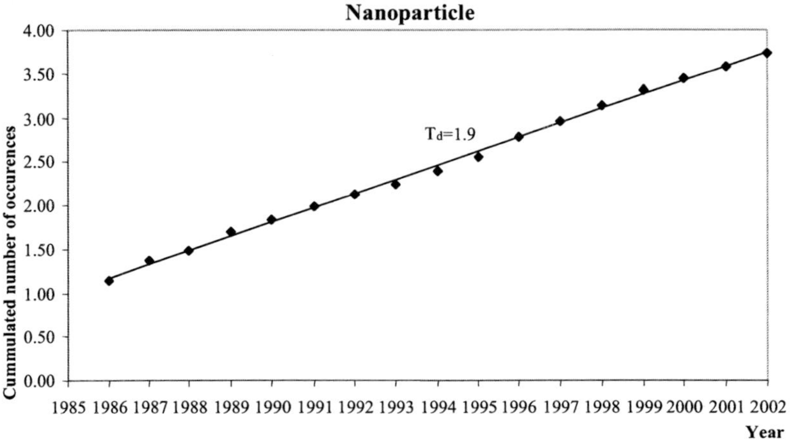


(2, f)

Figure 2 continued on next page.

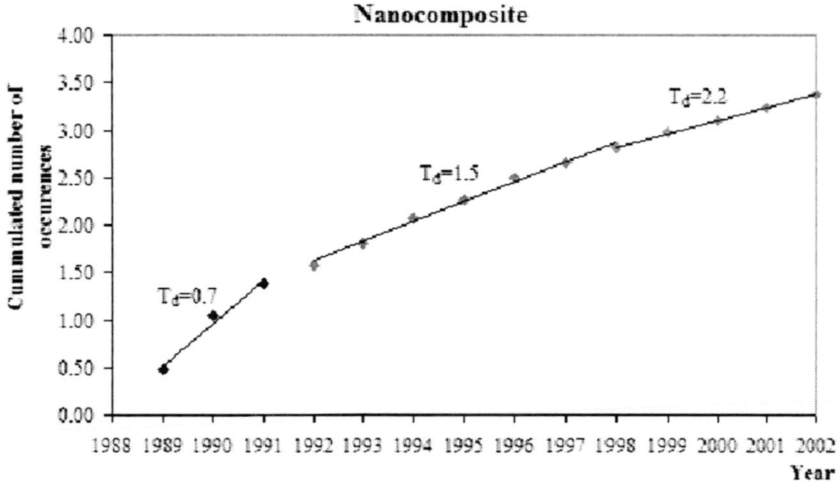


(2, g)

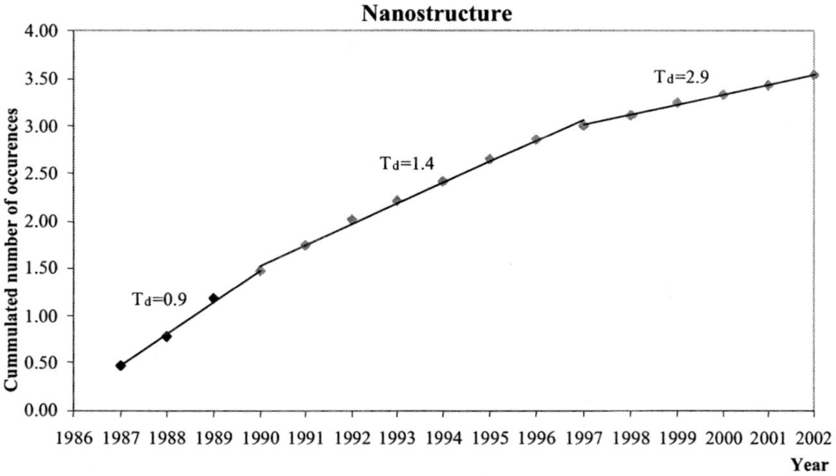


(2, h)

Figure 2: Nanoscience and nanotechnology. The growth rates of the most frequent nanoterms in the title of journal papers.

As visible, nanoscale chemistry, physics and material sciences are in the lead. Quite surprisingly, it seems that, as far as the journal literature is concerned, nanotechnology was not yet coming of age until 1995. This is also indicated by the relatively low citation rate of Reference 3 (a total of 66 citations during the 1989 – 2002 period This can be, *e.g.*, contrasted by the citation rate of Ref. 5 (3057 citations from 1991 to 2002), the path breaking paper describing the discovery of carbon nanotubes. This later topic seems to be presently one of the most active subfields of the nanoscience and nanotechnology international scene (Figure 3).

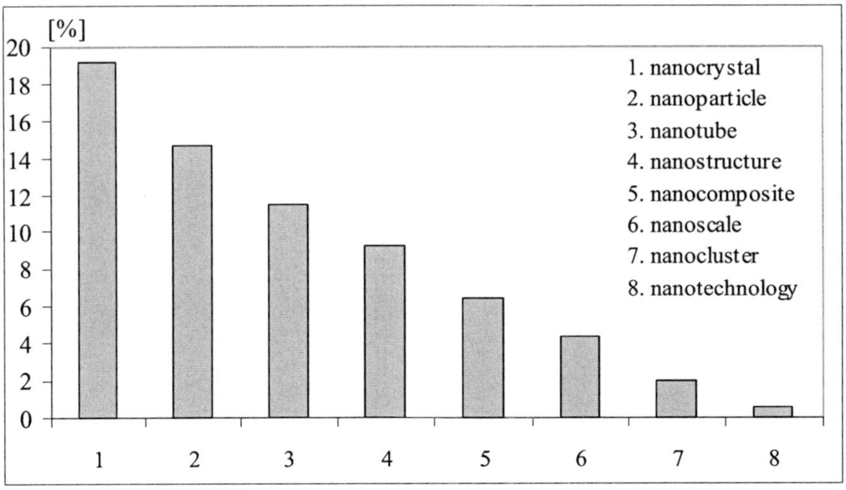


Figure 3: The relative frequency of the most frequent nanoterms in the title of journal papers

In spite of the fact that the whole field of nanoscience is as the title of this chapter suggests and Figure 1 testifies that it is in the skyrocketting phase, we should not forget Figures 2, a–h the classical observations by Derek Price[6]:

"In the real world things do not grow and grow until they reach infinity. Rather, exponential growth eventually reaches some limit, at which the process must slacken and stop before reaching absurdity. In its typical pattern, growth starts exponentially and maintains this pace to a point almost halfway between floor and ceiling, where it has an inflection. After this, the pace of growth declines so that the curve continues toward the ceiling in a manner symmetrical with the way in which it climbed from the floor to the midpoint. However, growths that have long been exponential seem not to relish the idea of being flattened. From this we are led to suggest a second basic law of the analysis of science: all the apparently exponential laws of growth must ultimately be logistic, and this implies a period of crisis extending on either side of the date of midpoint for about a generation. The outcome of the battle at the point of no return is complete reorganization or violent fluctuation. The detailed nature of this change, and any interpretation of it, must depend on what we are measuring and on how such an index is compiled. Separation seldom implies death, but rather that we have the beginning of new and exciting tactics for science, operating with quite new ground rules.[6]

Although some subfields and topics in the nanoscience field show a relative decreasing of their growth (as visible in Figures 2,a-h, we are sure that new topics will appear and their fast growth will sustain the overall growth of the whole field for quite a long time in the future.

At this moment all we can say is that the distribution of the literature on nanoscience and technology is as visible in Figure 3. A collection of the doubling times of the subfields is presented in Table 1. It has to be stressed that the data in this chapter have been derived from one basic source (database). Although interdisciplinary, it has only orientating value.

**Table 1: Growth rates (as shown by doubling time $T_d$ – in the periods given in Figure 2) of journal papers having the most frequent nanoterms in their title.**

| Nanoterm | $T_{d1}$ | $T_{d2}$ | $T_{d3}$ |
|---|---|---|---|
| nanotube | 0.5 | 1.6 | |
| nanocluster | 0.7 | 2.0 | |
| nanoscale | 0.6 | 1.6 | 2.9 |
| nanocomposite | 0.7 | 1.5 | 2.2 |
| nanostructure | 0.9 | 1.4 | 2.9 |
| nanocrystal | 0.4 | 1.1 | 2.8 |
| nanotechnology | 2.2 | | |
| nanoparticle | 1.9 | | |

## REFERENCES


[1] G. Ozin, *Adv. Mater.*, 1992, 4, 612.

[2] W. M. Tolles, *Nanotechnology*, 1996, *7*, 59.

[3] K. E. Drexler, *Engines of Creation; The Coming Era of Nanotechnology,* Anchor Press, 1989.

[4] T. Braun, E. Bujdosó, A. Schubert, *Literature of Analytical Chemistry. A Scientometric Evaluation,* CRC Press, 1987.

[5] S. Iijima, *Nature*, 1991, 354, 56.

[6] J. Derek de Solla Price, *Little Science, Big Science and Beyond,* Columbia University Press, New York, 1986.

In: Nanostructures: Novel Architecture
Editor: Mircea V. Diudea, pp. 9-24 ISBN: 1-59454-499-9


*Chapter 2*

# Fullerene Peapods and Related Nanomaterials: Synthesis, Structure, and Electronic Properties

***Viktoria V. Ivanovskaya*[a1], *Yurii N. Makurin*[b] and *Aleksander L. Ivanovskii*[a]**

[a] Institute of Solid State Chemistry
Ural Branch of the Russian Academy of Sciences,Ekaterinburg, 620219, Russia
[b] Ural State Technical University, Ekaterinburg, 620001, Russia

## Abstract

The discovery of new "hybrid" (0*D*+1*D*) nanostructures, *viz*. fullerenes incapsulated inside carbon nanotubes (the so-called fullerene peapods $C_{60}$@NT) has triggered recently intensive experimental and theoretical studies aimed at synthesis, investigation of properties, and simulation of the electronic structure of these unique nanosystems. The basic research findings along these lines have been analyzed.

## 1. Introduction

The discovery of fullerenes[1] and carbon nanotubes (NT)[2] determined large-scale investigations in the field of novel nanomaterials. Each class of these nanostructures, namely quasi-one-dimensional 1*D*-nanotubes and quasi-zero-dimensional 0*D*–fullerene-like molecules, possesses a complex of unusual properties. These nanostructures are capable of forming various associates. For example, fullerenes can be associated (self-assembled) into polymeric chains, films, molecular crystals, which may have radically different properties ranging from semiconducting, metallic-like to superconducting.[3-22]

Various associates (the so-called "ropes" or "bundles") and dense films ("nanopaper") were produced from carbon nanotubes. Therefore, fullerenes and NTs are considered as

[1] E-mail address: viktoria@ihim.uran.ru

"superatoms", which may be organized (self-organized) into condensed systems having an intrinsic structural hierarchy.

Along with exploration of "ideal" 0*D*- and 1*D*-nanostructures and their associates, of importance is the development of methods for purposeful control of their properties. For example, for single-wall or multi-wall nanotubes (SWNT, MWNT) this can be achieved by chemical "functionalization":[3-22]

- addition of atoms (or molecules) to "end" atomic rings of open NTs, "caps" of closed NTs or external NT walls (adsorption on the NT surface);
- substitution of NT atoms by atoms of a different species (NT "alloying");
- inculcation (encapsulation) of atoms (or molecules) between MWNT walls or adjoining NTs in "ropes", as well as into internal NT cavity.

The latter technique led to the discovery (1998)[23] of a new unique class of "hybrid" (0D+1D) nanomaterials, which are a combination of two types of nanostructures having different dimensions. These structures, namely nanotubes (1D) inside which an ordered one-dimensional ensemble $C_{60}$ of fullerenes (0D) is encapsulated, are frequently called peapods.

To date, a large number of carbon peapods with different fullerenes ($C_{36}$-$C_{82}$) and some metallofullerenes have been produced. Their properties were studied and growth scenarios were proposed. Electronic characteristics and chemical bonding were simulated.

A number of new peapods were predicted. After a brief description of synthesis techniques and some properties of carbon peapods in Section 2, we shall consider in greater detail the state-of-the-art modeling of these and related "hybrid" nanomaterials in Section 3.

## 2. Fullerenes and Metallofullerenes Inside Carbon Nanotubes: Synthesis and Properties

The filling of the interior space of carbon SWNTs with fullerenes was observed for the first time using high-resolution transmission electron microscopy (HRTEM) for the material obtained by pulsed laser vaporization (PLV) of a composite graphite target with metallic catalysts impurities, which was then annealed at 1100 °C and treated with acid.[23] As is known, PLV, a highly efficient process of producing SWNTs,[24] gives a certain amount of $C_{60}$ molecules. Then specimens containing SWNTs usually are purified to remove $C_{60}$.[25] As was noted earlier,[26] $C_{60}$ fullerenes ($D$ = 0.7 nm) may penetrate inner cavities of the majority of PLV-produced SWNTs thanks to their diameters ($D$ = 1.3-1.5 nm).

Findings[23] were confirmed in later experiments.[27-29] Clear-cut HRTEM images of "hybrid" structures ($C_{60}$,$C_{70}$)@NT (Figure 1) were obtained.[30] Moreover, molecule clusters and extended carbon "nanocapsules" (~ 2 nm long) were observed for some SWNT sections. Those "capsules" are likely formed due to spontaneous polymerization of fullerenes. The formation of two-layer NTs (TNT) was reported in ref. 28. The degree, to which the tubes are filled with molecules, and the tubes coalescence depend on the annealing temperature.

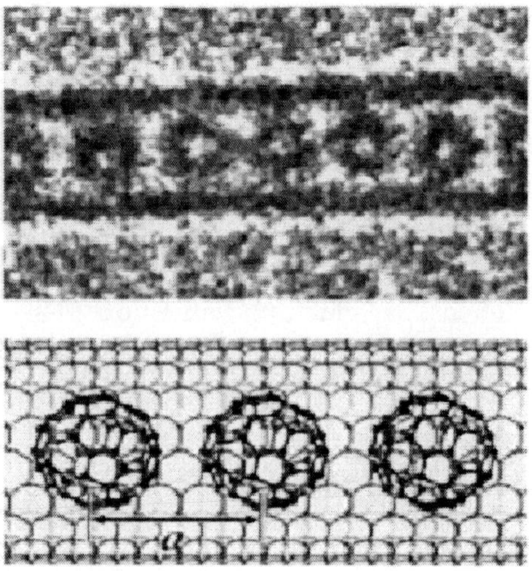


Figure 1: HRTEM image of $C_{60}$@NT peapod (*1*) and its structural model (*2*).[40]

Special experiments were performed to prove that peapods contain exactly $C_{60}$ fullerenes. According to estimates,[30] ~5.4% of total NT length in the examined material is filled with fullerenes. These structures remain stable until ~1200°C as distinct from fullerenes in gas and condensed phases, which decompose at 3000 and 1000°C respectively.[31],[32] Evidently, SWNTs of fixed diameters and chirality should be used when synthesizing peapods with reproducible properties (which is significant both for exploration of their properties and potential applications).

The incapsulation of fullerenes into SWNTs containing wall defects was observed in refs. 33, 34. Fullerenes forming regular one-dimensional motifs were found in 5-10% SWNTs.[34] Along with $C_{60}$, peapods contained certain quantities of higher ($C_N$, N > 60) and lower fullerenes (N < 60, up to $C_{36}$). Their content correlated with NT diameters: for example, the number of higher fullerenes increased with D. Fullerenes may coalesce under irradiation producing dimers, trimers, and more complicated structures up to "secondary" tubes. Conditions of the TNT growth from $C_{60}$@NT peapods were studied in ref. 35. The $C_{60}$@NT structure remained unchanged up to ~800°C. As the temperature increased, molecules coalesced to form extended nanocapsules and tubular cylinder fragments. At ~1200°C, separate $C_{60}$ disappeared almost completely, and peapods transformed to TNTs. In this case, $C_{60}$ did not leave the bulk of the peapods, a fact which presumably[35] attested to the energy stability of the hybrid structures.

Polymerization of $C_{60}$ inside NTs induced by introduction of an additional dopant (potassium) was reported in ref. 36. The doping (at atomic ratio C/K~6) caused transition of $C_{60}$@NT from the semiconducting to metallic state. The polymerization reaction yielded polymeric fullerene chains with covalent bonds. These linear polymers might be produced only in the NT cavity serving as a "nanoreactor".[36]

A high-efficiency gas-phase method for production of hybrid structures was developed in refs. 36, 37. At the first stage, the initial product was prepared,[37] which contained a considerable fraction of SWNTs with controlled diameters of 1.3-1.4 nm or larger. The product was placed into a quartz ampoule with a fullerene powder. The specimen was compacted and annealed at 650°C for 2 h. As a result, a sheet of "peapod paper" having NTs filled to 70% with fullerenes (mostly $C_{70}$) was obtained. Exploration of different modes of the gas-phase synthesis of peapods showed[38] that the most critical factors are the degree of NT purification and the annealing temperature.

Among the peapods properties, experimentally investigated are their structural[24],[28-30],[34-38] and optical[35],[37] characteristics, the basic techniques being HRTEM, Raman spectroscopy, and electron-energy-loss spectroscopy in transmission (EELS). EELS was used to examine $C_{60}$@NT films ~100 nm thick.[39] In particular, no covalent fullerene-tube interactions were found, whereas the electronic structure of SWNTs changed upon filling with fullerenes.

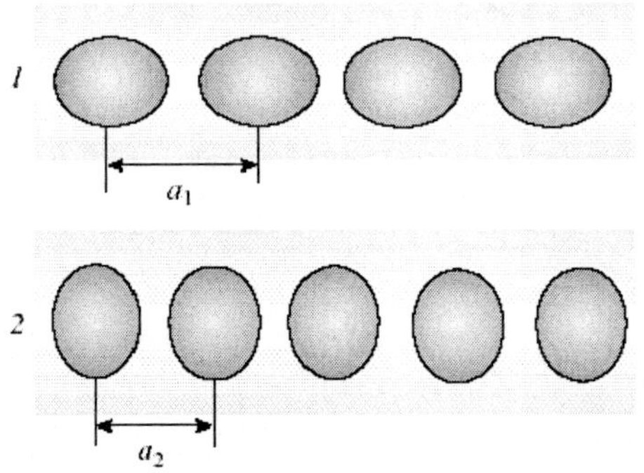


Figure 2: Models of $C_{70}$@NT peapods with different orientations of $C_{70}$ molecules: (*1*) $a_1$ = 1.10 and (*2*) $a_2$ = 1.00 nm.[40]

An effective method of studying structural features of peapods is electron diffraction, which was used to analyze a large number of hybrid structures.[40] Intermolecular distances in peapods are greater than in polymeric chains, but smaller than in molecular *3D*-crystals formed by the corresponding fullerenes. For example, in $C_{60}$@NT the inter-center distances of molecules (*a* being the "lattice period" of a quasi-one-dimensional crystal)[40] are ~0.97 nm. The distances between fullerenes and NTs are ~0.35 nm. This is the so-called Van der Waals gap of the third type. The former two types are formed between the walls of adjoining coaxial cylinders in multilayer NTs and between external walls of neighboring NTs in their compact "bundles" respectively.[3-7]

An examination of peapods with nonspherical isomers of higher fullerenes[40],[41] revealed two types of structures with $a_{1,2}$ = 1.00 and 1.10 nm in $C_{70}$@NT, which had different mutual orientations of $C_{70}$ molecules (see Figure 2). They were viewed[41] as independent phases. The structures possessed different thermal expansion coefficients. X-ray diffraction measurements were performed[42] to study the structural features of SWNTs with encapsulated $C_{70}$ (with $D \sim 1.37$ nm). In some tubes, two different domains co-existed which had different orientations of $C_{70}$ with respect to the NT axis. This long-range order was thermally destroyed by activation energy ~ 39±4 eV. For $C_{80}$@NTs, "mixed" orientations of molecules were also found.[40] It was suggested that diameters of the carbon tubes could control the mutual arrangement of nonspherical fullerenes (*i. e.* "lattice periods" of the corresponding one-dimensional crystals). The minimal tube dimensions providing encapsulation of various fullerenes were 1.37 ($C_{60}$), 1.45 ($C_{76}$, $C_{78}$), and 1.54 nm ($C_{84}$).[43]

The scanning tunneling microscopic (STM) examination[44] yielded interesting findings on structure visualization and electronic states of separate $C_{60}$@NTs. Current-voltage dependences (dI/dV) were determined for different peapod sections, which represented experimental densities of electronic states of the peapod shell, Figure 3. The electronic states of the peapod and the "pure" SWNT were drastically different. Moreover, an appreciable modulation of the electronic states distribution was observed along the $C_{60}$@NT axis. It was determined by the periodic location of encapsulated $C_{60}$ due to the interaction between fullerene MOs and NT energy bands. It was noted that *1D* resonance quantum structures could be created by controlling the period of $C_{60}$ distribution. The electrical properties of $C_{60}$@NT peapods forming bundles were measured in ref. 45. The resistance of the peapods was high and their temperature dependencies differed for two groups of specimens. This was explained[45] by disorder and defects in the peapods.

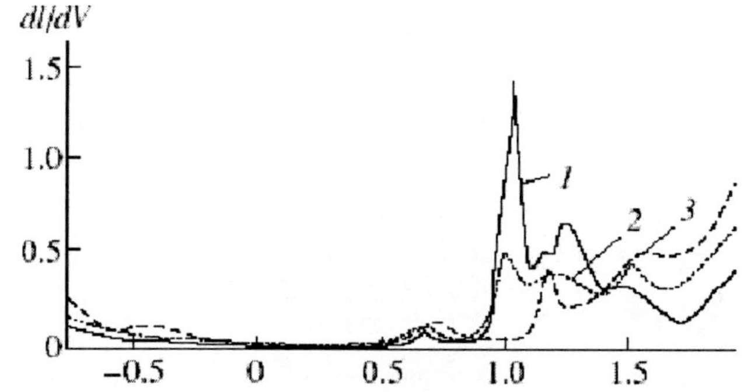


Figure 3: Current-voltage dependences for two segments of $C_{60}$@NT peapod: (*1*) near the fullerene and (*2*) between neighboring fullerenes as compared with (*3*) the pristine carbon tube; STM data.[44]

The hydrogen saturation of peapods was studied in ref. 46. It was due to $H_2$ sorption on the external surfaces of associated tubes and scarcely depended on filling of their intratubular region with fullerene molecules.

The diversity of the hybrid structures and their properties got more pronounced as the chemical composition of peapods was complicated, in particular, by incorporating metallofullerenes into the NT cavity. A series of carbon SWNTs doped with regular chains of endofullerenes Gd@$C_{82}$ was obtained[36] by gas-phase synthesis. HRTEM analysis revealed ordered Gd@$C_{82}$@NT structures called one-dimensional metallofullerene crystals in NTs. They were formed on the basis of NTs having $D$ ~1.4-1.5 nm with Gd@$C_{82}$ molecules located at a distance of ~1.10 nm. Gd@$C_{82}$@NTs were presumably formed due to thermally activated diffusion of molecules through "open" tube ends.

The effect of temperature on *p-n* transition was observed for Dy@$C_{82}$@NT peapods.[47] Dy@$C_{82}$ molecules served as electronic density donors with respect to NTs, the value of the transferred charge depending on temperature. At room temperature, Dy@$C_{82}$@NT was a semiconductor. As the temperature decreased, the conductivity (beginning at ~215 K) acquired the metallic type. Below 75 K, the transport properties suggested a series of quantum dots in Dy@$C_{82}$@NTs. This problem was the subject of special research[49] (using Gd@$C_{82}$@NT as an example) as part of the program dealing with materials alternative to silicon for production of super-miniature electronic devices (diodes, transistors, memory elements *etc.*). The forbidden gap (FG) of different sections of Gd@$C_{82}$@NT synthesized from semiconducting carbon nanotubes was determined at low temperatures by STM. It was found that the FG of Gd@$C_{82}$@NT (~0.5 eV) in the vicinity of inculcated endofullerenes diminished to ~0.1 eV. Presumably, such structures with a set of ordered quantum dots may be applied in electronics and optoelectronics.

Hybrid structures were produced containing SWNTs ($D$ ~1.4 nm) with incorporated endofullerenes $La_2$@$C_{80}$ ($D_{2h}$ isomers, $D$ ~0.8 nm) forming regular close-packed chains.[49] Atomic pairs La-La (interatomic distances ~0.47 nm) inside $C_{80}$ had different mutual orientations for neighboring molecules $La_2$@$C_{80}$ in $La_2$@$C_{80}$@NT.

The analyzed systems are ideal models of quasi-one-dimensional crystals.[36] Of particular interest is the opportunity to change considerably their composition (general formula $(M_l@C_n)_m$@NT, where $l$, $n$, $m$ are variables) and produce "mixed" systems with simultaneous participation of several types of different metallofullerenes.

## 3. Simulation of the Structure and Properties of Fullerene Peapods and Related "Hybrid" Nanomaterials

The discovery of peapods brought up a number of questions significant for the development of physics and chemistry of the "hybrid" nanostructures and prospects of their practical application:

- What are the nature and the mechanism of the peapod formation?
- What are specific features of the electronic structure, chemical bonding, and properties of these structures in comparison with their "constituent elements" (fullerenes and nanotubes)?
- Do hybrid structures related to carbon peapods exist and what are their properties?

The above questions seem to be effectively tackled by theoretical simulation approaches. The results obtained are summarized below.

### 3.1. Fullerene Peapods

To date, carbon peapods have been simulated both by various semi-empirical and ab-initio approaches of quantum theory (tight-bonding and local density approximation (LDA)) and molecular dynamics (MD) techniques,[50-56] see also below.

The tight-binding method was used[50] to analyze geometric characteristics of peapods. According to ref. 50, the minimum diameter of SWNTs capable of encapsulating $C_{60}$ is 1.29, while their optimum diameter is 1.36 nm. Electronic and transport properties of $C_{60}$@NTs were studied. They were shown to depend on the near-Fermi states of a regular chain of fullerenes.[51]

A large number of peapods containing *armchair* (n,n)- and *zigzag* (n,0)-like NTs was discussed[52] in the framework of the parametrized tight-binding method. The calculation was performed in the supercell model. For example, the cell used to model $C_{60}$@(10,10)NTs was a cylinder 3.637 nm long, including 600 atoms. The most stable systems were those with fullerenes centered on the NT axis. When the $C_{60}$–$C_{60}$ distance changed, the fullerene repulsion energy increased abruptly as the distance became less than ~0.95 nm.

The valence states of peapods with (10,10) and (17,0)NTs (diameters 1.3-1.4 nm) and those of $C_{60}$ did not overlap, whereas the overall spectrum represented a superposition of fullerenes and NTs.[52] Covalent interactions $C_{60}$–NT were practically absent.

The possibility that these interactions can take place was examined using model systems $C_{60}$@NT with metallic (8,8) or semiconducting (14,0) tubes of smaller diameters as an example. In $C_{60}$@(14,0)NTs, fullerenes served as donor "impurities": free (donor) states appeared in the FG and its value decreased. Since encapsulated fullerenes, like other donor impurities,[57],[58] may locally modify the electronic structure of NTs, peapods hold promise as nanodiodes.[59] When the $C_{60}$–$C_{60}$ distance in $C_{60}$@(8,8)NTs decreases (to 0.849 nm), fullerene states are hybridized, a set of levels appears in the pseudogap of the metallic tube, and also a very narrow (0.01-0.02 eV) FG "opens" as happens when metallic-like ONTs are associated into "ropes".[60]

Mechanical properties of peapods were evaluated.[52] By way of example, elastic properties of $C_{60}$@(10,10)NTs were examined: deformation effects of torsion and compression–extension. The calculated Young modulus ($Y$) approached $Y$ of the "pure" tube (51.86), but was smaller than $Y$ of the two-layer (10,10)@(5,5)NT (52.47 eV/atom = 0.94 TPa). The peapod torsion deformation modulus ($C_L$, 4.03) was also smaller than that of one- and two-layer NTs (4.07 and 4.62 eV A/deg$^2$ respectively). $Y$ and $C_L$ of the (9,9)NT-based peapod also decreased relative to their counterparts in the "pure" NT. The Poisson ratio ($\sigma$) was determined for two sections of $C_{60}$@(10,10)NT, namely near the ring of C atoms between two neighboring $C_{60}$ (1) and the ring concentrically "embracing" the fullerene (2). It was found that $\sigma$ decreased by 2.8 (1) and 5.3% (2) relative to the undoped tube. Therefore, cohesive properties of peapods were impaired as compared with "pure" NTs. On the other hand, doping of NTs with fullerenes was an efficient method of their "plasticization".

Mechanical properties of $C_{60}$@(10,10)NTs were simulated by the MD method as a function of temperature (in the range 140-1500 K).[61] When the NT was filled with fullerenes, deformation tensions increased; simultaneously, the presence of encapsulated molecules largely suppressed the temperature dependence of these tensions in the nanotube. The mobility of molecules in the tubes was studied using the MD simulation[62] of the interaction between $C_{60}$ and (n,n)NT (n = 5-10).

Potential energies of interactions of $C_{60}$ with carbon tubes of different diameters were calculated in refs. 63, 64. NTs capable of forming "hybrid" structures with a regular (as linear chains) arrangement of encapsulated fullerenes were found to be limited in size both by lower and upper boundaries. For example, the potential curve of the E($C_{60}$-NT) interaction for $C_{60}$@(10,10)NT had the only radially symmetrical minimum centered on the tube axis. For $C_{60}$@(15,15)NT ($D$ = 2.04 nm), the E($C_{60}$-NT) profile exhibited some minima near the tube walls and a local maximum on its axis. Consequently, the distribution periodicity of fullerenes will be impaired along the axis of tubes having larger diameters ($D$ > 1.46 nm).[63] Abrupt weakening of fullerene–NT interactions will be an additional factor of disordering: the bonding energy $E_{bond}$($C_{60}$-NT) having an extremum for the tube diameter ~1.356 nm successively decreased with increasing $D$ and approached asymptotically $E_{bond}$ of $C_{60}$ with a plane graphene net.

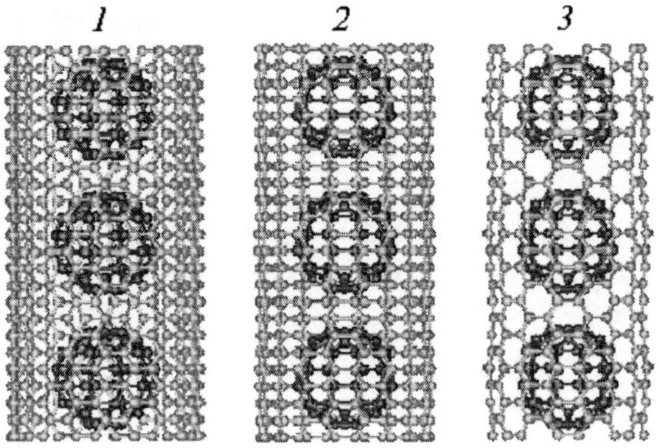


Figure 4: Optimized peapod structures: (*1*) $C_{60}$@(10,10)NT, (*2*) $C_{60}$@(9,9)NT, and (*3*) $C_{60}$@(8,8)NT. LDA calculation.[65]

Fundamental electronic properties of $C_{60}$@(n,n)NT (n = 8, 9, 10)[55],[65] and $C_{60}$@(14,7)NT peapods[66] were calculated in the LDA approximation.

For $C_{60}$@(10,10)NT ($C_{60}$–NT distance equals 0.331 nm), the geometry of the tube and fullerenes forming the system changed little. As the NT diameter decreased, deformations took place both in the NT (for instance, for $C_{60}$@(8,8)NT the difference in the cross-section diameters of the "widest" and "narrowest" parts of (8,8)NT was 0.06 nm) and the fullerenes, which experienced ellipsoid distortions, Figure 4.

The energy bands of $C_{60}$@(n,n)NT are displayed in Figure 5. The near-Fermi region of the $C_{60}$@(10,10)NT spectrum contains a system of π bands. Two bands having a considerable *k*-dispersion are referred to NT states, whereas two quasi-plane bands are related to π bands of the $[C_{60}]_\infty$ chain. As a result, $C_{60}$@(10,10)NT is a metal with a "mixed" type of carriers distributed along both the NT and the fullerene chain.

For the peapods with tubes of smaller diameters, the states (fullerene + NT) are not superimposed because of structural distortion effects modifying the band system.[65] For example, for $C_{60}$@(8,8)NT, the Fermi level intersects only two π bands of the NT, and the carriers are located on the cylinder peapod "shell". The energy distribution of the near-Fermi bands of such quasi-one-dimensional systems is affected by Peierls transitions. Probably, this effect will show up in peapods as the variation of distances between fullerenes in the chain.

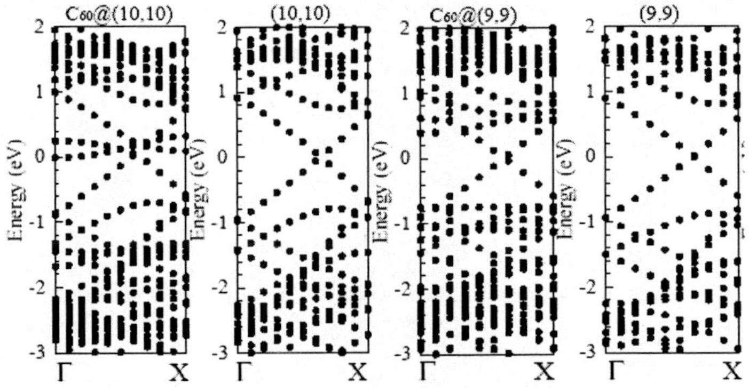


Figure 5: Energy bands of $C_{60}$@(10,10)NT and $C_{60}$@(9,9)NT as compared with "pure" (10,10) and (9,9) nanotubes.[65]

The analysis of spatial changes in the distribution of electronic states shows that $\pi$–electron density is redistributed partially between fullerenes and the NT during peapod formation. This effect becomes more pronounced with decreasing NT diameter, when the electron density in the system rises.

The peapod with a chiral SWNT ($C_{60}$@(14,7)NT) is a semiconductor with the FG of 0.2 eV, whereas the "pure" (14,7)NT has a gap of 0.5 eV. The bottom of the peapod conduction band is composed of $C_{60}$ $t_{1u}$-like states partly hybridized with nearly free-electron states of the carbon tube, while the top of the valence band is due to the (14,7)NT electronic bands.

The energy of peapod formation ($\Delta E$) was estimated as $E$((n,n)NT) + $E(C_{60})$ $\rightarrow$ $E(C_{60}$@(n,n)NT) – $\Delta E$.[65] The $\Delta E$ values were -0.51, 0.27, and 15.19 eV for n = 10, 9, and 8 respectively. Therefore, the encapsulation of $C_{60}$ into the (10,10)NT will proceed exothermally. On the contrary, for NTs of smaller diameters, this is an endothermal process ($\Delta E > 0$) explaining the presence of the experimentally detected lower "critical" diameter of the NTs capable of forming "hybrid" structures with $C_{60}$.

It is important to determine the peapod formation mechanism. Several phenomenological models of the process were proposed, which are based on experimental data. For example, using the concept of the capillary filling of SWNTs and MWNTs with intercalants (reviews 9-14), researchers[34] supposed that the peapod is formed in several stages. At the first stage, an "empty" tube grows on the face of a catalyst-metal (for example, Ni) particle. Under certain conditions, a "secondary" NT of smaller diameter may grow inside this tube. At some moment, the end of the inner tube "gets pinched" and is separated in the form of a molecule–fullerene $C_N$. The residual thermal energy moves the molecules inside the NT forming $C_N$ chains. Obviously, the model[34] is by far not universal and may hold only in a particular case of peapod growth in the presence of a catalyst.

Other mechanisms of tube filling have been advanced as well.[27-29],[35-37] By the first mechanism, $C_N$ is encapsulated via diffusion through "open" NT ends. By the second mechanism, $C_N$ molecules first deposit from the gas phase on the NT surface, penetrate into the inner tube cavity through the wall defects, and organize themselves in regular $[C_N]_\infty$ chains thanks to Van der Waals forces.

The above scenarios (illustrated by the example of $C_{60}$@(10,10)NT) were recently described at the microscopic level.[67] Potential surfaces were calculated when a fullerene was introduced into a single tube (or NT "bundle") through open ends or a large (diameter ~1.36 nm) "side" wall defect, Figure 6. The trajectory was most optimal when the $C_{60}$ molecule first was physsorbed on the NT surface and then moved freely along the tube walls in the direction of the tube axis. In both scenarios, the fullerene was encapsulated into the tube without activation.[67] Therefore, the most probable mechanism will be determined by the process conditions and the potential surface topology. When $C_{60}$ had initial speeds corresponding to thermal conditions of the peapod synthesis (350-400°C), the tubes most probably were filled through wall defects.

The theory of hybridization of 15 (n,n), (n,0), and (n,m) carbon NTs and encapsulated fullerenes was developed in ref. 68. It was used to interpret the electronic spectra of fullerene peapods, which were measured recently by scanning tunneling microscopy.[44] The Luttinger parameter $g$ of some carbon nanostructures was derived theoretically by the random phase approximation.[69] It was shown that the $C_{60}$@(10,10)NT peapod should have a smaller $g$ value as compared with a "pure" metallic-like (10,10) nanotube.

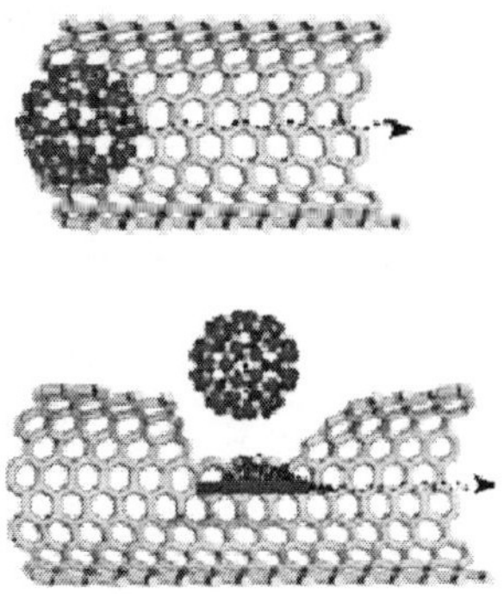


Figure 6: Peapod formation scenarios: diffusion of a fullerene through (1) an open end of a tube or (*2*) a defect in the tube wall.

## 3.2. Simulation of Related "Hybrid" Nanostructures

The class of "hybrid" nanostructures and their functional properties may be obviously expanded if they are formed:

- on the basis of non-carbon NTs or
- using different fullerene-like molecules as intercalants.

The former hypothetical objects were modeled in refs. 70-76.

LDA calculation of $C_{60}$ chains in (10,10) and (9,9) BN NTs, which are wide-band-gap semiconductors (FG $\geq$ 5 eV, see ref. 14,18, was performed in ref. 70. According to the inculcation energy estimates, this reaction is exothermal ($\Delta E$ = -0.12 and -1.27 eV for (9,9) and (10,10)BN-NTs respectively) and the energy gain is ~0.7 eV greater than in the formation of commensurate carbon peapods.[65] The energy translation and rotation barriers of $C_{60}$ in the BN tube are small (less than 0.04 eV/$C_{60}$) and cause a high mobility of molecules. This may lead to coalescence of $C_{60}$ inside the BN-HT and formation of two-layer tubes (carbon tubes inside boron-nitrogen tubes). The above system was proposed[70] as a prototype of a coaxial nanocable, where the conducting carbon tube is inside the "insulating shell", namely BN-NT. $C_{60}$@BN-NT peapods are semiconductors. The edge of the conduction band is formed by $\pi$ states of the fullerene chain. Electron doping may presumably render the system superconducting.[70]

As was noted above, diameters of the NTs and inculcated fullerenes correlate so that the "optimum" diameter of carbon NTs, which form peapods with $C_{60}$, should be 1.3–1.4 nm or larger. It was suggested[71] that potential intercalant molecules in NTs of smaller diameters might be fullerenes having minimum dimensions, for example $C_{20}$[77],[78] and $C_{28}$[79-82]. Specific features of the electronic structure of those molecules (in particular, dangling "external" bonds [79-82]) may probably cause their spontaneous polymerization leading to formation inside the tube of a large number of new nanoforms based on $C_{20}$ and $C_{28}$, such as dimers, trimers, various "nanocapsules" *etc.* Researchers[71] calculated hypothetical peapods with smaller fullerenes $C_{20}$ and $C_{28}$ intercalated into boron-nitrogen nanotubes ($C_{20}$,$C_{28}$)@(n,0)BN-NTs (n = 9, 10, 13), and also [$C_{20}$,$C_{28}$]$_{\infty}$ + (n,0)BN-NT systems, which are regular fullerene chains decorating the NTs on the outside (adsorption of $[C_{20},C_{28}]_\infty$ on the NT surface).

Researchers[71] examined the modification of the electronic structure of the hybrid nanosystem $C_{20}$@(9,0)BN-NT by its formation from "components": $[C_{20}]$ chain → @(9,0)BN-NT → adsystem $\{[C_{20}]$ + (9,0)BN-NT$\}$ → hybrid system $C_{20}$@(9,0)BN-NT with $C_{20}$–(9,0)BN-NT distances of ~0.155 nm. The spectrum of the $\{[C_{20}]$ + (9,0)BN-NT$\}$ adsystem differs from the superposition of the densities of states (DOS) of isolated [C@20] and (9,0)BN-NTs. New states appear in the BG of the tube. The tube–fullerene interactions are even stronger in $C_{20}$@(9,0)BN-NT, the most active orbitals being the outer (partially occupied) orbitals of $C_{20}$.

The above effects are absent upon formation of the adsystem and the peapod with distances of van der Waals bonds. The carrier density is concentrated in the outer states of fullerenes, which formed a narrow "impurity" band in the BG of BN-NTs, whose semiconducting properties persist in peapods. The intercalation of $C_{20}$ resulted in formation of anisotropic atomic bonds along the tube, the period of its "modulation" being determined by the period of the $C_{20}$ chain.

An efficient way to change the electronic properties of the aforementioned peapods is to vary the composition of the inculcated molecules. Therefore it is interesting to study peapods involving smaller endofullerenes M@$C_{28}$ (where M = Sc, Ti, Zr, Hf).[82]

By intercalation of the said fullerenes into NTs, it is possible to prepare new unique nanotubular materials including *d* metal atoms of the III-VI groups. Another possibility to produce such materials is discussed in refs. 72-75.

It is known that the inculcation of IV-VI-group *d*-metal atoms into carbon nanotubes is complicated.[3-7] While atoms of alkaline metals, namely Al, Pb, Zn, Ni, and Cu (under equilibrium conditions these elements either are inert to carbon or form unstable carbide phases),[83] easily penetrate into carbon NTs, these *d* metals are highly capable of forming stable carbides. They appear in NTs only in nanosized carbide phases, see ref. 18.

It was assumed[72-75] that novel nanostructures involving IV-VI-group *d*-metals may be prepared if they are inculcated into the NT bulk as "intrinsic" stable 0*D*–nanostructures. Recently synthesized molecular metal-carbon frame clusters, *viz.* the so-called metallo-carbohedrenes (metcars of the composition $M_8C_{12}$)[84-86] were proposed[72-75] as such structures, Figure 7.

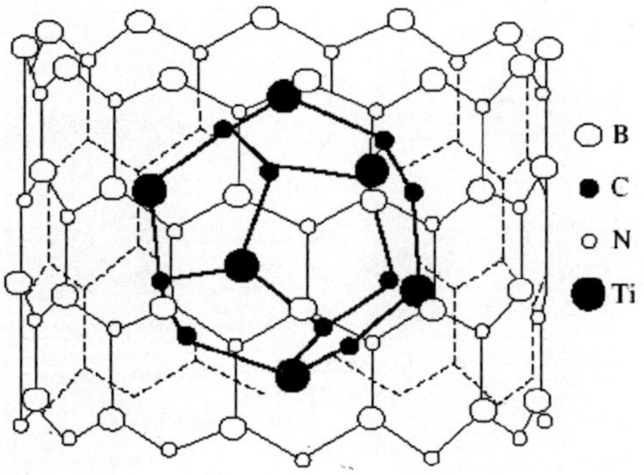


Figure 7: Model cell for calculation of a hybrid system "metcar in a nanotube" illustrated by the example of $Ti_8C_{12}$@BN-NT.[73]

Hypothetical hybrid nanosystems, which are metcar ($[M_8C_{12}]_\infty$, M = Sc, Ti, V, Zr, Nb) chains inside single-layer BN,[72] C[73], Si, GaN[74] or some borocarbonitride nanotubes of a variable composition ($M_8C_{12}@B_xN_yC_z$-NT),[75] were calculated. Properties of these systems were analyzed as a function of (1) the NT chemical composition, (2) the mutual arrangement of metcars and NTs, and (3) the electron concentration in the system determined by the type of the *d* metal and the chemical composition of the tube.

The valence band of these systems is close to the DOS superposition of the corresponding NTs and metcars. For example, the spectrum of $Ti_8C_{12}@$BN-NT contains five occupied bands of N2*s*, C2*s*,2*p*, C2*s*,2*p* +N2*p*, N2*p*-B2*s*,2*p* + Ti3*d*-C2*p*, and Ti3*d* states. The first four bands belong to $Ti_8C_{12}$ and BN-NT states. The most active states integrating $[Ti_8C_{12}]_\infty$ and BN-NT into $Ti_8C_{12}@$BN-NT are the near-Fermi states of Ti, which participate in bonds of neighboring metcars in the chain and metcar-NT bonds. In $Ti_8C_{12}@$BN-NT, Ti-Ti bonds in $[Ti_8C_{12}]_\infty$ increase insignificantly (with respect to an "isolated" chain, *i. e.* it stabilizes inside the NT), whereas bonds in the tube and metcars change maximum by 1-3%. As a result, the main types of bonds responsible for the invariance of these structures in the "isolated" state are preserved in hybrid structures.

A large number of configurations and compositions of hybrid systems "metcars inside NT" was analyzed.[72-75] The calculation of $Sc_8C_{12}@$BN-NT showed that bonds between individual metcars are lacking. Oppositely, in $V_8C_{12}@$(12,0)BN-NT they increased with the number of electrons, see Table 1. The population of interatomic bonds, electronic properties, and stability of the hybrid systems probably may be controlled by the chemical composition of metcars (by their doping).[87],[88] Another alternative is to use NTs having different chemical compositions. The electronic spectra of such systems were analyzed as a function of the tube composition using $Ti_8C_{12}@B_xN_yC_z$-NT as an example.[75]

Quite recently, one more type of the hybrid structure was proposed,[76] in which *d*-metal-doped silicon nanoclusters (namely, "magic" clusters $W@Si_{12}$) are inserted into carbon *arm-chair* NTs, Figure 8. Based on *ab initio* LDA calculations it was established that the (9,9) NT is the smallest nanotube to encapsulate clusters. Although the cluster-cluster and cluster-tube interactions are weak, the near-Fermi DOS of the hybrid system $W@Si_{12}@$(9,9)–NT is much higher than that of the "isolated" carbon tube.

**Table 1: Crystal orbital overlap populations (COOPs, e) for M-M (between metcars) and C-C (B-N) (in tubes) bonds for hybrid nanosystems $M_8C_{12}@$(12,0)C,BN-NT, one-dimensional $[M_8C_{12}]_\infty$ chains, and (12,0)C,BN-NTs.**

| Nanosystem | COOPs | | Nanosystem | COOPs | |
|---|---|---|---|---|---|
| | M - M | C - C (B - N) | | M - M | C - C (B - N) |
| (12,0)C–NT | - | 1.021 | $Sc_8C_{12}@$(12,0)C-NT | -0.017 | 1.009 |
| (12,0)BN–NT | - | 0.781 | $V_8C_{12}@$(12,0)C–NT | 0.375 | 0.996 |
| $[Ti_8C_{12}]_\infty$ | 0.138 | - | $Ti_8C_{12}@$(12,0)BN–NT | 0.236 | 0.776 |
| $[Sc_8C_{12}]_\infty$ | -0.018 | - | $Sc_8C_{12}@$(12,0)BN–NT | -0.021 | 0.801 |
| $[V_8C_{12}]_\infty$ | 0.334 | - | $V_8C_{12}@$(12,0)BN–NT | 0.364 | 0.785 |
| $Ti_8C_{12}@$(12,0)C-NT | 0.296 | 1.012 | | | |

Some important differences between $W@Si_{12}@(9,9)$–NT and "classical" fullerene peapods were also noted. Other recently synthesized metal-stabilized Si clusters, such as $(Hf,Zr)@Si_{16}$,[89] $(Ca,Sr,Ba)@Si_{20}$,[90] and $CrSi_{12}$[91] are likely to be candidates for encapsulating into NTs.

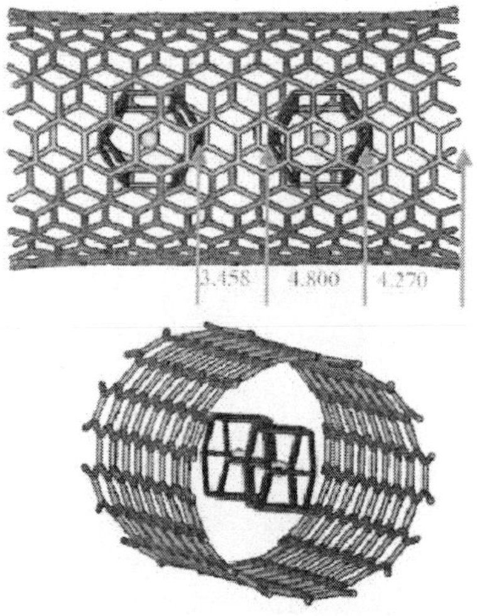


Figure 8: Preferable orientation of $W@Si_{12}$ cluster dimer inside a carbon (9,9) SWNT according to *ab initio* calculation.[76]

## 4. Conclusions

Let us note, in conclusion, that modern experimental investigations of the hybrid structures are determined by development of effective methods of synthesis and creation of a wide spectrum of peapods. The majority of the studies deal with synthesis procedures, product morphology, and some properties (chiefly structural, optical, thermal, and transport ones) of peapods. The theoretical simulation still faces a number of problems including an imperfect peapod morphology, reactivity, description of association processes, and prediction of the possible expansion of this class of unique hybrid structures.

Peapods are also viewed as promising materials for micro- and nanoelectronics (nanodiodes, transistors, memory elements, logic circuits)[44] and high-temperature superconductors.[92] However the quest for potential practical applications of peapods has a prognostic character and the obtained results are still far from technological solutions. Note also that peapods have great significance as unique model systems for the study of one-dimensional processes of diffusion and phase transformations and chemical reactions ("chemistry in tubes").

## Acknowledgement

The financial support from the Russian State Foundation of Leading Schools (grant 829.2003.3) and RFBR (grants 04-03-96117 (Ural) and 04-03-3211) has been gratefully appreciated.

## REFERENCES


[1] H.W. Kroto, J.R. Hearth, S.C. O'Brien *et al.*, *Nature*, 1985, 318, 162.
[2] S. Iijima, *Nature*, 1991, 347, 354.
[3] M.S. Dresselhaus, G. Dresselhaus, P. Eklund Eds. *Science of Fullerenes and Carbon Nanotubes*, San-Diego, Academic Press, 1996.
[4] R. Saito, G. Dresselhaus, M.S. Dresselhaus, *Physical Properties of Carbon Nanotubes*, London: Imperial College Press, 1998.
[5] The Science and Technology of Carbon Nanotubes (Eds. K. Tanaka, T. Yamabe, K. Fuku). Oxford: Elsevier, 1999.
[6] A L Ivanovskii, Quantum Chemistry in Materials Science: Nanotubular Forms of Matter (in Russian), Ekaterinburg, 1999.
[7] P. J. F. Harris, *Carbon Nanotubes and Related Structures: New Materials for the Twenty-first Century*. Cambridge: University Press, 1999.
[8] T. W. Ebbesen, *Phys. Today*, 1996, 49, 26.
[9] P. M. Ajayan, T. W. Ebbesen, *Rep. Progr. Phys.*, 1997, 60, 1025.
[10] A V Elezkii, *Uspecki Fiz. Nauk.*, 1997, 167, 945.
[11] Yu. E. Lozovik, A. M. Popov, *Uspecki Fiz. Nauk.*, 1997, 167, 751.
[12] A L. Ivanovskii, *Russ. Chem. Rev.*, 1999, 68, 511.
[13] M. Terrones, W. K. Hsu, H. W. Kroto, D. R. M. Walton, *Top. Curr. Chem.*, 1999, 199, 198.
[14] Fullerene Polymers and Fullerene Polymer Composites (Eds. P. C. Eklund, A. M. Rao) Springer, Berlin. 2000.
[15] E. G. Rakov, *Russ. Chem. Rev.*, 2000, 69, 41; 2001, 70, 934.
[16] R. Tenne, *Progr. Inorg. Chem.*, 2001, 50, 269.
[17] S. B. Sinnott, R. Andrews, *Crit. Rev. Solid State Mater. Sci.*, 2001, 26, 145.
[18] C. N. R. Rao, B. C. Satishkumar, A. Govindaraj, M. Nath, *Chem. Phys. Phys. Chem.*, 2001, 2, 78.
[19] A Loiseau, *Vide-Science Techn. Appl.*, 2001, 56,209.
[20] V.V.Pokropivny, *Powder Metall. Met.*, 2001, 40, 485.
[21] A L. Ivanovskii, *Russ. Chem. Rev.*, 2002, 71, 203.
[22] S. Iijima, *Physica.*, 2002, C.323, 1.
[23] B. M. Smith, M. Monthioux, D. E. Luzzi, *Nature*, 1998, 396, 323.
[24] A Thess, R. Lee, P. Nikolaev et al., *Science*, 1996, 273, 483.
[25] G. Rinzler, J. Liu, H. Dai et al., *Appl. Phys.*, 1998, *A*67, 29.
[26] P. Nikolaev, A. Thess, A. G. Rinzler et al., *Chem. Phys. Letters*, 1997, 266, 422.
[27] B. W. Smith, M. Monthioux, D. E. Luzzi, *Chem. Phys. Letters*, 1999, 315, 31.
[28] D. E. Luzzi, B. W. Smith, *Carbon*, 2000, 38, 1751.
[29] B. W. Smith, D. E. Luzzi, *Chem. Phys. Letters*, 2000, 321, 169.
[30] B. Burteaux, A. Claye, B. W. Smith et al., *Chem. Phys. Letters*, 1999, 310, 21.
[31] S. Von Gersum, T. Kruse, P. Roth, *Bunsenges. Phys. Chem.*, 1994, 98, 979.
[32] M. R. Stetzer, P. A. Heiney, J. E. Fischer, A. R. McGhie, *Phys. Rev.*, 1997, *B*55, 127.
[33] Y. Zhang, S. Iijima, Z. Shi, Z. Gu, *Philos. Mag. Letters*, 1999, 79, 473.
[34] J. Sloan, R. E. Dunkin-Borkowski, J. L. Hutchison et al., *Chem. Phys. Letters*, 2000, 316, 191.

[35] S. Bandow, M. Takizawa, K. Hirahara et al., *Chem. Phys. Letters*, 2001, 337, 48.
[36] H. Hirahara, K. Suenaga, S. Bandow et al., *Phys. Rev. Letters*, 2000, 85, 5384.
[37] H. Kataura, Y. Maniwa, T. Kodama et al., *Synthetic Metals*, 2001, 121, 1195.
[38] B. W. Smith, R. M. Russo, S. B.Chikkannanavar, D. E. Luzzi, *J. Appl. Phys.*, 2002, 91, 9333.
[39] X. Liu, T. Pichler, M. Knupfer et al., *Phys. Rev.*, 2002, *B*65, 5419.
[40] K. Hirahara, S. Bandow, K. Suenaga et al., *Phys. Rev.*, 2001, *B*64, 5420.
[41] H. Kataura, Y. Maniwa, M. Abe et al., *Appl. Phys.*, 2002, *A*74, 349.
[42] Y. Maniwa, H. Kataura, M. Abe et al., *J. Phys. Soc. Japan*, 2003, 72, 45.
[43] S. Bandow, M. Takizawa, H. Kato et al., *Chem. Phys. Letters*, 2001, 347, 23.
[44] D. J. Hornbaker, S.-J. Kahng, S. Misra et al., *Science*, 2002, 295, 828.
[45] H. Hongo, F. Nihey, M. Yudasaka et al., *Physica*, 2002, *B*323, 244.
[46] M. Shiraishi, T. Takenobu, A. Yamada et al., *Chem. Phys. Letters*, 2002, 358, 213.
[47] P. W. Chiu, G. Gu, G. T. Kim et al., *Appl. Phys. Letters*, 2001, 79, 3845.
[48] J. Lee, H. Kim, S. J. Kahng et al., *Nature*, 2002, 415, 1005.
[49] B. Smith, D. E. Luzzi, Y. Achiba, *Chem. Phys. Letters*, 2000, 331, 137.
[50] H. R. Palser, *Phys. Chem. Chem. Phys.*, 1999, 1, 4459.
[51] D. H. Kim, H. S. Sim, K. J. Chang, *Phys. Rev.*, 2001, *B*64, 5409.
[52] A. Farajian, M. Mikami, *J. Phys.: Condens. Matter.*, 2001, 13, 8049.
[53] D. Tomanek, *Physica*, 2002, *B*323, 86.
[54] M. M. Calbi, S. M. Gatica, M. W. Cole, cond-matter-0301362 (2003).
[55] Oshiyama, S. Okada, S. Saito, *Physica*, 2002, *B*323, 21.
[56] Roshefort, cond-matter – 0301310 (2003).
[57] K. Esfarjani, A. A. Faraijian, Y. Hashi, Y. Kawazoe, *Appl. Phys. Letters*, 1999, 74, 79.
[58] A. Faraijian, K. Esfarjaini, Y. Kawazoe, *Phys. Rev. Letters*, 1999, 82, 5084.
[59] G. W.Wang, K. Komatsu, Y. Murata, M. Shiro *Nature*, 1997, 387, 583.
[60] Y. K.Kwon, S. Saito, D. Tomanek, *Phys. Rev.* 1998, *B*58, R13314.
[61] Ni, S. B. Sinnott, P. Mikulski, J. A. Harrison, *Phys. Rev. Letters*, 2002, 88, 5505.
[62] Qian, W. K. Liu, R. S. Ruoff, *J. Phys. Chem.*, 2001, *B*105, 10753.
[63] L. A. Girifalco, M. Hodak, R. S. Lee, *Phys. Rev.*, 2000, *B*62, 13104.
[64] L. A. Girifalco, M. Hodak, *Phys. Rev.*, 2002, *B*65, 5404.
[65] S. Okada, S. Saito, A. Oshiyama, *Phys. Rev. Letters*, 2001, 86, 3835.
[66] T. Miyake, S. Saito, *Solid State Commun.*, 2003, 125, 201.
[67] S. Berber, Y.-K. Kwon, D. Tomanek. *Phys. Rev. Letters*. 2002, 88, 5502.
[68] C. L. Kane, E. J. Mele, A. T. Johnson et al., *Phys. Rev.*, 2002, *B*66, 235423.
[69] W. Que, *Phys. Rev.*, 2002, *B*66, 193405.
[70] S. Okada, S. Saito, A. Oshiyama, *Phys. Rev.*, 2001, *B*64, 1303.
[71] V. V. Ivanovskaya, A. A. Sofronov, Yu. N. Makurin *et al.*, *Russ. J. General Chem.*, 2004, 45, 243.
[72] V. V. Ivanovskaya, A. A. Sofronov, A. L. Ivanovskii, *Teor. Exper. Chem.*, 2002, 37, 331.
[73] A A Sofronov, V. V. Ivanovskaya, Yu. N. Makurin, A. L. Ivanovskii, *Phys. Chem. Letters*, 2002, 351, 35.
[74] V. V. Ivanovskaya, A. A. Sofronov, A. L. Ivanovskii, *Phys. Letters*, 2002, *A*297, 436.

[75] V. V. Ivanovskaya, A. A. Sofronov, Yu. N. Makurin, A. L. Ivanovskii, *J. Mol. Struct.* (THEOCHEM), 2002, 594, 31.
[76] Q. Sun, Q. Wang, Y. Kawazoe, P. Jena, *Phys. Rev.*, 2002, B66, 245425.
[77] Li, Q. Li, *J. Mol. Struct.* (THEOCHEM), 1998, 432, 115.
[78] M. Saito, Y. Miyamoto, *Phys. Rev. Letters*, 2001, 87, 5031.
[79] L. M. Balerieius, E. Stumbrys, A. Tamulis, *Fullerene Sci. Technol.*, 1997, 5, 85-96.
[80] K. Choho, G. Van de Woude, G. Van Lier, P. Geerlings, *J. Mol. Struct.* (THEOCHEM), 1997, 417, 265.
[81] R. K. Mishra, Y. T. Lin, S. L. Lee, *Intern. J. Quantum Chem.*, 2001, 84, 642.
[82] Yu. N. Makurin, A. A. Sofronov, A. I. Gusev., A. L. Ivanovskii, *Chem. Phys.* 2001, 270, 293.
[83] H. J. Goldschmidt, *Interstitial Alloys*. London: Butterworth, 1967.
[84] B. C. Guo, R. P. Kerns, A. W. Castleman, *Science*, 1992, 255, 1411.
[85] B. C. Guo, S. Wei, J. Purnell et al., *Science*, 1992, 256, 515.
[86] M. Rohmer, M. Bernard, J. Poblet, *Chem. Rev.*, 2000, 100, 495.
[87] V. V. Ivanovskaya, A. A. Sofronov, Yu. N. Makurin, A. L. Ivanovskii, *Russ. J. Coord. Chem.*, 2001, 27, 808.
[88] V. V. Ivanovskaya, A. A. Sofronov, Yu. N. Makurin, A. L. Ivanovskii, *Russ. J. Inorgan. Chem.*, 2002, 47, 972.
[89] V. Kumar, Y. Kawazoe, *Phys. Rev. Letters*, 2001, 87, 5503.
[90] Q. Sun, T. Wang, T. M. Breiere et al., *Phys. Rev.*, 2002, *B*65, 235417.
[91] S. N. Khanna, B. K. Rao, P. Jena, *Phys. Rev. Letters*, 2002, 89, 6803.
[92] R. F. Service, *Science*, 2001, 292, 45.

In: Nanostructures: Novel Architecture
Editor: Mircea V. Diudea, pp. 25-60
ISBN: 1-59454-499-9


*Chapter 3*

# REACTION PATHWAYS IN THE COALESCENCE OF FULLERENES

***Csaba L. Nagy,[a1] Mircea V. Diudea[a] and Teodor S. Balaban[b]***

[a]Faculty of Chemistry and Chemical Engineering
Babeş-Bolyai University, 400028, Cluj-Napoca, Romania
[b]Forschungszentrum Karlsruhe, Institute for Nanotechnology,
3640, D-76021 Karlsruhe, Germany

## ABSTRACT

The encapsulation of fullerene cages inside single-walled nanotubes SWNT can be performed very efficiently in both vapor phase and in solution. In the inner space of SWNT, zero dimensional molecules, like $C_{60}$, are ordered in one-dimensional arrays. Such structures are called nanopeapods. Annealing peapods over 800°C, results in double-walled carbon nanotubes DWNT. The growing of the inner tube is a multi-step process, within which the inner tube increases its diameter up to the van der Waals distance to the external parent tube. The coalescence of $C_{60}$ molecules is described by a sequence of Stone-Wales SW bond rotations. The process passes through the peanut-shaped intermediates and terminates with a capped (polyhex) nanotube, *i.e.*, tubulene. The strain relief from the spherical fullerene to the tubulene provides the driving force of the transformation. Tubulenes can further be transformed, by SW, in other tubular forms. The finally resulted DWNT could be the precursor of a distinct-walled torus DWT, with the apex zones decorated by non-hexagonal faces. Semiempirical calculations suggest that such tori are far more stable than the classical polyhex ones, at least at small and moderate size. Topology of fullerenes, tubulenes and related objects can be described by using the spiral code and/or highly discriminating topological indices operating on layer matrices, either on the parent associated graphs or on their transforms by medial and dual operations on maps.

[1] E-mail address: nc35@chem.ubbcluj.ro

# 1. Introduction

The merging process between carbon nanostructures, induced by thermal annealing or electron beam irradiation, is a way to the synthesis of more complex structures: double-walled nanotubes DWNT can be synthesized by merging fullerene molecules into a nanotube to form nanopeapod systems; two nanotubes coalesce into a diameter-doubled one; initially crossed nanotubes can form "Y" and "T" shaped junctions, etc. In the first part of this paper, we discuss the most important experimental data of this topic, reported in literature. The second part is devoted to a pertinent discussion, in the light of our own calculations.

## 1.1. Fullerene Dimers and their Derivatives

The first dimeric fullerene $C_{120}$ was synthesized by using light irradiation. Applying visible or ultraviolet light to a film of solid $C_{60}$ resulted in a polymer. The X-ray study of the photo-transformed film has shown a contraction in the fullerene lattice, found less than two van der Waals radii between the $C_{60}$ cages. It was supposed that a covalently bonded [2+2] cycloadduct (involving $sp^3$ hybridized carbon atoms in the junction) of two *quasi*-spherical molecules is formed.[1]

Later, by simultaneous application of high pressure and temperature, $C_{60}$ was converted into two different structures: a face-centered-cubic structure (300-400°C and 5GPa) and a rhombohedral structure (500-800°C and 5GPa). Further theoretical and experimental studies proved that the dimeric $C_{120}$ has a structure in which the two fullerene cages are connected through a four-membered ring, closed by a cycloaddition over two *h* bonds.[2] We call an *h*-bond the one shared by two hexagons and a *p*-bond that shared by pentagon-hexagons.

A synthesis of $C_{120}$, using KCN as a catalyst, has been described[3] and the resulting structure was assigned by X-ray analysis and $^{13}C$ NMR spectroscopy. It was suggested that $C_{120}$ can be taken as a subunit of the $C_{60}$ polymers.

Since KCN is a highly toxic substance, a different preparation method was elaborated, catalyzed by Li powder. Small amounts of LiOH or $Li_2CO_3$ increased considerably the yields. Using the same method for polymerization of $C_{70}$ failed, proving that $C_{70}$ is not as easily polymerized as $C_{60}$. Raman spectra of $C_{120}$ exhibit similar patterns with the dimeric oxides $C_{120}O$ and $C_{120}O_2$.[4]

Among the 1812 possible topological structures of $C_{60}$, the most stable is that in which all *p* edges are formally single bonds and all *h* edges are double. That is why, in $C_{120}$ dimer, the bridge formation is favored when it originates in a formal double bond, leaving the rest of the Kekulé structure unaltered. It was stressed that, the experimental factor deciding the preferable dimeric isomer is the mutual orientation of the two parent molecules, rather than the dimer stability.[5]

Keeping in mind the two types of bonds involved, the cycloaddition of $C_{60}$ will result in four different ($sp^3$-joined) isomers (Figure 1): (i) *hh* ; (ii) *hp*; (iii) *pp-c* and (iv) *pp-t*. In the last two isomers, the two pentagons lies on the same side (*cis*) or they are opposite (*trans*), with respect to the junction ring.

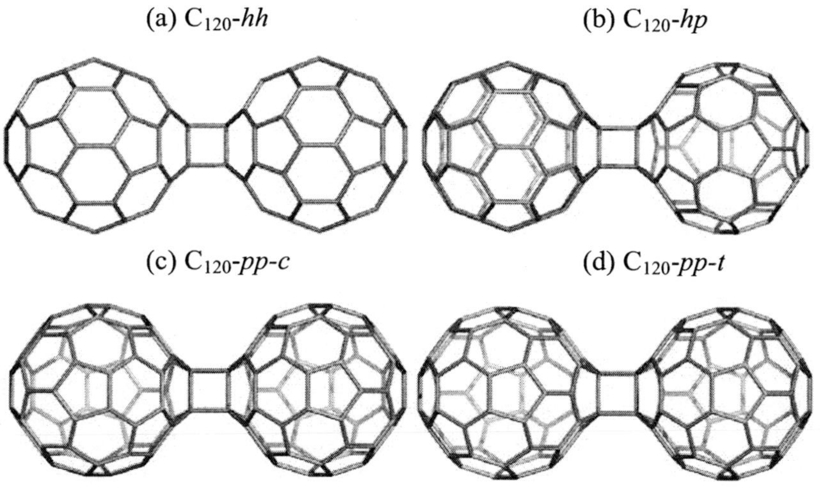


Figure 1: The four possible $C_{120}$ $sp^3$ dimers resulted by (2+2) cycloaddition.

Fullerene derivatives have attracted great attention of synthesists. Oxygen derivatives represent an important class of fullerenes, with particular applications (*e.g.*, precursors of odd-numbered fullerenes, production of singlet oxygen, etc.).

The first observed fullerene oxide was $C_{70}O$, an epoxide of $D_{5h}C_{70}$. The decay of $C_{70}O$ into $C_{70}$ suggested that the oxygen is located on the external surface of the cage. $^{13}C$ NMR data suggested a structure of $C_{2v}$ symmetry, with the epoxy group across an *h* double bond. Synthesis of $C_{60}O$ was performed by ozonolysis of $C_{60}$ solution in $CS_2$; its structure is analogous to that of $C_{70}O$ (Figure 2a).

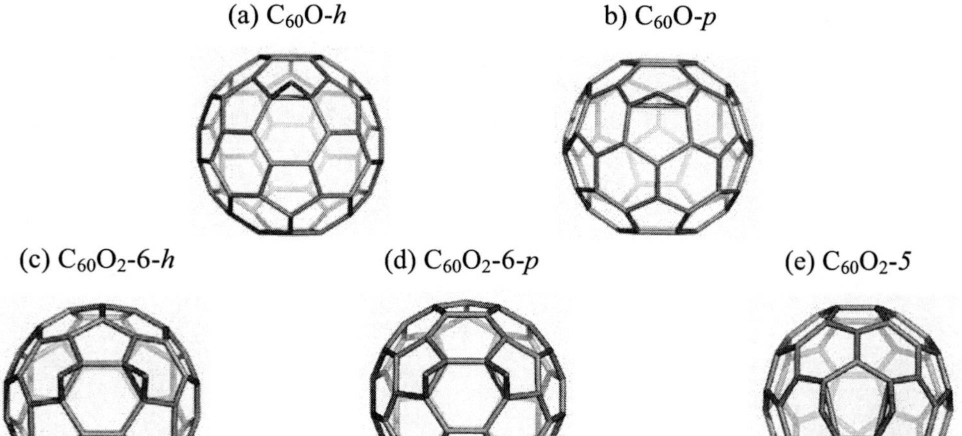


Figure 2: $C_{60}O$ and $C_{60}O_2$ fullerene epoxides.

The dimeric monoxide $C_{120}O$ was prepared, in solid phase, by heating a $C_{60}/C_{60}O$ mixture, at 200°C. This synthesis was also performed in solution, but the yields were much lower. Similarly, the ozonolysis of a $C_{60}$ solution, led to a mixture of $C_{60}O$ and higher oxides $C_{60}O_n$ ($n$ >1), next separated by the HPLC method. The reaction products were analyzed and characterized by UV-VIS, FT-IR and MALDI-TOF MS. In the mass spectrum of the isolated product, the peak of $C_{120}O$ appeared (together with the peaks of $C_{60}$ and $C_{60}O$, eventually resulted by the fragmentation of $C_{120}O$).[6]

The structure assignment for $C_{120}O$ was analogous to that for the dimer $C_{120}$; actually, a furan like bridge is formed between two *h* bonds of each cage (Figure 3a). For this, a [3+2] cycloaddition mechanism was proposed. CO and $CO_2$ were also detected, by heating the product over 400°C, and the peak of $C_{119}$ appeared in the mass spectrum as well. It was proposed that $C_{120}O$ is the precursor of the odd-numbered fullerene $C_{119}$.

The thermolysis of solid $C_{120}O$, in argon atmosphere at 400°C and normal pressure, led to the dimeric dioxide $C_{120}O_2$, along with other oligomeric oxides $(C_{60})_nO_m$ ($n$ = 3-5, $m$ = 3-9). $C_{120}O_2$ was separated by HPLC and characterized by MALDI-TOF MS, UV-VIS, and FTIR. The mass spectrum of $C_{120}O_2$ evidenced $C_{60}$ and $C_{60}O$ as major fragmentation ions. The $^{13}C$ NMR spectra suggested a structure of $C_{2v}$ symmetry.[7]

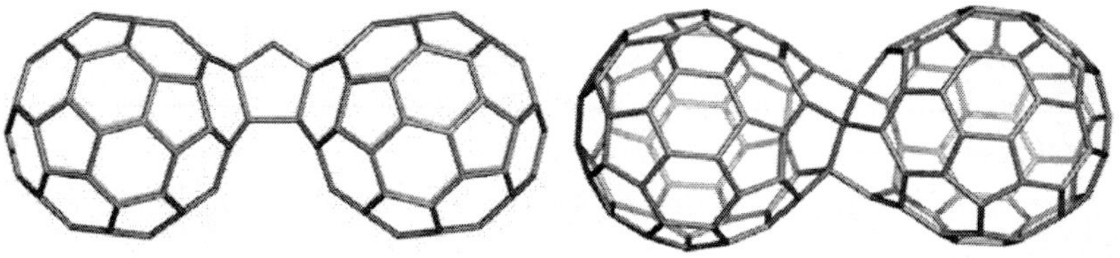


Figure 3: $C_{120}O$ (a) and the corresponding odd-numbered fullerene $C_{119}$ (b) resulted by the loss of CO or $CO_2$.

Two different structures, having the two cages bis-linked by adjacent furanoid rings, have been proposed. In the first case, the furanoid bridges are attached at the sites of neighboring *h* bonds in each parent cage (Figures 4a,b), while in the second one, *p* junctions are involved in the linkage (in this last case the carbon frame is that of $C_{120}$-*hh*).

Another method for preparing $C_{120}O_2$ is the thermal transformation of a $C_{60}/C_{60}O/C_{60}O_2$ system, in solid state.[8] $C_{60}O_2$ was prepared as a mixture of regio-isomers during ozonolysis of $C_{60}$. The mixture was heated at 200°C under argon, resulting, beside $C_{120}O$ and $C_{120}O_2$, a trimeric product, $C_{180}O_2$. The $^{13}C$ NMR analysis detected two isomers of $C_{120}O_2$, with $C_1$ symmetry, in 4:1 ratio, different from the one above discussed. Spectroscopic data suggested these isomers are closely related to $C_{120}O$, probably one of its epoxides. Again, the dimeric $C_{120}O_2$ can result by two ways: (i) the cycloaddition of one $C_{60}O$ molecule across a *h* double bond of another $C_{60}O$ one, or (ii) the cycloaddition of $C_{60}O_2$ to $C_{60}$ *via* one of the two epoxy groups. Regio-isomers of $C_{60}O_2$ and $C_{120}O_2$ have been modeled and the lowest energy isomers were found to have the epoxy group in the proximity of the furanoid bridge (Figure 4c,d). The *cis*-1 *anti* structure (nomenclature according to ref. [9]) can rearrange by opening the epoxy cycle to the proximal *h* bond of the other cage, resulting the bis-linked structure ($C_{2v}C_{120}O_2$). The other possible regio-isomers of $C_{120}O_2$ can easily react with molecules of $C_{60}$ to form $C_{180}O_2$, since the epoxy group is sterically accessible.

Similar to $C_{120}O_2$, the first sulfur containing dimeric fullerene, $C_{120}OS$, was prepared. The synthesis was performed either by the thermal reaction of a 1:1 mixture of solid $C_{120}O$ and elemental sulfur at 230°C under argon, or by heating of $C_{60}/C_{60}O$ system with sulfur.[10]

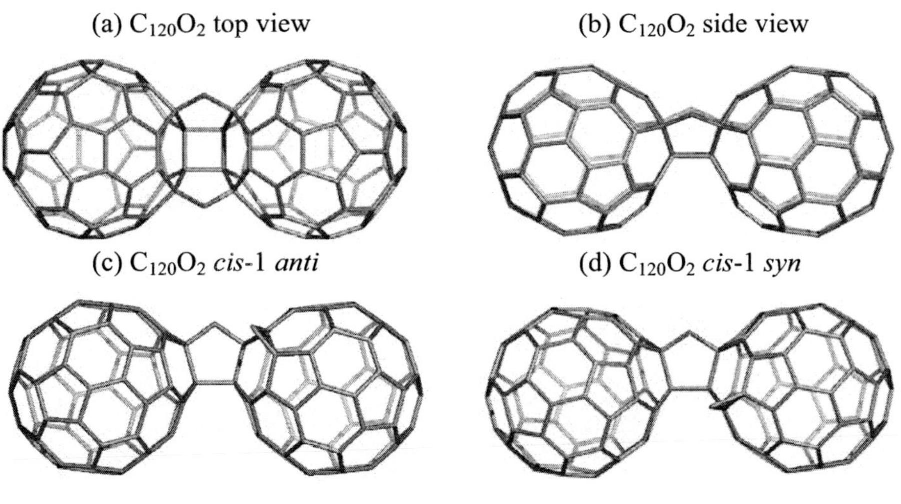


Figure 4: $C_{60}$ cages connected by furanoid ring(s) attached to *h* junctions in each cage.

The mass spectra of the reaction product also revealed $C_{60}$, $C_{60}O$ and $C_{60}S$. Molecular modeling has shown that the lowest energy structure is that having an oxygen atom in one of the furanoid rings in $C_{2v}$ $C_{120}O_2$ replaced by sulfur. This can result from $C_{120}O$ by addition of a sulfur atom to the *h* double bond next to the furanoid ring, followed by thiophene bridge formation.

As mentioned above, signals for odd numbered fullerenes ($C_{119}$, $C_{129}$ and $C_{139}$) were observed in the mass spectra of fullerene oxides. These species correspond to dimers of $C_{60}$, $C_{60}/C_{70}$ and $C_{70}$, respectively, less one carbon atom. Since the loss of a carbon atom in the reaction of two cages would be thermodynamically expensive, it was supposed that oxygen is involved, because loss of CO or $CO_2$ could be the driving force of the reaction. For this, fullerene solution was exposed to ozone; the experiments have confirmed the above hypothesis.[11]

On the ground of semiempirical calculations, a peanut structure of $C_2$ symmetry, where two equivalent $C_{58}$ cages are connected by a bridge of three four-connected carbons, was proposed. Later, calculated Raman spectra for a series of $C_{119}$ fullerenes with $C_2$ symmetry were compared with the experimentally obtained spectra. The conclusion was, the prepared isomer is indeed the thermodynamically most stable one, as proposed earlier (Figure 4b).[12]

## 1.2. Theoretical Considerations about Fullerene Fusion

At the pioneering age of fullerene science, it seemed impossible the merging of two (spherical) units to form a larger structure, by the reason that bond breaking requires high energy. Later, the coalescence of fullerenes has been reported as an experimental fact and modeled in view of understanding non-obvious details. The observed diameter doubling of carbon nanotubes was interpreted as a lateral coalescence of such suitably aligned objects.

The first coalescence experiment was performed by laser desorption of a $C_{60}$ film.[13] In another experiment, $C_{120}^{+}$ appeared by collision between $C_{60}$ and $C_{60}^{+}$. Further studies found that the fusion barrier lies in the region between 60 and 80 eV and increases as the number of atoms participating to the collision increases.[14],[15]

It is quite difficult to assign structures to such coalesced fullerenes. Theoretical studies have proposed several classes, including peanut-shaped dimers or cylindrical tubulenes, the most studied dimer being $C_{120}$. Atomically precise routes for complete coalescence of fullerene cages have been presented.

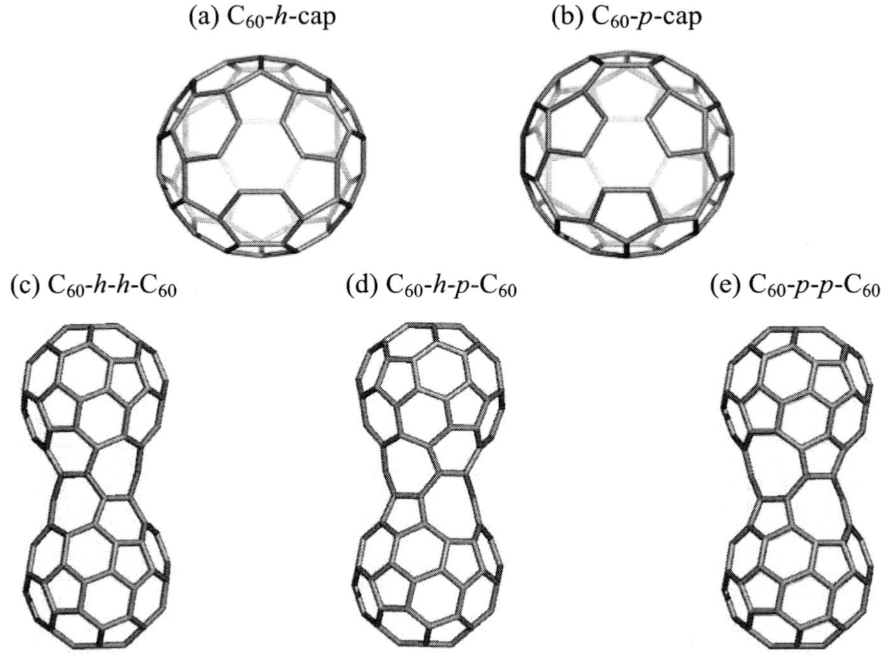


Figure 5: Caps derived from $C_{60}$ and the all $sp^2$ corresponding peanut dimers. The junction between the two caps is of armchair A type.

A class of all $sp^2$ peanut-shaped structures can be constructed by connecting the caps $C_{60(6\,(56)^3(665)^3(656)^3(*\,6)^3-A[12,0])}$ and $C_{60(6\,(56)^3(665)^3(656)^3(5\,*)^3-A[12,0])}$, derived from $C_{60}$ by deleting the three *p* and *h* edges, respectively, of a hexagonal polar ring (Figure 5a,b). The asterisk in the above symbols denotes the missing polygon(s).[16]

The free valences of the caps will be connected to another unit, avoiding the formation of four-membered rings. The three resulting peanut dimers are denoted by the complementary edges remaining in the cut polar ring (Figure 5c to 5e).

The construction of our all $sp^2$ peanut-shaped dimers makes use of quite different caps: $C_{60(56^5(56)^5(65)^5(1\,*)^5-Z[10,0])}$ and $C_{60(6(56)^3(665)^3(656)^3(*\,1)^6-Z[12,0])}$ with “1” denoting the dangling bonds (Figure 6a,b). The short symbols are $C_{60}$-*pd*-cap and $C_{60}$-*hd*-cap, respectively. It is easily seen that the number of dangling bonds equals the size of the polar ring, the mixing caps being thus prohibited. The isomers are named $C_{120(56^5(56)^5(65)^5 7^5 7^5-Z[10,1])}$ and $C_{120(6(56)^3(665)^3(656)^3(67)^3(76)^3-Z[12,0])}$, according to the tubulene nomenclature, proposed by us in ref.[17]. The proposed short symbols, specifying the junction covering, are $C_{120\text{-}77}$ and $C_{120\text{-}67}$, respectively (Figure 6c,d).

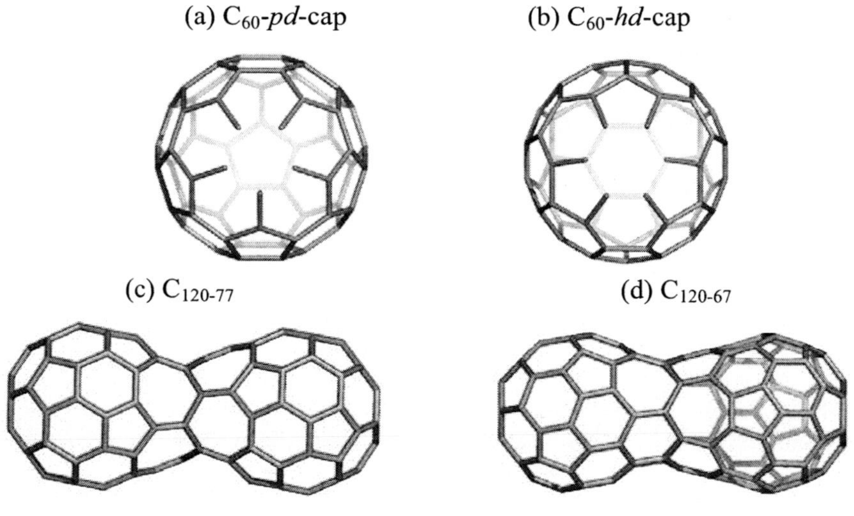


Figure 6: Caps resulted by elimination of all bonds in a pentagon or hexagon and the corresponding peanut dimers. The junction between the two caps is of zigzag Z type.

Topological analysis suggests that the merging process of two fullerene molecules can be achieved by a sequence of Stone-Wales SW bond rotations. The first step in the coalescence is the formation of a [2+2] ($sp^3$-joined) cycloadduct, *e.g.*, the $C_{120}$-*pp-t* dimmer (*vide supra*). An intermediate step is the formation of all $sp^2$ $C_{120-77}$ peanut structure; subsequent SW bond flipping, in a circumferential order, transforms the heptagon-pentagon junction into a hexagonal (6,3) covering (see Figure 15 below). In case when $C_{60}$ coalesced from a different orientation, the process was believed to stop at a peanut shaped structure. In view of the negative curvature relief (in the junction zone) a complete restructuring of the end caps is required. In this way, abutting pentagons could occur, which considerably raises the energy barrier.[18]

Similar coalescence paths were recently proposed for several tubulenes having a pentagonal polar ring: (a) two nanotubes A[10,*n*] capped by $C_{30(5\,6^5(5\,*)^5-A[10,0])}$, (see Figure 15a) with an energy barrier of 3.7eV; (b) two A[20,*n*] nanotubes capped by $C_{90(5\,6^5 6^{10} 6^{15}(5\,*)^5-A[10,0])}$, requiring 8eV; and (c) the fusion of $C_{60}$ with the tubulene resulted as described at point (b).[19]

T-shaped junctions can result by cap-to-wall coalescence, involving SW flipping. Recently, a process of $C_{60}$ penetration through the wall of a nanotube, to form a peapod, was described.[20]

The SW edge rotation is known as having a high energy barrier; however, the overall process is exothermic. The reason is that the strain energy of a spherical cage (*i.e.*, fullerene) is much higher than that of a (capped) tube (*i.e.*, tubulene), thus the energy gain, associated with a reduced local curvature, provides the driving force of the coalescence, up to the cylindrical tubulene.

Using a graphical search program, a 23 SW steps topological pathway, from the *hh* [2+2] cycloadduct to the $C_{120}$ tubulene, was found. The same tubulene is obtained when the SW cascade process starts from the $C_{120}$-*pp-t* cycloadduct.[21]

It was believed that, the initial [2+2] cycloaddition pays the most energy cost in the process of coalescence, probably being the rate-limiting step. For this first step, the proximity of two double bonds in each fullerene is necessary. At lower temperature, polygons rather than double bonds should preferentially face to each other, preventing the fusion. Only at high enough temperature, fullerene cages start to rotate, making effective the bond to bond addition. In bulk reaction, $C_{60}$ molecules are able to follow multiple coalescence with neighboring fullerenes. It is possible that the further SW transformation of one connection to interfere with another fusion process on the same fullerene cage, which would probably stop the fusion. This ability will be restrained in the case of one-dimensional space (*e.g.*, in nanopeapods, with the cages aligned in a chain), where polymerization is possible only on the opposite faces of a cage.[22]

Note that the activation barrier of a Stone-Wales transformation is significantly lower in non-planar structures in comparison to the graphite, due to the pyramidalization of $sp^2$ atoms. However, the presence of hydrogen atoms stabilizes the transition state in the SW rotation, and reduces the energy barrier to 4.1 eV.[23]

## 1.3. Coalescence in Peapods

The knowledge about the structural diversity of carbon allotropes has recently been enriched by the discovery of nanopeapods, a hybrid structure consisting of fullerene molecules encapsulated in single-walled nanotubes. The symbol for such structures is inspired from the endohedral metal doped fullerenes: for example $(C_{60})_n$@SWNT is for the $C_{60}$ nanopeapod.[24]

Nanopeapods represent a promise for new applications. For example, a positively charged $C_{60}^+$ molecule inside a nanocapsule could serve as a *memory element*, where the position of the ion encodes the information. An important aspect is that fullerene molecules can rotate freely in the space of a tube at room temperature, and this rotation will affect the peapod electronic properties significantly.[25]

A nanopeapod can be considered as a nano furnace in which the coalescence of fullerenes can be monitored (see below). Usually they are prepared in gas phase at 400°C or higher,[26] where $C_{60}$ molecules sublime and enter the SWNTs from the open ends or sidewall defects. In liquid phase, the affinity of the solvent to the tube and fullerenes dictates the direction of the process.[27]

By pulsed laser vaporization of graphite, in the presence of a metal catalyst, besides carbon nanotubes also $C_{60}$ molecules are formed. Although $C_{60}$ and other residues are separated from nanotubes, it was suggested that fullerene molecules might be encapsulated inside the nanotube. This was confirmed by high-resolution transmission electron microscopy HRTEM. It was observed that the fullerene molecules are arranged in a chain at a nearly uniform distance from each other. By exposure at an electron beam, fullerenes inside the nanotube coalesce into larger capsules, capped by $C_{60}$ halves. The length of such capsules corresponds to three-four fullerene units, and the diameter is constrained by the outer nanotube dimensions.[28]

After the insertion of fullerene molecules inside nanotubes in vapor phase, it is difficult to reverse the process by heat treatment. The fullerene array is stable up to 800°C and further heating induces coalescence, rather than dedoping. The coalescence process becomes more rapid as the temperature increases, and it completes around 1200°C. The inner tube diameter was found to be smaller by 0.71 nm than the outer nanotube. Thus, the growing of the inner tube is independent on the size of the doping fullerene.[29]

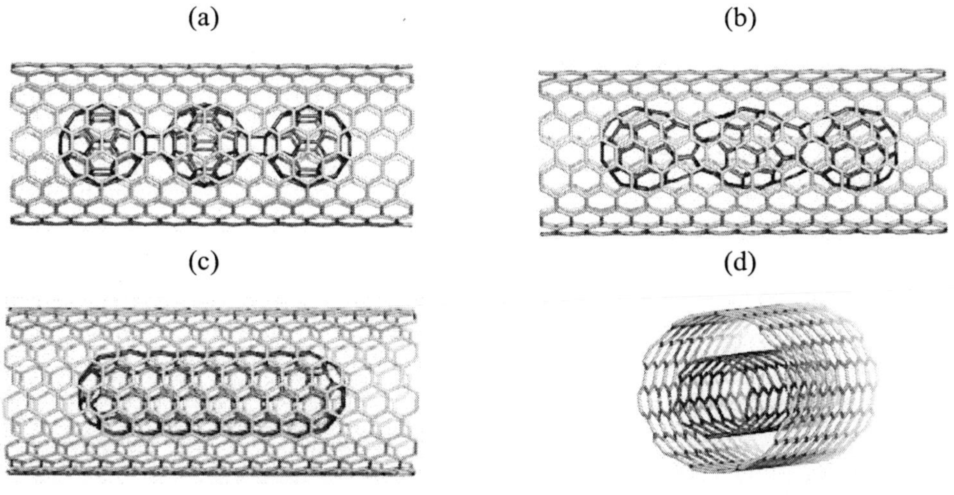


Figure 7: From the [2+2] cycloadduct of $C_{60}$, to peapods and double walled nanotube DWNT.

To elucidate intermediates in the transformation of nanopeapods to double-walled nanotubes DWNT, the heating was performed at low temperatures (800-900°C) and for a much longer time (200-300 hours). The growth of the inner tube was monitored by Raman study. After a short heating period the inner tube diameter corresponds to that of $C_{60}$. This is the result of the polymerization of adjacent $C_{60}$ molecules. Further heating induces the increase of the inner tube diameter until the difference between the inner and outer tube fits the van der Waals distance.[30]

In case of $C_{60}$, the polymerization takes place easily because the probability of having two fullerenes oriented in a preferential way is high, as expected by the free rotation of $C_{60}$. But when $C_{70}$ was heated at 800°C polymerization proceeded very slowly, and most of the $C_{70}$ molecules were found untransformed. This could be explained by that, the $C_{70}$ molecules must arrange in a zigzag way for the polymerization to take place. Indeed, the Raman spectrum indicated a larger inner tube than in the case of $C_{60}$.

As above mentioned, the structure resulted by the dimerization of $C_{60}$ is considered as a subunit in further polymerization. Therefore, the dimer must have *h* bonds at the tips of both ends, since it cannot freely rotate, as $C_{60}$ does. If the resulted dimer has a $C_{60}$ like cap, with a pentagon at the tip (see Figure 15a), no further polymerization is expected between two such adjacent units. It was proposed that, instead of $C_{120}$, the primary unit be $C_{114}$, which has $C_{60}$-like caps but with hexagons at the tip. For this, at every merging step, six carbon atoms have to be released, eventually moved around the inner tube. It is expected that tube diameter transformation occurs by the expense of these migrating carbon atoms.

Besides the thermal annealing, fullerenes can easily coalesce by electron irradiation. Under such conditions, carbon atoms are displaced from the fullerene cage by knock-on effects. The formation of vacancies and dangling bonds increase considerably the energy of the system, but the formation of fully coalesced products up to double walled nanotube will finally stabilize the system. Theoretical studies have proposed that such inside tubes contain heptagons and octagons, acting as links between fullerene cages, like in peanut dimers. These internal corrugate carbon nanotubes resemble to the Haeckelite structure. The topology of junction depends of the initial orientation of the molecules.[31]

# 2. RESULTS AND DISCUSSION

In the following, we present our own results, on nanostructure building and semiempirical and strain energy calculations, as: (i) a comparison between the previously proposed structures for fullerene epoxy dimers and our epoxides built up on coalesced fullerenes; (ii) novel dimeric structures possibly appearing in the coalescence of $C_{60}$ molecules, and cages that could be formed by loss of carbon atoms from the starting molecules; (iii) a new class of tori, constructed and compared with the polyhex tori;

(iv) discussion on nanotube diameter doubling with a remark on zigzag nanotubes; (v) a method of finding the vertex, edge, and face topological orbits in fullerenes.

## 2.1. The $sp^3$ Adducts: $C_{120}$, $C_{120}O$ and $C_{120}O_2$

It was discussed above that functionalized dimeric derivatives are synthesized starting from a mixture of $C_{60}/C_{60}O_n$. In the following we propose a pathway, analogous to that for the coalescence of fullerenes leading to epoxy tubulene.

First we have modeled the $C_{60}$ epoxides to find out which is the preferential position of the oxygen atoms. Of course if we consider the $C_{60}$ molecule as a Kekulé structure with alternating single and double bonds, the oxygen atom will be bound to one of these double bonds (*h* bond). Although the calculations presented in Table 1 did not confirm this, in the following it will be presented that the position of the oxygen atom will not influence very much the stability of the coalesced dimer.

**Table 1: PM3 data for $C_{60}$ and isomers of $C_{60}O_n$; n = 1, 2.**

| Name | *N* | $H_f / N$ | GAP (eV) | Sym |
|---|---|---|---|---|
| $C_{60}$ | 60 | 13.512 | 6.594 | $I_h$ |
| $C_{60}O$-*p* | 61 | 12.857 | 6.412 | $C_s$ |
| $C_{60}O$-*h* | 61 | 12.964 | 6.446 | $C_{2v}$ |
| $C_{60}O_2$-5 | 62 | 11.989 | 6.293 | $C_s$ |
| $C_{60}O_2$-6-*p* | 62 | 12.757 | 6.388 | $C_s$ |
| $C_{60}O_2$-6-*h* | 62 | 12.389 | 6.491 | $C_s$ |

**Table 2: PM3 data for the dimeric $C_{120}$ $sp^3$ cycloadducts.**

| Name | $N$ | $H_f / N$ | GAP (eV) | Sym |
|---|---|---|---|---|
| *hh* | 120 | 13.204 | 6.349 | $C_{2v}$ |
| *ph* | 120 | 13.373 | 5.854 | $C_S$ |
| *pp-c* | 120 | 13.544 | 5.749 | $C_{2v}$ |
| *pp-t* | 120 | 13.542 | 5.707 | $C_2$ |

Among the $C_{120}$ dimers presented in Figure 1 the most stable was found the *hh* isomer. The energy of other dimers increases with the number of *p* bonds involved in the cycloaddition. The data listed in Table 2 are comparable to those reported in literature.

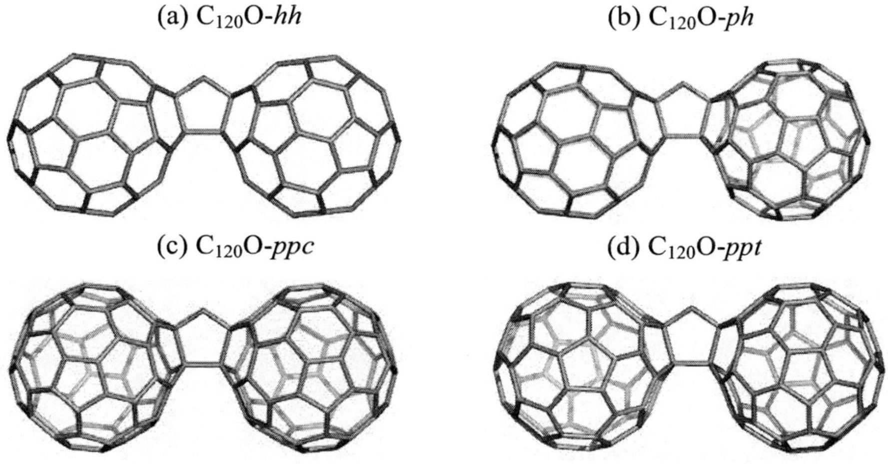


Figure 8: Structural isomers of $C_{120}O$.

The isomers of $C_{120}O$ (Figure 8) are constructed by analogy to the $C_{120}$ dimers; actually, the four-membered ring is replaced by a furanoid bridge. The isomers are named similarly to those of $C_{120}$ dimers.

Table 3 lists the results of semiempirical calculations performed on the four isomers in Figure 1. They are in agreement with the literature data, the most stable being the isomer $C_{120}O$-*hh*, with $C_{2v}$ symmetry. Note that their heat of formation increases as more *p* bonds are involved in the furanoid ring.

**Table 3: PM3 data for furanoid $C_{120}O$ $sp^3$ isomers.**

| Name | $N$ | $H_f / N$ | GAP (eV) | Sym |
|---|---|---|---|---|
| $C_{120}O$-*hh* | 121 | 12.793 | 6.356 | $C_{2v}$ |
| $C_{120}O$-*ph* | 121 | 12.944 | 5.888 | $C_1$ |
| $C_{120}O$-*ppc* | 121 | 13.095 | 5.837 | $C_s$ |
| $C_{120}O$-*ppt* | 121 | 13.094 | 5.832 | $C_1$ |

The construction of possible dimers, containing two oxygen atoms, can be done keeping in mind the following: there are four dimers of $C_{120}$, linked by a four-membered ring and the furanoid bridges are adjacent to this ring. If the two furanoid rings are on the same side of the four-membered ring, the isomer is denoted *cis*, while they are in the opposite sites, the structure is *trans*. Thus, for each $C_{120}$ dimer there are at least two possible structures of formula $C_{120}O_2$. In the case of *hp* and *ppc* dimmers, because there are different bond types on the two sides of the cyclobutane ring, one more dimer is possible. All together, ten structural isomers of $C_{120}O_2$ formula can be constructed (Figure 9). All carbon atoms involved in the junction are $sp^3$ hybridized.

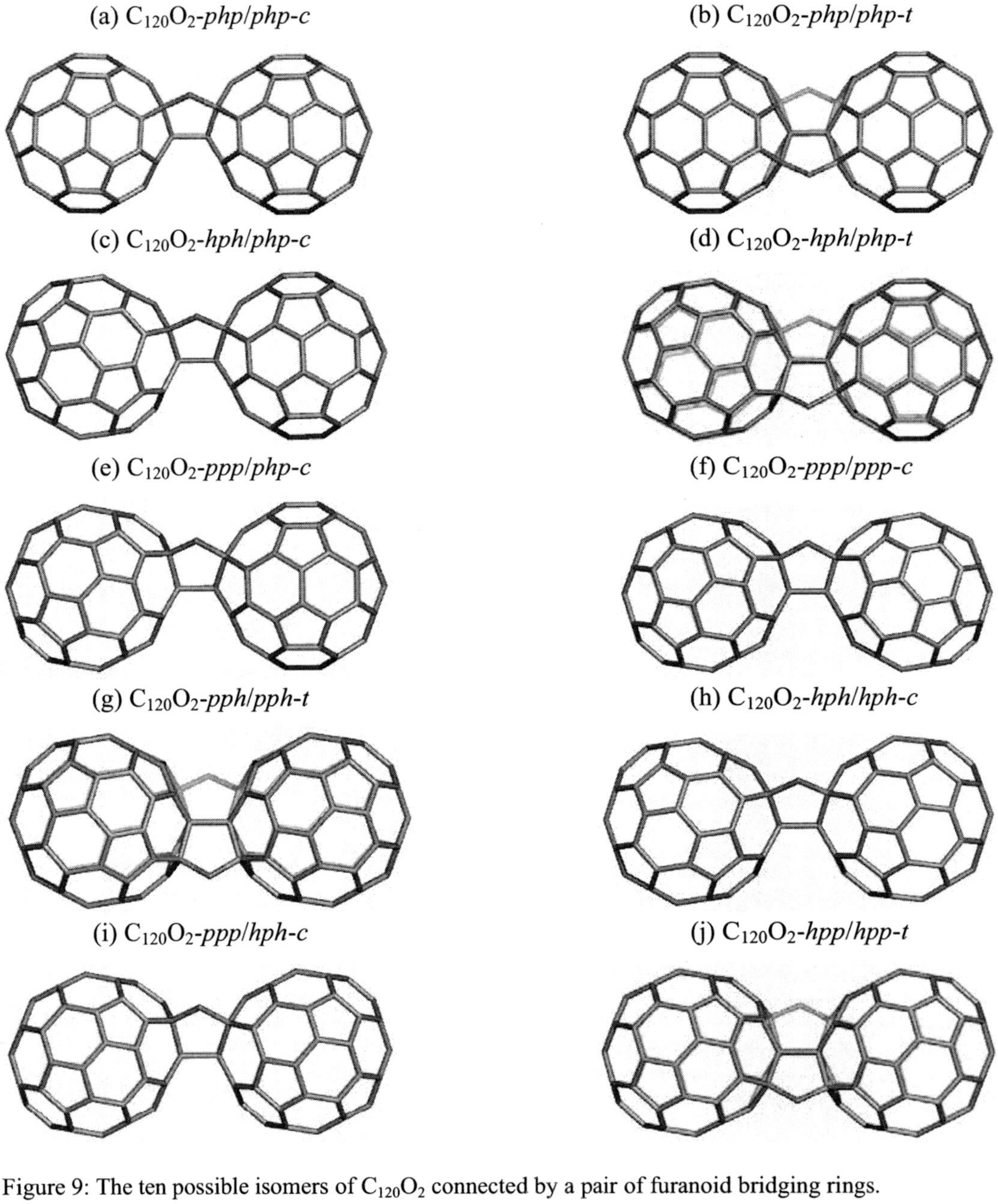


Figure 9: The ten possible isomers of $C_{120}O_2$ connected by a pair of furanoid bridging rings.

Calculations, carried out on the ten isomers (see figure 9 below) are presented in Table 4. Once again the heat of formation decrease with the number of *h* bonds involved in the furanoid bridges, the most stable being $C_{120}O_2$-*hph*/*hph-c*. Recall that the dimer $C_{120}$-*hh* was also the lowest energy isomer.

**Table 4: PM3 data for $C_{120}O_2$ $sp^3$ isomers**

| Dimer | Isomer | *N* | $H_f / N$ | GAP (eV) | Sym |
|---|---|---|---|---|---|
| *hh* | *php-php-c* | 122 | 12.257 | 6.414 | $C_{2v}$ |
| | *php-php-t* | 122 | 13.168 | 2.680 | $C_{2h}$ |
| *hp* | *hph-php-c* | 122 | 12.184 | 6.437 | $C_S$ |
| | *hpp-php-t* | 122 | 12.771 | 3.231 | $C_1$ |
| | *ppp-php-c* | 122 | 12.604 | 5.689 | $C_S$ |
| *ppc* | *ppp-ppp-c* | 122 | 12.828 | 5.645 | $C_{2v}$ |
| | *hph-hph-c* | 122 | 12.109 | 6.467 | $C_{2v}$ |
| | *pph-pph-t* | 122 | 12.433 | 5.878 | $C_S$ |
| *ppt* | *hph-ppp-c* | 122 | 12.533 | 5.695 | $C_S$ |
| | *hpp-hpp-t* | 122 | 12.486 | 5.866 | $C_1$ |
| | *cis-1 syn* | 122 | 10.148 | 6.332 | $C_1$ |
| | *cis-1 anti* | 122 | 10.170 | 6.318 | $C_1$ |

Below will be shone that the initial cycloaddition between $C_{60}$ and $C_{60}O_n$ molecules takes place in such manner, the further transformation to a $C_{120}O$ tubulene could be possible.

## 2.2. Coalescence through Stone-Wales Bond Flipping

Interconversion of fullerene isomers is possible by the rotation of atom pairs in the molecules about their mutual bond centre. Such atomic rearrangement, called Stone-Wales (SW) or pyracylenic transformation,[32] is suitable for describing the coalescence between fullerenes and/or nanotubes.

The initial orientation of the two molecular cages will affect the pathway of the coalescence. But, regardless of the starting cycloadduct, it was found that the very first steps are similar, leading to all $sp^2$ peanut-shaped dimers. Further, there are several ways for completing transformation up to the perfect tubulene (*i.e.*, tessellated by only hexagons and pentagons). It is also possible the interconversion among various coalescence appearing species. The energy diagram of the SW transformation of the four cycloadducts of $C_{60}$ to the corresponding peanut dimers is presented in Figure 10. The lowest energy peanut dimer is the $C_{120\text{-}77}$ (Figure 6c).

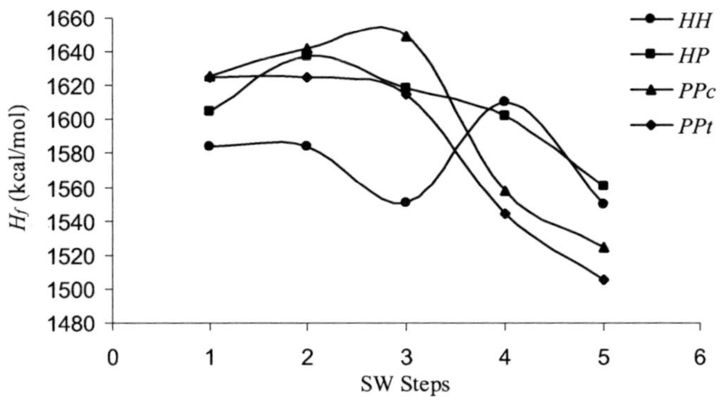


Figure 10: Energy diagram of the SW transformation from sp3 cycloadducts to sp2 peanuts, in case of C60 molecules.

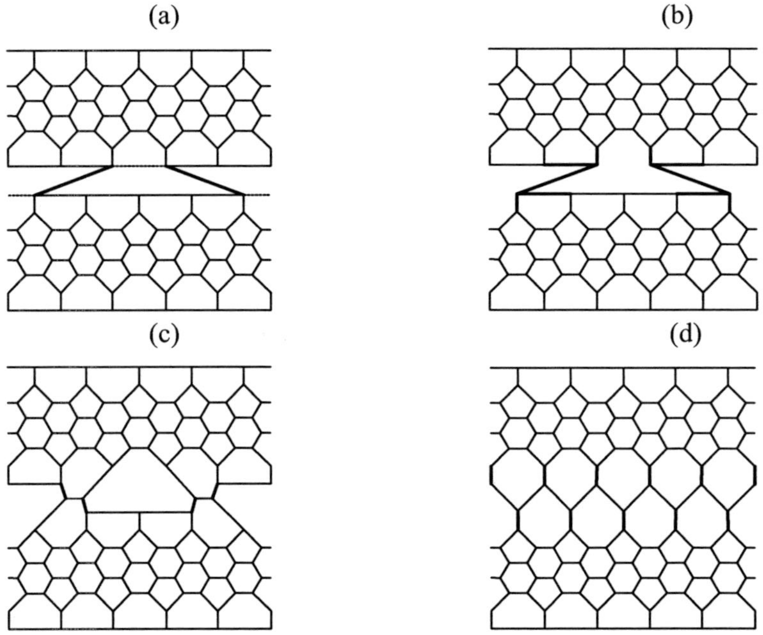


Figure 11: The complete lattice transformation, from [2+2] cycloadduct to peanut dimer, in case of $C_{60}$ molecules.

In a previous work,[33] we presented a series of $C_{120}$ epoxy derivatives, which energy varies with the location of the oxygen atom. The most stable isomers $C_{120}O$ and $C_{120}O_2$ are shown in Figure 12.

Figure 12: The most stable epoxy and di-epoxy dimeric tubulenes of $C_{60}$.

The energy difference between the $C_{120}$ dimer and $C_{120}O_n$ epoxy derivative is almost constant at every SW step, as shown in Figure 13.

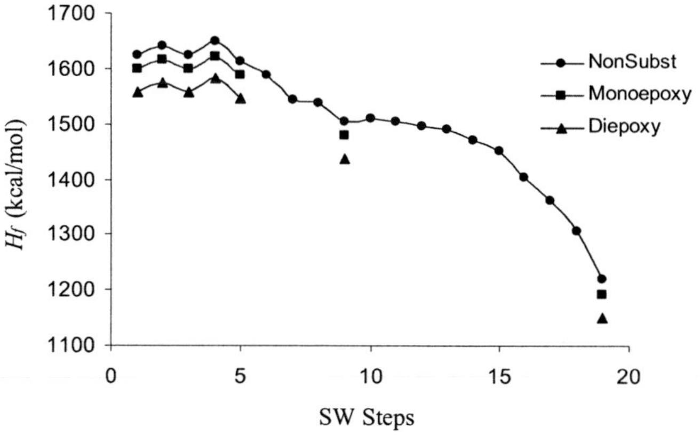


Figure 13: Comparative coalescence energy plot for $C_{60}$, $C_{60}O$, and $C_{60}O_2$ molecules.

A question arises about the proper route, if trimerization or oligomerization occurs. The transformation of a junction could terminate at a peanut shape, which results in an oligomer with multiple necks, or it could continue until a perfect tubule is formed. Figure 14 presents such possibilities in the coalescence of four $C_{60}$ molecules.

(a) $C_{240}$_1-2-3-4 (b) $C_{240}$_1234

(c) $C_{240}$_1-2-34 (d) $C_{240}$_1-23-4

(e) $C_{240}$_1-234 (f) $C_{240}$_12-34

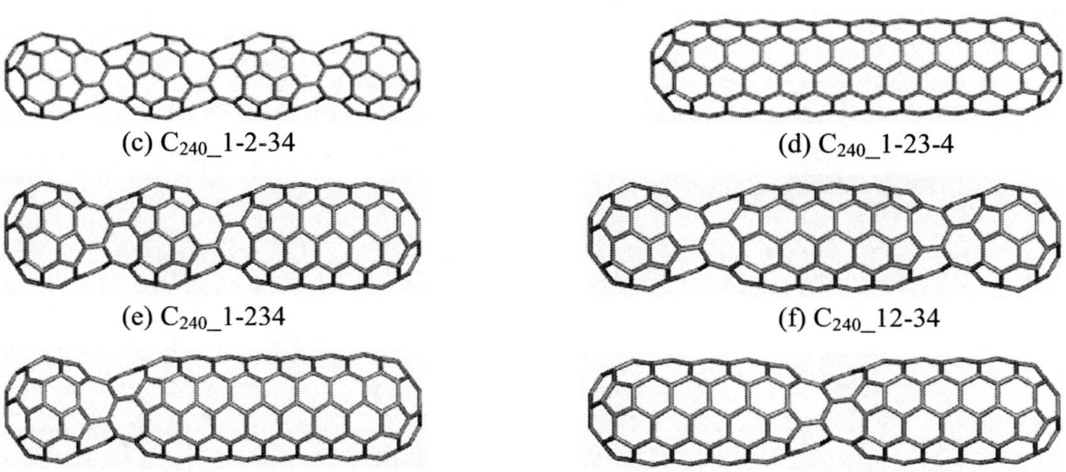

Figure 14: Multi-necked peanut and tubular tetramers of $C_{60}$.

Table 5 lists semiempirical data for both trimers and tetramers of $C_{60}$. It can be seen that, the heat of formation $H_f$ increases as the number of necks in oligomers increases. Thus, it seems that a new cycloaddition is favored only after the tubular shape was reached. The energy gain can be associated with the relief of strain originating in the negative curvature of the necks. However, in nanopeapods annealed at moderate temperature, corrugated tubules could be seen by TEM.[31] The strain relief is obtained at higher temperature, when DWNT is obtained.[30]

**Table 5: PM3 for all $sp^2$ $C_{60}$ trimers and tetramers**

| Name | $N$ | $H_f$ / $N$ (kcal/mol) | GAP (eV) | Sym |
|---|---|---|---|---|
| $C_{180}$_1-2-3 | 180 | 12.224 | 5.917 | $D_{5d}$ |
| $C_{180}$_12-3 | 180 | 10.602 | 4.830 | $C_5$ |
| $C_{180}$_123 | 180 | 9.052 | 4.057 | $D_{5d}$ |
| $C_{240}$_1-2-3-4 | 240 | 12.062 | 5.819 | $D_{5d}$ |
| $C_{240}$_12-3-4 | 240 | 10.846 | 4.831 | $C_5$ |
| $C_{240}$_1-23-4 | 240 | 10.826 | 4.870 | $D_{5d}$ |
| $C_{240}$_123-4 | 240 | 9.663 | 4.088 | $C_5$ |
| $C_{240}$_12-34 | 240 | 9.630 | 4.791 | $D_{5d}$ |
| $C_{240}$_1234 | 240 | 8.500 | 3.697 | $D_{5d}$ |

The $C_{120(56^5(56)^5-A[10,6])}$ dimeric armchair tubulene is the ultimate structure resulting from the all possible cycloadducts (Figure 15a).

Within the study of SW transformation of the peanut dimers we found pathways leading to armchair or zigzag tubulenes, having pentagons or hexagons as the polar ring. Such tubulenes have also been identified by molecular dynamics.[34]

Figure 15 illustrates several $C_{120}$ dimers, along with the corresponding cap (where the dimer is a tubulene).

These dimers have a properly closed shell, associated with a perfect Clar structure PCS (see below). Their $H_f$ is similar but there is a considerable difference in the HOMO-LUMO gap, the highest values corresponding to those structures which are PCS. Tubulenes with hexagonal polar rings have a more relaxed surface, as it can be seen by the strain energy values. Calculated data is summarized in Table 6.

**Table 6: PM3, strain and spectral data for $C_{120}$ cages in Figure 15**

| | $N$ | $H_f$ / $N$ (kcal/mol) | Gap (eV) | Strain (kcal/mol) | Spectral Data | | $D$ (nm) | Sym |
|---|---|---|---|---|---|---|---|---|
| | | | | | $\lambda_{N/2}$ | $\lambda_{N/2+1}$ | | |
| **(A)** | 120 | 10.156 | 4.793 | 5.162 | 0.310 | -0.061 | 0.649 | $D_{5d}$ |
| **(B)** | 120 | 10.271 | 3.706 | 4.664 | 0.319 | 0.319 | 0.792 | $D_{5h}$ |
| **(C)** | 120 | 10.175 | 5.789 | 4.903 | 0.521 | -0.099 | 0.110 | $D_{5d}$ |
| **(D)** | 120 | 9.8195 | 5.015 | 4.788 | 0.350 | 0.000 | 0.807 | $D_{6d}$ |
| **(E)** | 120 | 9.9375 | 4.587 | 4.504 | 0.402 | 0.336 | 0.954 | $D_{6d}$ |
| **(F)** | 120 | 10.295 | 4.419 | 4.928 | 0.271 | 0.137 | - | $D_{3d}$ |

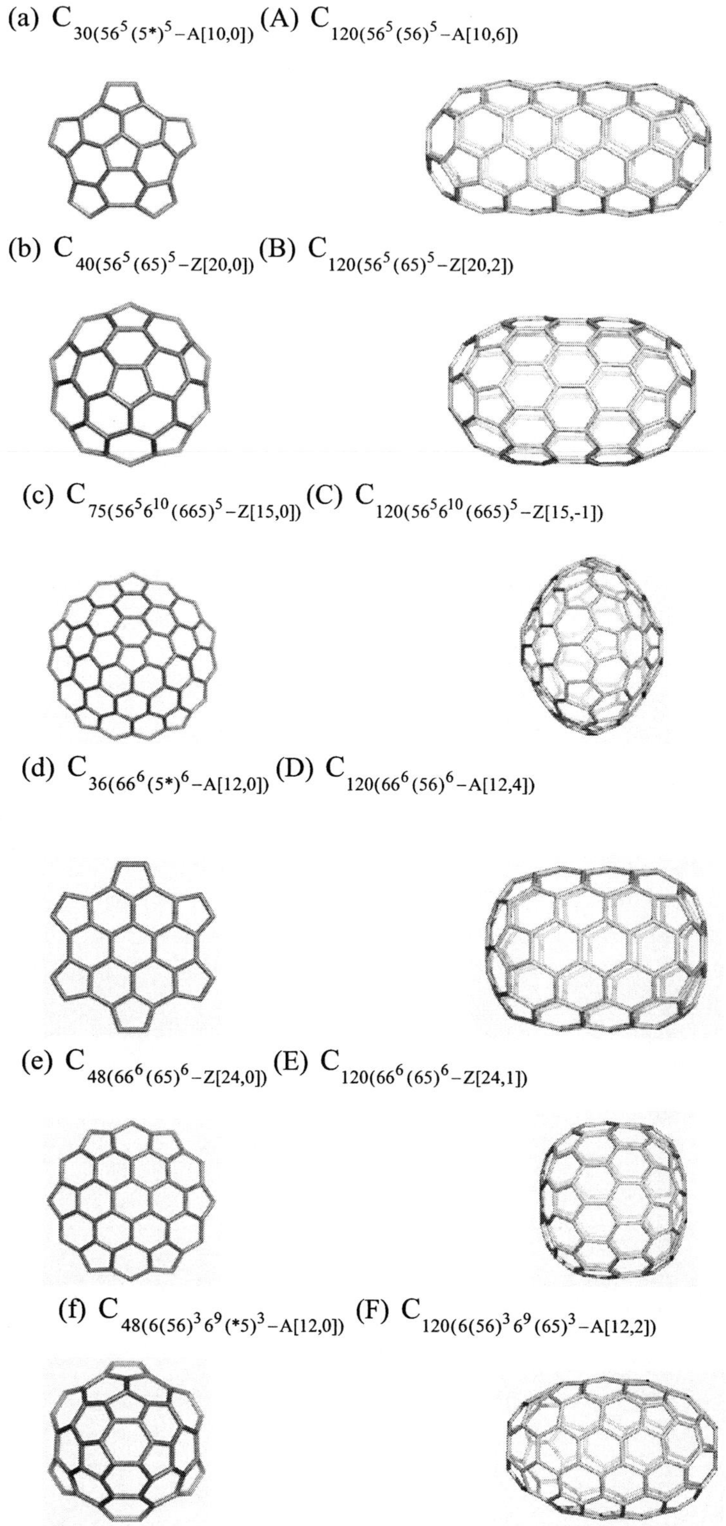


Figure 15: Different caps and corresponding dimeric cages possibly appearing in the coalescence of $C_{60}$.

Further transformation of $C_{120(56^5(56)^5-A[10,6])}$ is possible, leading to the other tubulenes. Because of similar $H_f$, a mixture of the molecules shown in Figure 15 is expected. Figure 16 gives the energetic plot of SW transformation, starting from $C_{120(56^5(56)^5-A[10,6])}$ and $C_{120(66^6(56)^6-A[12,4])}$

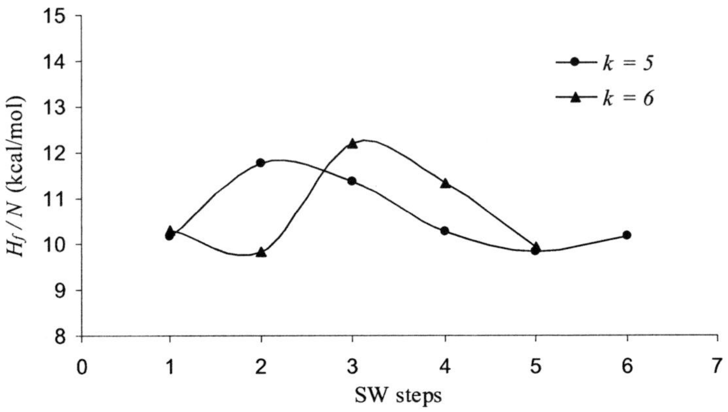


Figure 16: Energetic diagram of the SW interconversion of the objects in Figure 15.

## *2.2.1 Pathway to $C_{114}$ Tubulene*

It was mentioned above that, the primary coalescence intermediates can further polymerize only if the formation of a new cycloadduct between two *h* bonds is possible. In this respect, it is required that dimeric structures have hexagons at both ends. The tubulene that preserves the position of pentagons as in $C_{60}$ would be $C_{114}$, which corresponds to a polar face distance of ~1.3 nm, as measured by TEM in nanopeapods. The formation of such a unit involves the displacement of six carbon atoms per junction. Here we propose a possible way to $C_{114}$ which includes several SW steps (Figure 17).

(a) $C_{114(6(56)^3(665)^3(656)^3(86)^3-A[12,0])}$ (b) PCS

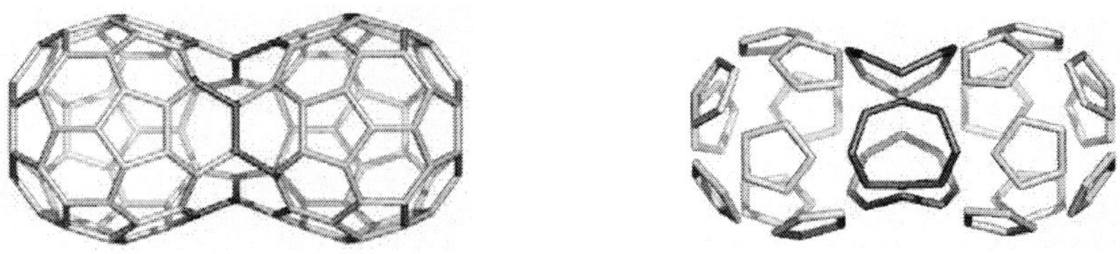

(c) $C_{39(6(56)^3(665)^3-Z[18,0])}$ (d) $C_{114(6(56)^3(665)^3-Z[18,2])}$

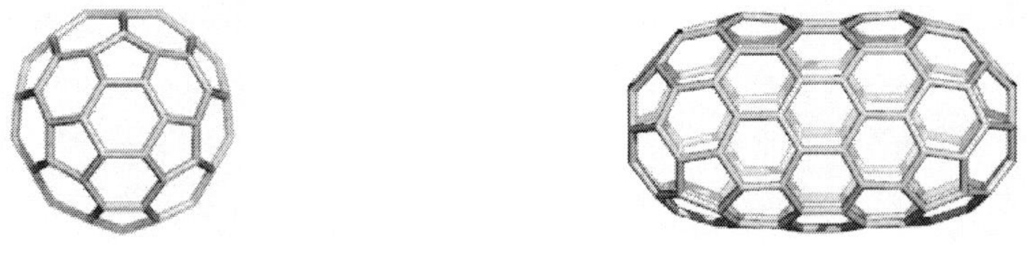

Figure 17: Pathway to $C_{114}$ cage.

Electron beam irradiation could result in defects of nano lattice so that the coalescence is promoted (see next section for merging nanotubes). By controlled irradiation, the displacement of three carbon atoms from a hexagonal face would be possible. The resulted units could increase their stability by dimerization (Figure 17a). This peanut-shaped intermediate is a PCS (Figure 17b); it further isomerizes to $C_{114}$, by SW edge rotation.

In $C_{114}$ peanut structure (Figure 17d) there are three bonds in the equatorial/joining zone. The removal of these bonds (together with their end-atoms) would generate two caps $C_{54(6(56)^3(665)^3(656)^3-Z[12,0])}$, perfectly oriented to form a $C_{108}$ peanut cage (having a PCS), with three hexagon-octagon pairs in the junction (Figure 18a).

This could isomerize to a cage with a fully heptagonal covering in the equatorial zone (Figure 18b). Such peanut structures were observed in annealed or irradiated peapods. Note that their length is identical to the $C_{114}$ cage. Figure 18c,d presents other two structures, which could result from $C_{114}$ by stepwise elimination of the six atoms.

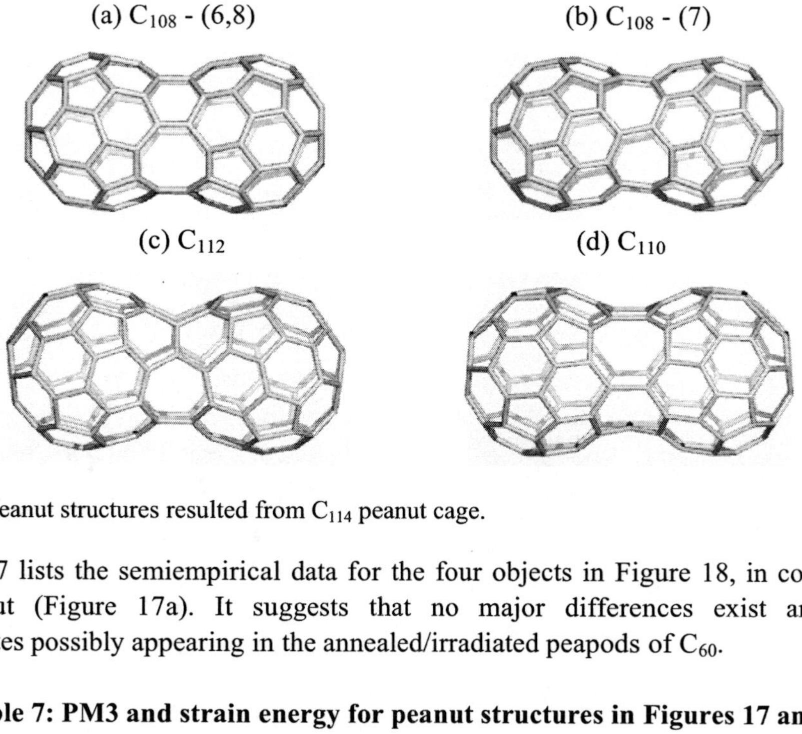


Figure 18: Peanut structures resulted from $C_{114}$ peanut cage.

Table 7 lists the semiempirical data for the four objects in Figure 18, in comparison to $C_{114}$ peanut (Figure 17a). It suggests that no major differences exist among these intermediates possibly appearing in the annealed/irradiated peapods of $C_{60}$.

**Table 7: PM3 and strain energy for peanut structures in Figures 17 and 18**

| Name | *N* | $H_f/N$ (kcal/mol) | GAP (eV) | POAV 1 (kcal/mol) | Sym |
|---|---|---|---|---|---|
| $C_{114}$ (peanut) | 114 | 12.7894 | 6.0814 | 6.711 | $D_{3h}$ |
| $C_{112}$ | 112 | 13.0186 | 5.3058 | 6.605 | $C_{2v}$ |
| $C_{110}$ | 110 | 13.2559 | 5.4234 | 6.476 | $C_{2v}$ |
| $C_{108}$-(68) | 108 | 13.0038 | 5.2419 | 6.380 | $D_{3h}$ |
| $C_{108}$-(7) | 108 | 12.9526 | 4.8702 | 6.491 | $D_{3d}$ |

## 2.3. DWNT as DWT Precursors

The final product in the thermal annealing of nanopeapods is a double-walled carbon nanotube DWNT. It was above discussed that coalescence is also induced by electron beam irradiation, when vacancies and dangling bonds appear in the carbon lattice and energy reduction is achieved by elimination of such defects. Field emission electron guns are available, which permit the controlled irradiation of a small region on the nanometer scale. The dangling bonds, generated by irradiation, could stabilize themselves by forming new bonds between the two interlaced tubular units. This would result in formation of an elongated torus, named hereafter "distinct-walled torus" DWT. Various types of DWTs, built up from only armchair nanotubes, are further presented.

The inner tube is shown in Figure 19 by the top hexagons. Breaking of the boldface bonds results in two dangling bonds. To remake the valence of three, each dangling bond has to connect with two carbon atoms from the outer tube (see dotted bonds in Figure 19). That is why twice as much carbon atoms in a row in the outer tube are needed. Of course, the growth of the inner tube depends on the outer tube diameter.

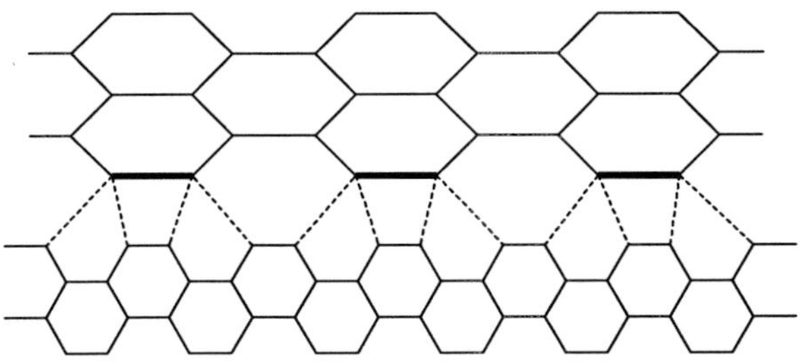

Figure 19: Joining scheme of the two distinct tubes in DWT formation.

The junction between the two tubes consists of alternating pentagon/octagon/ pentagon *POP* triplets, which ensure the complementary curvature and stability of the structure. The elimination of octagons is possible by the Stone-Wales rotation of the bonds connecting the two vicinal pentagons (see Figure 10, bond in boldface). However, the pentagon-heptagon *PH* containing junction appears more strained, after optimization, as shown by increased $H_f$ values (Table 10). It is easily seen (Figure 20) that both these junction coverings provide a PCS. Namely, the non-hexagonal faces are bounded only by single bonds, while the double bonds lye in the benzenoid hexagons.

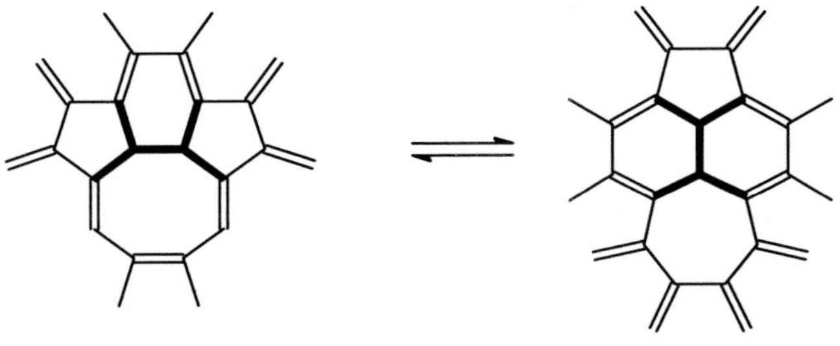

Figure 20: Transformation of the junction between the inner and outer tube of a DWT by Stone-Wales rotation of the boldface bond.

Considering an armchair nanotube with even or odd number of atom rows, the combination of different nanotubes leads to four distinct classes of DWTs, which resembles in their point group symmetry. The torus will have a horizontal plane of symmetry only in case when both tubes have an even number of rows. Figure 21 presents the repeat units of these tori. By copying the lines in boldface for $l_e / l_i = 0, 1, 2 \ldots$ the tube length increases. In case $l_e / l_i = 0$, the smallest possible torus, having abutting pentagons, is obtained.

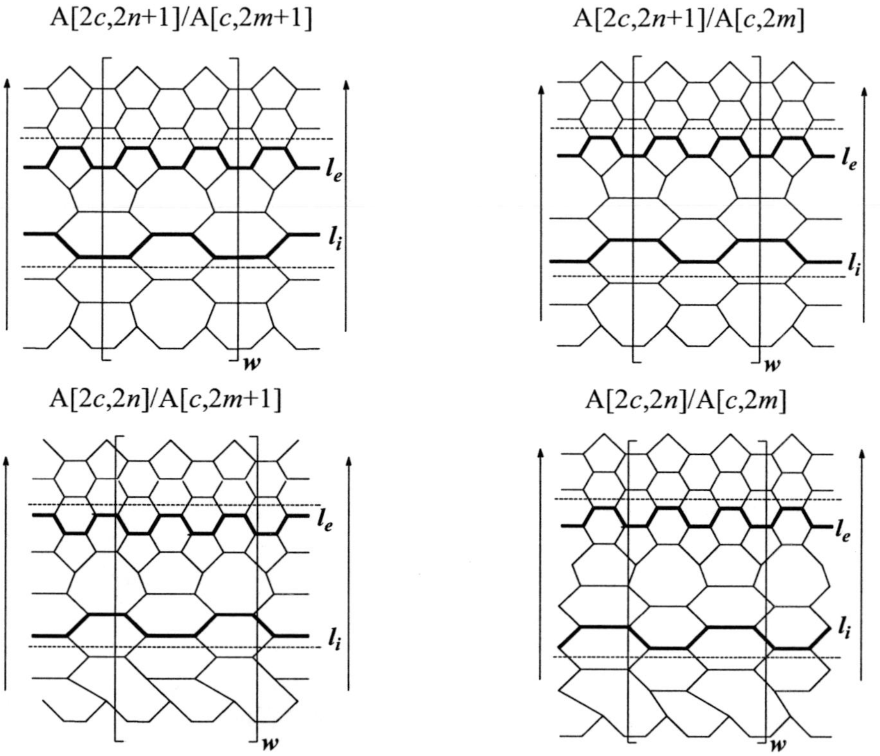


Figure 21: Units of the four *POP*-DWT types in geodesic projection.

The torus width *w* can be increased by adding more repeat units (in brackets). Although it is possible to construct a torus from one unit, the strain at the apex is reduced starting from $w \geq 4$. The name of these DWTs includes the specification of the two tubes. For example, A[2*c*,2*n*+1] corresponds to the outer tube with 2*c* atoms in a row and odd number of rows (2*n*+1) in length, while the inner tube A[*c*,2*m*+1] has half as many atoms in a row and again an odd number of rows (2*m*+1) in length. The length of two tubes can be either equal, when $n = m$, or different (Figure 21).

For every class of *POP*-DWTs an example is presented in Figures 22-23 both as top and side view; the non-hexagonal faces are shown in dark color. In all these examples, the inner tube is the armchair A[10,*n*], which is the tube that can result in the coalescence of two $C_{60}$ molecules. Figure 22 presents *POP*-DWTs with odd number of rows in the external tube. The structures in the left column have the highest point group symmetry, $D_{5h}$.

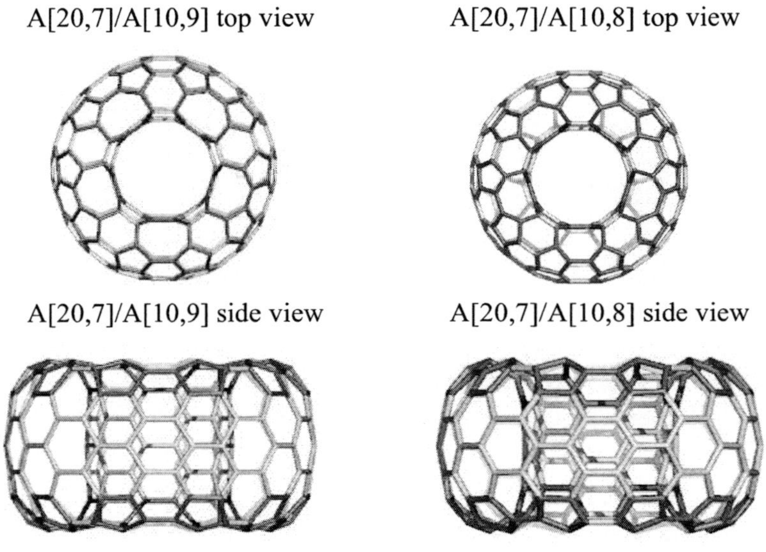


Figure 22: *POP*-DWTs with an odd number of rows in the outer tube.

Figure 23 presents tori with even number of rows in the outer tube.

It is well known that, the minimal energy, in case of polyhex tori, is reached only at a perfect circular section. This happens at very large central hollow of the torus.[35]

With DWTs the situation is different. If the length of both tubes is small enough, the cross section of the torus wall is near circular. As the tube length increases, the cross section becomes an elongated, elliptic-shaped one. The heat of formation decreases as the tube length increases, in all the four classes of DWTs (Table 8).

Note that similar tori have been previously proposed by several authors,[36-39] but no relation with peapods, coalescence and DWNT was suggested. Also note that hemitori, resembling with our DWTs, have been observed by TEM[40],[41] and inferred in multi walled nanotube growth.

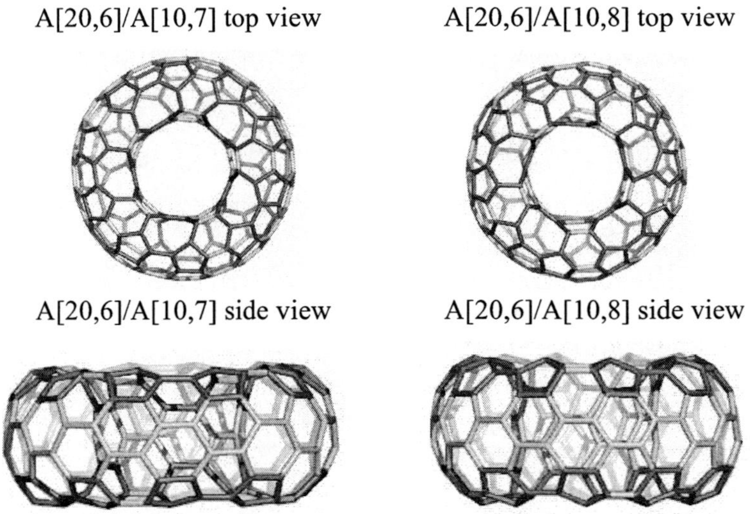


Figure 23: *POP*-DWTs with an even number of rows in the outer tube.

**Table 8: PM3 data for three classes of *POP*-DWTs**

| Name | *N* | $H_f/N$ (kcal/mol) | Gap (eV) | Sym |
|---|---|---|---|---|
| A[20,5]/A[10,7] | 170 | 15.159 | 5.031 | $D_{5h}$ |
| A[20,7]/A[10,9] | 230 | 12.122 | 5.685 | $D_{5h}$ |
| A[20,9]/A[10,11] | 290 | 10.827 | 4.218 | $D_{5h}$ |
| A[20,11]/A[10,13] | 350 | 9.890 | 3.767 | $D_{5h}$ |
| A[20,3]/A[10,4] | 100 | 32.213 | 6.274 | $D_{5d}$ |
| A[20,5]/A[10,6] | 160 | 15.978 | 5.134 | $D_{5d}$ |
| A[20,7]/A[10,8] | 220 | 12.678 | 5.589 | $D_{5d}$ |
| A[20,9]/A[10,10] | 280 | 11.376 | 4.092 | $D_{5d}$ |
| A[20,4]/A[10,6] | 140 | 20.306 | 5.967 | $D_5$ |
| A[20,6]/A[10,8] | 200 | 14.408 | 4.988 | $D_5$ |
| A[20,8]/A[10,10] | 260 | 12.319 | 4.051 | $D_5$ |
| A[20,10]/A[10,12] | 320 | 10.756 | 4.760 | $D_5$ |

The topology of the zone joining the two tubes of a DWT influences both the strain and heat of formation. Keeping constant one tube and varying the length of the other one enables construction of a series of tori. Semiempirical calculations are presented in Table and Figures 24-25. The minimal energy corresponds to the most relaxed structure. When there is a notable difference between the tubes length, the energy increases due to hexagons forced to lie at the apex. At extreme low/high values of *m*, the structure could not be optimized anymore. Figure 24 shows the energy curve for a series of *POP*-DWTs, where the inner tube length varied in the range *m* = 4,5... The structure with lowest energy was found A[20,5]/A[10,8].

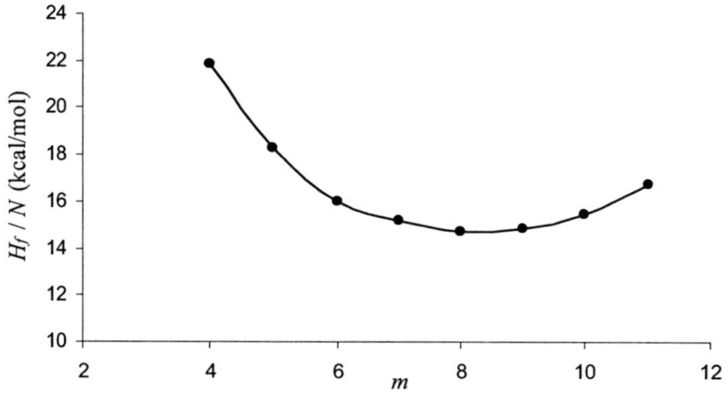


Figure 24: Heat of formation values for the series A[20,5]/A[10,*m*] *m* = 4,5…11. of *POP*-DWTs.

The plot in Figure 25 shows a series of *POP*-DWTs with the inner tube length maintained constant.

If the length of both tubes was increased by the same number of rows, the junction zone shoved an almost constant strain. The global strain energy decreases as the tube length increases (Figure 26).

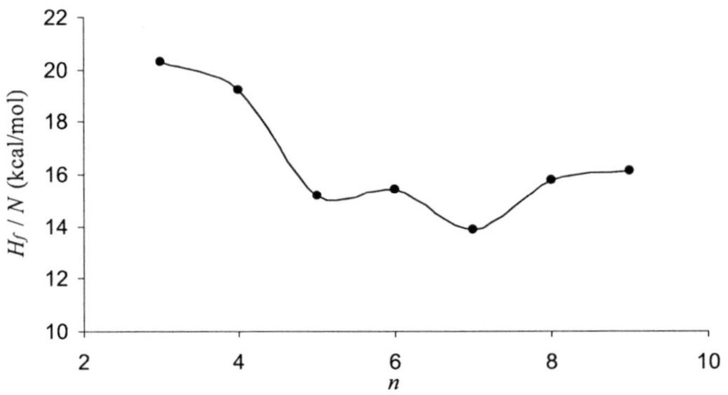


Figure 25: Heat of formation values for the series A[20,*n*]/A[10,7] $n$ = 3,5…9 of *POP*-DWTs.

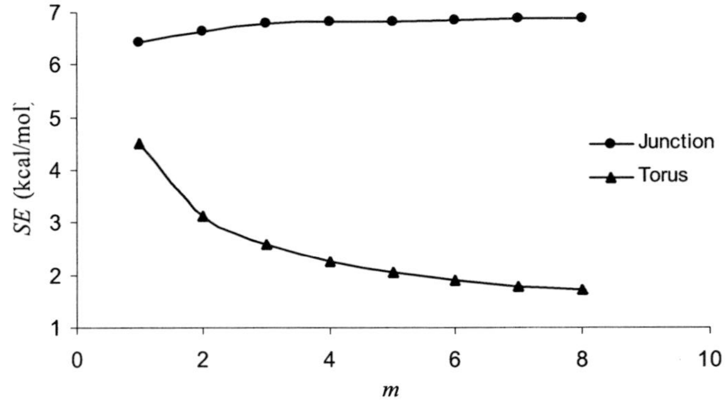


Figure 26: Strain energy for the series A[20,*n*]/A[10,*m*], n = 7, 13,…49; m = 11, 17, …53 of *POP*-DWTs.

We limited here to only A[20,*n*]/A[10,*m*] tori, because these DWTs could result in the coalescence of fullerene molecules. DWTs with larger central hollow have also been modeled, in studying the influence of the primary tube diameter on the structure stability. In this respect, a series of tori was constructed: the length was kept unchanged but the number of repeat units was varied. The tori consisting of one to three repeat units were very strained, they could not be optimized. Figure 27 presents tori with four and six repeat units, respectively.

The data in Table 9 show that both thermodynamic and kinetic stability increases as the tubes diameter increases. The strain of these structures decreases as the primary tube width increases.

Since semiempirical calculations are limited by the large number of atoms, we performed only strain energy calculations (Figure 28).

The Stone-Wales rotation can change the triple *POP* to the *PH* pair, in the junction zones. Figure 29 presents two structures resulted by such a transformation.

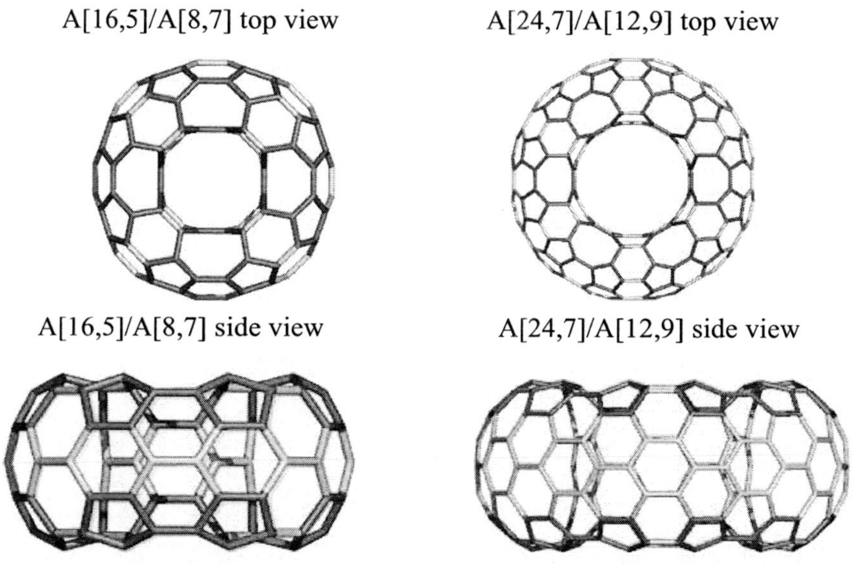


Figure 27: *POP*-DWTs with four/six repeat units.

**Table 9: PM3 data for a series of *POP*-DWTs having different number of repeat units**

| Name | $N$ | $HF / N$ kcal/mol | Gap (eV) | Sym |
|---|---|---|---|---|
| [12,5]/[6,7] | 102 | 28.146 | 4.059 | $D_{3h}$ |
| [16,5]/[8,7] | 136 | 18.542 | 4.687 | $D_{4h}$ |
| [20,5]/[10,7] | 170 | 15.159 | 5.031 | $D_{5h}$ |
| [24,5]/[12,7] | 204 | 14.635 | 5.053 | $D_{6h}$ |
| [28,5]/[14,7] | 238 | 15.647 | 5.278 | $D_{7h}$ |

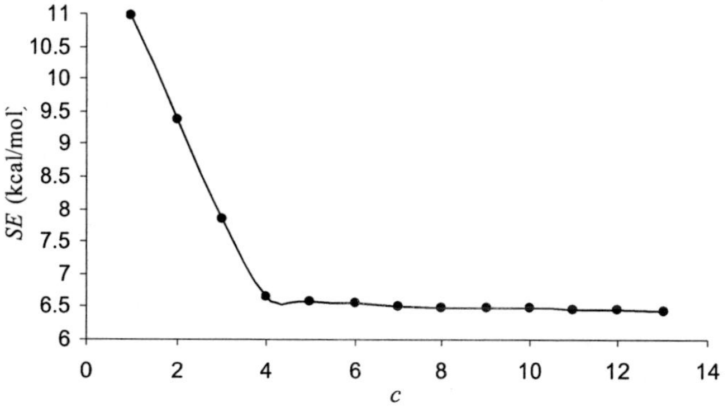


Figure 28: Strain energy for the series A[2*c*,7]/A[*c*,5], c = 4, 6, ...30 of *POP*-DWTs.

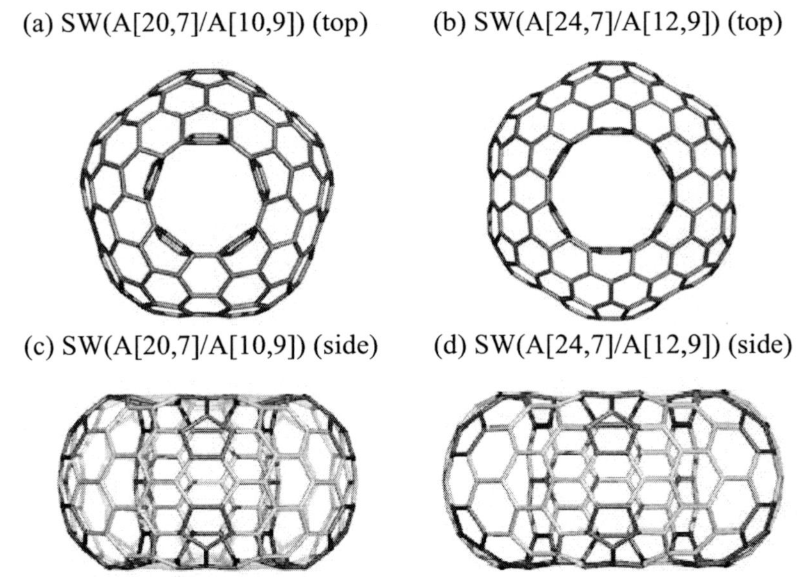


Figure 29: *PH*-DWTs resulted by the SW rotation of bonds joining two pentagons.

Table 10 lists data for sets of *PH*-DWT related by SW to tori in Table 8. The conclusion is that *PH*-junction rises the strain of the torus, as shown by the $H_f$ values.

**Table 10: PM3 data for three series of *PH*-DWTs**

| Name | $N$ | $H_f/N$ (kcal/mol) | Gap (eV) | Sym |
|---|---|---|---|---|
| SW(A[20,5]/A[10,7]) | 170 | 18.202 | 4.815 | $D_{5h}$ |
| SW(A[20,7]/A[10,9]) | 230 | 14.311 | 4.547 | $D_{5h}$ |
| SW(A[20,9]/A[10,11]) | 290 | 12.249 | 5.078 | $D_{5h}$ |
| SW(A[20,5]/A[10,6]) | 160 | 17.032 | 4.719 | $D_{5d}$ |
| SW(A[20,7]/A[10,8]) | 220 | 13.731 | 4.584 | $D_{5d}$ |
| SW(A[20,9]/A[10,10]) | 280 | 11.870 | 4.825 | $D_{5d}$ |
| SW(A[20,4]/A[10,6]) | 140 | 20.306 | 5.967 | $D_5$ |
| SW(A[20,6]/A[10,8]) | 200 | 14.408 | 4.988 | $D_5$ |
| SW(A[20,8]/A[10,10]) | 260 | 12.319 | 4.051 | $D_5$ |

### *2.3.1 Perfect Clar DWTs*

A molecule is more aromatic (eventually more stable) if it has a perfect Clar structure PCS.[42],[43] It is a disjoint set of faces, built up on all vertices in molecule, whose boundaries form a 2-factor (*i.e.*, a set of disjoint cycles). A PCS is associated with a Fries structure, which is a Kekulé structure having the maximum possible number of benzenoid faces. A Kekulé structure is a set of pairwise disjoint edges/bonds of the molecule (over all its atoms). Molecules, having PCS/Fries structures, represent *total resonant sextet* TRS benzenoid molecules and it is expected to be extremely stable, in the valence bond theory.[44] Leapfrog *Le* is the only operation on maps that provides PCS transforms.

In DWT, the apex junction of the two tubes looks as shown in Figure 20 (with no double bonds in the odd-fold cycles) and is suitable for a PCS. The criterion for PCS realization refers only to the polyhex tubular zone; for armchair A[$c$,$n$] nanotubes, it is:

*POP*-DWTs: A[$2c$,$3(n+1)+1$]/A[$c$,$3(m+1)-1$]
*PH*-DWTs: A[$2c$,$3(n+2)$]/A[$c$,$3(m+2)$]

In the above formulas, $2c/c$ represents the number of atoms in the outer/inner tube section, while $n/m$ denotes the length of the outer/inner tube.

A PCS object results by the *Leapfrog Le* operation on maps: $Le(M) = Du(P_3(M))$. Keeping in mind that dual of dual returns the original map, the *Retro-Leapfrog RLe,* leading to the parent torus, can be written as follows: $Du(Le(M)) = P_3(M)$ and $RLe(M) = P_{-3}(M)$, with $P_{-3}$ being the *retro*-operation of stellation $P_3$.

(a) *RLe*(A[20,7]/A[10,8]) (b) *RLe*(SW(A[20,9]/A[10,9]))

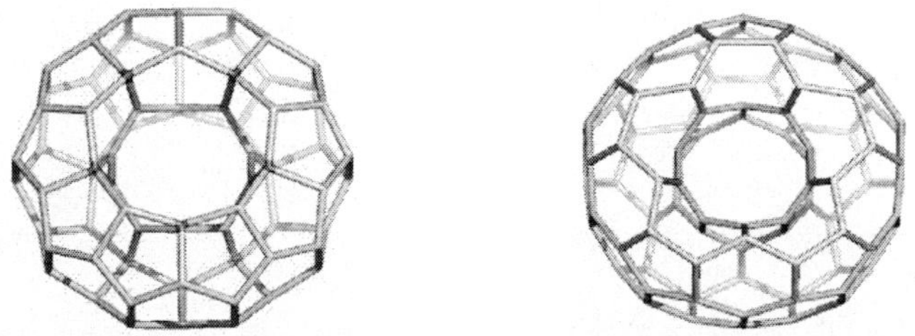

(c) A[20,7]/A[10,8] (d) SW(A[20,9]/A[10,9])

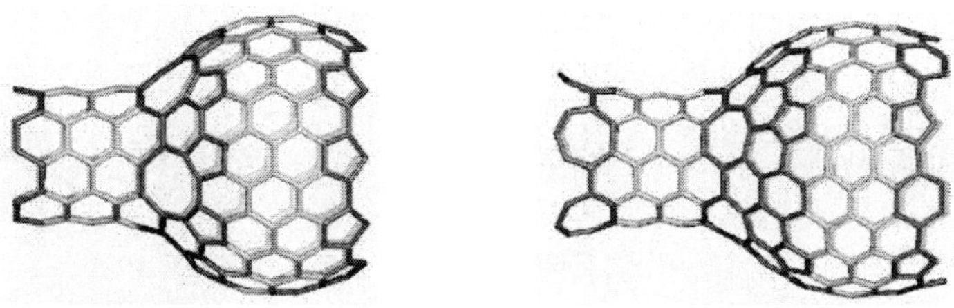

(e) PCS of A[20,7]/A[10,8] (f) PCS of SW(A[20,9]/A[10,9])

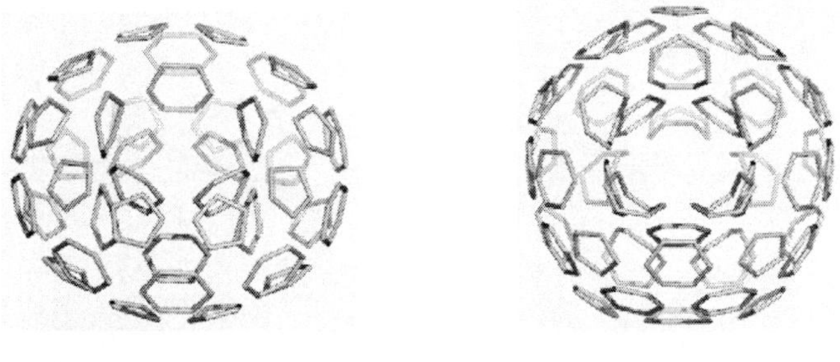

Figure 30: *Retro-Leapfrog* (a, b); PCS junctions (c, d) and Perfect Clar Structure DWTs (e, f).

The parent of a *POP*-DWT has the corresponding vertices of valence four (Figure 30a, in dark). The SW pair *PH*-DWT has an all trivalent parent map, as illustrated in Figure 30b. The corresponding junction zones are detailed in Figure 30c,d. The PCS of A[20,7]/A[10,8] and SW(A[20,9]/A[10,9]) tori are given in Figure 30e,f.

As above mentioned, DWTs with a *PH* junction are less stable than their corresponding SW *POP* pair. It is generally true, excepting the case they have a perfect Clar structure PCS; both the kinetic and thermodynamic stability of a PCS *PH*-DWT is higher than its PCS *POP*-DWT pair. Table 11 presents semiempirical data performed on PCS DWTs.

**Table 11: PM3 data for PCS DWTs**

| Name | *N* | $H_f / N$ (kcal/mol) | GAP (eV) | Sym | Junction type |
|---|---|---|---|---|---|
| A[20,4]/A[10,5] | 130 | 23.830 | 5.635 | $D_5$ | *POP* |
| A[20,4]/A[10,8] | 160 | 18.603 | 5.624 | $D_5$ | *POP* |
| A[20,4]/A[10,11] | 190 | 20.994 | 6.190 | $D_5$ | *POP* |
| A[20,7]/A[10,5] | 190 | 17.926 | 5.409 | $D_{5h}$ | *POP* |
| A[20,7]/A[10,8] | 220 | 12.678 | 5.589 | $D_{5d}$ | *POP* |
| A[20,7]/A[10,11] | 250 | 12.274 | 4.518 | $D_{5h}$ | *POP* |
| A[20,7]/A[10,14] | 280 | 14.917 | 5.181 | $D_{5d}$ | *POP* |
| A[20,10]/A[10,8] | 280 | 15.652 | 3.875 | $D_5$ | *POP* |
| SW(A[20,6]/A[10,6]) | 180 | 16.855 | 5.033 | $D_5$ | *PH* |
| A[20,9]/A[10,9] | 270 | 12.441 | 3.725 | $D_{5h}$ | *POP* |
| SW(A[20,9]/A[10,9]) | 270 | 12.051 | 4.266 | $D_{5h}$ | *PH* |

It was found that DWTs become more relaxed by the increase of the tube length. A supplementary stability is brought by the PCS. We limited here to the armchair achiral nanotubes, but it is also possible to construct tori from chiral nanotubes.

By extension, a perfect corannulenic structure *PCorS* was proposed by Diudea[45] (Figure 31). It is a *disjoint set* of (supra) faces covering all vertices in the molecular graph. The associated Fries-like structure is defined as above but its construction will avoid the corannulenic core vertices. A molecule has a PCorS if and only if it has a PCS. The reciprocal is not always true. The supra-organized corannulenic units are expected to contribute to the stability of the whole molecule and to a characteristic behavior in the magnetic field.

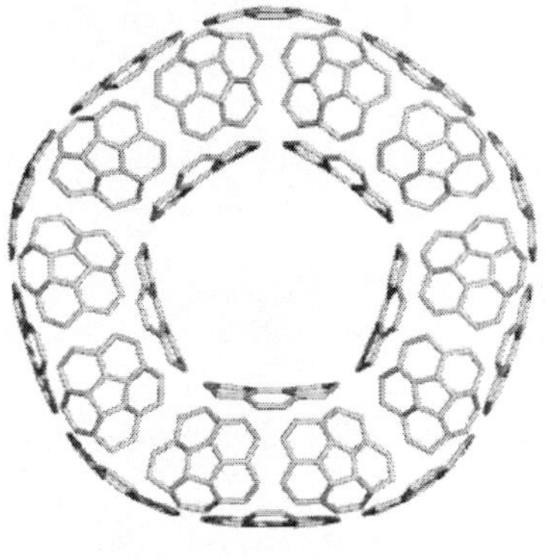

Figure 31: The PCorS of A[20,7]/A[10,5] *POP*-DWT.

## 2.4. Diameter Doubling of SWNT

Diameter doubling of single-walled carbon nanotubes has been observed experimentally. Armchair tubes were annealed at 1400 or 1500°C under vacuum, flowing argon, flowing hydrogen atmosphere, etc. As many as 60% of the nanotubes so treated coalesced with their neighbors, the proportion depending on the atmosphere. The coalesced nanotubes show twice (and occasionally three times) the diameter of the initial tubes. Only nanotubes with identical chirality are believed to undergo coalescence.[46,47]

Studies by high-resolution transmission electron microscopy HRTEM, allowed access to a variety of structural changes in carbon systems. Electron irradiation removes carbon atoms from their lattice sites, by knock-on displacements. The atoms can either be ejected from the tube, or migrate as interstitials along the inner or outer surface. The high temperature ensures a high mobility of interstitials and hence a rapid annealing of defects. Because of the vacancies, the system becomes energetically unstable. When tubes are assembled in bundles, the irradiated tubes will establish links in order to satisfy most of the dangling bonds. The fusion is driven by the minimization of strain energy so that tubes with larger diameter are created.[48]

Generation of interstitials at a graphite-diamond interface can lead to the transformation of graphite to diamond.[49] This phenomenon has also been discovered and studied by electron microscopy.

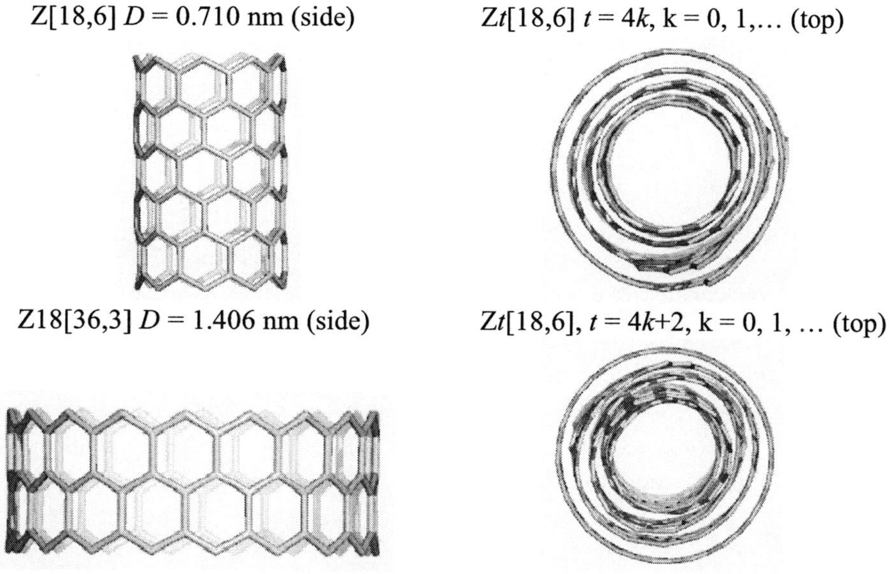


Figure 32: Zigzag nanotubes before and after twisting.

A zipper mechanism was proposed for armchair nanotube coalescence. The initial steps are similar to the formation of the local bridge between spherical fullerenes, followed by a sequence of SW transformations. The resulted Y-junction consists of an octagon and four heptagons, and it propagates axially, similar to a zipper. The transformation cycle requires nine SW steps, the process being exothermic. This transformation was evidenced experimentally by HRTEM. Nanotubes, lying close and parallel to each other, at 800°C start to merge fast after the initial connection. The main advantage of this mechanism is that such systems are vacancy free.[50]

In experiments with nanopeapods, the inner tube diameter growth, up to 0.71 nm distance from the host nanotube wall, was observed.[29] The authors proposed a model based on formation of achiral zigzag nanotubes in the beginning of fullerene coalescence. In the light of our previous results,[51] we propose the formation of chiral/twisted zigzag tubes, a process often appearing in the synthesis of nanotubes, which stepwise enlarges the diameter, up to the doubling one. Note that the final tube is again achiral (and zigzag as well), and its length is half of the initial tube. Figure 32 illustrates the diameter doubling by twisting (right hand column, top view) of a zigzag polyhex nanotube.

The twisting leads to more relaxed molecules, by decreasing the strain of their surface, as shown in Figure 33.

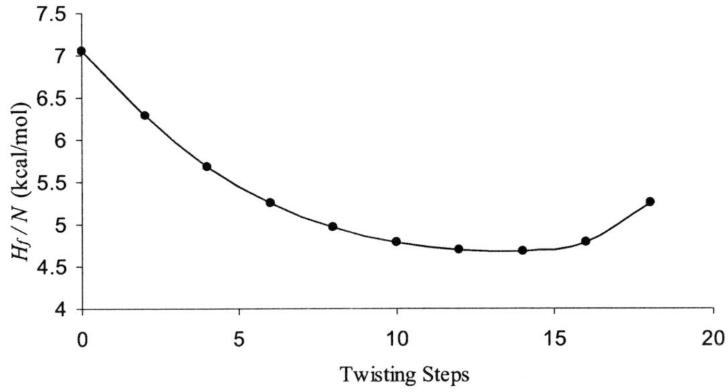


Figure 33: Energy curve of the twisting process of Z[18,6].

The armchair tubes change to zigzag ones at the complete twisting ($t = c$), but without diameter doubling. Such tubes can enlarge the diameter, either by the zipper mechanism, or by the above transformation of zigzag tubes.

Joining single-walled carbon nanotubes to form X, Y and T molecular junctions are experimentally achieved by electron beam irradiation of crossed tubes. The ultrasonic dispersion in solution of SWNTs leads to a random crisscrossing distribution. These arrangements were monitored under TEM, and the crossing points were irradiated for a short time. The two tubes merged together resulting an X shaped junction. By further careful irradiation it was possible to remove the arms of the junction in order to create a Y or T junction.[52]

The reason why the merging process takes place is the formation of dangling bonds, under electron beam. It is noticeable that, in the absence of irradiation, the process does not occur. The energy of the system is lowered by connecting the dangling bonds in both tubes, which results in the junction. In the new surface, heptagons and octagons appear to introduce a smooth negative curvature.

## 2.5. Topological Characterization of $C_{70}$ Dimers

In $D_{5h}$ $C_{70}$ there are five topologically distinct faces, two pentagons and three hexagons. Similar to the generation of the two caps from $C_{60}$, five caps (related to the five distinct faces) can be constructed (Figure 34). Connection of the dangling bonds between two caps results in peanut-shaped dimers.

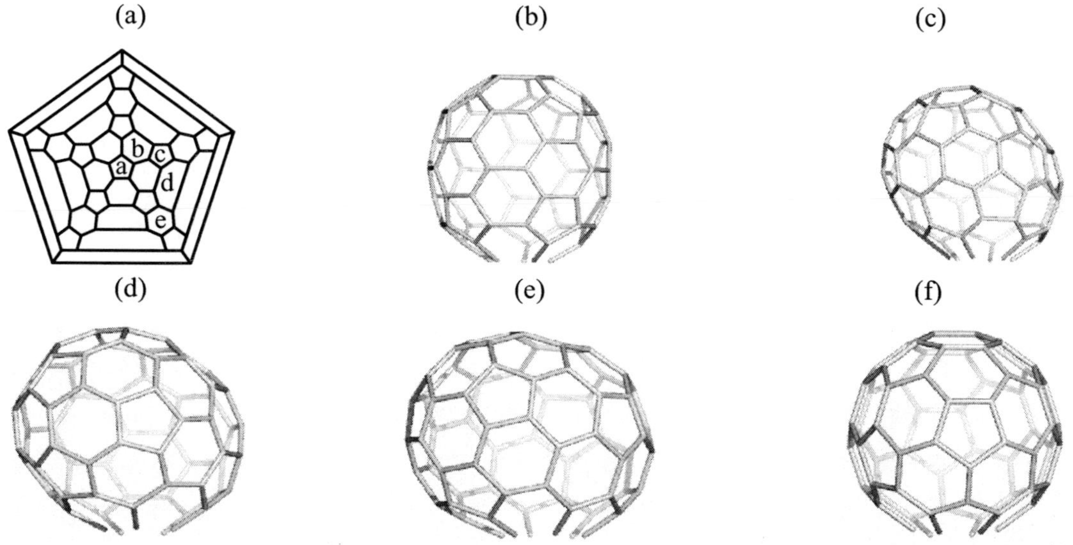


Figure 34: Geodesic projection of $C_{70}$ presenting the five face types (a), and the corresponding five caps (b-f)

Construction of all possible symmetric dimers was monitored by a cyclic permutation on the sequence of edges in the joining zone.[53] Among the resulted peanut shaped dimers several enantiomer pairs appear. Topological equivalence classes of the constitutive substructures were evaluated by calculating a topological index on the parent cages, their medial and dual, respectively.

Given a graph $G$ with $S = S(G)$ being the set of subgraphs of a given size and a group of automorphisms $Aut(G(S))$, two subgraphs $S_i$, $S_j$ of $G$ are called *equivalent* if there is a group element $aut(N_{S_i}) \in Aut(G(S))$ such that ( *i.e.*, there exists an automorphic permutation that transforms two subgraphs to each other).

The set of all subgraphs $S_j$ obeying the above equivalence relation is called the orbit (*i.e.*, class of equivalence) of the subgraph $S_i$,. The orbits represent disjoint automorphic partitions: $S = \bigcup S_{N_{S_i}}$ . We limited here to find the equivalence classes for the vertices, edges and faces of a fullerene cage. As an equivalence criterion we used two topological indices: of centrality $C(LM)$ and centrocomplexity $X(LM)$, respectively, defined on layer matrices **LM**.[54]

The vertex orbits are found by calculating the above indices on the layer matrix of distance sum **LDS** of the parent graph. The edge orbits are calculable with the same index but applied on the **LDS** matrix of the corresponding medial graph.[55]

The face orbits are obtained from the dual of the parent graph, the criterion being the index of centrocomplexity $X(L\ ^1W)$, where $\mathbf{L^1W}$ matrix is the layer matrix of valences. Each vertex in the dual has the valence equal to the parent face size. In this way one can select the orbits for each face type. Data are listed in Table 12.

(a) $C_{140}$aa the parent graph: $v = 140$; $e = 210$; $f = 72$

(b) medial of $C_{140}$aa, $v = 210$

(c) dual of $C_{140}$aa, $v = 72$

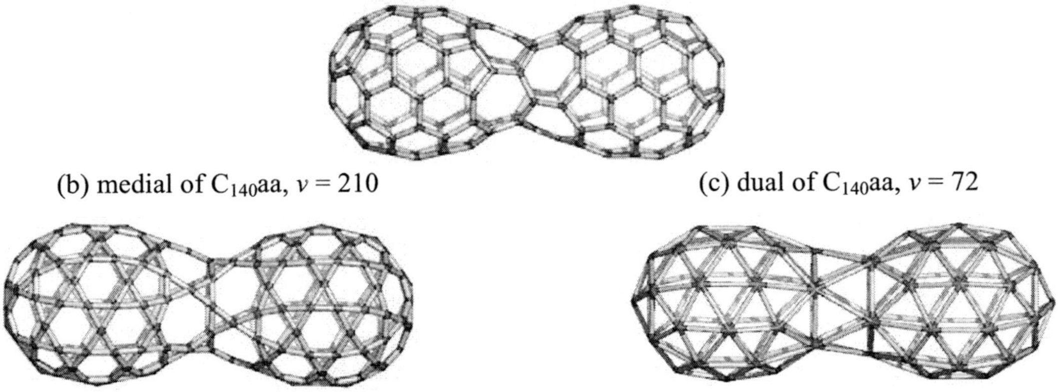

Figure 35: $C_{140}$aa peanut structure and its medial and dual graph.

**Table 12: Vertex, edge, and face orbits of C140 peanut dimmers**

| G | Vertices | Edges | Faces |
|---|---|---|---|
| c140aa | 4x{10};5x{20}; | 9x{10};6x{20}; | [5] 1{2};2{10}; [6] 4{10}; [7] 1{10}; |
| c140bb | 64x{2};3x{4}; | 2x{1};102x{2};1x{4}; | [5] 3{2};3{4}; [6] 10{2};7{4}; [7] 1{2};1{4}; |
| c140bc | 66x{2};2x{4}; | 2x{1};87x{2};7x{4};1x{6}; | [5] 3{2};3{4}; [6] 7{2};7{4};1{6}; [7] 3{2}; |
| c140bd | 70x{2}; | 2x{1};96x{2};4x{4}; | [5] 3{2};3{4}; [6] 10{2};7{4}; [7] 3{2}; |
| c140cc | 66x{2};2x{4}; | 2x{1};94x{2};5x{4}; | [5] 3{2};4{4}; [6] 1{2};8{4};1{6}; [7] 3{2};1{4}; |
| c140cd | 66x{2};2x{4}; | 2x{1};100x{2};2x{4}; | [5] 3{2};4{4}; [6] 1{2};8{4};1{6}; [7] 1{2};2{4}; |
| c140ce | 4x{2};33x{4}; | 9x{2};48x{4}; | [5] 3{2};4{4}; [6] 1{2};8{4};1{6}; [7] 1{2};2{4}; |
| c140dd | 68x{2};1x{4}; | 2x{1};104x{2}; | [5] 8{2};1{4}; [6] 9{2};5{4};1{6}; [7] 2{2};1{4}; |
| c140de | 68x{2};1x{4}; | 2x{1};98x{2};3x{4}; | [5] 6{2};2{4}; [6] 10{2};4{4};1{8}; [7] 4{2}; |
| c140df | 70x{2}; | 2x{1};102x{2};1x{4}; | [5] 6{2};2{4}; [6] 9{2};5{4};1{6}; [7] 1{2};1{6}; |
| c140ee | 35x{4}; | 3x{2};45x{4};3x{8}; | [5] 3{4};1{8}; [6] 5{4};3{8}; [7] 2{4}; |
| c140ef | 7x{4};14x{8}; | 1x{2};6x{4};23x{8}; | [5] 1{4};2{8}; [6] 5{4};3{8}; [7] 1{8}; |

## 3. CONCLUSIONS

The coalescence of fullerene molecules, experimentally observed by TEM in annealed/irradiated peapods, was described by a sequence of Stone-Wales SW bond rotations. Data reported recently in literature confirmed the apparition of peanut-shaped intermediates and capped (polyhex) nanotubes, *i.e.*, tubulenes. Tubulenes can further be transformed, by SW, in other tubular forms. The finally resulted double-walled nanotubes DWNTs could be precursors of distinct-walled tori DWTs, with the apex zones decorated by non-hexagonal faces. Semiempirical calculations proved that such tori are far more stable than the classical polyhex ones, at least at small and moderate size. Topology of some fullerenes and their dimers was described by the spiral code and some topological indices operating on layer matrices, either on the parent associated graphs or on their medials and duals, finally providing the equivalence classes (*i.e.*, orbits) of the constitutive substructures.

## REFERENCES

[1] M. Rao, P. Zhou, K-A Wang, G. T. Hager, J. M. Holden, Y. Wang, W.-T. Lee, X.-X. Bi, P. C. Eklund, D. S. Cornett, M. A. Duncan, and I. J. Amster, Photoinduced polymerization of solid $C_{60}$ films, *Science*, 1993, 259, 955-957.

[2] Y. Iwasa, T. Arima, R. M. Fleming, T. Siegrist, O. Zhou, R. C. Haddon, L. J. Rothberg, K. B. Lyons, H. L. Carter Jr., A. F. Hebard, R. Tycko, G. Dabbagh, J. J. Krajewski, G. A. Thomas, and T. Yagi, New phases of $C_{60}$ synthesized at high pressure, *Science*, 1994, 264, 1570-1572.

[3] G.-W. Wang, K. Komatsu, Y. Murata, M. Shiro, Synthesis and X-ray structure of dumb-bell-shaped $C_{120}$, *Nature*, 1997, 387, 583-586.

[4] S. Lebedkin, A. Gromov, S. Giesa, R. Gleiter, B. Renker, H. Rietschel, and W. Krätschmer, Raman scattering study of $C_{120}$, a $C_{60}$ dimer, *Chem. Phys. Lett.*, 1998, 285, 210-215.

[5] P. W. Fowler, D. Mitchell, R. Taylor, and G. Seifert, Structures and energetics of dimeric fullerene oxide derivatives, *J. Chem. Soc., Perkin Trans.* 2, 1997, 1901-1905.

[6] S. Lebedkin, S. Ballenweg, J. Gross, R. Taylor, and W. Krätschmer, Synthesis of $C_{120}O$: a new dimeric [60]fullerene derivative, *Tetrahedron. Lett.* 1995, 36, 4971-4974.

[7] Gromov, S. Lebedkin, S. Ballenweg, A. G. Avent, R. Taylor, and W. Krätschmer, $C_{120}O_2$: The first [60]fullerene dimer with cages bis-linked by furanoid bridges, *Chem. Comm.*, 1997, 209-210.

[8] Gromov, S. Lebedkin, W. E. Hull, and W. Krätschmer, Isomers of the dimeric fullerene $C_{120}O_2$, *J. Phys. Chem. A*, 1998, 102, 4997-5005.

[9] A Hirsch, Addition Reactions of Buckminsterfullerene ($C_{60}$), *Synthesis*, 1995, 139, 895-913.

[10] S. Giesa, J. H. Gross, W. E. Hull, S. Lebedkin, A. Gromov, R. Gleiter, and W. Krätschmer, C120OS: the first sulfur-containing dimeric [60]fullerene derivative, *Chem. Commun.*, 1999, 465-466.

[11] S. W. McElvany, J. H. Callahan, M. M. Ross, L. D. Lamb, and D. R. Huffman, Large odd-numbered carbon clusters from fullerene-ozone reactions, *Science*, 1993, 260, 1632-1634.

[12] S. Lebedkin, H. Rietschel, G. B. Adams, J. B. Page, W. E. Hull, F. H. Hennrich, H-J Eisler, M. M. Kappes, and W. Krätschmer, Quantum molecular dynamics calculations and experimental Raman spectra confirm the proposed structure of the odd-numbered dimeric fullerene $C_{119}$, *J. Chem. Phys.*, 1999, 110, 11768-11778.
[13] Yeretzian, K. Hansen, F. Diedrich, and R. L. Whetten, Coalescence rections of fullerenes, *Nature*, 1992, 359, 44-47.
[14] E. E. B. Campbell, V. Schyja, R. Ehlich, and I. V. Hertel, Observation of molecular fusion and deep inelastic scattering in $C^{+}_{60}$+$C_{60}$ collisions, *Phys. Rev. Lett.*, 1993, 70, 263–266.
[15] F. Rohmund, A. V. Glotov, K. Hansen, and E. E. B. Campbell, Experimental studies of fusion and fragmentation of fullerenes, *J. Phys. B: At. Mol. Opt. Phys.*, 1996, 29, 5143-5161.
[16] L. Strout, R. L. Murry, C. Xu, W. C. Eckhoff, G. K. Odom, and G. E. Scuseria, A theoretical study of buckminsterfullerene reaction products: $C_{60}$ + $C_{60}$, *Chem. Phys. Lett.*, 1993, 214, 576-582.
[17] M. V. Diudea, C.L. Nagy, I. Silaghi-Dumitrescu, A. Graovac, D. Janežić, and D. Vikic-Topić, Periodic cages, *J. Chem. Inf. Model.*, 2005, (in press).
[18] Y. Zhao, B. I. Yakobson, and R. E. Smalley, Dynamic topology of fullerene coalescence, *Phys. Rev. Lett.*, 2002, 88, 18, 185501(4).
[19] Y. Zhao, R. E. Smalley, and B. I. Yakobson, Coalescence of fullerene cages: topology, energetics, and molecular dynamics simulation, *Phys. Rev. B*, 2002, 66, 195409(9).
[20] Y. F. Zhao, Y. Lin, and B. I. Yakobson, Fullerene shape transformations via Stone-Wales bond rotations, *Phys. Rev. B*, 2003, 68, 233403(1)-(4).
[21] H. Ueno, S. Osawa, E. Osawa, and K. Takeuchi, Stone-Wales rearrangement pathways from the hinge-opened [2+2]C60 dimer to IRP C120 fullerenes. Vibrational analysis of intermediates, *Fullerene Sci. Technol.*, 1998, 6, 319-338.
[22] S. Han, M. Yoon, S. Berber, N. Park, E. Osawa, J. Ihm, and D. Tománek, Microscopic mechanism of fullerene fusion, *Phys. Rev. B*, 2004.
[23] B. R. Eggen, M. I. Heggie, G. Jungnickel, C. D. Latham, R. Jones, and P. R. Briddon, Autocatalysis during fullerene growth, *Science*, 1996, 272, 87-90.
[24] J. Lu, S. Nagase, S. Zhang, and L. Peng, Strongly size-dependent electronic properties in $C_{60}$-encapsulated zigzag nanotubes and lower size limit of carbon nanopeapods, *Phys. Rev. B*, 2003, 68, 121402(1)-(4).
[25] Y.-K. Kwon, D. Tománek, and S. Iijima, "Bucky shuttle" memory device: synthetic approach and molecular dynamics simulations, *Phys. Rev. Lett.*, 1999, 82, 1470-1473; J. Chen and J.Dong, *J. Phys. Condens. Matter*, Electronic properties of peapods: effects of fullerene rotation and different types of tube, 2004, 16, 1401-1408.
[26] M. Monthioux, Filling single-wall carbon nanotubes, *Carbon*, 2002, 40, 1809-1823.
[27] M. Yudasaka, K. Ajima, K. Suenaga, T. Ichihashi, A. Hashimoto, and S. Iijima, Nano-extraction and nano-condensation for $C_{60}$ incorporation into single-wall carbon nanotubes in liquid phases, *Chem. Phys. Lett.*, 2003, 380, 42-46.
[28] B. W. Smith, M. Monthioux, and D. E. Luzzi, Encapsulated $C_{60}$ in carbon nanotubes, *Nature*, 1998, 396, 323-324.

[29] S. Bandow, M. Takizawa, K. Hirahara, M. Yudasaka, and S. Iijima, Raman scattering study of double-wall carbon nanotubes derived from the chains of fullerenes in single-wall carbon nanotubes, *Chem. Phys. Lett.*, 2001, 337, 48-54.

[30] S. Bandow, T. Hiraoka, T. Yumura, K. Hirahara, and S. Iijima, Raman scattering study on fullerene derived intermediates formed within single-wall carbon nanotube: from peapod to double-wall carbon nanotube, *Chem. Phys. Lett.*, 2004, 384, 320-325.

[31] Hernández, V. Meunier, B. W. Smith, R. Rurali, H. Terrones, M. Buongiorno Nardelli, M. Terrones, D. E. Luzzi, and J.-C. Charlier, Fullerene coalescence in nanopeapods: a path to novel tubular carbon, *Nano Lett.*, 2003, 3, 1037-1042.

[32] A J. Stone and D. J. Wales, Theoretical studies of icosahedral $C_{60}$ and some related species, *Chem. Phys. Lett.*, 1986, 128, 501-503.

[33] M. V. Diudea, Cs. L. Nagy, O. Ursu, and T. S. Balaban, $C_{60}$ dimers revisited, *Fuller. Nanotub. Carbon Nanostruct.*, 2003, 11, 245-255.

[34] Y.-H. Kim, I.-H. Lee, K. J. Chang, and S. Lee, Dynamics of fullerene coalescence, *Phys. Rev. Lett.*, 2003, 90, 065501-065504.

[35] J. Han, Energetics and structures of fullerene crop circles, *Chem. Phys. Lett.*, 1998, 282, 187-191.

[36] S. Itoh, S. Ihara, and J.-I. Kitakami, Toroidal form of $C_{360}$, *Phys. Rev. B*, 1993, 47, 1703-1704.

[37] S. Ihara, S. Itoh, and J. Kitakami, Toroidal forms of graphitic carbon, *Phys. Rev. B*, 1993, 47, 12908-12911.

[38] S. Itoh and S. Ihara, Isomers of the toroidal forms of graphitic carbon, *Phys. Rev. B*, 1994, 49, 13970-13974.

[39] S. Itoh and S. Ihara, Toroidal forms of graphitic carbon. II. Elongated tori, *Phys. Rev. B*, 1993, 48, 8323-8328.

[40] M. Endo, K. Takeuchi, K. Kobory, K. Takahashi, H. W. Kroto, and A. Sarkar, in: M. Endo, S. Iijima and M. S. Dresselhaus, *Carbon Nanotubes*, Pergamon, 1996, pp. 1-9.

[41] A Sarkar, H. W. Kroto, and M. Endo, in: M. Endo, S. Iijima and M. S. Dresselhaus, *Carbon Nanotubes*, Pergamon, 1996, pp. 105-109.

[42] Clar, *The Aromatic Sextet*, Wiley, 1972, New York,.

[43] Clar, *Polycyclic Hydrocarbons*, Acad. Press, 1964, London,.

[44] J. R. Dias, The most stable class of benzenoid hydrocarbons and their topological characteristics - total resonant sextet benzenoids revisited, *J. Chem. Inf. Comput. Sci.*, 1999, 39, 144-150.

[45] M. V. Diudea, Covering forms in nanostructures, *Forma*, 2004 (submitted).

[46] P. Nikolaev, A. Thess, A. G. Rinzler, D. T. Colbert, and R. E. Smalley, Diameter doubling of single-wall nanotubes, *Chem. Phys. Lett.*, 1997, 266, 422-426.

[47] S. L. Fang, A. M. Rao, P. C. Eklund, P. Nikolaev, A. G. Rinzler, and R. E. Smalley, Raman scattering study of coalesced single walled carbon nanotubes, *J. Mater. Res.*, 1998, 13, 2405-2411.

[48] M. Terrones, H. Terrones, F. Banhart, J-C Charlier, and P. M. Ajayan, Coalescence of single-walled carbon nanotubes, *Science*, 2000, 288, 1226-1229.

[49] Banhart, The role of lattice defects in the formation of new carbon structures under electron irradiation, *J. of Electron. Microsc.*, 2002, 27, 189-194.

[50] M. Yoon, S. Han, G. Kim, S. B. Lee, S. Berber, E. Osawa, J. Ihm, M. Terrones, F. Banhart, J.-C. Charlier, N. Grobert, H. Terrones, P. M. Ajayan, and D. Tománek, Zipper

mechanism of nanotube fusion: theory and experiment, *Phys. Rev. Lett.*, 2004, 92, 075504-(1)-(4).

[51] M. V. Diudea, T. S. Balaban, E. C. Kirby, A. Graovac, Energetics and topology of polyhex nanotubes, *Phys. Chem. Chem. Phys.*, 2003, 5, 4210-4214.

[52] M. Terrones, F. Banhart, N. Grobert, J.-C. Charlier, H. Terrones, and P. M. Ajayan, Molecular junctions by joining single-walled carbon nanotubes, *Phys. Rev. Lett.*, 2002, 89, 075505(1)-(4).

[53] Cs. L. Nagy, M. Stefu, M. V. Diudea, A. Dress, and A. Müller, $C_{70}$ dimers - energetics and topology, *Croat. Chem. Acta*, 2004, accepted.

[54] M. V. Diudea and O. Ursu, Layer matrices and distance property descriptors, *Indian J. Chem.*, 2003, 42*A*, 1283-1294.

[55] M. V. Diudea, P. E. John, A. Graovac, M. Primorac, and T. Pisanski, Leapfrog and Related Operations on Toroidal Fullerenes, *Croat. Chem. Acta*, 2003, 76, 153-159.

In: Nanostructures: Novel Architecture
Editor: Mircea V. Diudea, pp. 61-84
ISBN: 1-59454-499-9


*Chapter 4*

# PERIODIC FINITE NANOSTRUCTURES

***Mircea V. Diudea,*[1a] *Csaba L. Nagy*[a] *and Ante Graovac*[b]**
[a]Faculty of Chemistry and Chemical Engineering
Babes-Bolyai University, 40028 Cluj, Romania
[b]The R. Bošković Institute, POB 180, HR-10002 Zagreb, Croatia
and Faculty of Natural Sciences, Mathematics and Education,
Nikole Tesle 12, HR-21000 Split, Croatia

## ABSTRACT

Tubulenes are capped nanotubes built up from two caps and a distancing nanotube/neck of diverse decoration. Ways of "in silico" construction and isomerization of several finite nanostructures, temptingly classified by their neck and/or caps, are presented. Periodicity is considered both as topology (constitutive typing enumeration) and $\pi$-electronic structure, in the Hückel model. Spectral regularities appearing in families of tubulenes are summarized in some rules for their $\pi$-electronic structure. The older and newer rules are (re)written in terms of the classical $C_{N(k6^k(56)^k - A[2k,n])}$ cylinders.

## 1. INTRODUCTION

The discovery of fullerenes and related nanostructures has generated an explosion of research in chemistry, physics, and materials science. Several books in this field have already been published.[1-7]

Fullerenes and nanotubes represent 0D and 1D novel carbon allotropes that have non-preceding physico-chemical properties.[8-11] They represent promising candidates for the development of nanodevices and super strong composites.

Apart from the well-known $C_{60}$ and $C_{70}$, other cages have been isolated in the solid state. Recently, the small cages $C_{36}$ and $C_{20}$ have been reported[12-14] and their halves used for modelling capped narrow nanotubes.[15]

[1] E-mail address: diudea@chem.ubbcluj.ro

Nowadays, the fullerene study shifted to real chemistry. Carbon allotropes with finite molecular cage structures have been functionalized or inserted in supramolecular assemblies.[16-27]

A fullerene is, according to a classical definition, an all-carbon molecule $C_N$ consisting entirely of pentagons (exactly 12) and hexagons (*N*/2-10).[28] Non-classical fullerene extensions to include rings of other sizes have been considered.[29],[30] Heterocyclic large cages have also been studied.[31]

Some basic relations in polyhedral graphs come out from the Euler's results:[32]

$$N - E + F = 2 \tag{1}$$

where *N, E,* and *F* are the number of vertices, edges and faces, respectively. Relation (1) is given here for a graph embedded on the sphere. Other basic relations are:

$$\sum_d d \cdot v_d = 2E \tag{2}$$

$$\sum_s s \cdot f_s = 2E \tag{3}$$

where $v_d$ and $f_s$ denote vertices of degree *d* and *s*-sized faces, respectively.

In the simple π-only Hückel theory,[33] the energy of the $i^{\text{th}}$ molecular orbital is calculated from the eigenvalues ($\lambda_i$) of the adjacency matrix associated with the molecular hydrogen- depleted graph as:

$$E_i = \alpha + \lambda_i \beta \tag{4}$$

The non-increasing sequence of $\lambda_i$ is called the spectrum of the graph. The HOMO-LUMO gap is taken as the absolute value of the difference $E_{\text{HOMO}}$-$E_{\text{LUMO}}$, in $|\beta|$ units. Shell typing[34] encountered in fullerenes is as follows: (i) *properly closed*, PC, when $\lambda_{N/2} > 0 \geq \lambda_{N/2+1}$; (ii) *pseudo-closed*, PSC, in case $\lambda_{N/2} > \lambda_{N/2+1} > 0$; (iii) *meta-closed*, MC, with $0 \geq \lambda_{N/2} > \lambda_{N/2+1}$ and (iv) *open*, OP, when the $N/2^{th}$ (HOMO) and $N/2{+}1^{th}$ (LUMO) molecular orbitals are degenerate, $\lambda_{N/2} = \lambda_{N/2+1}$.

Systematic studies of this matrix and its eigenvalue spectrum (*i.e.*, the non-increasing sequence of eigenvalues $\lambda_i$), have provided some magic number rules for the stability of molecules, such as the venerable Hückel 4*m* +2 rule for aromatic rings, and the more recent 60 + 6*m* ($m \neq 1$) *leapfrog* rule[35],[36] LER for properly closed-shell fullerenes. The other rules of thumb, enabling selection among thousands of topological isomers of the most stable fullerenes to be further submitted to calculations at a higher level of theory, will be presented in close connection with construction of specified structures.

The most frequent case is that of the pseudo-closed shell PSC, since, in fullerenes, the number of positive eigenvalues is, in general, larger than that of the negative ones,[37] $n_+ \geq n_-$. It implies electron deficiency, so that the charge-transfer from an electron-donating partner will stabilize their π-electronic structure.

A properly closed-shell PC represents the optimum, with all electrons in bonding orbitals and a definite gap between occupied and empty orbitals; in this case, $n_+ = n_-$. Such structures are relatively rare, and satisfy the leapfrog rule (see above) and the cylinder rules, as will be detailed in the following.

This paper is organised as follows: several caps are designed and named in Section 2, Capping Nanotubes. Tubulenes of *fa*- and *ta*- series are built up and their stability is discussed in the light of some rules of thumb, as shown in Section 3, *a*-Tubulenes. The next Section 4, *z*-Tubulenes, presents tubulenes constructed with zigzag distancing nanotubes, peanut-shaped, periodic fulleroids and some topology counting theorems. Section 5 deals with isomerization of some tubulenes while Section 6 draws conclusions. The paper is ended by references.

## 2. Capping Nanotubes

Covering transformation is one of the ways in understanding chemical reactions occurring in fullerenes.[38-41] A capped nanotube we call here a tubulene. Covering is different for the caps and for the nanotube/neck distancing the two caps. A constructive enumeration of nanotube caps the reader can find in ref. 42. Within this paper we limit at three types of caps: (1) *fa/z*-cap, derived from spherical fullerenes by deleting zones of various size and fitted to polyhex "armchair" (6,3)A and "zigzag" (6,3)Z nanotubes, respectively; (2) *kfz*-cap, obtainable by cutting off the polar ring (of size k) and fitted to Z nanotubes, and (3) ta/z-cap, suitable to variously covered nanotubes (Figure 1).

The name of a cap includes the number of atoms *N*, the *a*-spiral code[43] and the fitting tube specification, *e.g.*, A/Z[*c*,*n*], *c* being the number of atoms in the cross-section while *n* is the number of such sections along the tube. The number of atoms in the neck is $c \times n$. Note the value $n = 0$ for the tube length in a cap, *e.g.*, $C_{6k(k\,6^k(5*)^k-\mathrm{A}[2k,0])}$. When no ambiguity exists, the specification of the polyhex (6,3) covering for tubes is omitted.

Also note that, in two integer parameter (*k*,*l*), notation,[44] the armchair (single walled) nanotube SWNT is (*c*/2, *c*/2) while the zigzag one is (*c*/2,0).

Various caps and various junction zones could be used to construct tubulenes.[45- 48] Both the cap and nanotube covering can be changed by appropriate operations.

A tempting classification of tubulenes, by their distancing nanotube and/or caps, is presented in the following.

## 3. *A*-Tubulenes

### 3.1. *Fa*-Tubulenes

Tubulenes of general formula $C_{N(k6^k(56)^k-\mathrm{A}[2k,n])}$ are the most stable ones, being a natural extension of the spherical fullerenes, *e.g.*, $C_{60}$. The distancing "armchair" tube is capped by the *fa*-cap $C_{6k(k\,6^k(5*)^k-\mathrm{A}[2k,0])}$, where the asterisk stands for the missing polygons, in a cap (see also Figure 1). The number of atoms *N* is calculable by summing the number of atoms in the two caps and the neck.

*fa*-Cap $C_{6k(k\,6^k(5*)^k-A[2k,0])}$; $k=6$ *fz*-Cap $C_{13k/2(k\,(56)^{k/2}(665)^{k/2}-Z[3k,0])}$; $k=6$

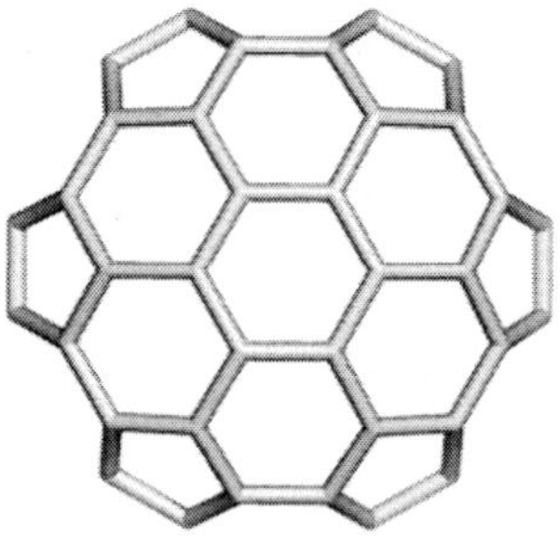

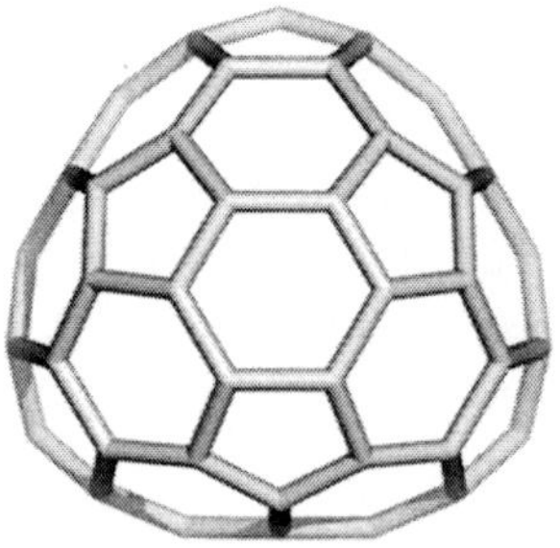

*kfz*-Cap $C_{11k(k\,6^k(56)^k(65)^k-Z[2k,0])}$; $k=5$ *kfz*-Cap $C_{9k(k\,(56)^{k/2}(665)^{k/2}(656)^{k/2}-Z[2k,0])}$; $k=6$

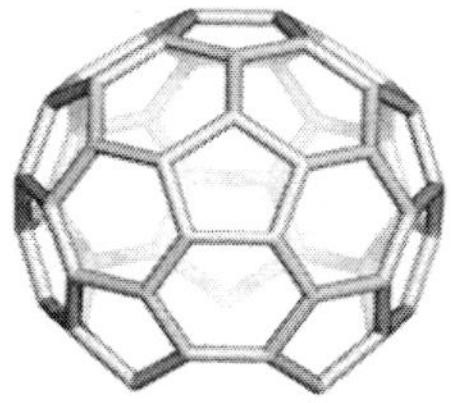

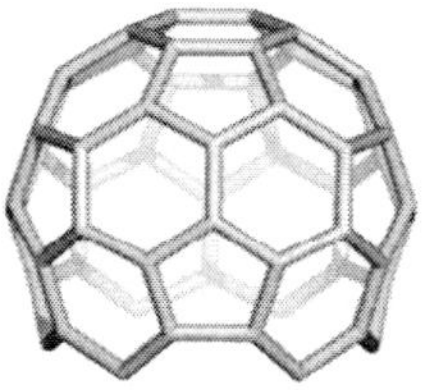

*ta*-Cap $C_{4k(k\,5^k1^k-A[2k,0])}$; $k=6$ *tz*-Cap $C_{3k(k\,5^k-Z[2k,0])}$; $k=7$

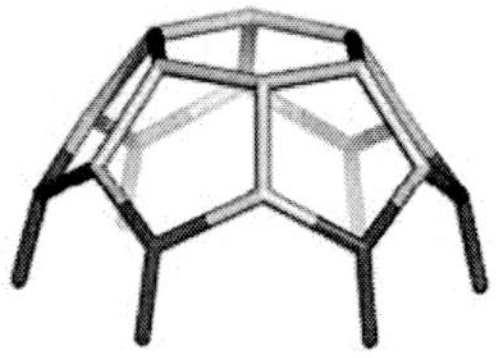

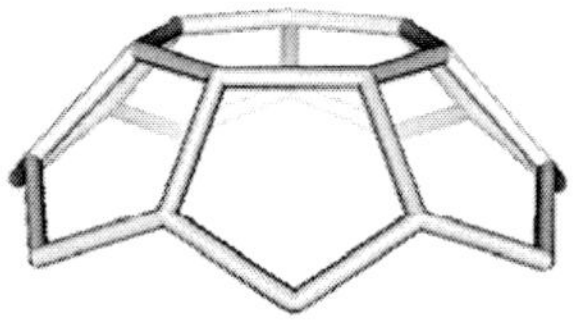

Figure 1: The three types of caps herein used for capping nanotubes.

The name of a tubulene continues the cap name by specifying the length (*e.g.*, taking $n \geq 0$) and tiling of the neck. The second cap is described by a string of numbers which is symmetrical to that of the first cap (*i.e.*, written from the end to the top). According to the fitted cap, these cages are called *fa*- or *ta*-tubulenes (see below). Figure 2 illustrates the most important of these structures, $C_{120(56^5(56)^5)-A[10,6])}$ being inferred in the coalescence of $C_{60}$ within $C_{60}$-peapods.[49],[50]

(a) $C_{120(56^5(56)^5)-A[10,6])}$ ($D_{5d}$) (b) $C_{168(66^6(56)^6)-A[12,8])}$ ($D_{6d}$)

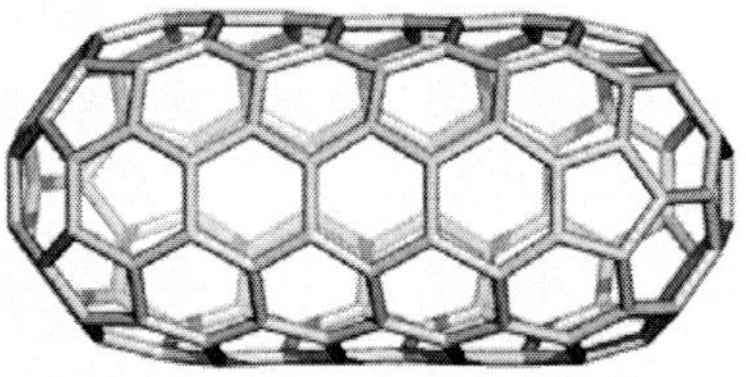

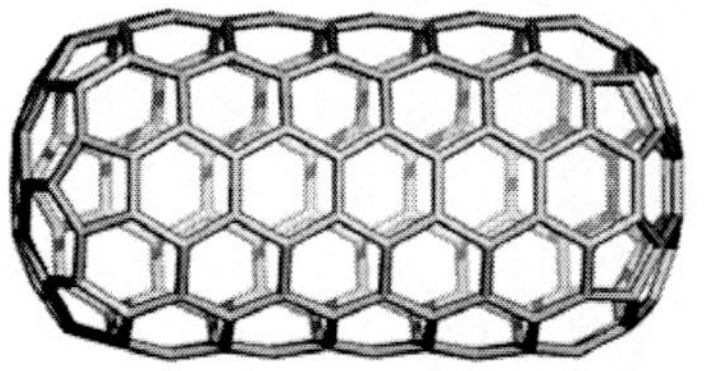

Figure 2: Examples of *fa*-tubulenes $C_{N(k\,6^k(56)^k-A[2k,n])}$, with $k=5$ (a) and 6 (b), respectively.

These tubulenes are far more stable than the corresponding open A[$c$,$n$] nanotubes, as shown in Figure 3. They include LER and ACR tubulenes (having properly closed PC $\pi$-electronic shells) as well as pseudoclosed PSC tubulenes, which appear in a periodic sequence: LER ($D_{kh}$), ACR($D_{kd}$), PSC($D_{kd}$), …, LER, ACR, PSC, …, as the $n$-dimension of (6,3)A-nanotube increases.

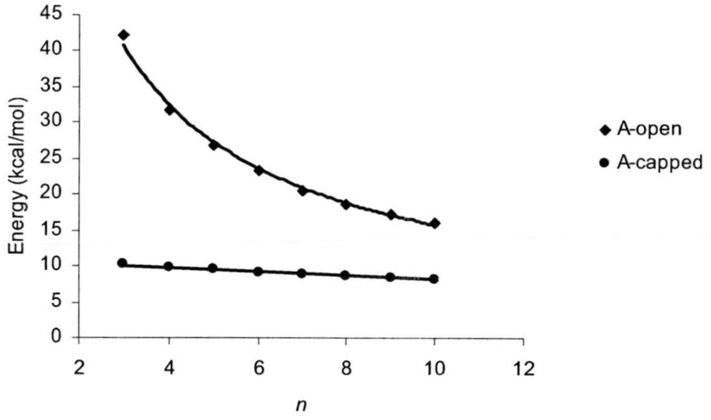


Figure 3: PM3 Heat of formation (per atom) of open and capped A-nanotubes of formula $C_{N(k6^k(56)^k-A[2k,n])}$; $k = 6$.

This periodicity is clearly illustrated by the plot of HOMO-LUMO gap vs. $n$, Figure 4.

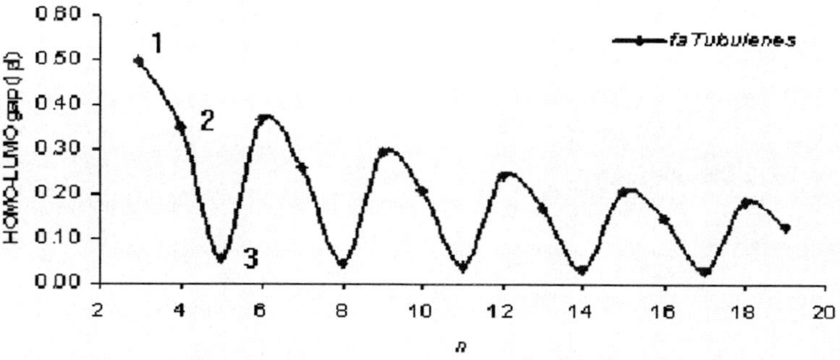


Figure 4: Periodicity of HOMO-LUMO gap *vs.* tube $n$-dimension of *fa*-tubulenes $C_{N(k6^k(56)^k-A[2k,n])}$; $k$ = 6. The points for LER (1), ACR(2) and PSC(3) in decreasing order of their gap.

Objects of general formula $C_{N(k6^k(56)^k-A[2k,n])}$, with $k$-fold cylindrical symmetry, obey the *armchair cylinder rule*[51] ACR, which predicts closed shell at each nuclearity $N = 12k + 2k(1+3m)$, $m$ = 0, 1, 2,…; $k$ = 4 to 7. In our notation, $n = 1 + 3m$, $n$ being the number of cross sections separating the two caps of a nanotube. These tubulenes have a non-bonding orbital (NBO) LUMO separated by a gap from HOMO. Exceptions have been observed. In this series of tubulenes, LER can be written as an armchair cylinder rule, A-LER: $N = 12k + 2k \cdot 3m$ , $m$ = 0, 1, 2,…; $k$ = 4 to 7.

Recall that stability of tubulenes is determined by the two main factors: the strain of the cap (attributed to the $\sigma$ frame) and the $\pi$-electronic structure. The aromaticity of these systems is claimed as a stability factor, directly related to the second one above. In this respect, Kekulé structure count,[52-54] Clar sextet count,[34],[55],[56] Topological resonance energy TRE,[57-60] Conjugated-circuit theory,[61-63] or NICS calculations [64-67] aimed at measuring the aromaticity of polycyclic conjugated systems, could help in understanding and predicting the stability and chemical behavior of such molecules.

## 3.2. *Ta*-Tubulenes

The "tubercule"-like *ta*-cap $C_{4k(k5^k1^k-A[2k,0])}$ ($k$ = 6, Figure 1), conceivable to evolve from the open end(s) of an A-nanotube by adding $C_2$ units up to closure, could provide a *ta*-tubulene $C_{N(k5^k7^k(56)^k-A[2k,n])}$, in our example $C_{96(6\,5^6 7^6 (56)^6 (65)^6 7^6 5^6\, 6)}$, with the equatorial zone tessellated as in $C_{60}$ (Figure 5).

$2 \times C_{24(6\,5^6 1^6-A[12,0])} + A[12,4]$ $C_{96(6\,5^6 7^6 (56)^6 (65)^6 7^6 5^6\, 6)}$ ($C_2$) Geodesic projection; $k$ =6

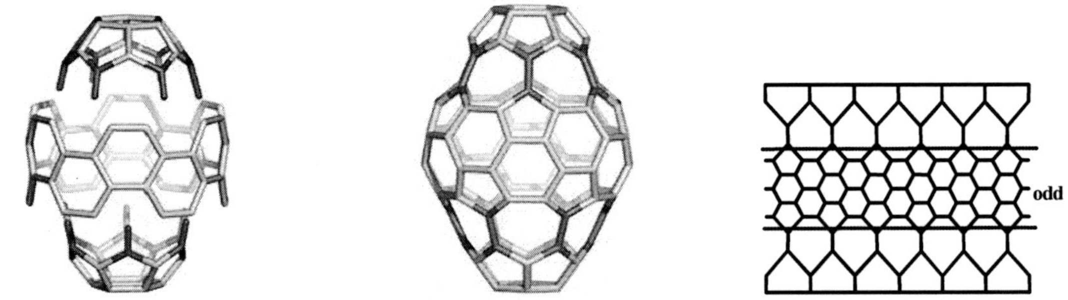


Figure 5: Capping an A-nanotube by a "tubercule" *ta*-cap.

The *ta*-tubulene may isomerizes[48] to *fa*-tubulene $C_{N(k6^k(56)^k-A[2k,n])}$ (*e.g.*, $C_{96(6\,6^6 (5\,6)^6 (6\,6)^6 (6\,5)^6 6^6\, 6)}$; $D_{6d}$ - Figure 6A) by cascade SW rotation of the marked edges. This transformation is general in *ta*-tubulenes (see Section 5).

$C_{96(66^6(56)^6(66)^6(65)^6 6^6 6)}$ ($D_{6d}$) Geodesic Projection

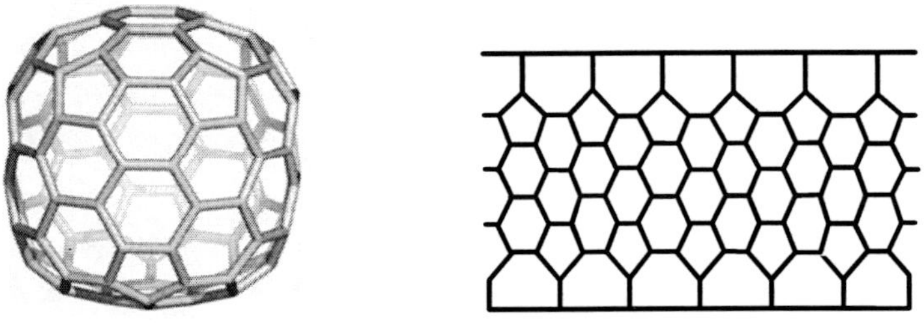

Figure 6A: *ta*-Tubulenes isomerize to *fa*-tubulenes by SW edge rotation.

The new series of *ta*-tubulenes obey the *t*-armchair cylinder rule[48] TACR, predicting the periodicity of closed shells at nuclearity $N = 8k + 2k(4+3m)$ ; $m$= 0, 1, 2,... Re-written to account for equivalent *fa*-tubulenes $C_{N(k6^k(56)^k - A[2k,n])}$, the rule becomes: $N = 12k + 2k(2+3m)$; $m$ = 0, 1, 2,...

In the same series, a second rule, called *tn*-armchair cylinder rule[48] TNACR, predicts closed shells (but LUMO being an NBO) at $N = 8k + 2k(11+3m)$ ; $m$= 0, 1, 2,....or, for equivalent *fa*-tubulenes, $N = 12k + 2k(3+3m)$; $m$ = 2, 3... Note that nuclearity given by this last rule superimposes to that provided by LER in the normal series $C_{N(k6^k(56)^k - A[2k,n])}$ (excepting for $m$ = 0 and 1).

The constitutive periodicity, by the polyhex net of the distancing tube, is obvious in both the *fa*- and *ta*-tubulenes. It superimposes over the electronic periodicity, as expressed by the above rules of thumb.

The *ta*-tubulenes have nuclearities that interlace[48] with those of *fa*-tubulenes and a similar periodicity of their HOMO-LUMO gap (Figure 6B).

Table 1 lists PM3 data for *fa*-tubulenes $C_{N(k6^k(56)^k - A[2k,n])}$ (a) and *ta*-tubulenes $C_{N(k5^k7^k(56)^k - A[2k,n])}$ (b and c). According to Fowler and Sandall,[68] fullerene stability can be expected at gap values higher than 5 eV, in semiempirical calculation, or 0.4 |β| in simple Hückel calculation. Note the quite good concordance between the Hückel gaps and those supplied by PM3.

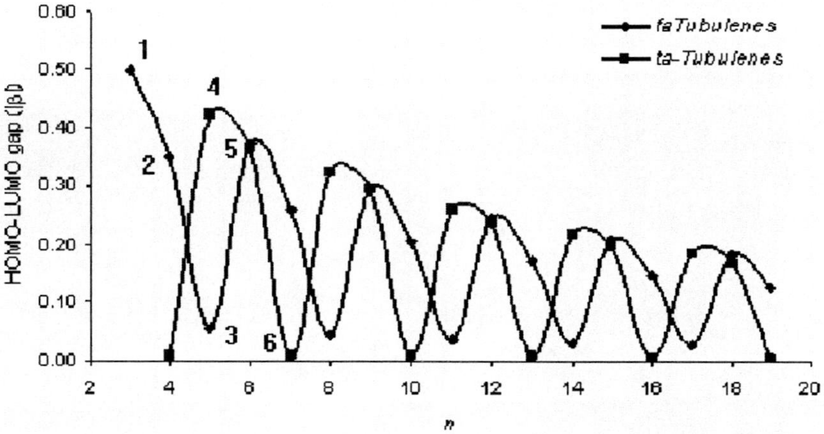


Figure 6B: Periodicity of HOMO-LUMO gap *vs.* tube *n*-dimension of *fa*-tubulenes (LER (1), ACR(2) and PSC(3)) and *ta*-tubulenes (TACR (4), TNACR(5) and PSC(6)); $k$ = 6. Observe that (5) superimposes to (1) at n ≥ 6 (see text).

**Table 1: PM3 data for *fa*-tubulenes $C_{N(k6^k(56)^k-A[2k,n])}$ (a) and *ta*-tubulenes $C_{N(k5^k7^k(56)^k-A[2k,n])}$ (b, c)**

| | Cage $k; n$ | $N$ | Sym. | PM3 HF/at. (kcal/mol) | PM3 Gap (eV) | Spectral Data $\lambda_{N/2}$ | $\lambda_{N/2+1}$ | Gap (\|β\| nits) | Shell |
|---|---|---|---|---|---|---|---|---|---|
| 1a | 5; 2 | 80 | $D_{5d}$ | 12.263 | 3.915 | 0.247 | 0.174 | 0.073 | PSC |
| 2a | 5; 5 | 110 | $C_s$ | 10.663 | 3.883 | 0.176 | 0.120 | 0.055 | PSC |
| 3a | 5; 9 | 150 | $C_s$ | 9.493 | 4.353 | 0.247 | -0.048 | 0.295 | PC |
| 4b | 5; 2 | 80 | $D_{5d}$ | 17.073 | 5.546 | 0.379 | -0.037 | 0.416 | PC |
| 5b | 5; 5 | 110 | $C_s$ | 14.144 | 5.130 | 0.414 | -0.009 | 0.423 | PC |
| 6c | 5; 9 | 150 | $C_s$ | 12.221 | 4.451 | 0.297 | 0.000 | 0.297 | PC |
| 7a | 6; 2 | 96 | $D_{6d}$ | 10.882 | 4.628 | 0.247 | 0.174 | 0.073 | PSC |
| 8a | 6; 5 | 132 | $C_{2h}$ | 9.592 | 3.713 | 0.176 | 0.120 | 0.055 | PSC |
| 9a | 6; 9 | 180 | $C_{2h}$ | 8.366 | 4.274 | 0.247 | -0.048 | 0.295 | PC |
| 10b | 6; 2 | 96 | $D_{6d}$ | 14.486 | 5.328 | 0.396 | -0.037 | 0.433 | PC |
| 11b | 6; 5 | 132 | $C_{2h}$ | 11.817 | 5.090 | 0.414 | -0.009 | 0.423 | PC |
| 12c | 6; 9 | 180 | $C_{2h}$ | 10.207 | 4.239 | 0.297 | 0.000 | 0.297 | PC |
| 13a | 7; 2 | 112 | - | 11.846 | 4.506 | 0.247 | 0.174 | 0.073 | PSC |
| 14a | 7; 5 | 154 | $C_s$ | 10.113 | 3.716 | 0.176 | 0.120 | 0.055 | PSC |
| 15a | 7; 9 | 210 | $C_s$ | 8.567 | 4.156 | 0.247 | -0.048 | 0.295 | PC |
| 16b | 7; 2 | 112 | - | 13.107 | 5.159 | 0.379 | -0.037 | 0.416 | PC |
| 17b | 7; 5 | 154 | $C_s$ | 10.492 | 5.005 | 0.405 | -0.023 | 0.428 | PC |
| 18c | 7; 9 | 210 | $C_s$ | 9.030 | 4.268 | 0.297 | 0.000 | 0.297 | PC |

Recall that HOMO-LUMO gap represents a rough measure of the kinetic stability while the heat of formation HF is related to the thermodynamic stability of molecules.

The largest gap follows the sequence: LER, ACR (*fa*-series) and TACR (*ta*-series), as shown in Table 2.

**Table 2: Sequence in *fa*- and *ta*-tubulenes π-electronic shell type**

| Cage / N | 60 | 70 | 80 | … | 180 | 190 | 200 |
|---|---|---|---|---|---|---|---|
| $C_{N(56^5(56)^5)-A[10,n])}$ | LER $m=0$ | ACR $m=0$ | PSC | | LER $m=4$ | ACR $m=4$ | PSC |
| $C_{N(5 5^5 7^5(56)^5-A[10,n])}$ | - | - | TACR $m=0$ | | TNACR $m=3$ | PSC | TACR $m=4$ |

Our hypothesis is as follows: a *ta*-tubulene is the first cage formed in the process of A-tube capping. It quickly isomerizes to the *fa*-tubulene (having the same N) if the product is a (more stable) LER and/or ACR.

This could be the case of TNACR objects that show quasi identical gap, (both in Spectral and Semiempirical approaches - Table 1, entries (3/6), (9/12) and (15/18)) but higher HF values in comparison with the LER *fa*-tubulenes (of identical *N*). The same is true for PSC *ta*-tubulenes, which would isomerize to ACR *fa*-tubulenes.

If *N* corresponds to a PSC in the normal series $C_{N(k6^k(56)^k-A[2k,n])}$, the cage would prefer to stay as *ta*-tubulene TACR (see Table 2). So far, no experimental evidence in the support of the above idea was found.

## 4. *Z*-Tubulenes

### 4.1. *Tz*-Tubulenes

Tubulenes which include a tube Z[*c*,*n*] are called *z*-tubulenes. The *tz*-Caps $C_{3k(k5^k-Z[2k,0])}$ (Figure 1) are derived as halves of some small fullerenes $k\ 5^k\ 6^k\ 5^k\ k$. Figure 7 illustrates two such *z*-tubulenes. Other *z*-tubulenes will be discussed in the next sections.

The *tz*-tubulenes of series $C_{N(k5^k-Z[2k,n])}$ are parents of LER *fa*-tubulenes of series $C_{3N(k6^k(56)^k-A[2k,n])}$. They have open shell electronic structure and surprisingly high HF values.[15]

$C_{80(55^5-Z[10,5])}$; $D_{5d}$ $C_{108(65^6-Z[12,6])}$; $D_{6h}$

Figure 7: Capping Z-tubes by fullerene halves.

As above mentioned, the stability of a tubulene is mainly conditioned by the strain of the cap (attributed to the $\sigma$ frame) and the $\pi$-electronic structure.

By examining the structure of *tz*-tubulenes, it seems that *tz*-cap $C_{3k(k5^k-Z[2k,0])}$, $k = 5; 6$ is not suitable for capping Z-tubes, despite their mutual fitting. One reason is the positive curvature induced by the pentagons, not relaxed by surrounding heptagons, as in case of *ta*-tubulenes. The other reason is the open shell electronic structure of the parent fullerene(s), hereditary in the corresponding *tz*-tubulene(s).

In this view, we calculated the strain energy, in terms of the POAV1 theory,[69-72] which defines the π-orbital axis vector is as the vector making equal angles $\theta_{\sigma\pi}$ to the three σ-bonds of the sp$^2$ carbon atom and a *pyramidalization* angle as:

$$\theta_p = \theta_{\sigma\pi} - 90^\circ \tag{5}$$

This angle is used for estimating the strain energy, induced by a pyramidalized carbon atom, by:

$$SE = 200 \cdot (\theta_p)^2 \tag{6}$$

with $\theta_p$ being measured in radians. The difference $120-(1/3)\cdot\sum\theta_{ij}$ gives the deviation to planarity.

Comparing the *tz*-tubulene $C_{96(65^6-Z[12,5])}$ and *ta*-tubulene $C_{96(6\,5^6 7^6 (56)^6 (6\,5)^6 7^6 5^6 6)}$, we found the following average strain energy values (in kcal/mol): [14.59 (cap); 4.74 (Z-tube); 8.43 (global)] and [10.53 (cap); 4.71 (A-tube); 6.89 (global)] respectively. Clearly, the *tz*-cap $C_{3k(k5^k-Z[2k,0])}$ is more strained in *tz*-tubulenes than in *ta*-tubulenes, while the joining tubes show a comparable strain.

In *tz*-tubulenes of $k = 6$, and $n = 0$ to 4, the PM3 HF values range from 23.5 to 17.5 kcal/mol while they are lower (14.5 to 11.8) for the *ta*-tubulenes.[48] Keeping in mind the average strain of 8.257 and HF of 13.5 for $C_{60}$, it seems that the difference could have an electronic origin (*e.g.*, the open $\pi$-electronic shell[48]), even though the electronic and geometric features are difficult to be separated. These data confirm the above suppositions about the stability of *tz*-tubulenes.

## 4.2. *Fz*-Tubulenes

Another cap, $C_{13k/2(k\,(56)^{k/2}(665)^{k/2}-Z[3k,0])}$ (k = 6, Figure 1), suitable for capping a Z-nanotube, is half of the isolated fullerene C78. The fz-tubulenes of general formula $C_{N(k\,(56)^{k/2}(665)^{k/2}-Z[3k,n])}$ ; k = 4, 6, 8,…, form a twin, odd- (Dk/2,d)/even- (Dk/2,h) n, series of properly closed PC shell objects (see Figure 8, and below). At n+1, the two series show members of PSC shell and lower symmetry ($D_{k/2}$ – Figure 9).

$C_{150(6\,(56)^3(665)^3)-Z[18,4])}$; $D_{3h}$; $k = 6$; PCS    $C_{168(6\,(56)^3(665)^3)-Z[18,5])}$; $D_{3d}$; $k = 6$; PCS/T

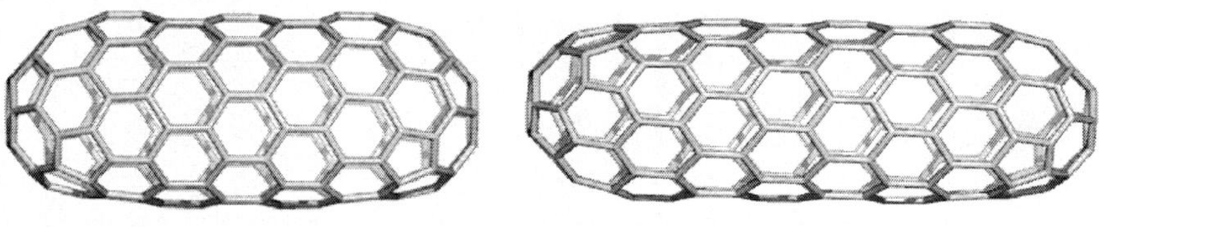

Figure 8: Capping Z-nanotubes by half fullerenes.

The PC shell *fz*-tubulenes (of highest symmetry) obey a true leapfrog rule, but written for zigzag cylinders, Z-LER: $N = 13k + 3km$; $k = 4,6,8,...$; $m = 1,2,...$. These objects show a perfect Clar system[55] PCS, which is not the case of PSC shell counterparts (of lower symmetry – see above). Note the wrong assessment[73] of this rule in a distinct zigzag cylinder rule, ZCR. For $k \neq 6$, the polar rings result by leapfrogging non-trivalent atoms in the parent cage.

Semiempirical calculations and spectral data for some *fz*-tubulenes $C_{N(k\,(56)^{k/2}(665)^{k/2}-Z[3k,n])}$ are given in Table 3.

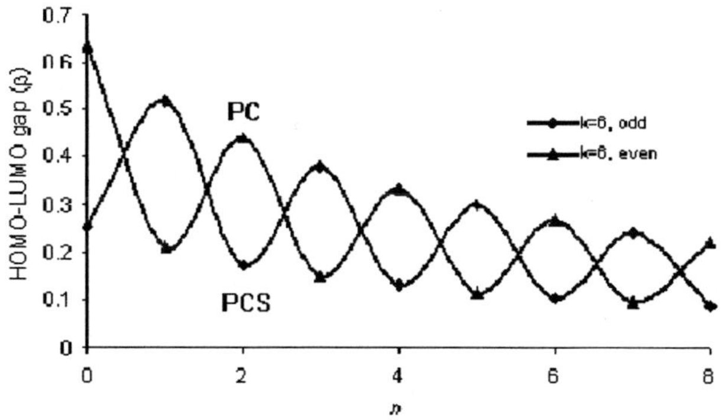


Figure 9: Periodicity of HOMO-LUMO gap *vs*. tube *n*-dimension of the twin odd-/even-*n* series of *fz*-tubulenes of general formula $C_{N(k\,(56)^{k/2}(665)^{k/2}-Z[3k,n])}$.

Out of $C_{78}$, other cages appear (pretty low HF and high gap, Table 3, entries 7-12) as candidates at real molecule status. Observe the quite good concordance between the Hückel gaps and those supplied by PM3, excepting cases of PC with LUMO being NBO (for 0(mod($k$,4)), entries 1, 7 and 13).

Finally, note that the cap $C_{13k/2(k\,(56)^{k/2}(665)^{k/2}-Z[3k,0])}$; $k = 6$, is $C_{60}$ deducible, by taking a hexagon as the polar ring.

**Table 3: Data for *fz*-tubulenes $C_{N(k\,(56)^{k/2}(665)^{k/2}-Z[3k,n])}$**

| | **Cage** | $N$ | **Sym** | **PM3** | **PM3** | **Spectral Data** | | | |
|---|---|---|---|---|---|---|---|---|---|
| | ***k*; *n*** | | **.** | HF/at. (kcal/mol) | **Gap (eV)** | $\lambda_{N/2}$ | $\lambda_{N/2+1}$ | Gap (\|β\| units) | Shell |
| 1 | 4; 0 | 52 | $D_{2h}$ | 21.585 | 5.533 | 0.2564 | 0 | 0.2564 | PC |
| 2 | 4; 1 | 64 | $D_{2d}$ | 19.271 | 5.783 | 0.3789 | -0.0774 | 0.4563 | PC |
| 3 | 4; 2 | 76 | $D_{2h}$ | 18.162 | 5.469 | 0.3260 | -0.0470 | 0.3731 | PC |
| 4 | 6; 0 | 78 | $D_{3h}$ | 12.294 | 6.083 | 0.5157 | -0.1176 | 0.6333 | PC |
| 5 | 6; 1 | 96 | $D_{3d}$ | 11.015 | 5.564 | 0.4329 | -0.0844 | 0.5173 | PC |
| 6 | 6;2 | 114 | $D_{3h}$ | 10.381 | 5.195 | 0.3688 | -0.0721 | 0.4409 | PC |
| 7 | 8; 0 | 104 | $D_{4h}$ | 11.453 | 5.730 | 0.2564 | 0 | 0.2564 | PC |
| 8 | 8; 1 | 128 | $D_{4d}$ | 10.158 | 5.497 | 0.3789 | -0.0774 | 0.4563 | PC |
| 9 | 8; 2 | 152 | $D_{4h}$ | 9.390 | 5.112 | 0.3260 | -0.0470 | 0.3731 | PC |
| 10 | 10; 0 | 130 | $D_{5h}$ | 12.518 | 5.998 | 0.4579 | -0.0874 | 0.5453 | PC |
| 11 | 10; 1 | 160 | $D_{5d}$ | 11.875 | 5.512 | 0.4329 | -0.0844 | 0.5173 | PC |
| 12 | 10; 2 | 190 | $D_{5h}$ | 10.715 | 5.115 | 0.3688 | -0.0721 | 0.4409 | PC |
| 13 | 12; 0 | 156 | $D_{6h}$ | 18.147 | 6.184 | 0.2564 | 0 | 0.2564 | PC |
| 14 | 12; 1 | 192 | $D_{6d}$ | 14.865 | 5.629 | 0.3789 | -0.0774 | 0.4563 | PC |
| 15 | 12; 2 | 228 | $D_{6h}$ | 13.083 | 5.210 | 0.3260 | -0.0470 | 0.3731 | PC |

## 4.3. *Kfz*-Peanut Tubulenes

By cutting off the polar ring $k$ of a spherical fullerene, a "$k$-deleted fullerene *kfz*-cap" of general formula $C_{11k(k\,6^k\,(56)^k\,(65)^k-Z[2k,0])}$ is obtained. Such caps fit to Z-nanotubes (possibly having various covering). Figure 10 illustrates two peanut-shaped *kfz*-tubulenes of the $C_{66(6\,6^6\,(56)^6\,(6\,5)^6-Z[12,0])}$ cap with different distancing nanotubes (observe the complete description of covering). Details on the stability (and isomerization) of these *kfz*-tubulenes will be given in Section 5.

$C_{168(6\,6^6\,(56)^6\,(65)^6\,7^6\,6^6\,6^6\,7^6)}$ $(D_{6d})$ $C_{168(6\,6^6\,(56)^6\,(65)^6\,7^6\,(57)^3\,(75)^3\,7^6)}$ $(S_6)$

Figure 10: Peanut-shaped *kfz* -tubulenes.

The cages in Figure 11 have even shorter distancing tube (one atom and zero atom rows, respectively) between the two caps. The name of such tubulenes includes the code for the cap and distancing zone up to the second cap, if the two caps are identical, or full description, if they are different.

Peanut tubulenes show negative curvature, induced by the presence of rings larger than hexagons. Cages such those in Figure 11 have been observed, by transmission electron microscopy TEM, in the coalescence reactions occurring in $C_{60}$ peapods.[49] Their energetic and topology are shown in the following sections.

(a) $C_{120(56^5\,(56)^5\,(65)^5\,7^5\,7^5-Z[10,1])}$ (b) $C_{108(6(56)^3\,(665)^3\,(656)^3\,7^6-Z[12,0])}$

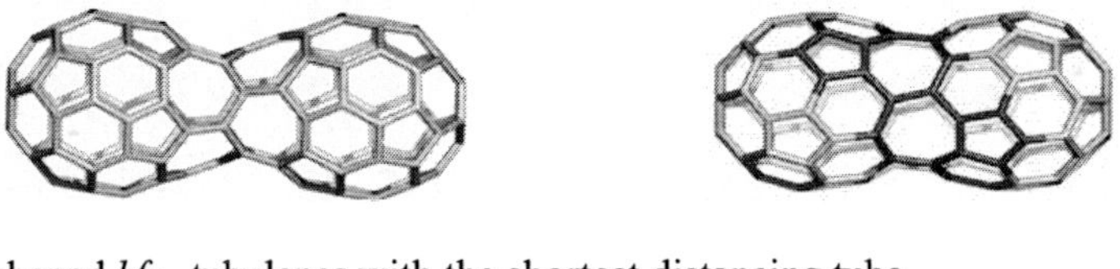

Figure 11: Peanut-shaped *kfz* –tubulenes with the shortest distancing tube

### *4.3.1. Periodic Fulleroids*

At moderate temperature and pressure spherical fullerenes arrange in one dimensional assemblies.[74] The primary $sp^3$ [2+2] adduct rearranges, either by Stone-Wales edge rotations[40] and/or atom exclusion,[39] to give "dimers", oligomers or just corrugated tubules (*e.g.*, having repeat units of peanut-shape, like the objects in Figure 11), as observed in fullerene peapods by thermal annealing or by electron irradiation.[49] Fullerene coalescence, the main molecular transformation occurring in such processes, is presented in detail elsewhere.[75]

Let consider the *kfz*-cap $C_{11k(k\,6^k\,(56)^k\,(65)^k-Z[2k,0])}$ (Figure 1) and the Z-tube having length $n = 1$; a repeat "dimerization" leads to periodic *kfz*-tubulenes of corrugated shape, of general formula $C_{N(k\,6^k\,(56)^k\,(65)^k\,7^k-Z[2k,1]-r)}$, just observed in annealed fullerene peapods.[49] The last subscript $r$ in the above formula means the number of sphere-like repeat moieties. $N$ is calculable by formula: $N = 12kr$ (see Table 4). Figure 12 illustrates such a periodic "fulleroid".[47]

$C_{N(k\,6^k\,(56)^k\,(65)^k\,7^k-Z[2k,1]-r)}$; $k = 5$; $r = 4$

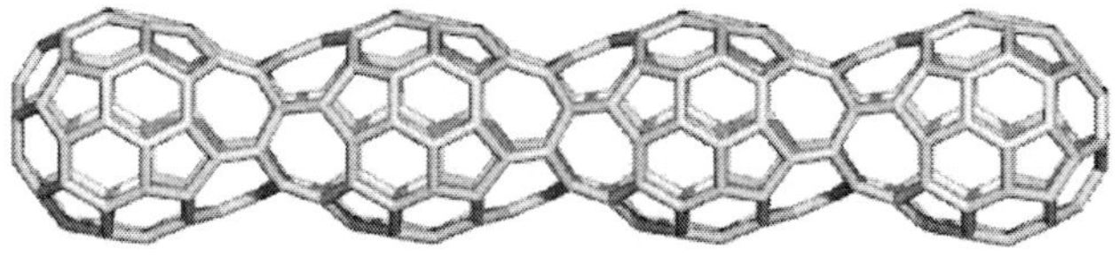

Figure 12: A periodic *kfz*-tubulene.

We were interested in topological characterization of such all $sp^2$ periodic objects. The following theorem states the counting formulas (*i.e.*, number of faces of size $s$, $f_s$, edges shared by faces of given size, $e_{s1,s2}$ and vertices shared by corresponding faces, $v_{s1,s2,s3}$) for objects with $k = 4$ to 7.

**Theorem 1**

For a periodic fulleroid, of general formula $C_{N(k\,6^k\,(56)^k\,(65)^k\,7^k-Z[2k,1]-r)}$, the number of faces, edges, and vertices of various types composing its associate graph, can be counted function of the repeat parameter r and polar ring size k (Table 4).

The theorem is demonstrated by construction.

**Table 4: Tiling counting formulas for periodic fulleroids of general formula $C_{N(k\,6^k\,(56)^k\,(65)^k\,7^k-Z[2k,1]-r)}$.**

| **k = 4, 5, 6, 7** | |
|---|---|
| **Faces** | **Vertices** |
| $f_{4k} = 2t_4$ * | $v_{4,6,6k} = 2kt_4$ |
| $f_{5k} = 2kr + 2t_5$ | $v_{5,6,6k} = 2k(2r+3+t_5)$ |
| $f_{6k} = 2k(r+1) + 2t_6$ | $v_{5,6,7k} = 4k(r-1)$ |
| $f_{7k} = 2k(r-1) + 2t_7$ | $v_{5,7,7k} = 2k(r-1)$ |
| **Edges** | $v_{6,6,6k} = 2kt_6$ |
| $e_{4,6k} = 2kt_4$ | $v_{6,6,7k} = 2kt_7$ |
| $e_{5,6k} = 2k(3r+2+t_5)$ | $v_{7,7,7k} = 2k(r-1)$ |
| $e_{5,7k} = 4k(r-1)$ | **Total No. atoms** |
| $e_{6,6k} = 2k(r+2+t_6)$ | $N_k = 12kr$ |
| $e_{6,7k} = 2k(r-1+t_7)$ | |
| $e_{7,7k} = 4k(r-1)$ | |

* $t_s = 1$ if $s = k$, and *zero* otherwise

Semiempirical calculations in series $C_{N(k\,6^k\,(56)^k\,(65)^k\,7^k-Z[2k,1]-r)}$, $k = 5$, for the first four terms (parent fullerene, with $r = 1$, included) show HF values in the range of C60, and large enough gaps (Table 5).

**Table 5: Semiempirical PM3 data for periodic fulleroids of the series $C_{N(k\,6^k\,(56)^k\,(65)^k\,7^k-Z[2k,1]-r)}$**

| | Cage (*k*; *r*) | N | Sym. | HF/atom | Gap |
|---|---|---|---|---|---|
| 1 | 5; 1 | 60 | $I_h$ | 13.512 | 6.593 |
| 2 | 5; 2 | 120 | $D_{5d}$ | 12.547 | 6.204 |
| 3 | 5; 3 | 180 | $D_{5d}$ | 12.224 | 5.917 |
| 4 | 5; 4 | 240 | $D_{5d}$ | 12.062 | 5.819 |

### *4.3.2. ((5,6,7)3) Kfz-Tubulenes*

Tubulenes capped with *kfz*-caps, like $C_{9k(k\,(56)^{k/2}(665)^{k/2}(656)^{k/2}-Z[2k,0])}$ (Figure 1), derived from $C_{60}$, and those cut off from $C_{78}$, are ((5,6,7)3) *kfz*-tubulenes with the shortest distancing tube. They can have the general formulas: $C_{N(k\,(56)^{k/2}(665)^{k/2}(656)^{k/2}7^k-Z[2k,0])}$ (Figure 11b), or $C_{N(k\,(56)^{k/2}(665)^{k/2}6^{3k/2}(566)^{k/2}7^k-Z[2k,0])}$ (Figure 13a), or $C_{N(k\,(56)^{k/2}(665)^{k/2}6^{3k/2}(656)^{k/2}7^k-Z[2k,0])}$ and even combinations of the three types (see Figure 13b). HF values, calculated with the PM3 Hamiltonian, are listed in Table 6 (entries 1 to 5). These values are in the range of that for $C_{60}$, lower for the peanuts derived from $C_{78}$. The strain energy per atom *S*, by POAV1 (see above), is lower than that calculated for $C_{60}$ (comparison made with the [2+2] adduct $sp^3$ dimer). The gaps are, however, with about 2 eV lower to that for $C_{60}$, suggesting a relative kinetic instability.

(a) $C_{144(6(56)^3(665)^3 6^9(566)^3 7^6-Z[12,0])}$ (b) $C_{126(6(56)^3(665)^3(656)^3 7^6\,(656)^3 6^9(665)^3(65)^3 6)}$

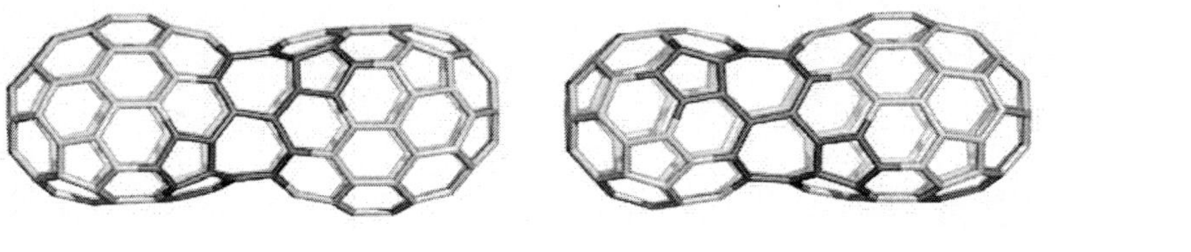

Figure 13: Peanut-shaped *kfz*–tubulenes with the shortest distancing tube.

Periodic tubulenes of corrugated shape, like the *kfz*–tubulene in Figure 14, were observed in $C_{60}$ peapods submitted to prolonged thermal annealing (the inner tube).[49] Oligomers with $r = 2$ to 4 show pretty low HF values (Table 6, entries 5 to 7). The corresponding of the $k = 5$ peanut dimer $C_{120(56^5(56)^5(65)^5 7^5 7^5-Z[10,1])}$ (Figure 11a), namely the $k = 6$ dimer, $C_{120(6(56)^3(665)^3(656)^3 7^6-Z[12,1])}$, was also inferred in the coalescence of $C_{60}$.[50]

$C_{204(6(56)^3(665)^3(656)^3 7^6-Z[12,0]-r)}$; $r = 4$

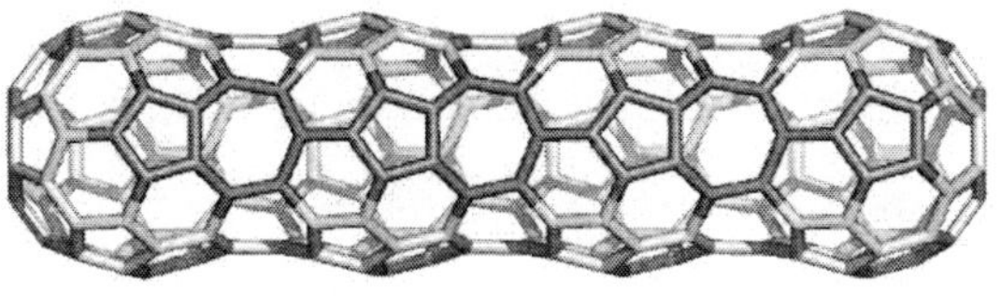

Figure 14: Periodic *kfz*–tubulene with the repeat spherical moiety derived from $C_{60}$ by cutting off the polar hexagons.

The ((5,6,7)3) covering pattern is easily derived, by Stone-Wales edge flipping, in polyhex (6,3) nanotubes.[76],[77] It can also be embedded in the torus (Figure 15).

**Table 6: Semiempirical PM3 data for periodic peanut *kfz*-tubulenes (Figure 13) and their relatives; for comparison, $C_{60}$ is included.**

| | Cage | Sym. | HF/atom (kcal/mol) | Gap (eV) | *S*/atom (kcal/mol) |
|---|---|---|---|---|---|
| 1 | $C_{144(6(56)^3(665)^3 6^k (566)^3 7^6-Z[12,0])}$ | $D_{3d}$ | 11.911 | 4.729 | 5.530 |
| 2 | $C_{144(k\ (56)^{k/2}(665)^{k/2} 6^k (656)^{k/2} 7^k-Z[2k,0])}$ | $S_6$ | 11.666 | 4.049 | 5.343 |
| 3 | $C_{144(6(56)^3(665)^3 6^k (566)^3 7^6 (665)^3 6^6 (665)^3 (65)^3 6)}$ | $C_3$ | 11.789 | 4.2664 | 5.436 |
| 4 | $C_{126(6(56)^3(665)^3(656)^3 7^6\ (656)^3(566)^3(65)^3 6)}$ | $C_{3v}$ | 12.349 | 4.796 | 5.848 |
| 5 | $C_{108(6(56)^3(665)^3(656)^3 7^6-Z[12,0]-2)}$ | $D_{3d}$ | 12.953 | 4.870 | 6.493 |
| 6 | $C_{108(6(56)^3(665)^3(656)^3 7^6-Z[12,0]-3)}$ | $D_{3d}$ | 12.878 | 4.502 | 5.687 |
| 7 | $C_{108(6(56)^3(665)^3(656)^3 7^6-Z[12,0]-4)}$ | $D_{3d}$ | 12.681 | 4.484 | 5.321 |
| 8 | $C_{60}$ | $I_h$ | 13.512 | 6.593 | 8.257 |

$T_{N((566)^{k/2}(665)^{k/2} 7^k-Z[2k,0]-r)}$; $k = 6$; $r = 30$; $N = 8kr$

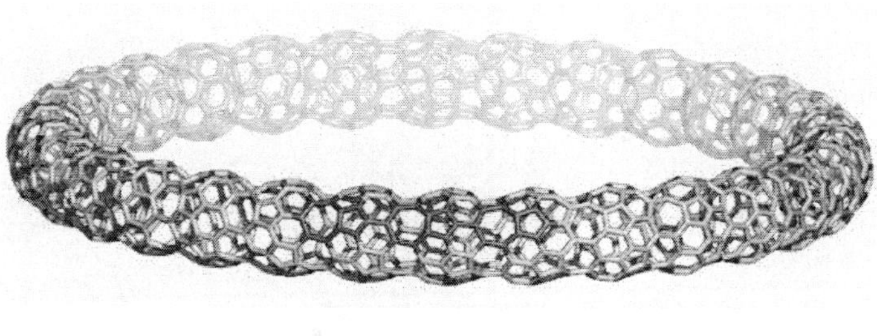

Figure 15: A toroidal embedding of the ((5,6,7)3) pattern.

The topology of ((5,6,7)3) decorated structures is calculable as stated by the following:[78]

**Theorem 2**

For a periodic ((5,6,7)3) covering, of local signature: t5j(0, 4, 1); t6j(2, 2, 2); and t7j(1, 4, 2), j = 5, 6, 7, the number of faces, edges, and vertices of various types composing its associate graph, embedded in the torus, can be counted function of the repeat parameter r and ring size k of the (equivalent) tube cross section (Table 7).

The theorem is demonstrated by construction.

**Table 7: Tiling counting formulas for toroids, of periodic ((5,6,7)3) covering, with general formula** $T_{N((566)^{k/2}(665)^{k/2}7^k-Z[2k,0]-r)}$

| $k = 4, 6,...$ | |
|---|---|
| **Faces** | **Vertices** |
| $f_5 = kr$ | $v_{5,6,6} = 3kr$ |
| $f_6 = 2kr$ | $v_{5,6,7} = 2kr$ |
| $f_7 = kr$ | $v_{6,6,7} = kr$ |
| **Edges** | $v_{6,7,7} = 2kr$ |
| $e_{5,6} = 4kr$ | **Total no. atoms** |
| $e_{5,7} = kr$ | $N = 8kr$ |
| $e_{6,6} = 2kr$ | |
| $e_{6,7} = 4kr$ | |
| $e_{7,7} = kr$ | |

A torus T((5,6,7)3)VA[$p$,$q$] derived, from an equivalent (6,3)Z[$c$,$n$] torus, by SW edge rotations, has the parameters: ($p$; $q$) = ($c$/4; $n$/4), with $c = 2f_7$ and the repeat unit parameter $r = q$ (see Table 7).

### *4.3.3. ((5,7)3) Kfz-Tubulenes*

The *tz*-Cap $C_{3k(k5^k-Z[2k,0])}$ fits to a suitable phenylenic ((4,6)3)H tube[79] (Figure 5), to give a cage $C_{N(k5^k7^k(46)^k7^k5^kk)}$, which further rearranges to a cage $C_{N(k5^k7^k5^{2k}7^k5^kk)}$, $k$ = 5,7, decorated with only pentagons and heptagons (see below).

The *kfz*-cap $C_{11k(k5^k7^k5^{2k}7^k-Z[2k,0])}$, derived from the above cage, leads, by identifying the *z*-boundary, to a "dimer" (Figure 16, $k$ = 7).

$2(C_{11k(k5^k7^k5^{2k}7^k-Z[2k,0])} - k); k = 7$ $\quad$ $C_{20k(k5^k(7^k5^{2k}7^k)^25^kk)}; k = 7$

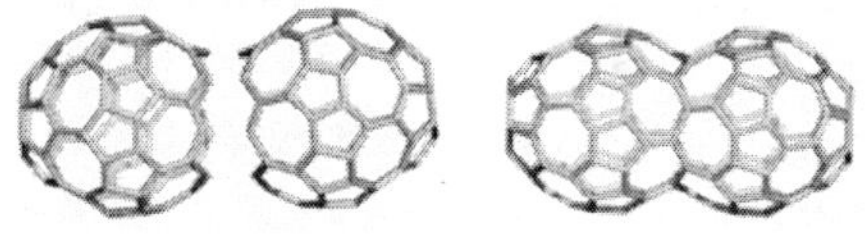

Figure 16: An *in silico* "dimerization" process.

The coupling process can continue to give periodic cages of formula $C_{N(k5^k(7^k5^{2k}7^k)^r5^kk)}$, with $N=4k(2r+1)$ (see Table 8). In the above formula, $r$ counts the spheroidal repeat units, $r$ = 2, 3,...

An example of such periodic cages is given in (Figure 17). It is a third cage having 260 points and ((5,7)3) decoration ($f_5=f_7+12;\ f_7=60$), in addition to the phantasmagoric fullerenes designed by Fowler and Dress, respectively.[80]

$$C_{260(k5^k(7^k5^{2k}7^k)^r5^kk);\quad k=5;\quad r=6}$$

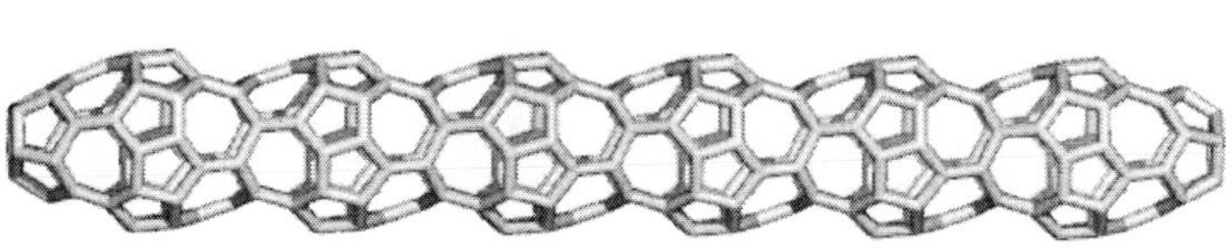

Figure 17: Diudea's $C_{260}$ cage with ((5,7)3) decoration.

The topology of ((5,7)3) decorated structures is calculable as shown by:[78]

**Theorem 3**

For a periodic cage with ((5,7)3) decoration, the number of faces, edges, and vertices of the various types can be counted function of the repeat unit r and polar ring size k (Table 8).

The theorem is demonstrated by construction.

**Table 8: Periodic cages with ((5,7)3) decoration - net counting**

| **Formulas for $k$ = 5; 7** | |
|---|---|
| $f_{5k}=2k(r+1)+2t_5$ | (1) |
| $f_{7k}=2kr+2t_7$ | (2) |
| $e_{5,5k}=2k(r+1+t_5)$ | (3) |
| $e_{5,7k}=2k(3r+2+t_7)$ | (4) |
| $e_{7,7k}=2k(2r-1)$ | (5) |
| $v_{5,5,5k}=2kt_5$ | (6) |
| $v_{5,5,7k}=2k(2r+1+t_7)$ | (7) |
| $v_{5,7,7k}=2k(r+1)$ | (8) |
| $v_{7,7,7k}=2k(r-1)$ | (9) |
| $N_k=4k(2r+1)$ | (10) |

*$t_s=1$ if $s=k$, and *zero* otherwise

A toroidal embedding of the ((5,7)3) pattern is shown in Figure 18.

$$\mathrm{T}_{N(7^k 5^{2k} 7^k - Z[2k,0] - r)};\ k = 6;\ r = 30;\ N = 8kr$$

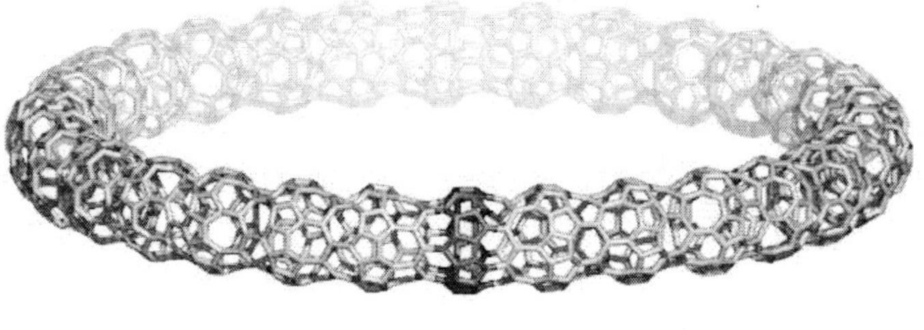

Figure 18: A toroidal embedding of the ((5,7)3) pattern.

PM3 calculations for some all ((5,7)3) periodic cages are given in Table 9. Only the dimeric tubulene $\mathrm{C}_{140(75^7(7^7 5^{17} 7^7)^2 5^7 7 - Z[14,1])}$ (entry 4) appears to have the HF closer to $C_{60}$ (entry 5), while the π-electronic structure is pseudo-closed or open.

**Table 9: PM3 data for data for some all ((5,7)3) periodic cages of general formula $\mathrm{C}_{N(k 5^k (7^k 5^{2k} 7^k)^r 5^k k - Z[2k,1])}$**

| | Cage N, k, r | Sym. | HF/at. | Gap |
|---|---|---|---|---|
| | Tube Z[10,1] | | | |
| 1 | 60, 5, 1 | $C_i$ | 21.158 | 5.623 |
| 2 | 100, 5, 2 | $C_i$ | 18.906 | 5.592 |
| | Tube Z[14,1] | | | |
| 3 | 84, 7, 1 | $C_i$ | 16.249 | 4.538 |
| 4 | 140,7, 2 | $C_i$ | 15.828 | 5.114 |
| 5 | $C_{60}$ | $C_i$ | 13.512 | 6.596 |

As a consequence, the all ((5, 7)3) tubulenes tend to isomerize to the more stable *fa*-tubulenes, as shown in the next section.

## 5. Isomerizing Tubulenes

The peanut $\mathrm{C}_{168(66^6(56)^6(65)^6 7^6 6^6 6^6 7^6)}$ ($D_{6d}$ - Figure 10a), can isomerize to the peanut $\mathrm{C}_{168(66^6(56)^6(65)^6 7^6 (57)^3 (75)^3 7^6)}$ ($S_6$ - Figure 10b), with azulenic tiling on the distancing tube, by (three) Stone-Walls SW edge rotations [40],[78] (see the necks and their geodesic projections in Figure 19). Semiempirical data for the above cages and their SW intermediates are given in Table 10.

(1) $C_{72(7^6 6^6 6^6 7^6)}$ (2) $C_{72(7^6 5 6^4 7 7 6^4 5 7^6)}$ (3) $C_{72(7^6 5 6^2 7 5 7^2 6^2 5 7 5 7^6)}$ (4) $C_{72(7^6 (57)^3 (75)^3 7^6)}$

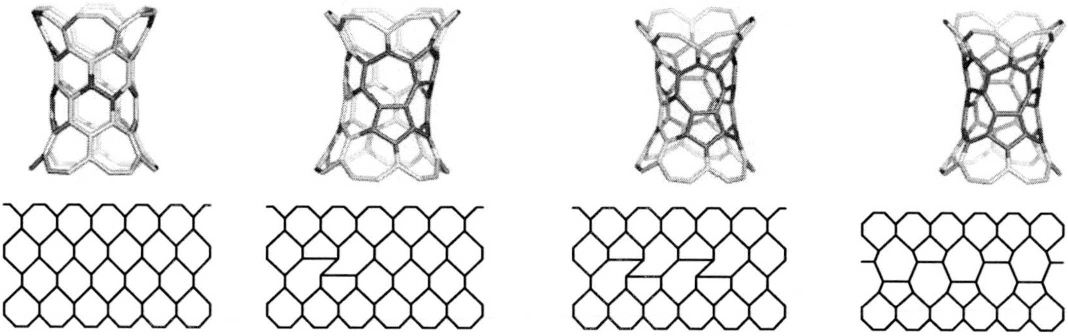

Figure 19: SW isomerization of the necks of the peanut cages from the polyhex $C_{72(7^6 6^6 6^6 7^6)}$ to azulenic $C_{72(7^6(57)^3(75)^3 7^6)}$ and their geodesic projections.

**Table 10: PM3 data for the peanut z-tubulenes in Figures 10 and 19**

| | Cage | *N* | Sym. | HF/atom | Gap |
|---|---|---|---|---|---|
| 1 | $C_{168(6\,6^6(56)^6(65)^6 7^6 6^6 6^6 7^6)}$ | 168 | $D_{6d}$ | 11.927 | 5.114 |
| 2 | $C_{168(66^6(56)^6(65)^6 7^6 5 6^4 7 7 6^4 5 7^6)}$ | 168 | $C_2$ | 12.261 | 5.157 |
| 3 | $C_{168(66^6(56)^6(65)^6 7^6 5 6^2 7 5 7^2 6^2 5 7 5 7^6)}$ | 168 | $C_2$ | 12.452 | 5.194 |
| 4 | $C_{168(66^6(56)^6(65)^6 7^6(57)^3(75)^3 7^6)}$ | 168 | $S_6$ | 12.553 | 5.425 |

The azulenic neck brings a PC shell of the π-electrons in the peanut structure (Table 10, entry 4), correlated with a higher gap and a better kinetic stability.

An SW process could transform the (4,6) pair in $C_{84(75^7 7^7 (46)^7 7^7 5^7 7)}$ into the (5,5) pair, in $C_{84(75^7 7^7 5^{14} 7^7 5^7 7)}$, Figure 20. Similarly, cages with ((5,7)3) covering and general formula $C_{12k(k\,5^k 7^k 5^{2k} 7^k 5^k\,k)}$ lead to the classical $C_{12k(k\,6^k (56)^k (65)^k 6^k\,k)}$ fullerenes, as shown in Figure 21.

$C_{84(75^7 7^7 (46)^7 7^7 5^7 7)}$ $(C_1)$ $C_{84(75^7 7^7 5^{14} 7^7 5^7 7)}$ $(C_i)$

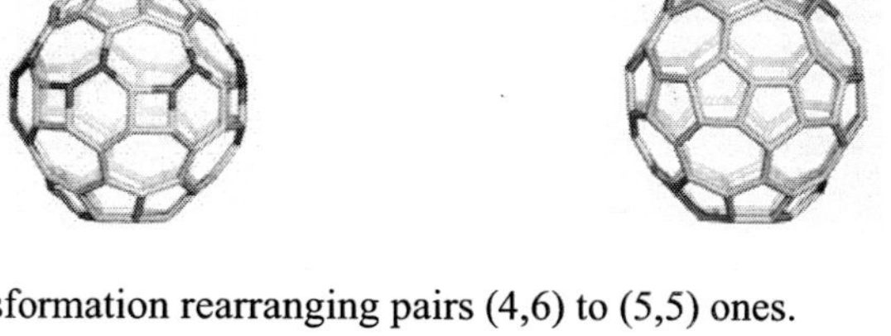

Figure 20: An SW transformation rearranging pairs (4,6) to (5,5) ones.

$C_{60(5 5^5 7^5 5^{10} 7^5 5^5 5)}$ $(C_i)$ $C_{60(5 6^5 (5 6)^5 (6 5)^5 6^5 5)}$ $(I_h)$

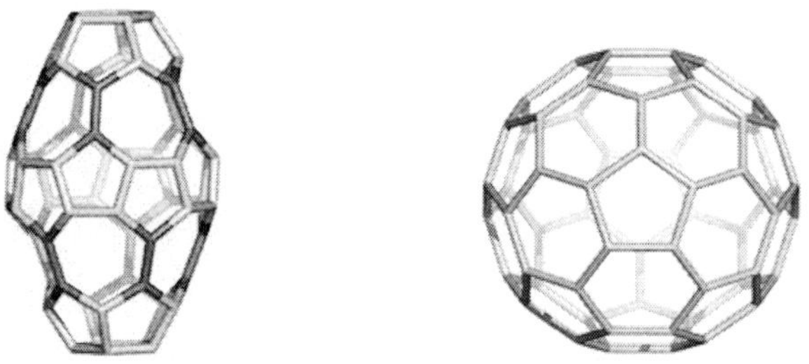

Figure 21: Isomerization of all ((5,7)3) cages leading to the classical $C_{12k}$ fullerenes.

Peanut cages with ((5,7)3) covering, like $C_{100(5\,5^5(7^5 5^{10} 7^5)^2 5^5 5)}$ (Figure 22a) can isomerize to the corresponding *fa*-tubulene, in our example $C_{100(5\,6^5(5\,6)^5-A[10,2])}$ (Figure 22b), by two major ways,[78] as the SW rotation process starts: (i) by the "red" bonds sharing two heptagons in the zone of joining the two repeat units or (ii) the "blue" bonds at the caps region. The stepwise process is illustrated in Figure 22 while the corresponding energetic data are listed in Table 11.

(a) $C_{100(k\,5^k(7^k 5^{2k} 7^k)^r 5^k k);\; k=5;\; r=2}$ (b) $C_{100(5\,6^5(5\,6)^5-A[10,2])}$

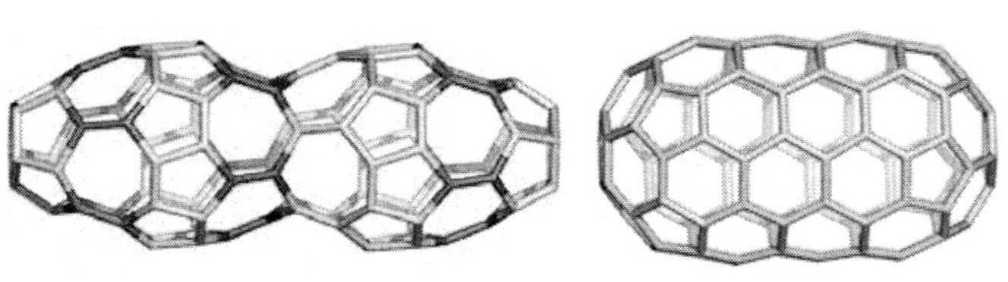

(c) $C_{100(k\,5^k 7^k 5^{2k} 8^k (56)^k (57)^k 5^k k);\; k=5}$ (d) $C_{100(5 5^5 7^5 (56)^5 (66)^5 (65)^5 7^5 5^5 5)}$

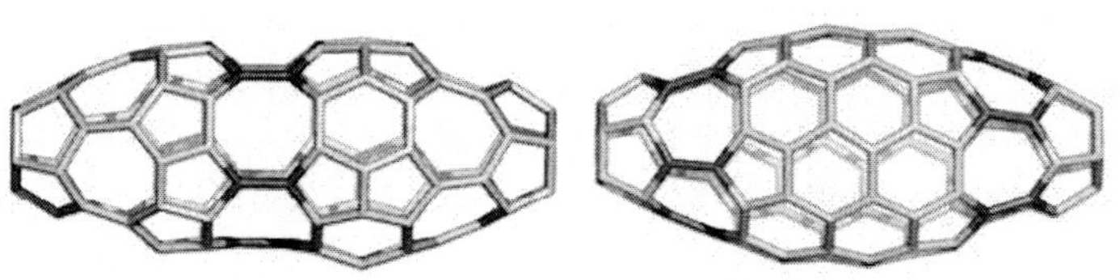

(e) $C_{100(k\,6^k (56)^k (57)^k 7^k (56)^k (56)^k k);\; k=5}$ (f) $C_{100(5 6^5 (56)^5 (58)^5 (56)^5 (65)^5 6^5 5)}$

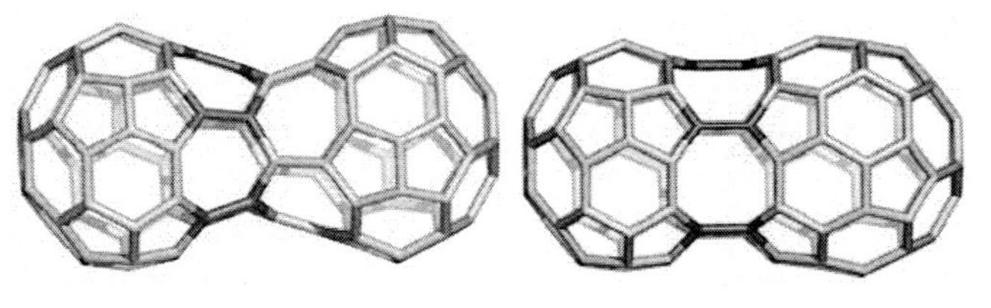

Figure 22: Isomerization of all (5, 7) cages leading to classical *fa*-tubulenes.

**Table 11: PM3 data for an all ((5, 7)3) cage and intermediates of isomerization to their corresponding *fa*-tubulene. The corresponding letters in Figure 22 are associated to the entry numbers**

| | Cage | Sym. | HF/atom | Gap |
|---|---|---|---|---|
| 1(a) | $C_{100(k\,5^k\,(7^k5^{2k}\,7^k)^r\,5^k\,k);\ k=5;\ r=2}$ | $C_1$ | 18.906 | 5.592 |
| 2(e) | $C_{100(k\,6^k\,(56)^k\,(57)^k\,7^k\,(56)^k\,(56)^k\,k);\ k=5}$ | $D_{5d}$ | 15.146 | 5.546 |
| 3(f) | $C_{100(56^5(56)^5(58)^5(56)^5(65)^5 6^5 5)}$ | $C_{5v}$ | 13.966 | 5.770 |
| 4(c) | $C_{100(k\,5^k7^k5^{2k}\,8^k\,(56)^k\,(57)^k\,5^k\,k);\ k=5}$ | $C_1$ | 18.260 | 5.404 |
| 5(d) | $C_{100(55^57^5(56)^5(66)^5(65)^5 7^5 5^5 5)}$ | $D_{5d}$ | 15.205 | 3.849 |
| 6(b) | $C_{100(5\,6^5(5\,6)^5-A[10,2])}$ | $D_{5d}$ | 10.876 | 5.045 |

Clearly, the corresponding *fa*-tubulene $C_{100(5\,6^5(5\,6)^5-A[10,2])}$ (Table 10, entry 6) is the far more stable structure.

The above presented isomerizations could be real routes in the synthesis and chemistry of nanostructures.[78]

Semiempirical calculations have been performed on a 2x1GHz Pentium III PC by using the PM3 Hamiltonian, in standard parametrization supplied by HyperChem (version 4.5, Hypercube, Inc.)[81] software. Structures were optimized by using the Polak-Ribier conjugate-gradient method, the energy minimization was terminated at an RMS gradient <0.01 kcal/(Å·mol) for all structures.

Spectral data (see Discussion) were calculated by TOPOCLUJ 2.0 Software Package.[82]

Stone-Wales rearrangements were performed by the CageVersatile 1.1 software package.[83]

## 6. Conclusions

Construction of tubulenes, from armchair and zigzag nanotubes, as "distancers", and caps mainly derived from the classical spherical fullerenes was presented. Their structure was rationalized by a tempting classification according to neck and/or caps covering. Periodicity of the presented tubulenes was considered both as topology (constitutive typing enumeration) and $\pi$-electronic structure.

The systematic search of spectral properties of various tubulenes has revealed some regularities in their $\pi$-electronic structure, rationalized in terms of the classical $C_{N(k6^k(56)^k-A[2k,n])}$ cylinders.

Some isomerizations, following the well-known Stone-Wales edge rotation, of the discussed tubulenes, could a7ppear in the real synthesis and chemistry of nanostructures.

Semiempirical calculations support the idea that new, relatively stable molecules, with tessellation including ((4,7)3) (and higher sized rings) to the classical ((5,6)3) one, may appear in the soot of vaporized graphite.

## REFERENCES

[1] P. W. Fowler and D. E. Manolopolous, *An atlas of fullerenes*, Oxford University Press, Oxford, U.K., 1995.

[2] R. Taylor, Ed., *The chemistry of fullerenes*, World Scientific: Singapore, 1995.

[3] T. Braun, A. Schubert, H. Maczelka, L. Vasvari, *Fullerene research* 1985-1993; World Scientific, Singapore, 1995.

[4] M. Endo, S. Iijima, and M. S. Dresselhaus, *Carbon Nanotubes*, Pergamon, 1996.

[5] M. S. Dresselhaus, G. Dresselhaus, and P. C. Eklund, Science of fullerenes and carbon nanotubes, Acad. Press, San Diego, 1996.

[6] A T. Balaban, (Ed.), *From Chemical Topology to Three - Dimensional Geometry*, Plenum Press, New York, 1997.

[7] K. Tanaka, T. Yamabe, and K. Fukui, The science and technology of carbon nanotues, Elsevier, 1999.

[8] H. F. Bettinger, B. I. Yakubson, and G. E. Scuseria, *J. Amer. Chem. Soc.*, 2003, 125, 5572-5580.

[9] Ge. G. Samsonidze, G. G. Samsonidze, and B. I. Yakobson, *Comput. Mater. Sci.*, 2002, 23, 67-72.

[10] D. Orlikowski, M. B. Nardelli, J. Bernholc, and Ch. Ronald, *Phys. Rev. B*, 2000, 61, 14194-14203.

[11] J. Hu, T. W. Odom, and C. M. Lieber, *Acc. Chem. Res.*, 1999, 32, 435-445.

[12] C. Piskoti, J. Yarger, and A. Zettl, *Nature*, 1998, 393, 771-773.

[13] H. Prinzbach, A. Woller, P. Landenberger, F. Wahl, J. Worth, L. T. Scott, M. Gelmont, D. Olevano, and B. Von Issendorff, *Nature*, 2000, 407, 60-63.

[14] Z. Wang, X. Ke, Z. Zhu, M. Ruan, H. Chen, R. Huang, and L. Zheng, *Phys. Lett. A*, 2001, 280, 351-356.

[15] Z. Slanina, F. Uhlik, and L. Adamowicz, *J. Mol. Graph. Modell.*, 2003, 21, 517–522.

[16] AHirsch, *The Chemistry of the Fullerenes*, Thieme Org. Chem. Monograph Ser., Stuttgart, 1994.

[17] H. Zorc, Lj. P. Tolić, S. Martinović, and D. Srzić, *Fuller. Sci. Technol.*, 1994, 2, 471.

[18] F. Diedrich and C. Thilgen, *Science*, 1996, 271, 317.

[19] I S. Neretin, K. A. Lyssenko, M. Yu. Antipin, Yu. L. Slovokhotov, O. V. Boltalina, P. A. Troshin, A. Yu. Lukonin, L. N. Sidorov, and R. Taylor, *Angew. Chem. Int. Ed.*, 2000, 39, 3273.

[20] W. Qian and Y. Rubin, *Angew. Chem. Int. Ed.*, 2000, 39, 3133-3137.

[21] K. Lee, Ch. H. Lee, H. Song, J. T. Park, H. Y. Chang, and M.-G. Choi, *Angew. Chem. Int. Ed.*, 2000, 39, 1801.

[22] T. F. Fässler, R. Hoffmann, S. Hoffmann, and M. Wörle, *Angew. Chem. Int. Ed.*, 2000, 39, 2091.

[23] J-. Y. Zheng, Sh. Noguchi, K. Miyauchi, M. Hamada, K. Kinbara, and K. Saigo, *Fullerene Sci. Technol.*, 2001, 9, 467-475.

[24] S. Lebedkin, A. Gromov, S. Giesa, R. Gleiter, B. Renker, H. Rietschel, and W. Krätschmer, *Chem. Phys. Lett.*, 1998, 285, 210 .
[25] S. Lebedkin, S. Ballenweg, J. Gross, R. Taylor, and W. Krätschmer, *Tetrahedron Lett.*, 1995, 36, 4971.
[26] A Gromov, S. Lebedkin, S. Ballenweg, A. G. Avent, R. Taylor, and W. Krätschmer, *Chem. Commun.*, 1997, 209.
[27] A Gromov, S. Lebedkin, W.E. Hull, and W. Krätschmer, *J. Phys. Chem. A*, 1998, 102, 4997.
[28] H. Kroto, *Fuller. Sci. Technol.*, 1994, 2, 333-342.
[29] Y. D. Gao and W. C. Herndon, *J. Amer. Chem. Soc.*, 1993, 115, 8459.
[30] P. W. Fowler, T. Heine, D. E. Manolopoulos, D. Mitchell, G. Orlandini, R. Schmidt, G. Seiferth and F. Zerbetto, *J. Phys. Chem.*, 1996, 100, 6984.
[31] A Müller, P. Kögerler, and Ch. Kuhlmann, *J. Chem. Soc., Chem. Commun.*, 1999, 1347-1358.
[32] L. Euler, *Comment. Acad. Sci. I. Petropolitanae*, 1736, 8, 128-140; *Novi Comment. Acad. Sci. I. Petropolitanae*, 1758, 4, 109-160.
[33] E. Hückel, *Z. Phys.*, 1931, 170, 204-286.
[34] P. W. Fowler and T. Pisanski, *J. Chem. Soc., Faraday Trans.*, 1994, 90, 2865-2871.
[35] P. W. Fowler, *Phys. Lett.*, 1986, 131, 444-450.
[36] P. W. Fowler and J. I. Steer, *J. Chem. Soc., Chem. Commun.*, 1987, 1403-1405.
[37] P. W. Fowler, *J. Chem. Soc., Faraday Trans.*, 1997, 93, 1-3.
[38] Y. Zhao, B. I. Yakobson, and R.E. Smalley, *Phys. Rev. Lett.*, 2002, 88, 185501.
[39] M. Endo and H. W. Kroto, *J. Phys. Chem.*, 1992, 96, 6941-6944.
[40] A J. Stone and D. J. Wales, *Chem. Phys. Lett.*, 1986, 128, 501-503.
[41] G. Brinkmann, P. W. Fowler, and C. Justus, *J. Chem. Inf. Comput. Sci.*, 2003, 43, 917-927
[42] G. Brinkmann, U. V. Nathusius, and A. H. R. Palser, *Discr. Appl. Math.*, 2002, 116, 55-71.
[43] D E. Manolopoulos, J. C. May, and S. E. Down, *Chem. Phys. Lett.*, 1991, 181, 105-111.
[44] N. Hamada, S. Sawada, and A. Oshiyama, *Phys. Rev. Lett.*, 1992, 68, 1579-1581.
[45] D J. Klein and H. Zhu, All-conjugated carbon species, in: A. T. Balaban, (Ed.), *From Chemical Topology to Three - Dimensional Geometry*, Plenum Press, New York, 1997, pp. 297-341.
[46] M. V. Diudea, *Bul. Stiint. Univ. Baia Mare Ser. B*, 2002, 18, 31-38.
[47] M. V. Diudea, *Int. J. Nanosci.*, 2003, 2, 171-183.
[48] M. V. Diudea, *Phys. Chem. Chem. Phys.*, 2004, 6, 332-339.
[49] E. Hernández, V. Meunier, B. W. Smith, R. Rurali, H. Terrones, M. Buongiorno Nardelli, M. Terrones, D. E. Luzzi, and J.-C. Charlierr, *Nano Lett.*, 2003, 3, 1037-1042.
[50] S. Bandow, T. Hiraoka, T. Yumura, K. Hirahara, H. Shinohara and S. Iijima, *Chem. Phys. Lett.*, 2004, 384, 320–325.
[51] P. W. Fowler, *J. Chem. Soc., Faraday Trans.*, 1990, 86, 2073-2077.
[52] D. J. Klein and H. Zhu, *Discrete Appl. Math.*, 1996, 67, 157-173.
[53] P. E. John, *Croat. Chem. Acta*, 1998, 71, 435-447.
[54] I Lukovits, A. Graovac, E. Kálmán, G. Kaptay, P. Nagy, S. Nikolić, J. Sytchev, and N. Trinajstić, *J. Chem. Inf. Comput. Sci.*, 2003, 43, 609-614.

[55] E. Clar, *The Aromatic Sextet*, Wiley, New York, 1972.
[56] D. J. Klein, *J. Chem. Ed.*, 1992, 69, 691-694.
[57] J. Aihara, *J. Am. Chem. Soc.*, 1976, 98, 2750-.
[58] J Aihara, *J. Mol. 0Struct. (Theochem)*, 1994, 311, 1-8.
[59] I Gutman, M. Milun, and N. Trinajstic, *MATCH-Commun. Math. Comput. Chem.*, 1975, 1, 171.
[60] I Gutman, M. Milun, and N. Trinajstic, *J. Am. Chem. Soc.*, 1977, 99, 1692-1704.
[61] M. Randić, *Chem. Phys. Lett.*, 1976, 38, 68-70.
[62] M. Randić, *J. Am. Chem. Soc.*, 1977, 99, 444-450.
[63] W. C. Herndon, *J. Am. Chem. Soc.*, 1973, 95, 2404-2406.
[64] P. v. R. Schleyer, C. Maerker, A. Dransfeld, H. Jiao, and N. J. R. van Eikema Hommes, *J. Am. Chem. Soc.*, 1996, 118, 6317-6318.
[65] H. Jiao and P. v. R. Schleyer, *Angew. Chem. Int. Ed. Engl.,* 1996, 35, 2383-286.
[66] A Hirsch, Z. Chen, and H. Jiao, *Angew. Chem. Int. Ed. Engl.,* 2000, 39, 3915-3917.
[67] Z. Chen, H. Jiao, and A. Hirsch, *J. Mol. Model.*, 2001, 7, 161-163.
[68] P. W. Fowler and J. P. B. Sandall, *J. Chem. Soc., Perkin Trans.* 2, 1994, 1917-1921.
[69] R.C. Haddon, *J. Am. Chem. Soc.*, 1990, 112*,* 3385- 3389.
[70] R.C. Haddon and K. Raghavachari, in: W.E. Billups and M. A. Ciufolini, Buckminsterfullerenes, VCH, 1993, p. 185-215.
[71] R.C. Haddon and S.-Y. Chow, *J. Am. Chem. Soc.*, 1998, 20, 10494-10496.
[72] R.C. Haddon, *J. Phys. Chem. A*, 2001, 105, 4164-4165.
[73] M. V. Diudea, *Studia Univ. "Babes-Bolyai*", 2003, 48, 31-40.
[74] S. Lebedkin, W. E. Hull, A. Soldatov, B. Renker and M. M. Kappes, *J. Phys. Chem. B*, 2000, 104, 4101.
[75] Cs. L. Nagy and M. V. Diudea, in: M. V. Diudea (Ed), *Nanostructures – Novel Architecture*, Nova, N.Y., 2004 (submitted).
[76] M. V. Diudea, *Studia Univ. "Babes-Bolyai"*, 2003, 48, 17-26.
[77] M. V. Diudea, *Forma*, 2004 (submitted).
[78] M. V. Diudea, Cs. L. Nagy, I. Dumitrescu-Silaghi, A. Graovac, D. Janežić; D. Vikic-Topić, *J. Chem. Inf. Model.*, 2004 (submitted).
[79] M. V. Diudea, *Fuller. Nanotub. Carbon Nanostruct.*, 2002, 10, 273-292.
[80] A Dress and G. Brinkmann, *MATCH, Commun. Math. Comput. Chem.* 1996, 33, 87-100.
[81] HyperChem™, Release 4.5 for SGI, © 1991-1995, HyperCube, Inc.
[82] O. Ursu and M. Diudea, TOPOCLUJ software package, "Babes-Bolyai" University, Cluj, 2003.
[83] M. Stefu and M. Diudea, CageVersatile 1.1 software package, "Babes-Bolyai" University, Cluj, 2003.

In: Nanostructures: Novel Architecture
Editor: Mircea V. Diudea, pp. 85-110

ISBN: 1-59454-499-9


***Chapter 5***

# THEORETICAL INVESTIGATIONS OF SINGLE-WALL NANOCONES

***Alexandru T. Balaban***[1]
Texas AandM University at Galveston,
5007 Avenue U, Galveston, TX 77551, USA

## ABSTRACT

The theoretical basis for remodeling a graphene sheet into the third dimension is briefly presented. Nanocones have been first observed in the 1960s, and discussed since the 1970s. They have been called by various names: graphite (graphene, graphitic) cones or molecular cones, but most authors now prefer *nanocones*. This review discusses predominantly single-wall nanocones (SWNCs) made from carbon atoms. However, SWNCs made from BN, Si, or from polymers are also known. Nanotubes may be considered to be a particular case of nanocones with a cap of $m$ = six 60° sectors of the graphene lattice. The other five possible graphitic SWNCs with positive curvature are discussed in detail. They can be obtained from $m$ = one through five 60°-sectors of the graphene lattice by linking the dangling bonds in various ways. A few additional cone-helix structures are also possible. Multi-wall graphitic nanocones may have much sharper tips than the SWNC with $m = 1$. The related *nanohorns* with less regular shape are also discussed. Various properties and possible applications of nanocones are summarily reviewed, and a comprehensive bibliography is attached.

## 1. INTRODUCTION

### 1.1. Plane Lattices

The bond angles φ associated with the regular polygons having $n$ sides (edges) which will be mentioned in the following paragraphs obey the simple relationship in which $s$ represents the number of sides of the polygon: $\varphi = 180(s - 2)/s$ [degrees] or $\varphi = \pi(s - 2)/s$ [radians]. Values of φ are 60° for triangles, 90° for squares, 108° for pentagons, 120° for hexagons, 128.6° for heptagons, 135° for octagons, and 150° for dodecagons. When $d$ polygons with $n$

[1] E-mail address: balabana@tamug.edu

sides meet at a vertex (meaning that the vertex has degree $d$), this is symbolized by $n^d$. There are five and only five regular (Platonic) polyhedra, namely: tetrahedron symbolized as $3^3$, octahedron $3^4$, hexahedron (cube) $4^3$, icosahedron $3^5$, and dodecahedron $5^3$. Their "sphericity" increases in the above order given by the number $V$ of their vertices or by their solid angle, and not by the number $F$ of their faces from which their name is derived.[1] The number $E$ of edges is the same for the two pairs of complementary polyhedra which are duals of each other (cube and octahedron; dodecahedron and icosahedron). The tetrahedron is self-dual. Strained valence isomers of annulenes with formulas $(CH)_{2k}$ are tetrahedrane ($k$ = 2), cubane ($k$ = 4) and dodecahedrane ($k$ = 10).[2] In addition, there are 13 semiregular Archimedean polyhedra (two of which are chiral), which have as faces more than one type of regular polygon. One of these, the truncated icosahedron $5,6^2$, has achieved prominence because when all vertices are carbon atoms, this represents buckminsterfullerene, $C_{60}$. It will be interesting to synthesize other annulene valence-isomeric polyhedral structures such as the truncated octahedron, $4,6^2$ i. e. $(CH)_{24}$.

There are three regular polygons that can each cover a plane: triangles, squares, and hexagons, because suitable sums of their bond angles add up to 360°. These three "regular planar tilings" are represented as $3^6$, $4^4$, and $6^3$, respectively. On adding polygonal bond angles, one has the following equalities for sums of bond angles (equivalence relationships denoted by ≈): $3^2 \approx 6$; $3 + 4 \approx 12$; $3^3 \approx 4^2 \approx 3 + 6$; $3^4 \approx 6^2 \approx 4 + 12$; $8^2 \approx 6 + 12$. Therefore there are, in addition to the three regular planar tilings, eight "semiregular plane tilings" symbolized as: $3^4,6$; $3^3,4^2$; $3^2,6^2$; $3^2,4,12$; $3,4^2,6$; $3,12^2$; $4,8^2$; 4,6,12.

Only the last three of these networks are three-connected, but the first two of these have considerable steric strain and antiaromatic character. One Kekulé structure of the last lattice, in which all bonds at the corners of the 4-membered rings are exocyclic to these rings, forming aromatic sextets in the 6-membered rings and yielding a continuous conjugation in the 12-membered rings, is comparable in many ways to the graphite lattice and could therefore be taken into consideration for forming analogs of fullerenes (e.g. "archimedene", $C_{120}$),[3] nanotubes, and nanocones. The results of Vollhardt's research group show that phenylenes can be stable, despite the antiaromaticity of the 4-membered rings.[4,5]

## 1.2. Polyhedra and Fullerenes

Vertices of planar or polyhedral networks are characterized by the number $d$ of edges meeting at each vertex, called the vertex degree. For the purpose of the present discussion, only the case of tricoordinated atoms (when $d$ = 3) is interesting. The arrangements that satisfy this condition are, among the regular plane tilings, only the hexagonal (honeycomb) lattice (known as graphene) $6^3$, and among the semiregular ones, the $4,8^2$; $3,12^2$; and 4,6,12 tilings. With $sp^2$-hybridized atoms, only the honeycomb lattice is strain-free, and this gives rise to the enormous number of possible chemical structures known as graphite, boon nitride, etc. (for plane tilings), fullerenes and heterofullerenes (for polyhedral systems), nanotubes and nanocones.

One of the earliest (possibly the first) analysis of alternative infinite networks of carbon atoms (different from graphite and diamond) was published by the present author in 1968.[6] Further reviews and papers on carbon and its nets were published by him later.[7-10] Prior to the discovery of fullerenes resulting in the award of the 1996 Nobel Prize for Chemistry to

Curl, Kroto and Smalley,[11-13] other authors had speculated on the possible existence of $C_{60}$.[14-17] The chemistry of fullerenes and their applications received a considerable boost after Krätschmer, Huffman and their coworkers discovered how to prepare, isolate, and purify fullerenes.[18]

The observation that $C_{60}$ and $C_{70}$ are formed predominantly was interpreted by Schmalz et al.[19-21] and by Kroto [22] as a proof that the most thermodynamically stable architecture of a cage formed from $sp^2$-hydridzed carbon atoms without dangling bonds contains 12 isolated (non-adjacent) pentagons and the corresponding numbers of hexagons. This led to the Isolated Pentagon Rule (IPR) postulating that pairs of adjacent pentagons (antiaromatic pentalenes) destabilize fullerenes and are avoided. However, Prinzbach et al. could obtain the smallest fullerene, the dodecahedral carbon $C_{20}$, in a spectacular achievement.[23]

## 1.3. Euler's Theorem

Euler's Theorem for polygon-faced closed surfaces of genus $g$ (with $g$ being the number of topological "handles") can be formulated as: $V - E + F = 2 - 2g$. A convex closed surface has $g = 0$, and a torus has g = 1. When all vertex degrees $d$ are equal to three ("cubic graphs"), $3V = 2E$, so that for a plane or an ellipsoid one obtains an equivalent expression:

$$3n_3 + 2n_4 + n_5 + 0n_6 - n_7 - 2n_8 ... = 12 \quad (1)$$

One can see that the number of hexagons ($n_6$) in this equation cannot influence the result, so that this number can be varied *ad libitum*. If only hexagons and pentagons are allowed as faces, then Euler's equation for such systems reduces to $n_5$ = 12. In other words, for closed structures such as single-wall fullerenes or closed nanotubes one must have twelve pentagons and any number of hexagons (except $n_6 = 1$).

One can generalize this theorem [24] in order to include polyhedral systems with vertices of different degrees $d$ (even when $d = 2$) in terms of the numbers $v_d$ of vertices with degree $d$ and of the number of faces with $n$ sides as follows:

$$\sum_i n_i(6-i) + 2\sum_d v_d(3-d) = 12 \quad (2)$$

or

$$3n_3 + 2n_4 + n_5 + 0n_6 - n_7 - 2n_8 ... + 2v_2 + 0v_3 - 2v_4 - 4v_5 - ... = 12 \quad (3)$$

Thus, one can see that just as the number of hexagons has coefficient zero, so also does the number of trivalent vertices. One vertex of degree 2 has the same contribution as one square or two pentagons, as we shall see later in this review.

A closed structure has a total disclination (the whole solid angle around a point equal to the surface of a sphere divided by the square of its radius) of $4\pi$ radians or 720°. Another name for disclination is "polygonal defect".[25] If 12 pentagons are needed to close a polyhedron containing only regular pentagons and hexagons, this means that each pentagon contributes a disclination of 720°/12 = 60°. A square or two pentagons would contribute with

120°, a triangle or three pentagons with 180°, and a heptagon with an opposite disclination (–60°). Such negative curvatures lead to fluted or crenelated saddle-shaped geometries and are not common, because extra atoms are needed in addition to those present in a planar net.

## 2. Carbon Single-Wall Nanocones (Carbon SWNCs)

### 2.1. Naming Nanocones and Nanohorns

Most of the present review will be dedicated to nanocones formed from carbon atoms. Such structures have received various names in the published literature, and this diversity creates difficulties for computer-assisted documentation. Whereas most chemists and physicists use names such as nanocones, graphitic (graphite, graphene) cones, or molecular cones, Sattler's group used occasionally the term "buckycones". On looking up in *Google* or *SciFinder* for the syntagm "molecular cone", most citations that come out refer to retinal cones, which along with retinal rods, are responsible for vision (therefore this syntagm is not recommended). Probably "nanocone" should be the universally accepted name.

Ijima's research group uses the term "nanohorns" [26-43] for related structures which have less regular shape, but are mostly conical, with occasional tubular portions or buds. A particularity of such nanohorns is the fact that they form aggregates of about 100 nanometers. As a consequence, the specific surface area is extremely large. Because of the ease with which gases and liquids permeate the inside of such nanohorn aggregates that may be prepared with high purity (unlike nanotubes), they are believed to be of high technological significance for fuel cells, for hydrogen-based engines, and for platinum catalyst support in the automotive industry. Some authors such as Berber et al.[44] consider that nanocone and nanohorn are synonyms, but this is debatable.

The observed impurity-driven cone formation during laser sputtering of graphite is something quite different, involving hundreds of layers of overlapping graphene sheets, and resulting in *micro*cones (not *nano*cones !).[45]

### 2.2. History of Nanocones

In 1968 Gillot and coworkers (Battelle Institute, Geneva) converted amorphous carbon (that had been dissolved in martensitic iron and had been set free by removing the iron electrolytically) into crystalline graphite under argon at 2,800 °C obtaining cigar-shaped crystals with conical structure at the ends.[46] Along with round graphitic planar disks, their figure 9 has the spectacular shape of a 300° sector corresponding to a crushed blunt cone with $m = 5$ of 60°-sectors. In 1972, Haanstra and coworkers reported that the graphitic whiskers (formed by decomposition of carbon monoxide above 1800 °C and deposited on a β-SiC substrate with diameters of 3–6 μm and lengths of about 0.5 mm) had conical ends with an apex angle of about 140°.[47]

Double and Hellawell rationalized in 1974 these and other observations of graphitic filaments or whiskers with conical ends by recalling that graphite from metal solutions adapt to the metallic crystals by adopting layered low-energy stacking faults involving rotations (about the axis orthogonal to the graphene plane) by angles 13.2°, 21.8° and 27.8° from the

60° symmetry positions.[48] These authors then described how a cone-helix can be formed from a long and narrow graphene strip rolled helically at an angle around the axis of a cylinder. Thus different forms of columnar pyrolytic carbon result: scroll-layer, stacked layer, or screw-layer. The above authors discussed the overlap angles θ and the resulting apex angles α of the pencil-shaped graphitic cylinder ending with a conical tip (see below). The self-assembly of sharp nanocones can give rise to various morphologies: spherulites with the center formed by all cone apexes, "herring-bone", "coral", "vermiculite" or "rosette" patterns. However, before the discovery of fullerenes, research work on graphitic forms was not widely known or publicized.

A poster on predicted properties of the five single-walled nanocones was presented by the present author (shortly after he had come to the Texas A and M University Galveston on a part-time basis, at the invitation of Professor D. J. Klein) on July 19–24, 1992 at the Seventh International Symposium on Novel Aromatic Compounds held in Victoria, Canada. The corresponding paper was published jointly with D. J. Klein and X. Liu after a considerable delay in 1994.[49] At that time we were not aware of the publications mentioned in the previous paragraph. Soon afterwards, the experimental detection of such single-wall graphitic cones was reported.[50] In rapid succession, a numerous series of papers appeared both on experimental and on theoretical aspects of such nanocones.

## 2.3. Topology of Single-Wall Nanocones

If a graphene sheet is divided into six sectors, each with an angle of 60° (Fig. 1), and if $m$ of these sectors (with $m$ varying from 1 to 5) are selected sequentially with the dangling bonds being fused together (see below), a series of five single-walled nanocones is obtained, with a linear angle α at the cone apex equal to 112.9°, 83.6°, 60.0°, 38.9°, and 19.2° (Table 1). These values correspond to the formula:

$$\alpha = 2\arcsin(m/6) \tag{4}$$

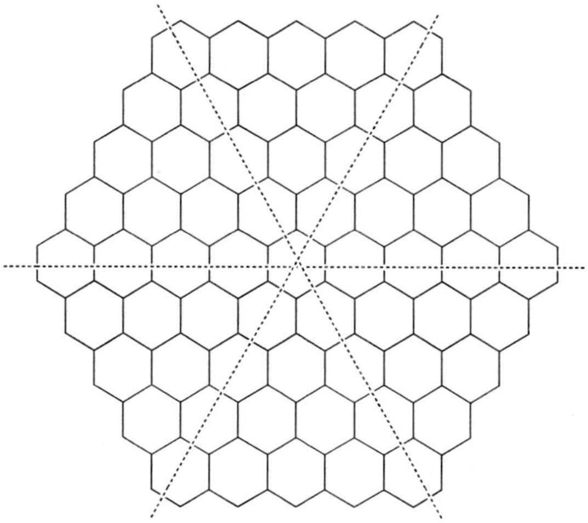

Figure 1: The six sectors on a honeycomb lattice.

**Table 1: The five standard SWNCs and their characteristics as a function of the number *m* of sectors, the disclination $P = (6 - m) \times 60°$, and the apex angle α between two opposite generators of the standard nanocone considered to have a circle as director.**

| Sectors *m* | *P* (°) | α (°) | Pentagons* | Nanocone or congener |
|---|---|---|---|---|
| 0 | 360 | 0.0 | 6 | Nanotube + cap (SWNC) |
| 1 | 300 | 19.2 | 5 | First (sharpest) SWNC |
| 2 | 240 | 38.9 | 4 | Second SWNC |
| 3 | 180 | 60.0 | 3 | Third SWNC |
| 4 | 120 | 83.6 | 2 | Fourth SWNC |
| 5 | 60 | 112.9 | 1 | Fifth (bluntest) SWNC |
| 6 | 0 | 180.0 | 0 | Graphene |

* Equivalent *n*-membered rings or vertices of other degrees than 3 may replace pentagons

One can add two extreme cases: (i) the graphene sheet, with all $m = 6$ sectors being involved, corresponding to a "cone" with an angle of 180°; and (ii) when $m = 6$, one obtains a "cone angle" equal to zero, corresponding to a nanotube capped at one end with any combination of hexagons and six pentagons, e.g. a "half-buckminsterfullerene". Thus *nanotubes capped at one end can be considered to be a particular case of nanocones,* and indeed Ebbesen has observed by transmission electron microscopy how a blunt nanocone with $m = 5$ on adding a sixth 60° sector becomes converted into a nanotube.[51,52] Because of this relationship between nanocones and nanotubes, in the next section we will intercalate a brief digression on relevant aspects of nanotubes. A nanotube capped at both ends is a long extended fullerene.

The linking of dangling bonds can proceed in several ways: the simplest of these is to connect corresponding numbers, i. e. 1 with 1', etc. (Fig. 2). Alternatively, one may decalate the bonding, and then more complicated geometries result both at the apex and at the more distant regions, but the apical angle is the same in all these isomers. The experimentally obtained carbon nanocones have usually appreciable radii (*r*), but of course these are finite, and the dangling bonds at the base of the cone are C–H bonds because of the hydrogen atmosphere in which these nanocones are obtained. Thus the chemical formula of such a "standard nanocone hydrocarbon" is $C_{mr^2}H_{mr}$.

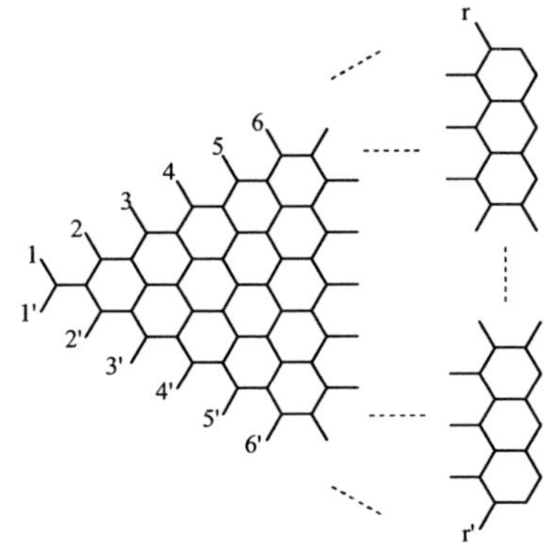


Figure 2: Formation of the sharpest nanocone with $m = 1$ and radius *r* by connecting the dangling bonds labeled with numbers 1, 2, 3, …, r with their primed counterparts.

The Gaussian curvature κ of a sphere with radius $R$ is $1/R^2$. According to the Gauss-Bonnet theorem, the total integral of the Gauss curvature for the whole spherical surface around a point is $4\pi = 720°$, and for a cone from $m$ sectors the total positive curvature (disclination $P$) is $\pi(6 - m)/3$ radians, as seen in Table 1 for an infinite cone (with $r = \infty$).

In addition to these five standard nanocones, a few other apex angles have been found in naturally occurring graphite, and they have been explained by the cone-helix model which assumes a positive wedge disclination in a wrapped-around graphene sheet.[48,53] The resulting apex angles α are given by the formula:

$$\alpha = 2\arcsin(1 - \theta / 360^\circ) \quad (5)$$

where the overlap angle is $\theta = m\times60° \pm \omega$, and ω can take the values 13.2°, 21.8°, or 27.8°. The resulting angles (one increases, the other decreases) are presented in Table 2.

**Table 2: Apex angles for standard nanocones (boldface) and cone-helix systems**

| | **Overlap angle θ (°)** | **Apex angle α (°)** |
|---|---|---|
| 1 | 13.2 | 148.9 |
| 2 | 21.8 | 139.9 |
| 3 | 27.8 | 134.7 |
| 4 | 60.0 | 112.9 |
| 5 | 120.0 | 83.6 |
| 6 | 180.0 | 60.0 |
| 7 | 240.0 | 38.9 |
| 8 | 300.0 | 19.2 |
| 9 | 300 + 13.2 | 14.9 |
| 10 | 300 + 27.8 | 10.3 |
| 11 | 360 – 21.8 | 6.9 |

## 2.4. Intermezzo on Nanotubes

The discovery of nanotubes by Ijima in 1991[54] was preceded by a long series of investigations on how to improve the quality of carbon fibers. These are normally made by thermo-oxidative treatment of either (i) polymer fibers which are based on rayon (cellulose acetate) or on polyacrylonitrile, or (ii) of pitch under certain conditions of timing at the right temperature to harden sufficiently. However, in 1889, a patent described the formation of "hair-like carbon filaments" by pyrolyzing an old feedstock of methane and hydrogen.[55] Later advances with feedstocks using higher aliphatic hydrocarbons were able to look at the structure of the filaments by electron microscopy.[56-58] In the 1970s, Koyama and Endo investigated vapor-grown carbon filaments from benzene,[59,60] and Endo's collaboration with Mildred S. Dresselhaus and her husband is well reviewed in their book on carbon fibers.[61] Some chemists believe that Tibbets deserves to be considered the discoverer of multiple-wall carbon nanotubes (MWNTs) in the early 1980s by pyrolyzing natural gas in stainless steel tubes.[62]

Single-wall carbon nanotubes (SWNTs) and their multi-wall congeners (MWNTs) are reviewed in several chapters of the present volume, and it seems likely that they will have many applications due to their fantastic mechanical and electrical properties. A few reviews and monographs on fullerenes and nanotubes are listed here.[63–84] In the first chapter of this book, T. Braun and coworkers provide scientometric evidence that the bibliography and applications of nanotubes have rapidly overtaken those of fullerenes. The same research group edits a bibliography on fullerene research.[85,86]

## 2.5. Geometry of Single-Wall Nanocones

Since a nanocone with positive curvature is situated on one and the same side of a plane, the Euler formula that has to be satisfied is similar to relationships (2) and (3) but one has to replace the right-hand term 12 by one of the numbers $6 - m = 0$ to 6, depending on the type of the nanocone. The formula (3) then becomes:

$$3n_3 + 2n_4 + n_5 - n_7 - 2n_8 ... + 2v_2 - 2v_4 - 4v_5 - ... = 6 - m \qquad (6)$$

The "bluntest" nanocone with $m = 5$ has a single pentagon at the apex corresponding to a total disclination $P = 60.0°$ (Figure 3).

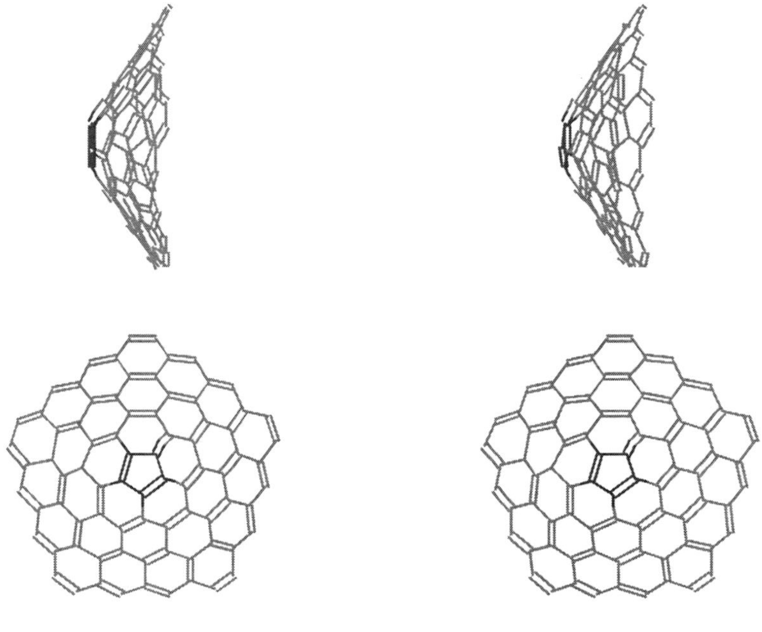

Figure 3: Stereo-view of the bluntest nanocone with m = 5 (side and front).

The next nanocones (Table 1) can be obtained if there are two to five pentagons close to the apex, but there are other possibilities too: many isomeric nanocones may be generated by placing in different reciprocal positions the $6 - m$ pentagons and/or the equivalent $n_3$- or $n_4$-membered rings in agreement with formula (6). According to the isolated pentagon rule mentioned in Section 1.2, Dresselhaus et al. suggested in their books that in $C_{60}$ the nearest pentagons (*meta* to each other with respect to the hexagon adjacent to both) are to be denoted as (1,1), whereas in $C_{70}$ there are in addition (2,0) pentagons (*para* to each other).[61,76,77]

If the apex has a vertex of degree two, or one square, or two pentagons (which need not be contiguous) the total disclination is $P = 120°$ and the second type of cone with $m = 2$ is obtained (Figures 4 and 5). All stereo-views in this chapter were generated by molecular mechanics optimization using the Chem3D Pro CambridgeSoft program.

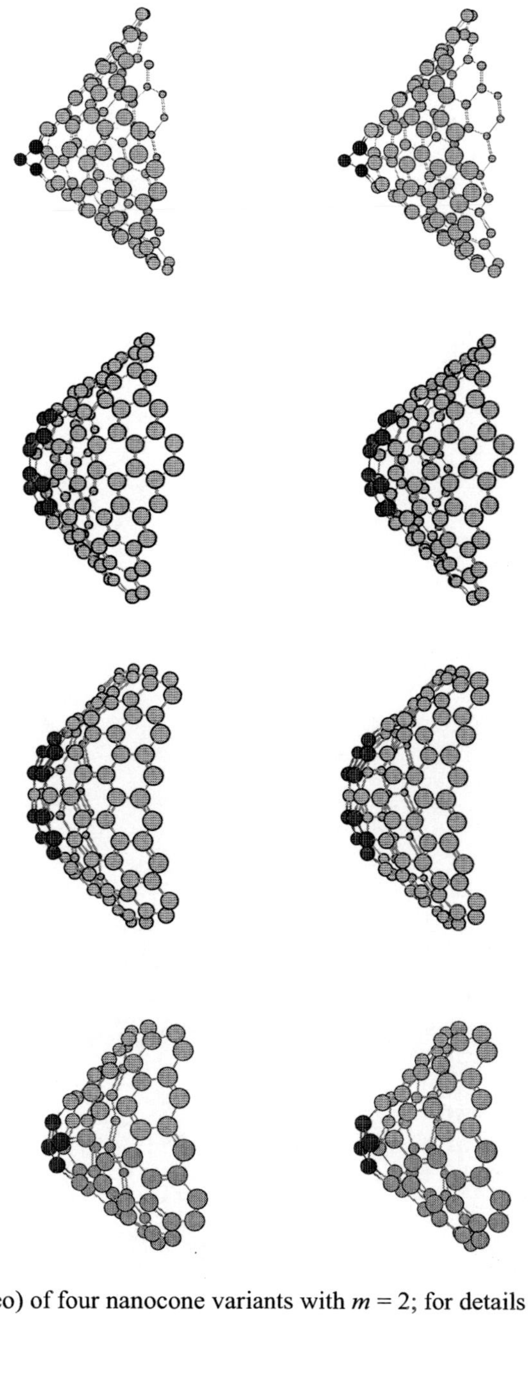

Figure 4: Side views (stereo) of four nanocone variants with $m = 2$; for details see Figure 5.

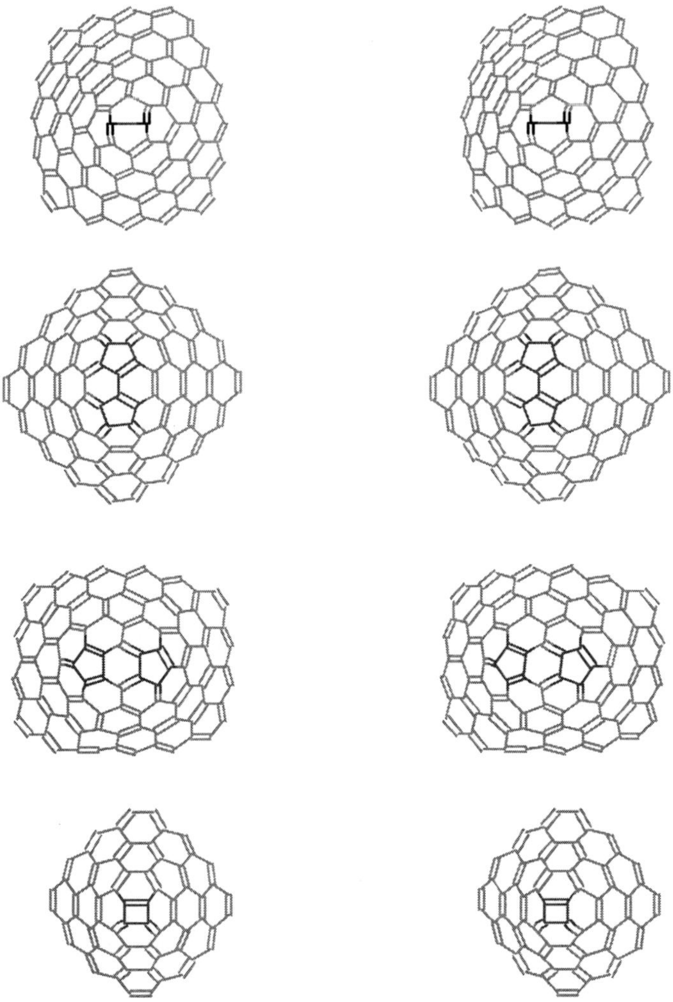

Figure 5: Front view (stereo) of the same four nanocones with $m = 2$ as in Fig. 4: starting from top, the apex has (i) a vertex of degree 2; (ii) and (iii) two pentagons in (1,1) and (2,0) geometries, respectively; (iv) a square.

The third type of standard nanocone (with $m = 3$, total disclination $P = 180°$ and angle at the apex $\alpha = 60.0°$) is shown as an example with three non-contiguous pentagons in Figures 6 and 7 as stereo-views. The last type (the sharpest standard nanocones) with $m = 1$ is represented in the same figure by two representatives: one has a square and a triangle, and the other two squares and a pentagon.

In papers by Ebbesen [51,52] photos of electron microscope images of these five types of single-walled standard nanocones obtained experimentally are shown. In practice, however, in addition to these single-walled nanocones, other (sharper) nanocones have been observed. The apical angle of 6.9° is fairly frequent (see Table 2).[87] When even shaper cones are detected, it is supposed that they are multiple-walled with a "herring-bone" architecture.[88,89]

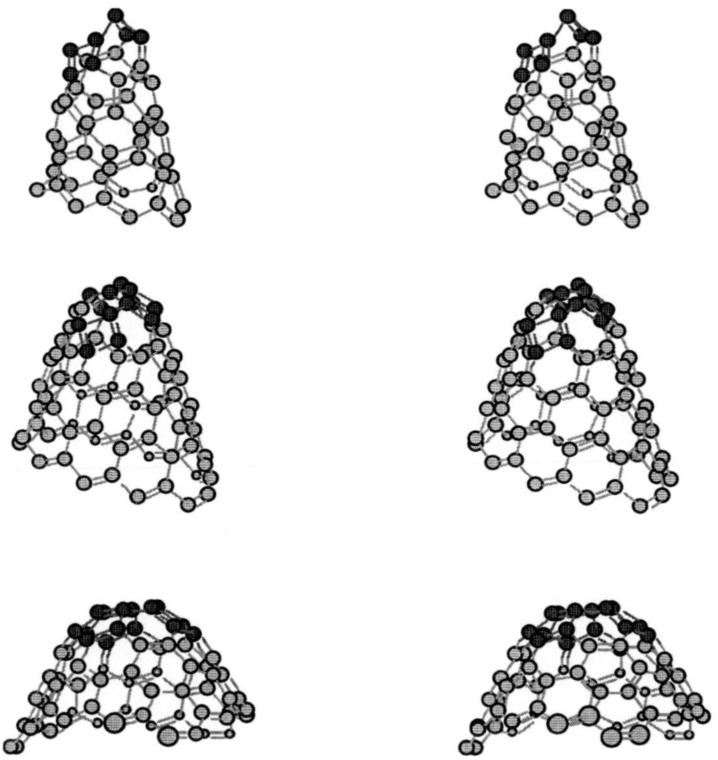

Figure 6: Side stereo-views of two nanotubes with $m = 1$, and a nanotube with $m = 3$.

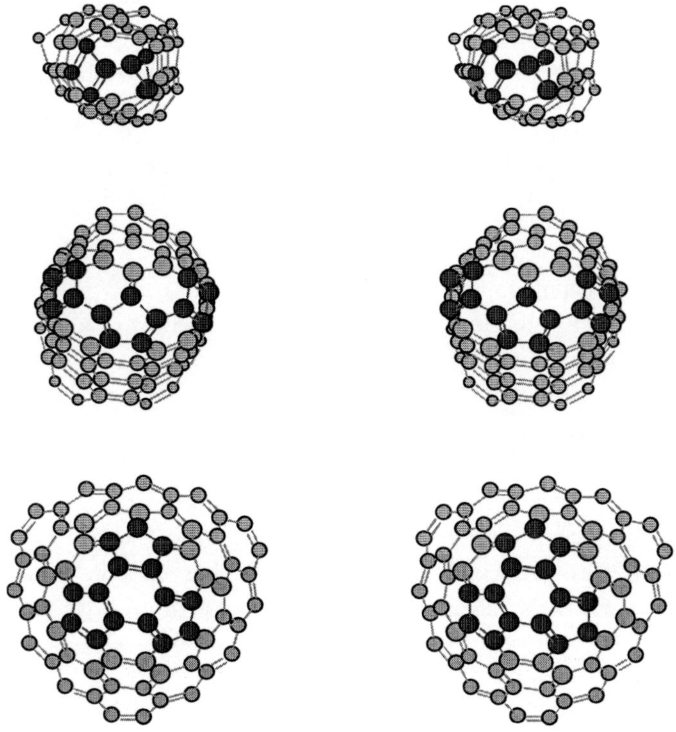

Figure 7: Frontal stereo-views of two nanotubes with $m = 1$, and a nanotube with $m = 3$.

A different way of arranging three pentagons around the apex of a nanocone with m = 3 from that presented in the last structure of Figures 6 and 7 is shown in Figure 8.

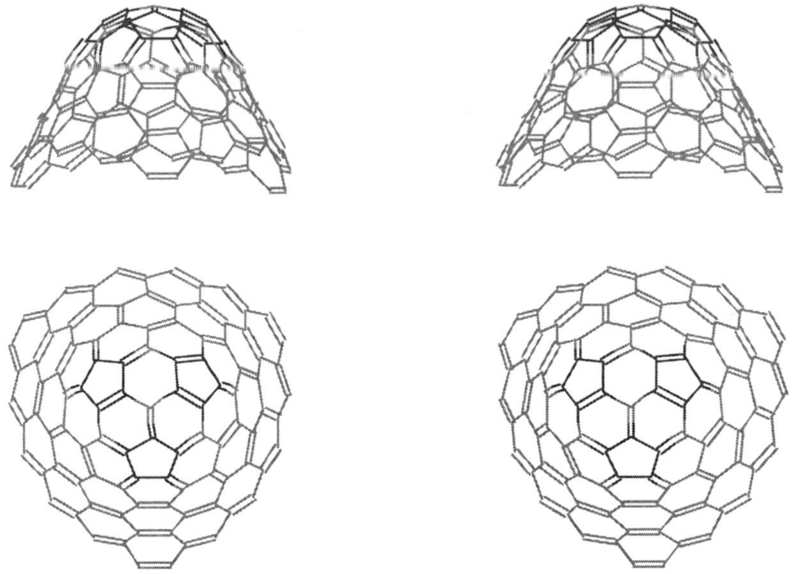

Figure 8: Another nanocone with $m = 3$ with side and front stereo-views

Whereas the pentagons in Fig. 8 are in (2,0) coordinates, the last structure of Figures 6 and 7 had the three pentagons in (1,1) coordinates. As in all preceding stereo-views, the non-hexagonal rings close to the apex of the nanocone are highlighted in red. Figures 3–8 contained four standard nanocones; the fifth one with $m = 4$ will be shown in Fig. 11.

## 2.6. Hydrocarbons $C_{mr^2}H_{mr}$ with a Conical Structure and Their Relevance to Fullerenes

The blunt and short nanocone with $m = 5$ and $r = 2$ affords the hydrocarbon $C_{20}H_{10}$ (corannulene) by attaching hydrogen atoms at the remaining dangling bonds. It has a pentagon surrounded by condensed benzenoid rings. The synthesis of this stable bowl-shaped compound (Fig. 9) was improved by Scott and coworkers.[90] They found that the bowl-to-bowl inversion barrier is only about 10 kcal/mol at –64 °C, approximately the same as for the ring flip of cyclohexane, and less than the tub-to-tub inversion barrier for cyclooctatetraene.[91] Interestingly, three isomeric $C_{20}H_{10}$ dicyclopentapyrenes were found in ethylene flames, but they are absent in fullerene-producing flames.[92] However, the $C_{60}$ buckminsterfullerene was produced under a wide range of conditions by pyrolysis of corannulene and of a few other systems whose molecular graphs are subgraphs of this $C_{60}$ molecule.[93,94]

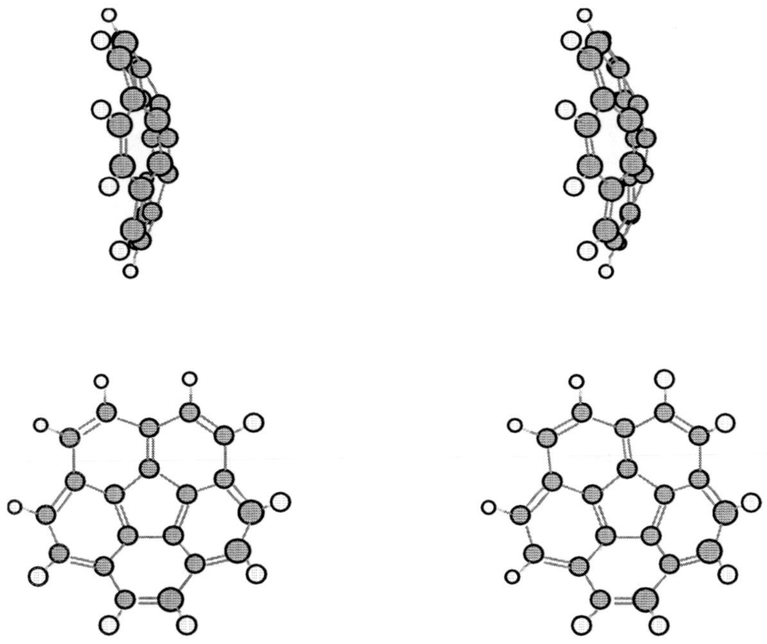

Figure 9: Profile and face stereo-views of corannulene.

The analogous hydrocarbon with a heptagon surrounded by condensed benzenoid rings (the [7]circulene) was synthesized by Yamamoto and his coworkers.[95,96] The saddle shape of this molecule (Fig. 10) reflects the negative curvature that is expected to appear in all nanocones with $m > 6$.[25] The [7.7]circulene resembles even more the figure of 8.[97]

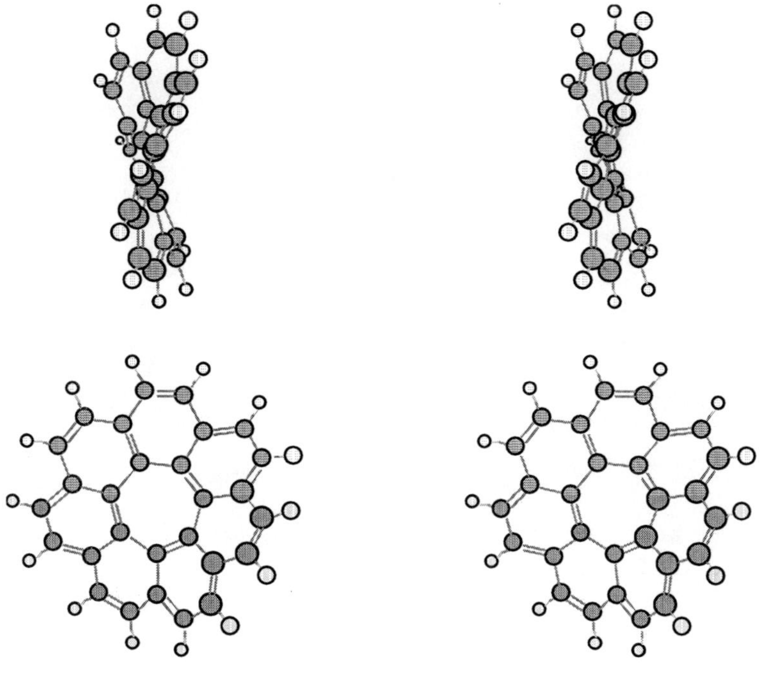

Figure 10: Profile and face stereo-views of 7-circulene.

## 3. Properties of SWNCs

For the π-electronic structure within the approximation of $C_{2v}$ symmetry, Table 2 inspired by ref.[49] displays the symmetry under reflection for irreducible representations of molecular orbitals. There are $r^2$ MOs, and the HOMO–LUMO gap decreases fast with increasing $r$ values in a semilogarithmic representation. For more details, ref.[49] should be consulted.

**Table 2. π-Electronic characteristics of regular (standard) nanocones**

| r | m | Symmetry | Gap | Nanocone |
|---|---|---|---|---|
| even | all | $^1A_+$ | $\neq 0$ | stable |
| | 4k | $^1A_-$ | 0 | biradical, distorts |
| odd | 4k + 2 | $^1A_+$ | $\neq 0$ | stable |
| | 4k – 1 | $^2E$ | 0 | radical, distorts |
| | 4k + 1 | $^2E$ | $\neq 0$ | radical, distorts |

More recent contributions have been performed by several authors. Thus, Klein introduced a topo-combinatoric classification of nanocones based on their connectivity.[98,99] The tight-binding Local Density of States (LDOS) and charge for infinite cones have been calculated by Lammert et al.[100,101] and by Tamura et al.,[25,102-104] but Osipov et al.[105-107] have obtained different results. Berber et al.[44] investigated the LDOS using a parametrized Linear Combination of Atomic Orbitals (LCAO).

Compernolle *et al.*[108] have published very recently an *ab initio* study of small graphitic real nanocone molecules of the *m*th type $C_{mr^2}H_{mr}$ with $m$ = 3, 4, and 5 having a triangle, square, and pentagon apex, respectively. They used the DFT with the B3LYP functional and explained some features of Table 2 and of the actual geometry, such as Δα (Table 3), which depends on the radius $r$ and the type $m$ of the nanocone according to the formula

$$\Delta\alpha = \alpha_{\text{fit}} - \alpha_{\text{theor}} \tag{7}$$

where the $\alpha_{\text{theor}}$ values are those from Table 2, and $\alpha_{\text{fit}}$ is the angle:

$$\alpha_{fit} = 180^o - 2\arctan(t) \tag{8}$$

with $t$ being the slope of the angle between the principal axis of the cone and the line of the calculated cone that fits best the carbon atoms.

**Table 3. Diference Δα (°) between "best fitting cone" angle $\alpha_{fit}$ and $\alpha_{theor}$**

| $r \backslash m$ | 3 | 4 | 5 |
|---|---|---|---|
| 2 | 5.59 | 8.49 | 21.90 |
| 3 | –0.95 | 0.86 | 4.86 |
| 4 | –0.52 | 0.07 | 2.21 |

One can see from Table 3 that the deviation decreases fast with increasing $r$ values; the deviation is maximal (21.9°) for corannulene, which is much more planar than the predicted value based on the ideal cone angle. Pentagons allow much more flexibility than triangles or squares, and this is borne out by the low bowl-to-bowl barrier of corannulene discussed earlier. It was found that the local densities of states cones that are locally different in the proximity of the apex but belong to the same topo-combinatoric class, share common features. The HOMO-LUMO gap was also explained in detail by the above authors.[108]

Han and Jaffe computed energies and geometries of carbon nanoconic tips by DFT semiempirical MO calculations,[109] finding that (1,1) pentagon geometries are preferred over (2,0) tip structures. This result was interpreted as indicating that tips containing (1,1) pentagon pairs induce less in-plane and out-of-plane strain than (2,0) pentagon pairs. A similar contrast between these two kinds of nanocones with (1,1) and (2,0) pentagon pairs was recently found by Lammert and Crespi: when the difference $a - b$ in such an $(a,b)$ pair is not a multiple of three, the behavior is different than when $a \equiv b \pmod 3$.[100,101] This finding is reminiscent of the rule differentiating metallic from semiconducting nanotubes.[110] The local density of states vanishes at the Fermi energy level in one class, but in the other it is nonzero and decreases inversely with distance from the apex.

Krishnan et al.[51] reported that in a sample obtained by pyrolysis of methane in a plasma torch the predominating structures were graphitic disks and, among the five classes of nanocones, those with $m = 2$ and 3; nanotubes and the other classes of nanocones were much less abundant. A similar statistical result was reported by Gogotsi and coworkers when exploring 554 naturally occurring graphite nanocones.[53] According to Treacy and Kilian, the formation enthalpy should favor disks, whereas the sharpest cone and the cylinder should be the rarest. Therefore the entropy contribution (the number of reaction pathways) provides the explanation for the preferred nanocone classes, which are more “designable” than the rarer ones.[111] This situation is reminiscent of the fact that in Hasse diagrams of partially ordered sets (posets) representing the halogenation of annulenes, the more symmetric halogenated systems have lower connectivities than the less symmetric ones. The above-mentioned authors concluded that Smalley’s [112] and Klein and Schmalz’s [113] “pentagon road” for the formation of fullerenes should also take into consideration the designability mechanism.

## 4. Non-Carbon Nanocones

Although most literature data on nanocones and nanohorns involves graphitic carbon, the hexagonal boron nitride (closely similar to graphite) has been shown to be able to form nanocones. However, because five-membered rings lead to a non-alternant system, any pentagonal ring of B and N atoms must contain a homonuclear bond (B–B or N–N) which is considerably weaker than the heteronuclear B–N bond. An analysis of BN nanocones with 240° disclination (with four adjacent pentagons, Fig. 11) [114] via the density functional theory with the B3LYP exchange-correlation functional revealed that when the top apical bond was a B–N bond, a doubly occupied level appeared close to the top of the valence band.

By contrast, when the cone terminates in a B–B bond a non-occupied level appears above the Fermi energy (acceptor state), whereas when the cone terminates in an N–N bond, a doubly occupied level (donor state) appears close to the Fermi energy level. The same group of Brazilian authors determined charge distributions, electronic properties, and electron field emission properties of BN nanocones.[115-117]

Although most non-carbon nanocones examined experimentally and theoretically have been obtained from boron nitride,[118-123] other elements and compounds have also been observed to afford nanocones under suitable conditions: silicon,[124] molybdenum sulfide ($MoS_2$),[125,126] zinc oxide,[127] $Bi_{20}TiO_{32}$,[128] and polymers such as poly(3-methylthiophene).[129,130]

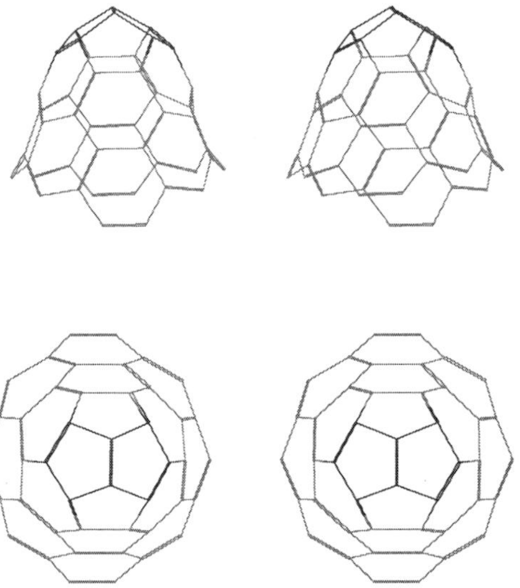

Figure 11: A nanocone with m = 4, relevant to the discussion of BN nanocones. From the four pentagons, only the top two are highlighted in red to evidence the critical B–N bond.

## 5. Nanocone and Nanohorn Formation and their Potential Applications

Without entering into details, one can obtain from the bibliography of this chapter (titles of papers have been included on purpose) a comprehensive coverage of the preparative methods for preparing carbon nanocones.[131—140] For the synthesis of nanohorn aggregates the numerous and important publications of Ijima and his coworkers should be consulted.[26-43] Various methods have been used for creating a high-energy carbon vapor atmosphere, among which one may cite arc evaporation, laser irradiation, electron or ion impact. The role of catalysts cannot be underestimated, because in most cases metal atoms have been located at tips of nanocones. It is not yet certain which defect in the graphite network that becomes deformed into a fullerene, nanotube or nanocone comes first: pentagon formation or disclination.

Among the most obvious potential applications one may cite: field emitters (only one patent is cited),[141] tips for tunnel effect and atomic force microscopy, "sponges" for hydrogen in fuel cells.

## 6. CONCLUSION

From the two well-known allotropic forms of carbon, diamond requires special conditions for its nucleation and formation, because it is not the thermodynamically favored form (under ordinary conditions). Its kinetically-conditioned existence is due to the high energy of carbon–carbon bonds that leads to a very high barrier for bond rearrangement and conversion into the graphite allotrope. On the other hand, graphite is the usual form in which elemental carbon is found in nature and is formed in chemical reactions, or is obtained after its alloys with (or solutions in) metals are decomposed.

Ugarte considered that the various forms in which graphene sheets may appear are due to the tendency of avoiding dangling bonds.[142] In an atmosphere of noble gas, carbon atoms in a plasma link together in $C_2$ pairs (one has to remember that the acetylide dianion is isoelectronic with dinitrogen and has therefore a closed electronic shell), then in $C_n$ rings ($n < 30$), and then in fullerenes. Metal catalysts act as nucleation centers for the formation of nanotubes (SWNTs and MWNTs) that can be capped at both ends. In the presence of hydrogen, the dangling bonds at the margins of the graphitic structure will be saturated with hydrogen, so that graphitic disks, nanotubes, nanohorns, and nanocones can be formed. An open question is what happens in a gaseous atmosphere with high-energy carbon atoms, and atoms of an element such as Cl, Br, or I (F probably forms too strong bonds), whose combined steric and electronic effects may influence the outcome.

This review ends with a few other open questions, the first of which concerns the detailed mechanism(s) allowing this plurality of three-dimensional architectures to be formed from such a simple structural element as carbon. Would it be possible to devise a diamond synthesis based on fitting interatomic distances on the graphite and diamond lattices, or inserting certain catalytic atoms into holes of these lattices?[143,144] The superstructure induced by topological defects in graphite was explained by computations,[145] but supergiant lattices of graphite still raise questions.[146] Terrones, Terrones, and Hsu published ten years ago an inspiring review entitled "Beyond $C_{60}$: graphite structures for the future", raising some questions that still await their answer.[147] Gogotsi demonstrated experimentally that graphite-based structures may be formed under "hydrothermal" conditions, and he cited data proving that at pressures of a few GPa and temperatures between 650 °C and 750 °C amorphous filaments of $sp^2$-hybridized carbon are formed, whereas above 800 well-ordered graphite is obtained.[148] The "hydrothermal" formation of graphite and diamond is well documented.[148,149]

It is likely that carbon and other elements will reveal many other interesting structures that will certainly lead to useful applications.

## ACKNOWLEDGEMENT

Helpful discussions with Professor D. J. Klein are gratefully acknowledged.

## NOTE ADDED IN PROOF

It should be noted that when the number *m* of sectors is 2, 3, and 4, each of such nanocone types has two distinct classes (so that the five types of standard nanocones give rise to eight classes of nanocones) as indicated in refs. [98,99] and in a recent paper: D. Klein and A. T. Balaban, The eight classes of positive-curvature graphitic nanocones, *J. Chem. Inf. Model.* 2005, 45, in press.

## REFERENCES

[1] A. T. Balaban and D. Bonchev, Complexity, sphericity, and ordering of regular and semiregular polyhedra, *MATCH, Commun. Math. Comput. Chem.*, 2005, 54, 137-152 .

[2] A.T. Balaban, M. Banciu and V. Ciorba, *Annulenes, Benzo-, Hetero-, Homo-Derivatives and Their Valence Isomers*, CRC Press, Boca Raton, Florida, 1986, 3 volumes.

[3] J. M. Schulman and R. L. Disch, Theoretical studies of the [*N*]phenylenes, *J. Am. Chem. Soc.,* 1996, 118, 8470-8474.

[4] R. Diercks and K. P. C. Vollhardt, Tris(benzocyclobutadieno)benzene, the triangular [4]phenylenes with a completely bond-fixed cyclohexatriene ring:cobalt-catalyzed synthesis from hexaethynylbenzene and thermal ring opening to 1,2:5,6:9,10-tribenzo-3,4,7,8,11,12-hexadehydro[12]annulene, *J. Am. Chem. Soc.*, 1986, 108, 3150-3152.

[5] K. P. C. Vollhardt, The phenylenes, *Pure Appl. Chem.* 1993, 65, 153-156; K. P. C. Vollhardt *et al.,* Physical, chemical and some biological properties of strained ring hydrocarbons, *Pure Appl. Chem.* 1980, 52, 1645-1667. O. S. Ognjen and K. P. C. Vollhardt, [*N]*Phenylenes: a novel class of cyclohexatrienoid hydrocarbons. In: *Carbon-Rich Compounds* (eds. M. Haley and R. Tykwinski), Wiley-VCH, Weinheim, 2006, pp. 140-197.

[6] A. T. Balaban, C. C. Rentea and E. Ciupitu, Chemical graphs. VI. Estimation of the relative stability of several planar and tridimensional lattices for elementary carbon, *Rev. Roum. Chim.*, 1968, 13, 231-247; (Erratum, *ibid.*, p. 1233).

[7] A. T. Balaban, Theoretical investigation of carbon nets and molecules, in: *Theoretical Organic Chemistry* (ed. C. Parkanyi), Elsevier, Amsterdam, 1998, pp. 381-404.

[8] K. M. Merz Jr., R. Hoffmann and A. T. Balaban, 3,4-Connected carbon nets : through-space and through-bond interactions in the solid state, *J. Am. Chem. Soc.,* 1987, 109, 6742-6751.

[9] A. T. Balaban, Carbon and its nets. *Computers Math .Applic.*, 1989, 17, 397-416. Reprinted in "*Symmetry II"* (ed. I. Hargittai, Pergamon Press, Oxford, 1989, 416).

[10] H. Zhu, A. T. Balaban, D. J. Klein and T.P. Zivkovic, Conjugated-circuit computation on two-dimensional carbon networks. *J. Chem. Phys.*, 1994, 101, 5281-5292.

[11] R. F. Curl, Dawn of the fullerenes: conjecture and experiment, *Angew. Chem. Internat. Ed. Engl.,* 1997, 36, 1567-1576.

[12] H. W. Kroto, Symmetry, space, stars, and $C_{60}$, *Angew. Chem.Internat. Ed. Engl.,* 1997, 36, 1578-1593.

[13] R. E. Smalley, Discovery of the fullerenes, *Angew. Chem.Internat. Ed. Engl.,* 1997, 36, 1594-1601.

[14] D. E. H. Jones (under 'Daedalus' pseudonym), Ariadne, *New Scientist*, 1966, 35, 245.

[15] E. Osawa, *Kagaku (Kyoto)* 1970, 25, 854; Z. Yoshida and E. Osawa, *Aromaticity*, Kagakudojin, Kyoto, 1971.
[16] R. A. Davidson, Unified theory of graph spectral reduction networks. 4. Spectral analysis of graphs by cyclic automorphisms groups, *Theor. Chim. Acta*, 1981, 58, 193-231.
[17] D. A. Bochvar and E. G. Gal'pern, Truncated dodecahedron and truncated icosahedron hypothetical structures, *Dokl. Akad. Nauk SSSR*, 1973, 209, 610-612; I. V. Stankevich, M. V. Nikerov, and D. A. Bochvar, The structural chemistry of crystalline carbon: geometry, stability and electronic spectrum, *Usp. Khim.* 1984, 53, 1101-1121 (in Russian).
[18] W. Krätschmer, L. D. Lamb, K. Fostiropoulos, and D. R. Huffman, Solid $C_{60}$: a new form of carbon. *Nature*, 1990, 347, 354-358.
[19] T. G. Schmalz, W. A. Seitz, D. J. Klein, and G. E. Hite, $C_{60}$ carbon cages, *Chem. Phys. Lett.*, 1986, 130, 203-207.
[20] T. G. Schmalz, W. A. Seitz, D. J. Klein, and G. E. Hite, Elemental carbon cages, *J. Am. Chem. Soc.*, 1988, 110, 1113-1127.
[21] D. J. Klein, W. A. Seitz, and T. G. Schmalz, Icosahedral symmetry carbon cage molecules, *Nature*, 1986, 323, 703-706.
[22] H. W. Kroto, The stability of the fullerenes $C_n$, with $n$ = 24, 28, 32, 36, 50, 60, and 70, *Nature*, 1987, 329, 529-531.
[23] H. Prinzbach, A. Weiler, P. Landenberger, F. Wahl, J. Wörth, L. T. Scott, M. Gelmont, D. Olevano, and B. v. Isasendorff, Gas-phase production and photoelectron spectroscopy of the smallest fullerene, $C_{20}$, *Nature*, 2000, 407, 60-63.
[24] A.T. Balaban, Theoretical examination of electrically-charged buckytubes, *Bull. Soc. Chim. Belges*, 1996, 105, 383-389.
[25] S. Ihara, S. Itoh, K. Akagi, R. Tamura, and M. Tsukada, Structure of polygonal defects in graphitic carbon sheets, *Phys. Rev. B*, 1996, 54, 14713-14719.
[26] S. Ijima, M. Yudasaka, R. Yamada, S. Bandow, K. Suenaga, F. Kokai, and K. Takahashi, Nano-aggregates of single-wall carbon nanohorns, *Chem. Phys. Lett.*, 1999, 309, 165-170.
[27] S. Bandow, F. Kokai, K. Takahashi, M. Yudasaka, L. C. Qin, and S. Ijima, Interlayer spacing anomaly of single-wall carbon nanohorn aggregate, *Chem. Phys. Lett.*, 2000, 321, 514-519.
[28] M. Murata, K. Kaneko, F. Kokai, K. Takahashi, M. Yudasaka, and S. Ijima, Pore structure of single-wall carbon nanohorns aggregates, *Chem. Phys. Lett.*, 2000, 331, 14-20.
[29] S. Bandow, A. M. Rao, G. U. Sumanasekera, P. C. Eklund, F. Kokai, K. Takahashi, M. Yudasaka, and S. Ijima, Evidence for anomalously small charge transfer in doped single-wall carbon nanohorn aggregates with Li, K and Br, *Appl. Phys. A*, 2000, 71, 561-564.
[30] K. Murata, K. Kaneko, W. A. Steele, F. Kokai, K. Takahashi, D. Kasuya, M. Yudasaka, and S. Ijima, Porosity evaluation of intrinsic intraparticle nanopores of single wall carbon nanohorns, *Nanolett.*, 2001, 1, 197-199.
[31] K. Murata, K. Kaneko, W. A. Steele, F. Kokai, K. Takahashi, D. Kasuya, K. Hirahara, M. Yudasaka, and S. Ijima, Molecular potential structures of heat-treated single-wall carbon nanohorn assemblies, *J. Phys. Chem. B*, 2001, 105, 10210-10216.

[32] T. Ohba, K. Murata, K. Kaneko, W. A Steele, F. Kokai, K. Takahashi, D. Kasuya, M. Yudasaka, and S. Ijima, $N_2$ Adsorption in an internal nanopore space of single-walled carbon nanohorn: GCMC simulation and experiment, *Nanolett.*, 2001, 1, 371-373.

[33] K. Murata, K. Kaneko, H. Kanoh, D. Kasuya, K. Takahashi, F. Kokai, M. Yudasaka, and S. Ijima, Adsorption mechanism of supercritical hydrogen in internal and interstitial nanospaces of single-wall carbon nanohorn assembly, *J. Phys. Chem. B*, 2002, 106, 11132-11138.

[34] K. Murata, K. Hirahara, M. Yudasaka, and S. Ijima, Nanowindow-induced molecular sieving effect in a single-wall carbon nanohorns, *J. Phys. Chem. B*, 2002, 106, 12668-12669.

[35] F. Kokai, K. Takahashi, D. Kasuya, M. Yudasaka, and S. Ijima, Growth dynamics of single-wall carbon nanotubes and nanohorn aggregates by CO2 laser vaporization at room temperature, *Appl. Surface Sci.*, 2002, 197-198, 650-655.

[36] D. Kasuya, M. Yudasaka, K. Takahashi, F. Kokai, and S. Ijima, Selective production of single-wall carbon nanohorn aggregates and their formation mechanism, *J. Phys Chem. B*, 2002, 106, 4947-4951.

[37] J.-M. Bonard, R. Gaal, S. Gara, L. Thien-Nga, L. Foro, K. Takahashi, F. Kokai, M. Yudasaka, and S. Ijima, Field emission prepreties of carbon nanohorn films, *J. Appl. Phys.*, 2002, 91, 10107-10109.

[38] E. Bekyarova, K. Kaneko, D. Kasuya, K. Murata, M. Yudasaka, and S. Ijima, Oxidation and porosity valuation of budlike single-wall carbon nanohorn aggregates, *Langmuir*, 2002, 18, 4138-4141.

[39] E. Bekyarova, K. Kaneko, D. Kasuya, K. Takahashi, F. Kokai, M. Yudasaka, and S. Ijima, Pore structure and adsorption properties of single-walled carbon nanohorn bud-like aggregates treated in different atmospheres, *Physica B*, 2002, 323, 143-145.

[40] S. Bandow, T. Omori, H. Kanoh, M. Yudasaka, S. Ijima, and K. Kaneko, Interstitial nanopore change of single-wall carbon nanohorn assemblies with high temperature treatment, *Chem. Phys. Lett.*, 2004, 389, 332-336.

[41] T. Yamaguchi, S. Bandow, and S. Ijima, Synthesis of carbon nanohorn particles by simple pulsed arc discharge ignited between pre-heated carbon rods. *Chem. Phys. Lett.*, 2004, 389, 181-185.

[42] C. Yang, D. Kasuka, M. Yukasada, S. Ijima, and K. Kaneko, Microporosity development of single-wall carbon nanohorns with chemically indices coalescence of the assembly structure, *J. Phys. Chem. B.*, 2004, 108, 17775-17782.

[43] A. Hashimoto, H. Yurimitsu, K. Ajima, K. Suenaga, H. Isobe, J. Miyawaki, M. Yudasaka, S. Ijima, and E. Nakamura, Selective deposition of gadolinium(III) cluster in a hole opening of single-wall carbon nanohorns, *Proc. Natl. Acad. Sci. USA*, 2004, 101, 8527-8530.

[44] S. Berber, Y. Kwon, and D. Tomanek, Electronic and structural properties of carbon nanohorns, *Phys. Rev. B*, 2000, 62, R2291-R2294.

[45] D. J. Krajnovich, J. E. Vasquez, and R. J. Savoy, Impurity-driven cone formation during laser sputtering of graphite, *Science*, 1993, 259, 1590-1592.

[46] J. Gillot, W. Bollmann and Lux, Cristaux de graphite en forme de cigare et à structure conique, *Carbon*, 1968, 6, 381-387.

[47] H. B. Haanstra, W. F. Knippenberg, and G. Verspui, Columnar growth of carbon, *J. Cryst. Growth*, 1972, 16, 71-79.

[48] D. D. Double and A. Hellawell, Cone-helix growth forms of graphite, *Acta Metallurg.* 1974, 22, 481-487.

[49] A. T. Balaban, D. J. Klein and X. Liu, Graphitic cones, *Carbon*, 1994, 32, 357-359.

[50] M. Ge and K. Sattler, Observation of fullerene cones, *Chem. Phys. Lett.*, 1994, 220, 192-196.

[51] A. Krishnan, E. Dujardin, M. M. J. Treacy, J. Hugdahl, S. Lynum, and T. W. Ebbesen, *Nature*, 1997, 388, 451-454.

[52] T. W. Ebbesen, Cones and tubes: Geometry in the chemistry of carbon, *Acc. Chem. Res.*, 1998, 31, 558-566.

[53] J. A. Jaszczak, G. W. Robinson, S. Dimovski, and Y. Gogotsi, Naturally occurring graphitic cones, *Carbon*, 2003, 41, 2085-2092.

[54] S. Ijima, Helical microtubules of graphitic carbon, *Nature*, 1991, 354, 56-58.

[55] T. V. Hughes and C. R. Chambers, Manufacture of carbon filaments, U. S. Pat. 405, 480 (1889).

[56] W. R. Davis, R. J. Slawson, and G. R. Rigby, An unusual form of carbon, *Nature,* 1953, 171, 756.

[57] R. Iley and H. L. Rile, The deposition of carbon on vitreous silica, *J. Chem. Soc.* 1948, II, 1362-1636

[58] R. T. K. Baker and P. S. Harris, The formation of filamentous carbon, in: *Chemistry and Physics of Carbon. A Series of Advances* (P. L. Walker and P. A. Thrower, Eds.), Vol. 14, Marcel Dekker Publ., New York, 1978, p. 83 (review of early work on carbon filaments).

[59] T. Koyama, Formation of carbon fibers from benzene, *Carbon*, 1972, 10, 757-758.

[60] A. Oberlin, M. Endo, and Y. Koyama, Filamentous growth of carbon through benzene decomposition, *J. Cryst. Growth*, 1976, 32, 335-349.

[61] M. Endo, S. Ijima, and M. S. Dresselhaus, *Carbon Nanotubes*, Pergamon, Oxford, 1996.

[62] G. C. Tibbets, Vapor-grown carbon fibers. In: *Carbon Fibers, Filaments, and Composites* (J. L. Figueredo *et al.,* Eds.), Kluwer, The Netherlands, 1990, pp. 73-94.

[63] H. W. Kroto and D. R. M. Walton (Eds.), *The Fullerenes: New Horizons for the Chemistry, Physics and Astrophysics of Carbon.* Cambridge University Press, Cambridge, 1993.

[64] H. W. Kroto, J. E. Fischer, and D. E. Cox (Eds.), *The Fullerenes.* Pergamon Press, Oxford, 1993.

[65] J. Baggott, *Perfect Symmetry. The Accidental Discovery of Buckminsterfullerene,* Oxford University Press, Oxford, 1994.

[66] J. Cioslowski, *Electronic Structure Calculations on Fullerenes and Their Derivatives,* Oxford University Press, New York and Oxford, 1995.

[67] W. Weltner, Jr. and R. J. Van Zee, Carbon molecules, ions, and clusters, *Chem. Rev.* 1989, 89, 1713-1747.

[68] H. W. Kroto, A. W. Alaf, and S. P. Balm, $C_{60}$: buckminsterfullerene, *Chem. Rev.* 1991, 91, 1213-1235.

[69] F. Diederich and R. L. Whetten, Beyond $C_{60}$: the higher fullerenes, *Acc. Chem. Res.* 1992, 25, 119-126.

[70] W. E. Billups and M. A. Ciufolini (Eds.), *Buckminsterfullerene*, Wiley, New York, 1993.

[71] P. W. Fowler and D. E. Manolopoulos, *An Atlas of Fullerenes*, Clarendon Press, Oxford, 1995.
[72] P. W. Stephens (Ed.), *Physics and Chemistry of Fullerenes*, World Scientific, Singapore.
[73] R. Taylor (Ed.), *The Chemistry of Fullerenes*, World Scientific, Singapore.
[74] A. Hirsch, *The Chemistry of the Fullerenes*, Thieme Organic Chemistry Monograph Series, Stuttgart, 1994; A. Hirsch and M. Brettreich, *Fullerenes. Chemistry and Reactions*, Wiley-VCH, New York, 2004.
[75] M. S. Dresselhaus, G. Dresselhaus, and P. Eklund, *The Science of Fullerenes and Carbon Nanotubes*, Academic Press, New York, 1996.
[76] M. S. Dresselhaus, G. Dresselhaus, and P. Avouris (Eds.), *Carbon Nanotubes: Synthesis, Structure, Properties, and Aplications*, Springer-Verlag, Berlin, 2000.
[77] R. Saito, G. Dresselhaus, and M. S. Dresselhaus, *Physical Properties of Carbon Nanotubes*, World Scientific, Singapore, 1998.
[78] T. W. Ebbesen (Ed.), *Carbon Nanotubes, Preparation and Properties*, CRC Press, Boca Raton, 1996.
[79] P. J. F. Harris, *Carbon Nanotubes and Related Structures: New Materials for the Twenty-First Century*, Cambridge University Press, Cambridge, 1999.
[80] K. Tanaka, T. Yamabe, and K. Fukui, *The Science and Technology of Carbon Nanotubes*, Elsevier, Amsterdam, 1999.
[81] P. M. Ajayan, Nanotubes from Carbon, *Chem. Rev.* 1999, 99, 1787-1799.
[82] T. W. Ebbesen, Carbon nanotubes, *Phys. Today*, 1996, 49, 26-32.
[83] T. W. Ebbesen and P. M. Ajayan, Large-scale synthesis of carbon nanotubes, *Nature*, 1992, 358, 220-222.
[84] E. G. Rakov, Methods for preparation of carbon nanotubes, *Russ. Chem. Revs.* 2000, 69, 35-52.
[85] O. Zhou, C. Bower, L. Jin, S. Suzuki, and K. Tanigaki, Carbon nanotubes: synthesis, processing and intercalation, in: *Recent Advances in the Chemistry and Physics of the Fullerenes and Related Materials* (K. M. Khadish and R. S. Ruoff, Eds.), *Electrochemical Society Proceedings*, 1998, vol. 6, pp. 885-896.
[86] T. Braun, A. Schubert, H. Maczelka, and L. Vasvari, *Fullerene Research* 1985-1993. *A Computer-Generated Cross-Indexed Bibliography of the Journal Literature*. World Scientific, Singapore, 1995; T. Braun, A. Schubert, G. Schubert, and L. Vasvari, *Fullerene Research* 1994-1996. *A Computer-Generated Cross-Indexed Bibliography of the Journal Literature*, World Scientific, Singapore, 1997.
[87] N. Muradov and A. Schwitter, Formation of conical carbon structures on vapor-grown carbon filaments, *Nanolett.* 2002, 2, 673-676.
[88] V. I. Merkulov, A. V. Melechko, M. A. Guillorn, D. H. Lowndes, and M. L. Simpson, Sharpening of carbon nanocone tips during plasma-enhanced chemical vapor growth, *Chem. Phys. Lett.*, 2001, 350, 381-385.
[89] G. Zhang, X. Jiang, and E. Wang, Tubular graphite cones, *Science*, 2003, 300, 472-474.
[90] L. T. Scott, P. Cheng, M. Hashemi, M. S. Bratcher, D. T. Meyer, and H. B. Warren, Corannulene. A three-step synthesis, *J. Am. Chem. Soc.*, 1997, 119, 10963-10968.
[91] L. T. Scott, M. M. Gashemi, and M. S. Bratcher, Corannulene bowl-to-bowl inversion is rapid at room temperature, *J. Am. Chem. Soc.*, 1992, 114, 1920-1921.

[92] A. L. Lafleur, J. B. Howard, K. Taghizadeh, E. F. Plummer, L. T. Scott, A. Necula, and K. C. Swallow, Identification of $C_{20}H_{10}$ dicyclopentapyrenes in flames: correlation withy corranulene and fullerene formation, *J. Phys. Chem.*, 1996, 100, 17421-17428.
[93] C. Crowley, R. Taylor, H. W. Kroto, D. R. M. Walton, P. Cheng, and L. T. Scott, Pyrolytic production of fullerenes, *Synthetic Metals*, 1996, 77, 17-22.
[94] C. Crowley, H. W. Kroto, R. Taylor, D. R. M. Walton, M. S. Bratcher, P. Cheng, and L. T. Scott, Formation of [60]fullerene by pyrolysis of corannulene, 7,10-bis(2,2'-dibromovinyl)fluoranthene, and 11, 12-benzofluoranthene, *Tetrahedron Lett.*, 1995, 36, 9215-9218.
[95] K. Yamamoto, T. Harada, Y. Okamoto, H. Chikamatsu, M. Nakazaki, Y. Ka, T. Nakao, M. Tanaka, S. Harada, and N. Kasai, Synthesis and molecular structure of [7]circulenes, *J. Am. Chem. Soc.*, 1988, 110, 3578-3584.
[96] H. Hopf, *Classics in Hydrocarbon Chemistry. Syntheses, Concepts, Perspectives*, Wiley-VCH, Weinheim, 2000, chapters 5 (platonic hydrocarbons) and 12 (non-planar aromatic compounds).
[97] K. Yamamoto, Y. Saitho, D. Iwaki, and T. Ooka, [7.7]Circulene, a molecule shaped like a figure of eight. *Angew. Chem. Int. Ed. Engl.*, 2003, 30, 1173-1174.
[98] D. J. Klein, Topo-combinatoric categorization of quasi-local graphitic defects, *Phys. Chem. Chem. Phys.*, 2002, 4, 2099-2110.
[99] D. J. Klein Quasilocal defects in regular plnar networks: Categorization for molecular cones, *Int. J. Quantum Chem.*, 2003, 95, 600-616.
[100] P. E. Lammert and V. H. Crespi, Topological phases in graphitic cones, *Phys. Rev. Lett.*, 2000, 85, 5190-5193.
[101] P. E. Lammert and V. H. Crespi, Graphene cones: Classification by fictitious flux and electronic properties, *Phys. Rev. B*, 2004, 69, 035406-1 – 035406-13.
[102] R. Tamura and M. Tsukada, Electronic states of the cap structure in the carbon nanotube, *Phys. Rev. B*, 1995, 52, 6015-6026.
[103] R. Tamura and M. Tsukada, Disclinations of monolayer graphite and their electronic states, *Phys. Rev. B*, 1994, 49, 7697-7708.
[104] R. Tamura, K. Akagi, M. Tsukada, S. Itoh, and S. Ihara, Electronic properties of polygonal defects in graphitic carbon sheets. *Phys. Rev. B*, 1997, 56, 1404-1411.
[105] V. A. Osipov and E. A. Kochetkov, Dirac fermions on graphite cones, *J. Exp. Theor. Phys. Lett.*, 2001, 73, 562-565.
[106] V. A. Osipov, E. A. Kochetkov, and M. Pudlak, Electronic structure of carbon nanoparticles, *J. Exp. Theor. Phys.*, 2003, 96, 140-149.
[107] R. Pincak and V. A. Osipov, Localized electron states near pentagons in variously shaped carbon nanoperticles, *Phys. Lett. A*, 2003, 314, 315-321.
[108] S. Compernolle, B. Kiran, L. F. Chibotaru, M. T. Nguyen, and A. Ceulemans, *Ab initio* study of small graphitic cones with triangle, square, and pentagon apex, *J. Chem. Phys.*, 2004, 121, 2326-2336.
[109] J. Han and R. Jaffe, Energetics and geometries of carbon nanoconic tips, *J. Chem. Phys.*, 1998, 108, 2817-2823.
[110] D. J. Klein, W. A. Seitz, and T. G. Schmalz, Symmetry of infinite tubular polymers: application to buckytubes, *J. Phys. Chem.*, 1993, 97, 1231-1236.
[111] M. M. J. Treacy and J. Killian, Designability of graphitic cones, *Mat. Res. Soc. Symp. Proc.*, 2001, 675, 1-6.

[112] R. E. Smalley, Self-assembly of the fullerenes, *Acc. Chem. Res.*, 1992, 25, 98-105.

[113] D.J. Klein and T. G. Schmalz, Buckminsterfullerene. Part A: Introduction. In: *Quasicrystals, Networks, and Molecules with Fivefold Symmetry* (I. Hargittai, Ed.), VCH Publishers, New York, 1990, pp. 239-275.

[114] R. Mota, M. Machado, and P. Piquini, Structural and electronic properties of 240° nanocones, *Phys. Stat. Sol. (c),* 2003, 799-802.

[115] M. Machado, P. Piquini, and R. Mota, Electronic properties of selected BN nanocones, *Mater. Charact.*, 2003, 50, 179-182.

[116] M. Machado, R. Mota, and P. Piquini, Electronic properties of BN nanocones under electric fields, *Microelectronics J.*, 2003, 34, 545-547.

[117] M. Machado, P. Piquini, and R. Mota, Charge distribution in BN nanocones: electric field and tip termination effects, *Chem. Phys. Lett.*, 2004, 392, 428-432.

[118] P. W. Fowler, K. M. Rogers, G. Seifert, M. Terrones, and H. Terrones, Pentagonal rings and nitrogen excess in fullerene-based BN cages and nanotube caps, *Chem. Phys. Lett.*, 1999, 299, 359-367.

[119] F. F. Xu, Y. Bando, and D. Goldberg, The tubular conical helix of graphitic boron nitride, *New J. Phys.*, 2003, 5, 118.1-118.16.

[120] L. Bourgeois, Y. Bando, K. Kurashima, and T. Sato, Co-produced carbon and boron nitride helical cones and the nucleation of curved BN sheets, *Phil. Mag. A*, 2000, 80, 129-142.

[121] S. Azevedo, Effect of substitutional atoms in carbon nanocones, *Phys. Lett. A*, 2004, 325, 283-286.

[122] S. Azevedo, M. S. C. Mazzoni, H. Chacham., and R. W. Nunes, Electron states in boron nitride nanocones, *Appl. Phys. Lett.*, 2003, 82, 2323-2325.

[123] M. Terauchi, M. Tanaka, K. Suzuli, O. Ogino, and K.Kimura, Production of zigzag-type BN nanotubes and BN cones by thermal annealing, *Chem. Phys. Lett.*, 2000, 324, 359-364.

[124] Self-organization of silicon nanocone array induced by pulsed $CO_2$ laser irradiation, *Appl. Surface Sci.*, 2003, 208-209, 148-152.

[125] M. Remskar, A. Mrzel, Z. Skraba, A. Jesih, M. Ceh, J. Demsar, P. Stadelmann, F. Levy, and D. Mihailovic, Self-assembly of subnanometer-diameter single-wall $MoS_2$ nanotubes, *Science*, 2001, 292, 479-481.

[126] L. Margulis, G. Saltra, R. Tenne, and M. Tallanker, Nested fullerene-like structures, *Nature*, 1993, 365, 113-114.

[127] J. Jie, G. Wang, X. Han, J. Fang, B. Xu, Q. Yu, F. Li, and J. G. Hou, Growth and properties of well-aligned ZnO hexagonal cones prepared by carbothermal reaction, *J. Crystal Growth*, 2004, 267, 223-230.

[128] Y. Hou, M. Wang, X. Xu, H. Wang, S. Shang, D. Wang, and W. Yao, $Bi_{20}TiO_{32}$ nanocones prepared from Bi–Ti–O mixture3 by metalorganic decomposition method, *J. Cryst. Growth*, 2002, 240, 489-494.

[129] B. Rajesh, K. Ravindranathan Thampi, J.-M. Bonnard, A. J. McEvoy, N. Xanthopoulos, H. J. Mathieu, and B. Viswanathan, Pt particles supported on conducting polymer nanocones as electro-catalysts for methanol oxidation, *J. Power Sources*, 2004, 133, 155-161.

[130] B. Rajesh, K. Ravindranathan Thampi, J.-M. Bonnard, N. Xanthopoulos, H. J. Mathieu, and B. Viswanathan, Template synhesis of conducting polymeric nanocones of poly(3-methylthiophene), *J. Phys. Chem. B*, 2004, 108, 10640-10644.

[131] Y. Gogotsi, S. Dimovski, and J. A. Libera, Conical crystals of graphite, *Carbon*, 2002, 40, 2263-2267.

[132] P. Milani, C. E. Bottani, A. Parisini, and G. P. Banfi, Photochemical and thermal effects in laser irradiated fullerite: From periodic exfoliation to cone formation, *Appl. Phys. Lett.*, 1998, 72, 293-295.

[133] Z. Sun, Z. Liu, J. Du, Y. Wang, B. Han, and T. Mu, Synthesis of tubular graphite cones through a catalytically thermal reduction route, *J. Phys. Chem. B*, 2004, 108, 9811-9814.

[134] H. Terrones, T. Hayashi, M. Munoz-Navia, M. Terrones, Y. A. Kim, N. Grobert, R. Kamalakaran, J. Dorantes-Davila, R. Escudero, M. S. Dresselhaus, and M. Endo, Graphitic cones in palladium-catalysed carbon nanofibres, *Chem. Phys. Lett.*, 2001, 343, 241-250.

[135] H. Terrones, T. Hayashi, M. Munoz-Navia, M. Terrones, T. Hayashi, Y. A. Kim, M. Endo, N. Grobert, R. Kamalakaran, J. Dorantes-Davila, R. Escudero, and M. S. Dresselhaus, Graphitic cones in carbon nanofibres, *Molec. Cryst. Liq. Crist.*, 2002, 387, 39-50.

[136] K. Sattler, Scanning tunneling microscopy of carbon nanotubes and nanocones, *Carbon*, 1995, 33, 915-920.

[137] T. Kawai, Y. Miyamoto, O. Sugino, and Y. Koga, Nanotube and nanohorn nucleation from graphitic patches: Tight-binding molecular dynamics simulation, *Phys. Rev. B*, 2002, 66, 033404-1 – 033404-4.

[138] X. Li, Z. Lei, R. Ren, J. Liu, X. Zuo, Z. Dong, H. Wang, and J. Wang, Characterization of carbon nanohorn encapsulated Fe particles, *Carbon*, 2003, 41, 3063-3072.

[139] P. J. F. Harris, S. C. Tang, J. B. Claridge, and M. L. H. Green, High-resolution electron microscopy of a microporous carbon produced by arc-evaporation, *J. Chem. Soc. Faraday Trans.*, 1994, 90, 2799-2802.

[140] S. Amelinckx, W. Luyten, T. Krekels, G. Van Tendeloo, and J. Van Landuyt, Conical, helically wound, graphite whiskers: a limiting member of the "fullerenes"? *J. Cryst. Growth*, 1992, 121, 543-558.

[141] S. I. U. Shah, G. B. Blanchet-Fincher, and W. L. Holstein (to Du Pont), Carbon cone and carbon whisker field emitters, U. S. Pat. 6,020,677, Eur. Pat., 0938739, WO 9,821,736 (1998).

[142] D. Ugarte, Curling and closure of graphitic networks under electron-beam irradiation, *Nature*, 1992, 359, 707-709.

[143] A. T. Balaban and D. J. Klein, Local interconversions between graphite and diamond structures, *Carbon*, 1997, 35, 247-251.

[144] A. T. Balaban, D. J. Klein and W. A. Seitz, Holes in diamond or carbon nitride lattices, *Internat. J. Quantum Chem.*, 1996, 60, 1065-1068.

[145] K. Kobayashi, Superstructure induced by a topological defect in graphitic cones, *Phys. Rev. B*, 2000-II, 61, 8496-8500.

[146] J. Xhie, K. Sattler, M. Ge, and N. Venkateswaran, Giant and supergiant lattices of graphite, *Phys. Rev. B*, 1993, 47, 15835-15841.

[147] H. Terrones, M. Terrones, and W. K. Hsu, Beyond $C_{60}$: graphite structures for the future, *Chem. Soc. Revs.*, 1995, 24, 341-350.
[148] Y. G. Gogotsi and K. G. Nickel, Formation of filamentous carbon from paraformaldehyde under high temperatures and pressures, *Carbon*, 1998, 36, 937-942.
[149] Y. Gogotsi, Designing carbon crystals for nanotechnology application, *Cryst. Growth and Design*, 2001, 1, 179-181.

In: Nanostructures: Novel Architecture
Editor: Mircea V. Diudea, pp. 111-126
ISBN: 1-59454-499-9


*Chapter 6*

# POLYHEX TORI ORIGINATING IN SQUARE TILED TORI

***Mircea V. Diudea*[1]**
Faculty of Chemistry and Chemical Engineering
"Babeş-Bolyai" University, 400084 Cluj, ROMANIA

## ABSTRACT

Construction of polyhex (6,3) tori from square tiled (4,4) tori is given in detail. Topological description by $\alpha$-spiral code was adapted for polyhex toroidal nets. Strain energy calculation, in terms of the POAV1 theory, revealed a dependence by $D^{-2}$ (or $h_n^{-2}$) at constant tube cross-section, with $D$ and $h_n$ being the large diameter and the number of hexes on the $n$-dimension (counting the cross-sections around the large hollow) of the torus, respectively. Varying the tube section showed dependence by $h_c^{-1}$, where $h_c$ is the number of hexes on the $c$-dimension, counting the number of atoms around the tube. The $\pi$-electronic structure is evaluated at the simple Hückel level of theory. Criteria for metallic character in (6,3) tori are given in terms of the $[c,n]$ dimensions of the net. These criteria superimpose over those for perfect Clar structure and support the newly proposed perfect Corannulenic structure, in a valence bond representation of total resonant sextet molecules. Resonance energy calculations support the idea of further increasing stability/aromaticity by highly organized corannulenic polyhex systems.

## 1. INTRODUCTION

Among the carbon allotropes, intensively studied in the last decade,[1-5] the only orientable closed surface $S$ entirely coverable by a benzenoid lattice is the torus. A polyhedral (combinatorial) torus obeys the Euler theorem:[6]

$$N - E + F = \chi(S) \qquad (1)$$

[1] E-mail address: diudea@chem.ubbcluj.ro

where $\chi(S)$ is the Euler characteristic: $\chi(S) = 2 - 2g$ (*N, E, F,* and *g* being respectively the number of vertices, edges, faces, and genus - unity, in case of the torus).

"Circle crops" structures were first observed by Liu *et at.*[7] and then by other groups.[8-10] Martel *et al.*[10] argued that the observed rings were coils rather than perfect tori, but these structures have continued to attract a multitude of theoretical studies, dealing with construction, mathematical and physical properties of graphitic tori.[11-19]

Covering a torus by hexagons is most often achieved by the graphite zone-folding procedure.[12,17-19] The method defines an equivalent planar parallelogram on the graphite sheet and identifies a pair of opposite sides to form a tube. Finally the two ends of the tube are glued in order to form a torus. The resulting polyhex torus is completely defined by four independent integer parameters,[16-18,20] reducible to three parameters.[17,19,21]

A second procedure uses the so-called topological coordinates, extracted from the adjacency matrix eigenvectors.[22-25]

The paper is further structured as: Construction of polyhex tori on a square tiled toroidal surface (section 2); Topology of polyhex tori (section 3); Strain energy calculation in large polyhex tori (section 4); π-Electronic structure (section 5); and Conclusions (section 6).

## 2. Construction of Polyhex Tori From Square Tiled Tori

A third way to polyhex tori starts from the tetragonal (4,4) net embedded in the torus, as (4,4)[*c,n*] objects.[26,27] The embedding of the (4,4) net is made by circulating a *c*-fold cycle, circumscribed to the toroidal tube cross-section of radius *r*, around the large hollow of the torus, of radius $R > r$ (Figure 1). Its subsequent *n images*, equally spaced and joined with edges, point by point, form a polyhedral torus tiled by a tetragonal pattern. The position of each of the *n* images of the "circulant" around the central hollow is characterized by angle $\theta$ while angle $\varphi$ locates the *c* points across the tube. In all, *c* x *n* points are generated. The parameters *R* and *r* are not directly involved in the topological characterization of the lattice.

Toroidal parametersA (4,4) square tiled torus

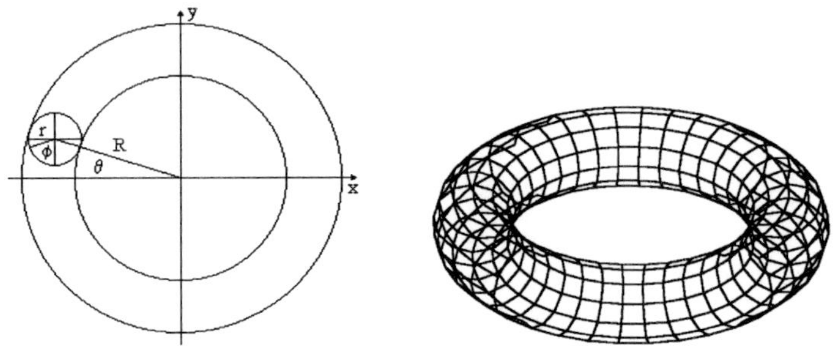


Figure 1: Embedding of the (4,4) net in the torus.

The parameters are calculated by the following formulas:

$$P(x, y, z): \quad (2)$$

$$x = \cos(\theta)(R + r\cos\varphi)$$

$$y = \sin(\theta)(R + r\cos\varphi)$$

$$z = r\sin\varphi$$

$$\theta_i = \frac{2\pi}{n}i \quad ; \quad i = 0,...,n-1$$

$$\varphi_j = \frac{2\pi}{c}j \quad ; \quad j = 0,...,c-1$$

The squares are changed to hexagons (or other tiling patterns, suitable from chemical point of view) by appropriate edge cutting.[27-32] To obtain the (6,3) lattice, each second edge is cut off. Two embedding isomers could result at each given $[c,n]$ pair, as the cut edges lye either horizontally or vertically (*i.e.*, perpendicularly and parallel to the Z axis of the torus). The two isomers are called "zigzag" Z (or H) and "armchair" A (or V), respectively, according to their cross-section shape (or cut edge location). The name of such a torus is a string of characters including the tiling, type of embedding and tube dimensions $[c,n]$.

Note that each hexagon consumes exactly two squares in the (4,4) net. By construction, the number of hexagons in the (6,3)Z pattern is half the number of squares on dimension $c$ of the (4,4) torus. The same is true for the (6,3)A, but on dimension $n$. Thus, (6,3)Z$[2c,n]$ has the same number of hexes as its embedding isomer (6,3)A$[c,2n]$. However, they represent topologically distinct objects (see below), and correspond to two different classes of aromatic chemical compounds: phenacenes and acenes, respectively.

After optimization, by a Molecular Mechanics procedure, the polyhex tori look like in Figure 2. Tori in these examples represent non-chiral objects and their embedding shows $c < n$.

(6,3)Z[20,100] (6,3)A[20,100]

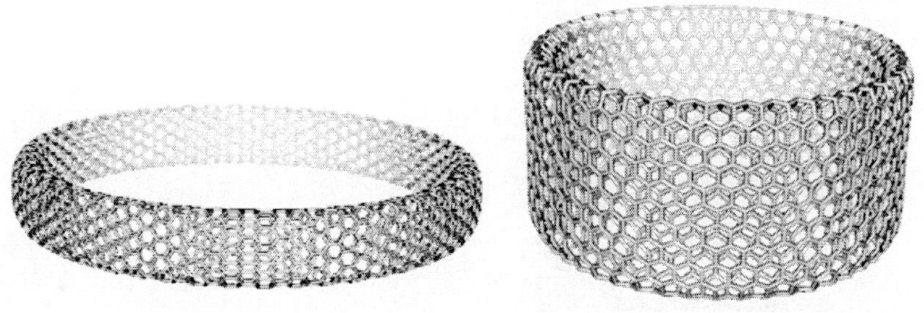

Figure 2: Optimized polyhex tori on 20 × 100 atoms, in the Z (zigzag) and A (armchair) embedding, respectively.

The shape of polyhex tori tends to that of the ideal torus (*i.e.*, a circular tube cross-section) at a sufficiently large ratio $n/c$. Figure 3 gives some cross-sections on polyhex tori along with their geometric eccentricity (in %, taken as the ratio of the difference of the two diameters of the cross section to the largest one).

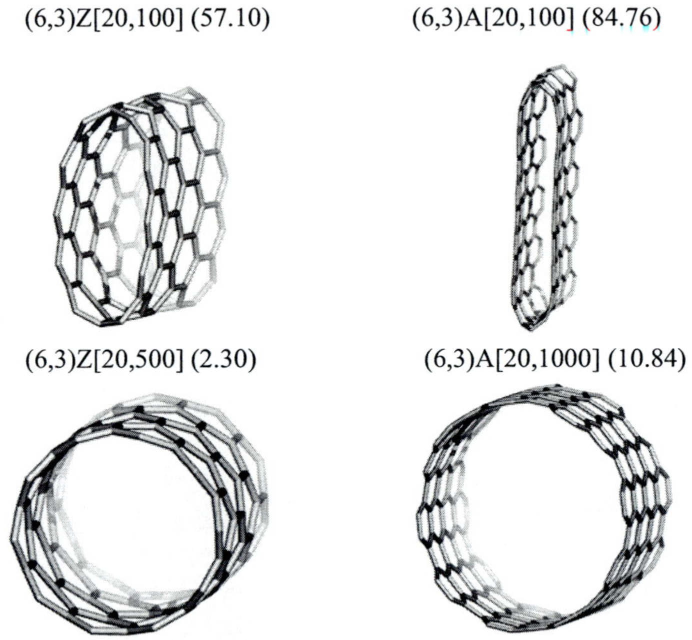


Figure 3: Cross-sections on polyhex tori and their geometric eccentricity (in %).

When $c > n$, the objects show a more elongated cross-section, as illustrated in Figure 4. Such elongated tori[33] have been inferred in multi walled nanotube growth.[34,35]

Successive torus opening and tube closing, as in the sequence: (6,3)Z[$c$,$n$]; TU(6,3)Z[$c$,$n$]/ TU(6,3)V[$n$,$c$]; (6,3)V[$n$,$c$], leads to pair tori of different embedding but having the same topology.

Figure 4: Elongated polyhex tori, in Z and A embedding, respectively.

Twisted, chiral, tori can be generated by the following two procedures: (1) twisting the horizontal layer connections (Figure 5a) and (2) twisting the vertical layer (offset) connections (Figure 5b).

(a) (4,4)H2[8,24] (b) (4,4)V2[8,24]

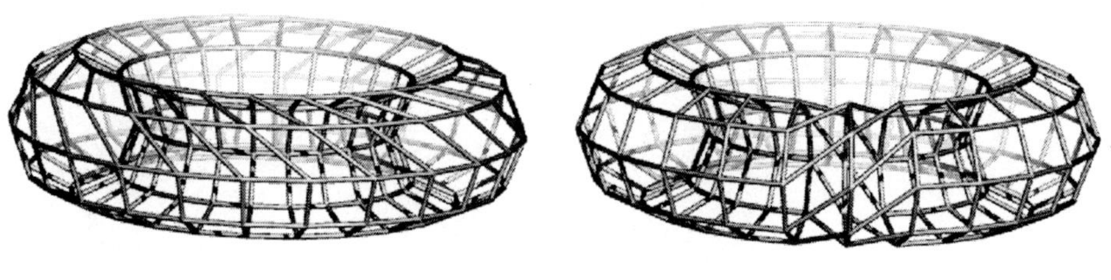

Figure 5: An H-twisted (4,4) net (a) and a V-twisted (offset) (4,4) one (b).

## 3. Topology of Polyhex Tori

Each of the above twistings superimpose on the two basic cuttings, thus resulting four classes of twisted tori: (i) H-twist, H-cut HH*t*[*c*,*n*]; (ii) V-twist, H-cut, VH*t*[*c*,*n*]; (iii) H-twist, V-cut, HV*t*[*c*,*n*] and (iv) V-twist, V-cut, VV*t*[*c*,*n*]. The type of cutting will dictate the type of embedding and, ultimately, the shape of objects. Figure 6 illustrates some (non-optimized) twisted polyhex tori.

The twist number *t* is just the deviation (in number of hexagons) of the chiral (*i.e.*, rolling-up) vector to the zigzag line, in the graphite sheet representation.[12,16-18,36] Accordingly, a toroidal object is drawn as an equivalent planar parallelogram, that needs the specification (in two integer parameter notation[36]) of the two involved tubes (Figure 7). One tube is built up by taking as circumference the rolling-up vector *R*, given in terms of the primitive lattice vectors of graphene:

$$R = ka_1 + la_2 \tag{3}$$

The second tube is formally defined on the translating vector *T*:

$$T = pa_1 + qa_2 \tag{4}$$

Going from a torus, generated as above, to its composing vectors, the first tube can be identified by cutting the object across the tube while the second one results by cutting it around the large hollow. Anyway, a four integer parameter description (*k*,*l*,*p*,*q*) is obtained.

The coordinates for the HH4[14,6] torus depicted in Figure 7 are: (5, -4, 3, 3). Note that this representation is not unique and is reducible to three parameter notation, theorized by Kirby *et al.*[17,21]

In our procedure, H-twisting involves an intrinsic chiral (HH or HV) *R*-based tube and a non-twisted *T*-based one (Table 1, entries 3 and 4). Conversely, in V-twisting, the *R*-based tube is a non-chiral (H/Z or V/A) one while the second *T*-based tube is twisted (leading to "offset" closed toroids - Table 1, entries 5 and 6). The resulting tori are all different, at least as 3D structures, because of distinct embedding. Note that the HV lattice is essentially a HH one, the objects differing, however, as embedding and electronic structure. We stress that, in generating chiral tori, the above procedure leaves untwisted one of the two constitutive vectors (see above and Figure 7). The correspondence between the two notations: X*t*[*c*,*n*] and (*k*,*l*,*p*,*q*) is detailed in Table 1.

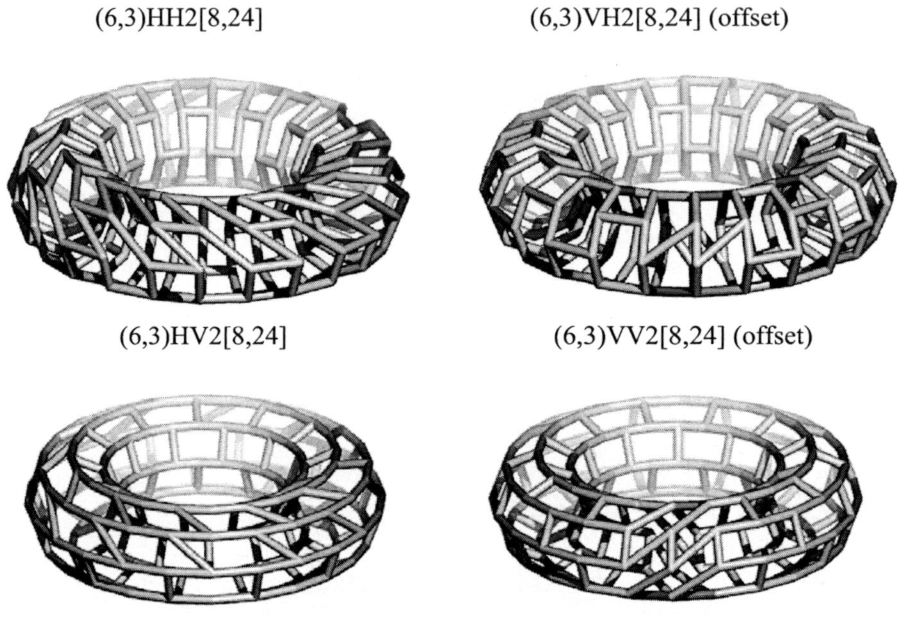


Figure 6: The four classes of twisted polyhex tori (non-optimized geometry).

Encoding the type of tessellation can be done, for example, by the spiral code.[37-39] It was first proposed for coding and constructing spherical fullerenes. We adapted the spiral code for tubular structures.[31] In a periodic tubular net, the spiral code brings information on size and sequence of faces and embedding the actual object on the parent (4,4)[*c*,*n* ] lattice. The $\alpha$-spiral code for the polyhex toroids is given in Table 2.

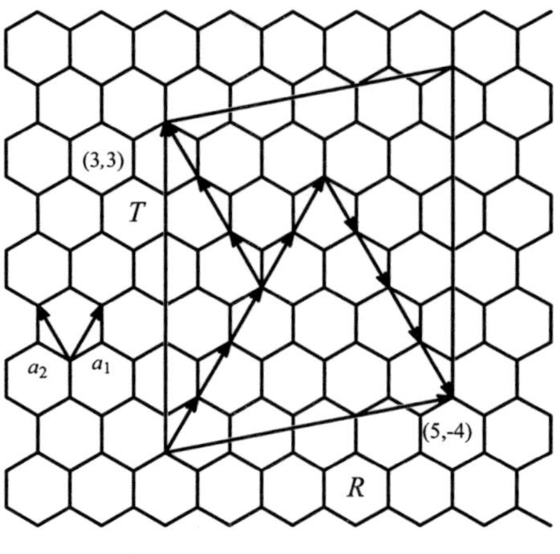


Figure 7: Representation of the torus HH4[14,6] by an equivalent parallelogram involves specification of two tubes: one defined on the rolling-up vector *R* (with integer coordinates (*k*,*l*)) and the other on the translating vector *T* (given by pair (*p*,*q*)). The four parameter specification of the depicted torus is (5, -4, 3, 3).

**Table 1: Correspondence between the X*t*[*c*,*n*] and (*k*,*l*,*p*,*q*) notations**

| | Torus [*c*,*n*] | Tube *R* | Tube *T* | Torus (*k*,*l*,*p*,*q*)* | *N* |
|---|---|---|---|---|---|
| 1 | H | H/Z | V/A | $(c/2,-c/2,n/2,n/2)$ | $2(kq-lp)=cn$ |
| 2 | V | V/A | H/Z | $(c/2,c/2,n/2,-n/2)$ | $2(lp-kq)=cn$ |
| 3 | HH*t* | HH (tw) | V/A | $[(c-t)/2,-t,n/2,n/2]$ | $2(2kq-lp)=cn$ |
| 4 | HV*t* | HV (tw) | H/Z | $[(c+t)/2,(c-t)/2,n/2,-n/2]$ | $2(lp-kq)=cn$ |
| 5 | VH*t* (offset) | H/Z | HV (tw) | $[c/2,-c/2,(n+t)/2,(n-t)/2]$ | $2(kq-lp)=cn$ |
| 6 | VV*t* (offset) | V/A | HH (tw) | $[c/2,c/2,(n-t)/2,-t)]$ | $2(2lp-kq)=cn$ |

*First pair (*k*,*l*) denotes the rolling-up vector *R* while last pair (*p*,*q*) specifies the translating vector *T*; (*m*,-*m*) = (*m*,0), is a Z-tube; (*m*,*m*) is an A-tube (see Figure 7).

**Table 2: Ring spiral code of tori**

| | Series | Ring spiral code |
|---|---|---|
| 1 | H/Z[*c*,*n*] | $[6^{c/2}]^{n}$ |
| 2 | HH[*c*,*n*] | $[(6^{c/2})^{t}]^{n/t}$ |
| 3 | HV[*c*,*n*] | $[(6^{c})^{t}]^{n/2t}$ |
| 4 | VH[*c*,*n*] | $[(6^{n})^{t}]^{c/2t}$ |
| 5 | VV[*c*,*n*] | $[(6^{n/2})^{t}]^{c/t}$ |
| 6 | V/A[*c*,*n*] | $[6^{c}]^{n/2}$ |

The number *t* inside the brackets equals the helicity[40] while the number out the brackets gives the steps of a helix. Note that the helicity could be less than *t*, if an integer number of steps appear.

## 4. Strain Energy Calculation

Our TORUS 3.0 software package[41] enables the generation of huge tori, up to 20,000 atoms, which could be optimized by a Molecular Mechanics procedure. As shown in Figure 3, the cross-section of a sufficiently large torus (*i.e.*, a large ratio $h_n/h_c$ of the number of hexes on the two dimensions) tends to the circular shape. Our objects approach that shape at the ratios 50 and 25, in case of H/Z[20,*n*] and V/A[20,*n*] series, respectively.

In the POAV1 theory,[42-45] the π-orbital axis vector makes equal angles $\theta_{\sigma\pi}$ to the three σ-bonds of the $sp^2$ carbon atom and the *pyramidalization* angle is obtained as:

$$\theta_p = \theta_{\sigma\pi} - 90^{\circ} \quad (5)$$

This angle is used for estimating the strain energy, induced by a pyramidalized carbon atom, by:

$$SE = 200 \cdot (\theta_p)^2 \tag{6}$$

with $\theta_p$ being measured in radians. The difference $120-(1/3)\cdot\sum\theta_{ij}$ gives the deviation to the planarity. POAV analysis provides a reliable description of electronic structure of non-planar conjugated molecules.

In a first test, we varied the *n*-dimension and found a dependence of the strain energy by $D^{-2}$, with D being the large diameter of the torus. A similar result was reported by Han.[46] This is true in large tori built up on both thick ($h_c = 20$, Figure 8) and thin ($h_c = 10$, Figure 9) tubes. In both cases, the V/A tori were found more strained. Observe the proportionality between D and *n*.

(a) (b)

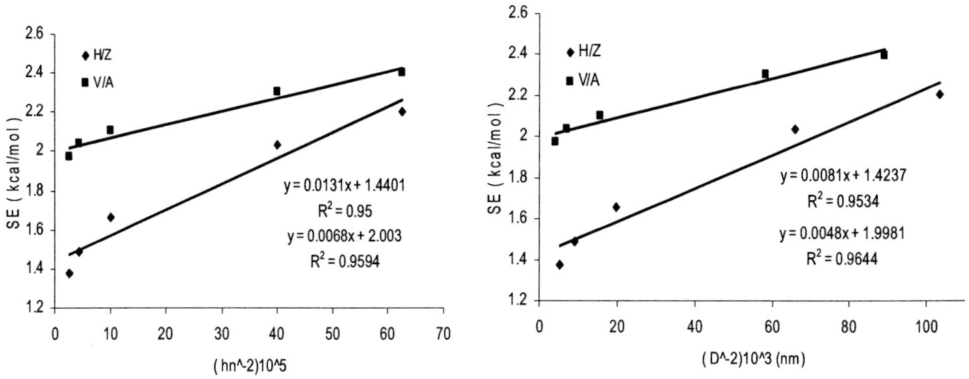


Figure 8: Strain energy per atom of thick tori, H/Z[40,*n*], *n* = 40,…, 500 and V/A[20,*n*], *n* = 80,…, 1000: (a) function of the number of hexes $h_n$ on dimension *n* and (b) function of the large diameter , in nm, respectively.

(a) (b)

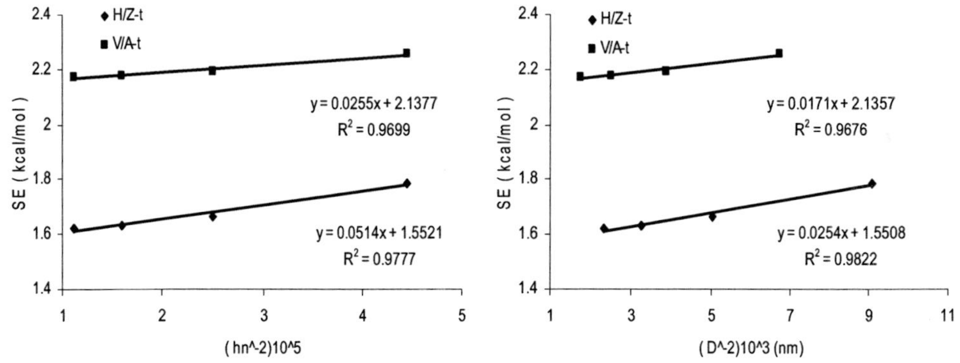


Figure 9: Strain energy per atom of thin tori, H/Z[20,*n*], *n* = 100,…, 300 and V/A[10,*n*], *n* = 200,…, 600: (a) function of the number of hexes $h_n$ on dimension *n* and (b) function of the large diameter , in nm, respectively.

When the $c$-dimension, at $n$ = 100, was varied a different trend appeared for the two series. Dependence by $h_c^{-2}$ in thin tori (the most strained subset) and by $h_c^{-1}$ in thick tori of the H/Z series was found (Figure 10). The series V/A appear split in two subseries, 0mod($c$,4) (having at apex a strain up to 11 times the average value) and 2mod($c$,4) (with up to 17 times the average strain at apex), respectively. Strain of both series depends on $h_c^{-1}$ (Figure 11). The two series are, however, not directly comparable.

(a) (b)

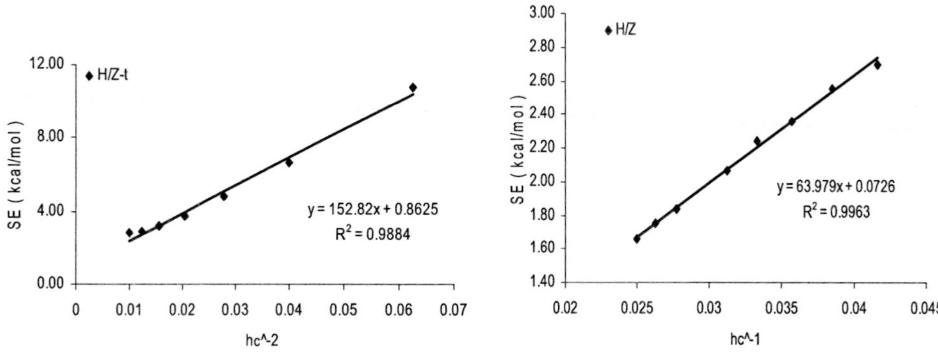


Figure 10: Strain energy per atom of: (a) thin tori, H/Z[$c$,100], $h_n$ = 100, $c$ = 8,..., 20 and (b) thick tori H/Z[$c$,100], $c$ = 24,...,40, function of the number of hexes $h_c$ on dimension $c$.

(a) (b)

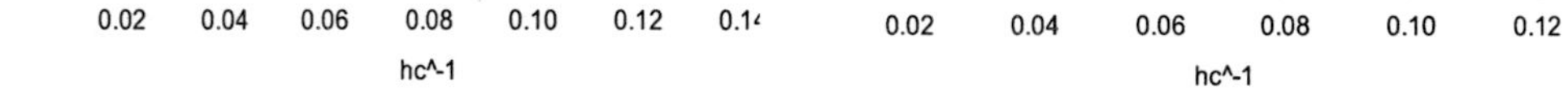


Figure 11: Strain energy per atom of tori V/A[$c$,100], $h_n$ = 50, $c$ = 8,..., 40, series 0mod($c$,4) (a) and 2mod($c$,4) (b), function of the number of hexes $h_c$ on dimension $c$.

Compare the strain energy per atom of tori with that of $C_{60}$ (8.26 kcal/mol) and find the polyhex tori as more relaxed structures. Among twisted tori, only the objects with $t$ = 2 could be optimized and their strain was in the range of corresponding non-twisted tori.

## 5. Π-ELECTRONIC STRUCTURE OF POLYHEX TORI

In the Spectral Theory, at the simple π-only Hückel[47] level of theory, the energy of the $i^{th}$ molecular orbital $E_i = \alpha + \lambda_i \beta$ is calculated on the grounds of the adjacency matrix associated to the molecular hydrogen depleted graph.

The π-electronic shells of neutral graphitic finite cages are classified, function of their eigenvalue spectra, as:[48,49] (i) *properly closed*, PC, when $\lambda_{N/2} > 0 \geq \lambda_{N/2+1}$; (ii) *pseudo-closed*, PSC, in case $\lambda_{N/2} > \lambda_{N/2+1} > 0$; (iii) *meta-closed*, MC, with $0 \geq \lambda_{N/2} > \lambda_{N/2+1}$ and (iv) *open*, OP, when the $N/2^{th}$ (HOMO) and $N/2+1^{th}$ (LUMO) molecular orbitals are degenerate, $\lambda_{N/2} = \lambda_{N/2+1}$.

The *metallic* character involves the existence of a zero HOMO-LUMO gap (a particular case of the open shell) and the degeneracy of some non-bonding orbitals[50] (NBOs) favoring the spin multiplicity, cf. the Hund rule. In polyhex tori, the metallic behavior is ensured by *four* NBOs, also present in the graphite sheet. The gap (in $\beta$ units) is taken as the absolute value of the difference $E_{HOMO}$ - $E_{LUMO}$.

A perfect Clar structure [51,52] PCS (Figure 12a) is a vertex disjoint set of faces whose boundaries form a 2-factor. A *k*-factor is a regular *k*-valent spanning subgraph. A PCS will include each vertex of *G* once and only once. It is associated with a Fries structure,[53] which is a Kekulé structure drawn over the maximum possible number of benzenoid (alternating single-double edge) hexagonal faces. A Kekulé structure is a set of pairwise disjoint edges of *G* (covering all its vertices) that coincides with a perfect matching and a 1-factor in Graph Theory.[54] A trivalent polyhedral graph has a PCS if and only if it has a Fries structure.[55] Such structures represent *total resonant sextet* TRS benzenoid molecules and is expected to be maximally stable, in the valence bond theory.[55,56] There exist some coverings by large numbers of conjugated 10-circuits (*i.e.*, naphthalenoids and azulenoids), only slightly less favorable, in the valence-bond based conjugated-circuit theory.[57,58]

(a) (6,3)H[24,36] PCS (b) (6,3)H[24,36]) PCorS

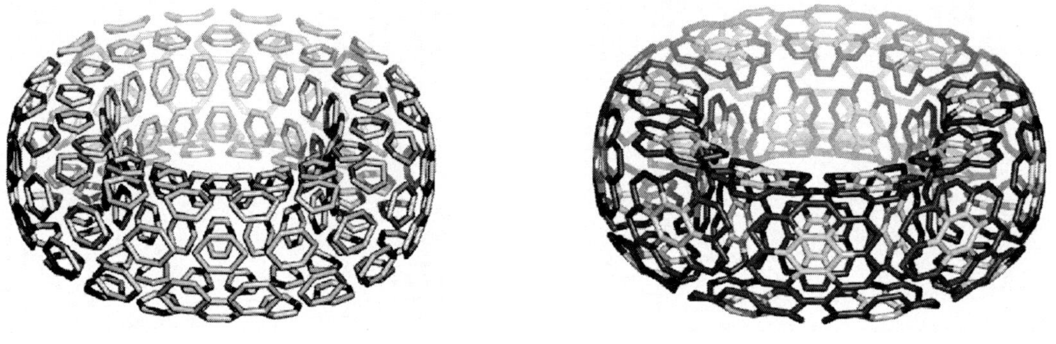

Figure 12: The perfect Clar and Cor structures of a H/Z polyhex torus (see text).

By extension, a perfect corannulenic system PCorS can be imagined. The operation[59] sequence *Leapfrog-Quadrupling*, applied on trivalent maps, provides a perfect Clar-like corannulenic structure PCorS, with the associated Fries-like structure defined on all the vertices of graph, excepting the corannulenic core ones (Figure 12b, in black).

Table 3 gives the criteria for metallic shell in (6,3) tori of various types. These criteria superimpose over those of perfect Clar structure PCS: divisibility by 6 of the net dimensions (see also [50,55,56,60]). In the opposite, Clar fullerenes (available by Leapfrog operation) show closed-shell structure.[50,55]

The last column in Table 3 gives the criteria for the perfect corannulenic structure PCorS. Any PCorS is necessarily a PCS. The supplementary condition is divisibility by four of the net dimensions.

At this stage, it is not clear if the supra organized corannulenic system would bring a supplementary stability in possible molecules, with respect to fully benzenoid (Clar) structures. Out of the energetic aspects, it is possible that such molecules behave similarly to planar corannulenes, in the magnetic field. Confirmation of this hypothesis, by evaluating the ring currents appearing in polycyclic molecular systems, *e.g.*, by the CTOCD method[61] is, however, needed.

**Table 3: Covering criteria for metallic shell/perfect Clar and perfect Corannulenic structure in polyhex tori originating in square tiled tori**

| No | Torus | Criterion for metallic shell and perfect Clar structure PCS | Criterion for perfect Corannulenic structure PCorS |
|---|---|---|---|
| | Non-Twisted | | |
| 1 | H/Z[$c$,$n$] | 0 mod ($c$,6) | 0 mod ($c$,12); 0 mod ($n$,4) |
| 2 | V/A[$c$,$n$] | 0 mod ($n$,6) | 0 mod ($c$,4); 0 mod ($n$,12) |
| | H-Twisted | | |
| 3 | HH$t$[$c$,$n$] | 0 mod ($c$,6) | 0 mod ($c$,12); 0 mod ($n$,4); 0 mod ($t$,4) |
| 4 | HV$t$[$c$,$n$] | 0 mod ($n$,6) and 0 mod($t$,6) | 0 mod ($c$,4); 0 mod ($n$,12) 0 mod ($t$,12) |
| | V-Twisted | | |
| 5 | VH$t$[$c$,$n$] | 0 mod ($c$,6) and 0 mod ($t$,6) | 0 mod ($c$,12); 0 mod ($n$,4); 0 mod ($t$,12) |
| 6 | VV$t$[$c$,$n$] | 0 mod ($n$,6) | 0 mod ($c$,4); 0 mod ($n$,12); 0 mod ($t$,4) |

*Identical Graphs*

Toroidal objects, correctly named to reflect different embeddings, could represent one and the same graph. In the following, the most frequent cases of identical graphs, in terms of our nomenclature, are given:

Non-twisted/achiral tori: Z[$c$,$n$] = A[ $n$, $c$]; A[$c$,$n$] = Z[ $n$, $c$]

Twisted HH$t$[$c$,$n$]: For [$c$, $c + 2m$]; $m = 0, 1, \ldots, (c - 2)/2$

Twisted HV$t$[$c$,$n$]: For {$t(2r + 2m)$, $t(c - 2r)$}; $r = 0, 1, \ldots, [c/4]$ ); $(2r + 2m) < c$

{HV$c$[$c$, $pc$]); HH$c$[$c$,$pc$],}; $p$ = 1, 2, 3,...

Twisted VH$t$[$c$,$n$]: For {$t$(2$r$), $t$($c$-2$r$)}; $r$ = 0, 1,..., [$c$/4]);

VH$t$[$c$, $c$] = HV$t$[$c$,$c$];

Twisted VV$t$[$c$,$n$]: VV$t$[$c$, $c$] = HH$t$[$c$, $c$];

*Non-Twisted H[c,n] and V[c,n] Tori*

In non-twisted tori, the HOMO-LUMO gap remains constant, at a given $c$ (in H/Z[$c$,$n$] series) and $n$ (V/A[$c$,$n$]), respectively. The same is true for twisted tori, according to the second capital letter in their name. Also note the identity of graphs H/Z[$c$,$n$] = V/A[ $n$,$c$] but having different embedding.

*H-Twisted HHt[c,n] and HVt[c,n] Tori*

Condition for all distinct spectra (implying all distinct objects) within a family of H-twisted tori (*i.e.*, a set of objects of given dimensions [$c$,$n$] and variable integer $t$, in the range [0,$c$]) is: $n \geq 2c$. In case [$c$, $c$ + 2$m$]; $m$ = 0, 1, ..., ($c$ - 2)/2, there exist one or more pairwise different embeddings (having different twisting $t$-values) with identical spectra at: ($t$(2$r$ + 2$m$), $t$($c$ – 2$r$); $r$ = 0, 1, ..., [$c$/4] ) and (2$r$ + 2$m$) < $c$. In general, HH$t$[$c$,$n$] $\neq$ HV$t$[$n$,$c$] (*i.e.*, different graphs, different embeddings). Pairs of different embeddings appear at maximum twisting $t = c$ with $n$ being multiple of $c$: (HH$c$[$c$, $pc$], HV$c$[$c$, $pc$]); $p$ = 1, 2, 3,... At $c = n$, HH$t$[$c$,$c$] = VV$t$[$c$,$c$].

*V-Twisted VHt[c,n] and VVt[c,n] Tori*

Within a family of V-twisted tori there exist pairwise different embeddings with identical spectra at: ($t$(2$r$), $t$($c$-2$r$); $r$ = 0, 1, ..., [$c$/4] ). In general, VH$t$[$c$,$n$] $\neq$ VV$t$[$n$,$c$], (different graphs, different embeddings). At $c = n$, VH$t$[$c$,$c$] = HV$t$[$c$,$c$].

*Resonance Energy in Polyhex Tori*

Resuming to the aromaticity of polyhex tori, recall that "resonance energy" $RE$ is defined as the difference $E_{\pi} - E_{\pi,ref}$ between the total π-energy of a conjugated (aromatic) system and the energy of some reference structure. Several approaches have been developed so far.[62]

A system having $RE > 0$ is classified as "aromatic", that having $RE < 0$ as "antiaromatic" while that for which RE $\approx$ 0 as "non-aromatic".

There exists a theorem[63] enabling the calculation of resonance energy from the number of Kekulé structures $K$, by means of eigenvalues of the molecular graph:

$$K = \prod_{i=1}^{N/2} \lambda_i \tag{7}$$

$$RE = A \cdot \ln K \tag{8}$$

with *A* being 1.185 (eV). We used this approach to evaluate the aromaticity/stability of the X*t*[12,14] series of toroidal objects, the results being listed in Table 4.

**Table 4: Tori of X*t*[12,14] series, topology, π-electronic shell and resonance energy *RE*.**

| | Torus (6,3)X*t*[12,14] | Tube *R* | Shell | RE (eV/atom) | Observation |
|---|---|---|---|---|---|
| | | | **HH series** | | |
| 1 | HH0 | (0,6) | M | 0.2142 | H/Z; (0,6) = (6,0) = (6,-6) |
| 2 | HH2 | (2,5) | M | 0.2143 | - |
| 3 | HH4 | (4,4) | M | 0.2150 | - |
| 4 | *HH6* | (6,3) | *M* | *0.2161* | (6,-3,7,7) (see Table 1) |
| 5 | *HH8* | (8,2) | *M* | *0.2161* | (8,-2,7,7) (see Table 1) |
| 6 | HH10 | (10,1) | M | 0.2150 | - |
| 7 | HH12 | (12,0) | M | 0.2143 | H/Z; non-twisted tube |
| | | | **HV series** | | |
| 8 | HV0 | (6,6) | PC | 0.1947 | V/A; non-twisted tube |
| 9 | HV2 | (7,5) | PC | 0.1960 | twisted tube |
| 10 | HV4 | (8,4) | PC | 0.1966 | twisted tube |
| 11 | HV6 | (9,3) | PC | 0.1951 | twisted tube |
| 12 | HV8 | (10,2) | PC | 0.1951 | twisted tube |
| 13 | HV10 | (11,1) | PC | 0.1966 | twisted tube |
| 14 | HV12 | (12,0) | PC | 0.1960 | H/Z; non-twisted tube |
| | | | **VV series** | | |
| 15 | VV*t*, *t* =0,2,4,..12 | (6,6) | PC | 0.1942 | Average<br>V/A; non-twisted tube |
| | | | **VH series** | | |
| 16 | **VH0** | (6,0) | **M** | **0.2142** | H/Z; non-twisted tube |
| 17 | VH2 | (6,0) | PC | 0.1847 | |
| 18 | VH4 | (6,0) | PC | 0.1847 | |
| 19 | **VH6** | (6,0) | **M** | **0.2142** | |
| 20 | VH8 | (6,0) | PC | 0.1847 | |
| 21 | VH10 | (6,0) | PC | 0.1847 | |
| 22 | **VH12** | (6,0) | **M** | **0.2142** | |

Following Table 3, the π-electronic shell of the toroids, in the above series, is either properly closed PC (HV series, entries 8-14, and in part VH series, entries 17,18, 20,21 – average *RE* < 0.196 eV/atom) or metallic M (HH series, entries 1-7, and VH series, entries 16,19 and 22). The value of *RE* per atom in metallic shell objects is larger than 0.21 eV, suggesting a possible higher stability of the associated molecules. Objects in entries 4 and 5 (in italics) represent one and the same graph.

## 6. Conclusions

Construction of polyhex (6,3) tori starting from the square tiled (4,4) tori proved to be a robust and versatile procedure. The topology of polyhex tori can be described by the α-spiral code, adapted for tubular objects. Strain energy calculation, in terms of the POAV1 theory, revealed toroidal polyhex objects more relaxed than the well-known $C_{60}$. The π-electronic structure was evaluated at the Hückel level of theory, criteria for metallic character in (6,3) tori being given in terms of the $[c,n]$ net dimensions. These criteria superimpose over those of perfect Clar and Cor structure, envisaging the metallic polyhex tori as fully-resonant-benzenoid/corannulenoid molecules. Resonance energy calculations provided support of the aromaticity and supplementary stability of metallic tori.

## References

[1] M. Endo, S. Iijima, and M. S. Dresselhaus, *Carbon Nanotubes*, Pergamon, 1996.
[2] P. W. Fowler and D. E. Manolopolous, *An atlas of fullerenes*, Oxford University Press, London, 1994.
[3] M. S. Dresselhaus, G. Dresselhaus, and P. C. Eklund, *Science of fullerenes and carbon nanotubes*, Acad. Press, San Diego, 1996.
[4] K. Tanaka, T. Yamabe, and K. Fukui, *The science and technology of carbon nanotubes*, Elsevier, 1999.
[5] A T. Balaban, (Ed.), *From Chemical Topology to Three - Dimensional Geometry*, Plenum Press, New York, 1997.
[6] L. Euler, Solutio Problematis ad Geometriam Situs Pertinentis. *Comment. Acad. Sci. I. Petropolitanae* 1736, 8, 128-140.
[7] J. Liu, H. Dai, J. H. Hafner, D. T. Colbert, R. E. Smalley, S. J. Tans, and C. Dekker, *Nature*, 1997, 385, 780-781.
[8] M. Ahlskog, E. Seynaeve, R. J. M. Vullers, C. Van Haesendonck, A. Fonseca, K. Hernadi, and J. B. Nagy, *Chem. Phys. Lett.*, 1999, 300, 202-206.
[9] R. Martel, H. R. Shea, and Ph. Avouris, *Nature*, 1999, 398, 299-299.
[10] R. Martel, H. R. Shea, and Ph. Avouris, *J. Phys. Chem. B,* 1999, 103, 7551-7556.
[11] D. Babić, D. J. Klein, and T. G. Schmalz, *J. Mol. Graphics Modell.*, 2001, 19, 222-231.
[12] A Ceulemans, L. F. Chibotaru, and P. W. Fowler, *Phys. Rev. Lett.*, 1998, 80, 1861-1864.
[13] J. K. Johnson, B. N. Davidson, M. R. Pederson, and J. Q. Broughton, *Phys. Rev., B*, 1994, 50, 17575-17582.
[14] V. Meunier, Ph. Lambin, and A. A. Lucas, *Phys. Rev., B*, 1998, 57, 14886-14890.
[15] M. F. Lin and D. S. Chuu, *Phys. Rev. B*, 1998, 57, 6731-6737.
[16] E. C. Kirby, *Croat. Chem. Acta*, 1993, 66, 13-26.
[17] E. C. Kirby, R. B. Mallion, and P. Pollak, *J. Chem. Soc. Faraday Trans.*, 1993, 89, 1945-1953.
[18] S. A. Bovin, L. F. Chibotaru, and A. Ceulemans, *J. Molec. Catalys.*, 2001, 166, 47-52.
[19] D. Marušić and T. Pisanski, *Croat. Chem. Acta*, 2000, 73, 969-981.
[20] D. J. Klein, *J. Chem. Inf. Comput. Sci.*, 1994, 34, 453-459.
[21] E. C. Kirby and P. Pollak, *J. Chem. Inf. Comput. Sci.*, 1998, 38, 66-70.

[22] T. Pisanski and J. Shawe-Taylor, *J. Chem. Inf. Comput. Sci.*, 2000, 40, 567-571.
[23] A Graovac, D. Plavšić, M. Kaufman, T. Pisanski, and E. C. Kirby, *J. Chem. Phys.*, 2000, 113, 1925-1931.
[24] Laszlo, A. Rassat, P. W. Fowler, and A. Graovac, *Chem. Phys. Lett.*, 2001, 342, 369-374.
[25] Laszlo and A. Rassat, *J. Chem. Inf. Comput. Sci.*, 2003, 43, 519-524.
[26] M. V. Diudea and A. Graovac, *MATCH, Commun. Math. Comput. Chem.*, 2001, 44, 93-102.
[27] M. V. Diudea, I. Silaghi-Dumitrescu, and B. Parv, *MATCH, Commun. Math. Comput. Chem.*, 2001, 44, 117-133.
[28] M. V. Diudea and E. C. Kirby, *Fullerene Sci. Technol.*, 2001, 9, 445-465.
[29] M. V. Diudea, *Bull. Chem. Soc. Japan*, 2002, 75, 487-492.
[30] M. V. Diudea, *Fullerenes, Nanotubes, Carbon Nanostruct.*, 2002, 10, 273-292.
[31] M. V. Diudea, *Phys. Chem., Chem. Phys.*, 2002, 4, 4740-4746.
[32] M. V. Diudea, B. Parv, and E. C. Kirby, *MATCH, Commun. Math. Comput. Chem.*, 2003, 47, 53-70.
[33] S. Itoh and S. Ihara, *Phys. Rev., B*, 1993, 48, 8323-8328.
[34] M. Endo, K. Takeuchi, K. Kobory, K. Takahashi, H. W. Kroto, and A. Sarkar, in: M. Endo, S. Iijima and M. S. Dresselhaus, *Carbon Nanotubes*, Pergamon, 1996, pp. 1-9.
[35] A Sarkar, H. W. Kroto, M. Endo, in: M. Endo, S. Iijima and M. S. Dresselhaus, *Carbon Nanotubes*, Pergamon, 1996, pp. 105-109.
[36] N. Hamada, S. Sawada, and A. Oshiyama, *Phys. Rev. Lett.*, 1992, 68, 1579.
[37] D. E. Manolopoulos, J. C. May and S. E. Down, *Chem. Phys. Lett.*, 1991, 181, 105-111.
[38] G. Brinkmann and P. W. Fowler, *J. Chem. Inf. Comput. Sci.*, 1998, 38, 463-468.
[39] P. W. Fowler, T. Pisanski, A. Graovac, and J. Žerovnik, *DIMACS Ser. Discrete Maths. Theor. Comput. Sci.*, 2000, 51, 175-187.
[40] D. J. Klein, W. A. Seitz, and T. G. Schmalz, *J. Phys. Chem.*, 1993, 97, 1231-1236.
[41] M. V. Diudea, B. Parv and O. Ursu, TORUS 3.0, "Babes-Bolyai" University, 2003.
[42] R.C. Haddon, *J. Am. Chem. Soc.*, 1990, 112, 3385-3389.
[43] R.C. Haddon and K. Raghavachari, in: W.E. Billups and M. A. Ciufolini, Buckminsterfullerenes, VCH, 1993, pp. 185-215.
[44] R.C. Haddon and S.-Y. Chow, *J. Am. Chem. Soc.*, 1998, 120, 10494-10496.
[45] R.C. Haddon, *J. Phys. Chem. A*, 2001, 105, 4164-4165.
[46] Han, *Chem. Phys. Lett.*, 1998, 282, 187-191.
[47] E. Hückel, *Z. Phys.*, 1931, 70, 204-286
[48] P. W. Fowler, *J. Chem. Soc., Faraday Trans.*, 1997, 93, 1-3.
[49] P. W. Fowler, *J. Chem. Soc., Faraday Trans.*, 1990, 86, 2073-2077.
[50] M. Yoshida, M. Fujita, P. W. Fowler, and E. C. Kirby, *J. Chem. Soc., Faraday Trans.*, 1997, 93, 1037-1043.
[51] E. Clar, *Polycyclic Hydrocarbons*, Acad. Press, London, 1964.
[52] E. Clar, *The Aromatic Sextet*, Wiley, New York, 1972.
[53] Fries, *J. Liebigs Ann. Chem.*, 1927, 454, 121-324.
[54] W-Ch. Shiu, P. Ch. B. Lam, and H. Zhang, *Theochem*4, 2000, p 0210.
[55] P. W. Fowler and T. Pisanski, *J. Chem. Soc., Faraday Trans.*, 1994, 90, 2865-2871.
[56] R. Dias, *J. Chem. Inf. Comput. Sci.*, 1999, 39, 144-150.

[57] D. J. Klein and H.Y. Zhu, in: *From Chemical Topology to Three - Dimensional Geometry*, A. T. Balaban, (Ed.), Plenum Press, New York.1997, pp. 297-341.

[58] E. C. Kirby, *MATCH, Commun. Math. Comput. Chem.*, 1996, 33, 147-156.

[59] V. Diudea, P. E. John, A. Graovac, M. Primorac, and T. Pisanski, *Croat. Chem. Acta*, 2003, 76, 153-159.

[60] P. W. Fowler, *Chem. Phys. Lett.*, 1986, 131, 444-450.

[61] T. A. Keith and R. F. W. Bader, *Chem. Phys. Lett.*, 1993, 210, 223-231.

[62] Trinajstić, *Chemical Graph Theory*, CRC Press, Boca Raton, FL, 1992, pp. 125-160.

[63] Trinajstić, *Chemical Graph Theory*, CRC Press, Boca Raton, FL, 1992, pp. 161; 188.

In: Nanostructures: Novel Architecture
Editor: Mircea V. Diudea, pp. 127-165

ISBN: 1-59454-499-9


*Chapter 7*

# DISTANCE COUNTING IN TUBES AND TORI: WIENER INDEX AND HOSOYA POLYNOMIAL

***Monica Ştefu and Mircea V. Diudea***[1]
Faculty of Chemistry and Chemical Engineering, "Babeş-Bolyai"
University, 400084, Cluj, Romania

## ABSTRACT

Counting of all distances in a molecular graph can be expressed as a single number, known as the Wiener index. This number equals the first derivative of the Hosoya polynomial (in $x = 1$), defined as a distance-based increasing power sequence. Analytical formulas and/or recursions for calculating Wiener index and Hosoya polynomial, in several classes of tubular and toroidal nets, are given.

## 1. INTRODUCTION

Wiener[1] has defined his path number $W$ as "the sum of the distance between any two carbon atoms in the molecule, in terms of carbon-carbon bonds." Wiener index $W$ equals the sum of all distances in a graph and can be obtained by summing all the entries above the main diagonal of the distance matrix **D**, as shown by Hosoya:[2]

$$W = (1/2) \sum_i \sum_j [\mathbf{D}]_{ij} \quad (1)$$

Recall that the non-diagonal entries in **D** count the number of edges on the shortest path joining any two atoms $i$ and $j$ in the molecular graph, while the diagonal entries are zero. These entries represent just the topological distance between $i$ and $j$.

The *Hosoya polynomial* was defined[3] as a distance-based increasing power sequence, in any connected graph:

[1] E-mail address: diudea@chem.ubbcluj.ro

$$H(G,x) = \sum_{k \geq 0} d(G,k) x^k \tag{2}$$

where *d*(*G*,*k*) is the number of pair vertices lying at distance *k* of each other. Then *d*(*G*,0) = *N* and *d*(*G*,1) = *e*, where *N* is the number of vertices in the graph *G* and *e* is the number of edges.

The *Wiener Index* can be obtained from the first derivative of the Hosoya polynomial (in *x* = 1):

$$\frac{dH(G,x)}{dx}\Big|_{x=1} = W(G) \tag{3}$$

The Hosoya polynomial can be expressed as a function of the vertex contributions:[4]

$$H(i,x) = \sum_{k=0}^{d(G)} d(i,k) x^k \tag{4}$$

where *d*(*i*,*k*) is the number of vertices at distance *k* from the vertex *i*. The summation runs from zero to *d*(*G*), which is the *diameter* of G or the longest distance in the graph. Since each path has two endpoints (*i.e.*, each path is counted twice), it becomes clear that, in a vertex transitive graph, the following relation holds:

$$N \cdot H(i,x) = 2H(G,x) - N \tag{5}$$

## 2. Wiener Index in (6,3) Polyhex Nanotubes

Conceptually, a single walled nanotube SWNT can be obtained by rolling up a graphite sheet (see[5] and therein refs.). In the constructive version of Diudea *et al.*,[6-8] the (4,4) net, embedded in the cylinder (or torus), is appropriately cut to give the polyhex (6,3) net. Tubes are named TU(6,3)A[*c*,*n*] and TU(6,3)Z[*c*,*n*] (Figure 1). They correspond to (*c*/2,*c*/2) "armchair" and (*c*/2, 0) "zig-zag" classically named nanotubes.[5] In the above, *c* counts the number of atoms across the tube while *n* is the number of cross-sections along the tube. The product *c* x *n* gives the total number of atoms in the molecule.

(a) TU(6,3)A[20,30] (b) TU(6,3)Z[20,30]

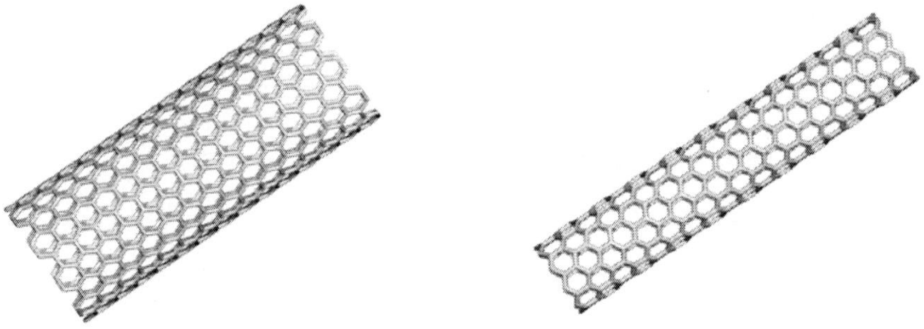

Figure 1: An "armchair" and a "zig-zag" polyhex nanotube.

## 2.1. Armchair (6,3) Polyhex Nanotubes

Let us consider a hexagonal armchair lattice, embedded in the cylinder, as illustrated in Figure 2. We choose a reference vertex $v$, from which the topological distance to al other vertices are evaluated.[9] The sum of such distances, on each level, is given as $S_i$ (Figure 2).

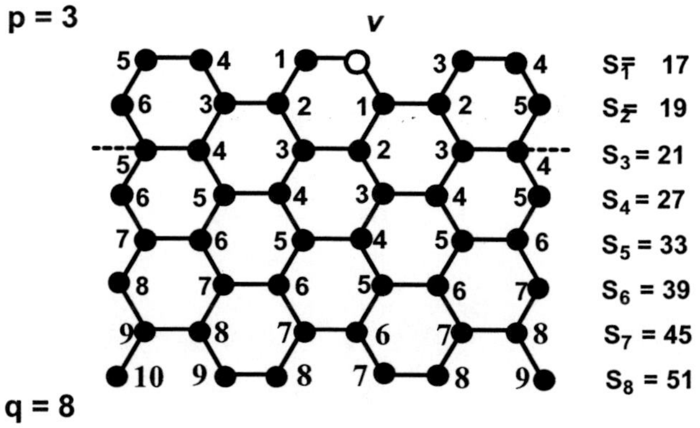


Figure 2: An "armchair" polyhex net embedded in the cylinder.

### *2.1.1. Method a*

The sum from $v$ to all vertices lying at level $m = 1$ is given by:

$$s_1(p,z) = 2p^2 - z \tag{6}$$

where $z = \mathrm{mod}(p,2)$. Here, $2p = c$ and $q = n$, in a tube TU(6,3)A[$c$,$n$].

For levels in the range $1 < k \leq p$, (see the dashed line in Figure 2) the increment to $s_1$ is calculated as:

$$s_n(k,z) = 2k + 2z - 3 - (-1)^k \tag{7}$$

and the distance sum:

$$s_m(p,s,z)=\sum_{k=1}^{s-z} s_n(k,z)+(2p^2-z) \tag{8}$$

The total distance sum up to level $m$ is given by:

$$st_m(p,m,z)=(2p^2-z)+\sum_{s=2}^{m} s_m(p,s,z) \tag{9}$$

The distance sum at level $m=p$ is:

$$s_p(p,z)=\sum_{k=1}^{p-z} s_n(k,z)+(2p^2-z) \tag{10}$$

Calculate now the sum at levels $m>p$ as:

$$s_c(p,s,z)=s_p(p,z)+2p(s-p) \tag{11}$$

and the total sum up to $m$:

$$st_c(p,m,z)=\sum_{s=p+1}^{m} s_c(p,s,z) \tag{12}$$

The total sum from $v$ located at level $m=1$ to all vertices in TU(6,3)A will be:

$$s_v(p,m,z)=st_m(p,p,z)+st_c(p,m,z) \tag{13}$$

Now we are ready to calculate the Wiener index of a TU(6,3)A[2$p$,$q$] as:

$$W(p,q,z)=\sum_{m=p+1}^{q} 2p\cdot s_v(p,m,z)+\sum_{m=2}^{p} 2p\cdot st_m(p,m,z)+2p\cdot s_1(p,z)-(q/2)\cdot 2p\cdot s_1(p,z) \tag{14}$$

The subtraction of the last term in the above equation is reasoned as follows: the reference vertex $v$ may be located at any level $1<m<q$, each time looking at TU(6,3)A as being obtained by two smaller tubes sharing a common level, namely that containing the vertex $v$. It is obvious that the actual level of $v$ is counted twice.

Expansion of functions in (14) leads to the final formula for calculating $W$, in **case** $q\geq p$:

*2.1.2. Method B*

The sum from $v$ to all vertices lying at level “1” is:[9]

$$s_1(p,z) = 2p^2 - z \qquad (21)$$

where $z = \mathrm{mod}(p,2)$.

For levels in the range $1 < k \le p$, the sum is:

$$sv_k(p,k,z) = 2p^2 + z \cdot (-1)^k + (k-1)^2 - \mathrm{mod}((k-1),2) \qquad (22)$$

The total distance sum up to level $n$ is given by:

$$st_n(p,n,z) = 2p^2 - z + \sum_{k=2}^{n} sv_k(p,k,z) \qquad (23)$$

and, after calculations:

$$st_n(p,n,z) = 2p^2 n - \frac{1}{2} \cdot z \cdot [1-(-1)^n] + \frac{1}{3} \cdot n^3 - \frac{1}{2} \cdot n^2 - \frac{1}{3} \cdot n + \frac{1}{4} \cdot [1-(-1)^n] \qquad (24)$$

The distance sum at level $n = p$ is:

$$st_p(p) = \frac{7}{3}p^3 - \frac{1}{2}p^2 - \frac{1}{3}p - \frac{1}{4}(1-(-1)^p) \qquad (25)$$

Calculate now the sums at levels $k > p$ as follows:

$$svn_p(p,k) = 3p^2 + 2p(k-p-1) \qquad (26)$$

The total distance sum from $v$ located at level “1” to all vertices in TUA(6,3), for $n > p$, is:

$$stn_p(p,n) = st_p(p) + \sum_{k=p+1}^{n} svn_p(p,k) \qquad (27)$$

and, after calculations:

$$stn_p(p,n) = st_p(p) + p(2p+m-1)(m-p) \qquad (28)$$

$$W_{\mathrm{TU(6,3)A}}(p,q,z)=\frac{p}{12}\begin{bmatrix}12(-1)^{(p-z)}pq+3(-1)^{(p-z+1)}+3(-1)^{-z}-12q^2z^2+12q^2z+\\12(-1)^{(p-z)}p+12z^2q+6(-1)^{(p-z+1)}p^2+8pq^3-28pq+6q^2+\\8p^3q+12p^2q^2-12(-1)^{(p-z)}q+6(-1)^{(p-z+1)}q^2-24qz-12p+\\18q+14p^2+6(-1)^{(1-z)}q-2p^4\end{bmatrix} \quad (15)$$

Keeping in mind the following:

I. $(-1)^{(p-z)}=1$, because: if $p$ is even, z = 0 and $(-1)^{(p-z)}=(-1)^p=1$

i. if $p$ is odd, z = 1, p-1 is even and $(-1)^{(p-z)}=(-1)^{p-1}=1$

II. $(-1)^{(p-z+1)}=(-1)\cdot(-1)^{(p-z)}=(-1)$ since, as calculated above, $(-1)^{p-z}=1$

III. $(-1)^{-z}=(-1)^p$ because:if $p$ is even and z = 0, $(-1)^{-z}=(-1)^0=1=(-1)^p$

1. if $p$ is odd and z = 1, $(-1)^{-z}=(-1)^{-1}=-1=(-1)^p$

IV. $z=(1-(-1)^p)/2$

relation (10), for $p = q$ becomes:

$$W_{\mathrm{TU(6,3)A}}(p,q)=\frac{p}{12}\cdot\left[p^2(26q^2-8)+3(-1+(-1)^p)\right] \quad (16)$$

In **case** $q\le p$, (*i.e.*, short tubes, sTU(6,3)A), *W* is calculated by formula:

$$W_{s\mathrm{TU(6,3)A}}(p,q,z)=\sum_{m=2}^{q}2p\cdot st_m(p,m,z)+2p\cdot s_1(p,z)-(q/2)\cdot 2p\cdot s_1(p,z) \quad (17)$$

Expansion of functions in (17) leads to the final formula for calculating W, in case of short tubes:

$$W_{s\mathrm{TU(6,3)A}}(p,q,z)=\frac{p}{12}\left[3(-1)^z+24p^2q^2+2q^4-8q^2+3(-1)^{(q-z+1)}\right] \quad (18)$$

A similar procedure as used for relation (15) leads to the final formula:[9]

$$W_{s\mathrm{TU(6,3)A}}(p,q)=\frac{p}{12}\cdot\left[24p^2q^2+2q^4-8q^2+3(-1)^p(1-(-1)^q)\right] \quad (19)$$

For $q = 2,\ p\ge 2$, the formula for simple cycles on 4*p* vertices is recovered:

$$W(C_{4P})=\frac{p}{4}\left[(-1)^z+32p^2+(-1)^{(1-z)}\right]=8p^3 \quad (20)$$

The Wiener index of a TU(6,3)A[2$p$,$q$], $q \geq p$ is given by:[9]

$$W_{\mathrm{TU(6,3)A}}(p,q,z) = p \cdot \left[ 2 \cdot \sum_{n=1}^{p} st_n(p,n,z) + 2 \cdot \sum_{n=p+1}^{q} stn_p(p,n) - q \cdot s_1(p,z) \right] \tag{29}$$

In **case** $q \leq p$ (*i.e.*, short tubes, sTU(6,3)A), the formula is:

$$W_{\mathrm{sTU(6,3)A}}(p,q,z) = p \cdot \left[ 2 \cdot \sum_{n=1}^{q} st_n(p,n,z) - q \cdot s_1(p,z) \right] \tag{30}$$

Expansion of the above functions leads to the final formulas for calculating $W$.
**Case** $q \geq p$ (long tubes, TU(6,3)A):

$$W_{\mathrm{TU(6,3)A}}(p,q,z) = p \cdot \left[ \frac{13}{6} p^4 - \frac{2}{3} p^2 + \frac{(1-2z)}{4} \cdot (1-(-1)^p) + \frac{1}{6}(q-p)[14p^3 + 10p^2 \cdot q + 4p \cdot q^2 - 8p - 3 + 3 \cdot (-1)^p + 6z] \right] \tag{31}$$

With $z = (1-(-1)^p)/2$, relation (31) transforms into relation (16).
**Case** $q \leq p$ (short tubes, sTU(6,3)A); expansion of relation (30) leads to:

$$W_{s\mathrm{TU(6,3)A}}(p,q,z) = p \cdot [2p^2q^2 + \frac{1}{6}q^4 - \frac{2}{3}q^2 + \frac{1}{4}(1-(-1)^q) - \frac{1}{2}z(1-(-1)^q)] \tag{32}$$

Substituting $z$ as above, (32) transforms into relation (19).

Tables 1 and 2 list some numerical values for the Wiener index of long TU(6,3)A[2$p$,$q$] and short tubes sTU(6,3)A[2$p$,$q$], respectively.

**Table 1: Wiener index of long tubes, TU(6,3)A[2$p$,$q$], $q \geq p$.**

| p | q | W | p | q | W |
|---|---|---|---|---|---|
| 3 | 3 | 507 | 4 | 4 | 2,176 |
| 3 | 8 | 5,112 | 4 | 8 | 10,624 |
| 3 | 16 | 32,136 | 4 | 16 | 62,336 |
| 5 | 5 | 6,685 | 6 | 6 | 16,704 |
| 5 | 10 | 32,560 | 6 | 12 | 81,216 |
| 5 | 15 | 89,685 | 6 | 18 | 223,488 |
| 5 | 20 | 190,560 | 6 | 24 | 474,624 |
| 7 | 7 | 36,183 | 8 | 8 | 70,656 |
| 7 | 14 | 175,784 | 8 | 16 | 343,040 |
| 7 | 21 | 483,455 | 8 | 24 | 943,104 |
| 7 | 28 | 1,026,424 | 8 | 32 | 2,001,920 |

**Table2: Wiener index of short tubes, sTU(6,3)A[2*p*,*q*], $q \le p$.**

| p | q | W | p | q | W |
|---|---|---|---|---|---|
| 9 | 9 | 127,449 | 10 | 10 | 216,000 |
| 9 | 8 | 99,072 | 10 | 8 | 134,400 |
| 9 | 7 | 74,745 | 10 | 6 | 73,920 |
| 9 | 6 | 54,216 | 10 | 5 | 50,880 |
| 9 | 5 | 37,233 | 10 | 4 | 32,320 |
| 9 | 4 | 23,616 | 10 | 3 | 18,080 |
| 9 | 2 | 5832 | 10 | 2 | 8,000 |

## 2.2. Zigzag (6,3) Polyhex Nanotubes

*2.2.1. Method A*

We applied method A in the case of "zig-zag", TU**(6,3)Z**[2*p*,*q*] tubes (Figure 3), as follows:[10]

The sum from *v* to all vertices on level $m = 1$ is:

$$s_1(p,z) = 2p^2 \tag{33}$$

For levels in the range $1 < k \le p$, the distance sum is now:

$$s_m(p,s) = \sum_{k=1}^{s-1} 2(p+k) + p^2 \tag{34}$$

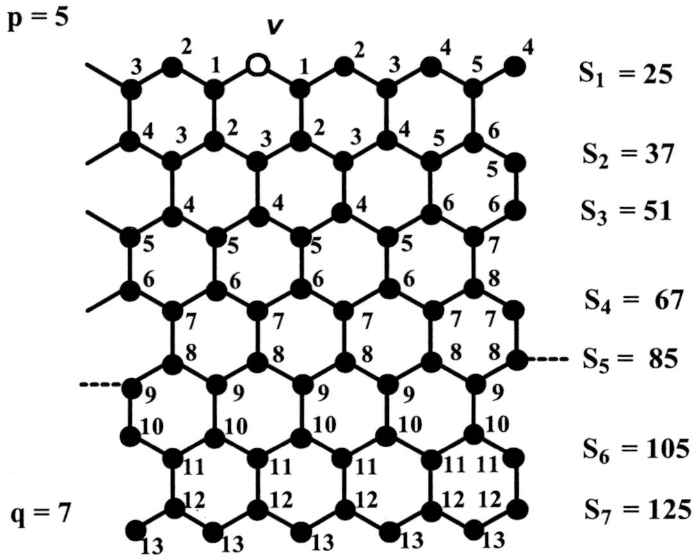


Figure 3: A "zig-zag" polyhex lattice embedded in the cylinder.

The total distance sum up to level $m$ is given by:

$$st_m(p,m) = p^2 + \sum_{s=2}^{m} s_m(p,s) \tag{35}$$

The distance sum at level $m = p$ is:

$$s_p(p) = \sum_{k=1}^{p-1} 2(p+k) + p^2 \tag{36}$$

Calculate the sum at levels $m > p$ as:

$$s_c(p,s) = s_p(p) + 4p(s-p) \tag{37}$$

and the total sum up to $m$:

$$st_c(p,m) = \sum_{s=p+1}^{m} s_c(p,s) \tag{38}$$

The total sum from $v$ located at level $m = 1$ to all vertices in TU(6,3)Z will be:

$$s_v(p,m) = st_m(p,p) + st_c(p,m) \tag{39}$$

The Wiener index of a long TU(6,3)Z[$2p,q$] is now:[10]

$$W(p,q) = \sum_{m=p+1}^{q} 2p \cdot s_v(p,m) + \sum_{m=2}^{p} 2p \cdot st_m(p,m) + 2p \cdot s_1(p,z) - d_a(p) - d_b(p,q) \tag{40}$$

The last two negative terms in the above equation have mainly the same reason as in the case of TU(6,3)A and account for the modulo ($p$,3) and $q$-dimension, respectively.

The first difference has the form:

$$d_a(p) = d_0(p)[(1-\text{mod}(p,3))(2-\text{mod}(p,3))/2] + d_1(p)[(2-\text{mod}(p,3))(\text{mod}(p,3))] + d_2(p)[(1-\text{mod}(p,3))(\text{mod}(p,3))/(-2)] \tag{41}$$

where:

$$d_0(p) = 4p^2 + (\text{trunc}(p/3))(p^3 - p) \tag{42}$$

$$d_1(p) = p^2 \lfloor 4p^2 + (\text{trunc}(p/3))(p+1) \rfloor \quad (43)$$

$$d_2(p) = p^2 \lfloor 4p^2 + (1 + \text{trunc}(p/3))(p-1) \rfloor \quad (44)$$

Evaluation of $d_a(p)$, in (41) leads to:

$$d_a(p) = (p^2/3)(13p^2 - 1) \quad (45)$$

The second difference $d_b(p,q)$, in (40) is:

$$d_b(p,q) = p^2(q-2p)[4p + (q-2p)] \quad (46)$$

Expansion of all the functions in (40) leads to the final formula for calculating *W*, in case of long tubes $q \geq p$:

$$W_{\text{TU}(6,3)\text{Z}}(p,q) = \frac{p^2}{6}\left[8q^3 + 4p^2q - 6q - p^3 + p\right] \quad (47)$$

which is identical to the formula reported in ref.[10]

Formula for short tubes (**case** $q \leq p$, TU(6,3)Z), is:

$$W_{s\text{TU}(6,3)\text{Z}}(p,q,m) = \sum_{m=2}^{q} 2p \cdot st_m(p,m) + 2p \cdot s_1(p) - (q/2)(2p \cdot s_1(p)) - d_c(p,q) \quad (48)$$

where:

$$d_c(p,q) = p\left(\sum_{k=2}^{q} k(k-1)\right) \quad (49)$$

Expansion of the above functions leads to the final formula for *W*, in case of short tubes:

$$W_{s\text{TU}(6,3)\text{Z}}(p,q) = \frac{pq}{6}\left[q^3 + 4pq^2 + 6p^2q - q - 4p\right] \quad (50)$$

For $q = 1,\ p \geq 2$, the formula for simple cycles on $2p$ vertices is recovered:

$$W(C_{2P}) = p^3 \quad (51)$$

# 3. Distance Counting in (6,3) Polyhex Tori

## 3.1. Polyhex Tori

Gluing the two ends of nanotubes results in corresponding (6,3)A and (6,3)Z polyhex tori.[8] The toroidal objects in Figure 4 represent embedding isomers, with different topology: (6,3)Z has 10 and 100 hexagons while (6,3)A, 20 and 50 hexagons, on their $c$ and $n$ dimensions, respectively.

The general form of the vertex Hosoya polynomials, in tori of type (6,3)Z[$c,n$] and (6,3)A[$c,n$], (regular graphs, having $N = c$ x $n$), in going from normal (6,3)Z to normal (6,3)A tori, are:[8],[11]

$$(6,3)\mathrm{Z}[c,n];\ n > c:$$

$$H(i,x) = 1 + 3kx^{k}{}_{,k=1,2,\ldots,(c/2-1)} + (3c/2-1)x^{c/2} + \\ + (3c/2-k)x^{(c/2+k)}{}_{,k=1,2,\ldots,(c/2-1)} + cx^{k}{}_{,k=c,c+1,\ldots,(n-1)} + (c/2)x^{n} \tag{52}$$

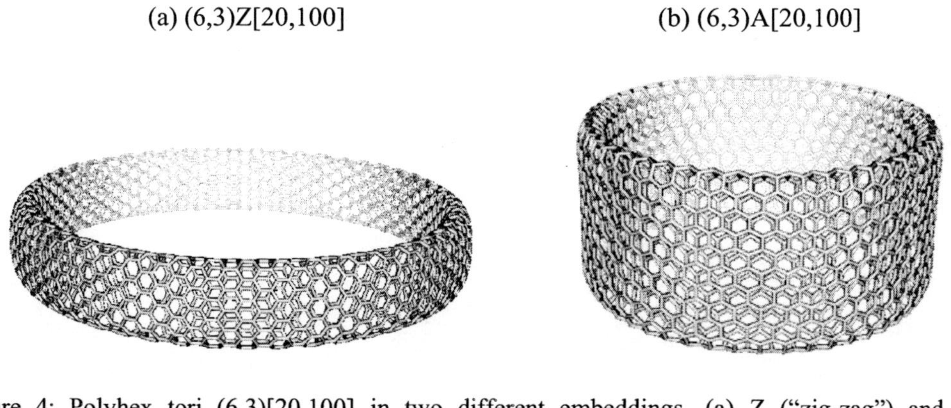

Figure 4: Polyhex tori (6,3)[20,100] in two different embeddings, (a) Z ("zig-zag") and (b) A ("armchair"), according to their cross section shape.

$$(6,3)\mathrm{Z}[c,n] = (6,3)\mathrm{A}[c,n];\ n = c:$$

$$H(i,x) = 1 + 3kx^{k}{}_{,k=1,2,\ldots,(c/2-1)} + (3c/2-1)x^{c/2} + (3c/2-k)x^{(c/2+k)}{}_{,k=1,2,\ldots,(\frac{c}{2}-1)} + \\ + (c/2)x^{n} \tag{53}$$

$$(6,3)\mathrm{A}[c,n];\ c + 2 \le n \le 2(c-1):$$

$$H(i,x) = 1 + 3kx^{k}{}_{,k=1,2,\ldots,(n/2-1)} + (3n/2-1)x^{n/2} + \\ + (3n/2-k)x^{(n/2+k)}{}_{,k=1,2,\ldots,(c-1-n/2)} + (2n-3c/2-1)x^{c} + \\ + [2(n-c)-4k]x^{(c+k)}{}_{,k=1,2,\ldots,((n-c)/2-1)} + x^{(c+n)/2} \tag{54}$$

(6,3)A[$c$,$n$]; $n = 2c$:

$$H(i,x) = 1 + 3kx^k{}_{,k=1,2,...,(c-1)} + (5c/2 - 2)x^c + (2c - 4k)x^{(c+k)}{}_{,k=1,2,...,(c/2-1)} + \\ + x^{(c+n)/2} \quad (55)$$

(6,3)A[$c$,$n$]; $n \geq 2(c+1)$:

$$H(i,x) = 1 + 3kx^k{}_{,k=1,2,...,(c-1)} + (5c/2 - 1)x^c + 2cx^k{}_{,k=c+1,c+2,...,(n/2-1)} + \\ + (2c-1)x^{n/2} + (2c - 4k)x^{(n/2+k)}{}_{,k=1,2,...,(c/2-1)} + x^{(c+n)/2} \quad (56)$$

"Normal" torus, in the above relations, means a toroidal net having the number of hexes across the tube smaller than around the torus. The "normal" status is already reached at $n > c$, in (6,3)Z tori, while $n \geq 2(c+1)$ is needed in case of (6,3)A tori.

Note that the coefficients of the vertex Hosoya polynomial are just the entries in the LC matrix (*i.e.*, Layer matrix of Counting)[12],[13] or the (vertex) Distance Degree Sequence DDS($i$) (*i.e.*, the number of vertices lying at distance $k$ form the vertex $i$).[13] Clearly, the vertex decomposition of $H(G,x)$ would be more complicated in vertex non-transitive graphs.

The polynomial coefficients can be viewed as a "distance degree" spectrum, useful in topological characterization of graphenes. In case of the (normal) (6,3)[20,$n$] series, the spectra (per vertex d($i$,$k$) values) are shown in Figures 5 and 6.

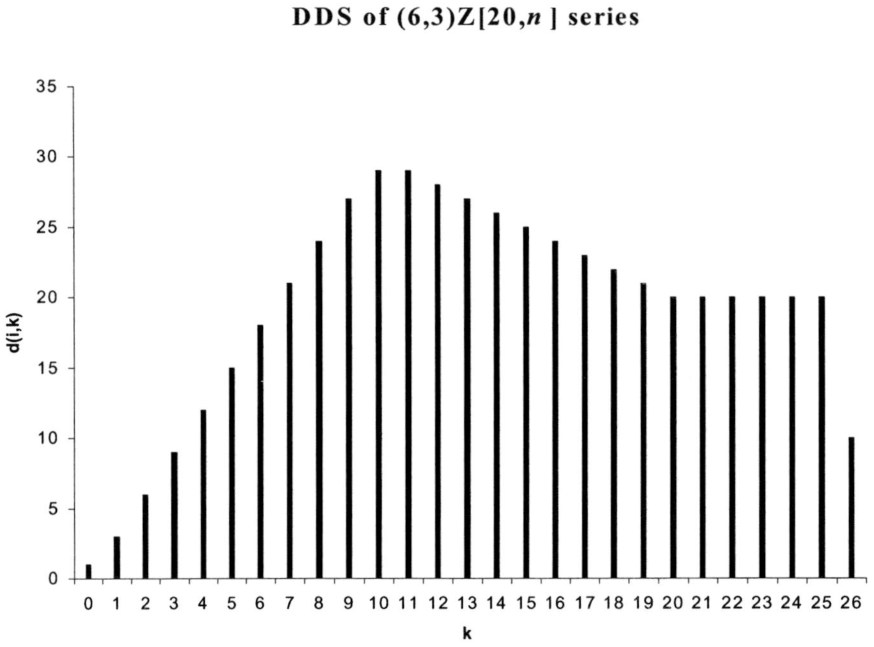


Figure 5: The distance degree spectrum of the (6,3)Z[20,$n$] tori.

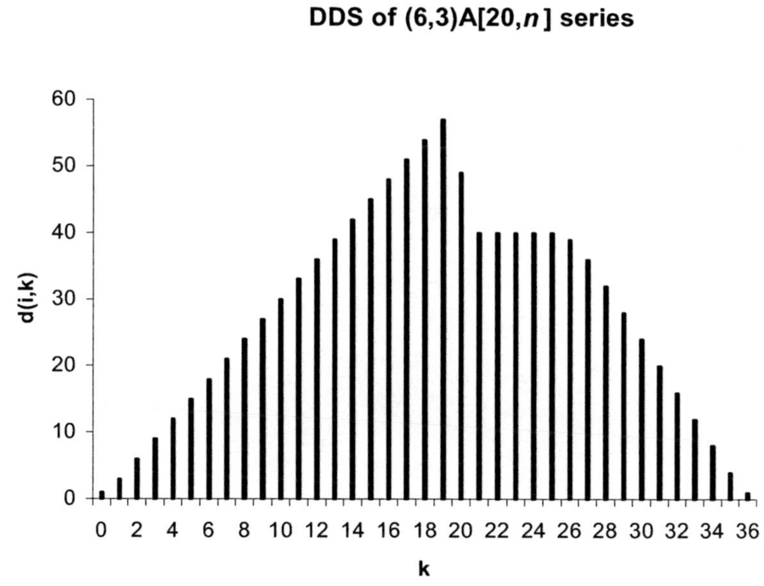


Figure 6: The distance degree spectrum of the (6,3)A[20,*n*] tori.

The repeat terms: $cx^k, _{k=c,c+1,\dots,n-1}$ and $2cx^k, _{k=c+1,c+2,\dots,n/2-1}$, respectively, are the only changes, as *n* increases, in the spectrum of a given series (*i.e.*, a series of fixed *c*). By changing the series, the spectrum will change drastically, according to the general formulas (52) -(56).

The first derivative[14] of the Hosoya polynomial (in $x = 1$) enables the calculation of the well-known Wiener[1] number *W* (eq 3).

In case of the normal series of tori, the first derivative of the Hosoya polynomial leads to:[8],[11]

(6,3)Z[*c*,*n*]:

$$W = \frac{nc}{2}\left[\sum_{k=1}^{c/2-1} 3k^2 + (3c/2-1)c/2 + \sum_{k=1}^{c/2-1}(3c/2-k)(c/2+k) + \sum_{k=c}^{n\text{-}1} ck + nc/2\right] \quad (57)$$

(6,3)A[*c*,*n*]:

$$W = \frac{nc}{2}\left[\begin{array}{l}\sum_{k=1}^{c-1} 3k^2 + (5c/2-1)c + \sum_{k=c+1}^{n/2\text{-}1} 2ck + (2c-1)n/2 + \sum_{k=1}^{c/2-1}(2c-4k)(n/2+k) + \\ +(n/2+c/2)\end{array}\right] \quad (58)$$

By expanding the sums one obtains:

$$(6,3)Z[c,n]:\ W = \frac{1}{24} nc^2 (6n^2 + c^2 - 4) \tag{59}$$

$$(6,3)A[c,n]: W = \frac{1}{24} nc^2 (3n^2 + c^2 + 3nc - 4) \tag{60}$$

Expansion of eq. 54 (case (6,3)A; $c + 2 \le n \le 2(c-1)$ ) also leads to relation (60). Moreover, the formulas for the other two cases:

$$n = c: W = \frac{c^3}{24}(7c^2 - 4) \tag{61}$$

and

$$n = 2c:\ W = \frac{c^3}{12}(19c^2 - 4) \tag{62}$$

can be deduced from the first derivative of the corresponding polynomials (eqs. 53 and 55, respectively), as well as from eq. 60. Relation (61) is also a particular case of eq (59), proving the self-consistency of the formulas (59) and (60) for calculating the Wiener index in polyhex tori.

As examples, the (vertex) Hosoya polynomials and Wiener numbers for the pair of isomers ((6,3)Z[24,72]; (6,3)A[24,72]) are given below:

(6,3)Z[24,72] (12x72 hexes):

$$H((6,3)Z[24,72]\_i,x) = 1 + 3x + 6x^2 + 9x^3 + 12x^4 + 15x^5 + 18x^6 + 21x^7 + 24x^8 + \\ 27x^9 + 30x^{10} + 33x^{11} + 35x^{12} + 35x^{13} + 34x^{14} + 33x^{15} + 32x^{16} + 31x^{17} + \\ 30x^{18} + 29x^{19} + 28x^{20} + 27x^{21} + 26x^{22} + 25x^{23} + 24x^{24-71} + 12x^{72}$$

$$W((6,3)Z[24,72]) = 5.4736128*10^7$$

(6,3)A[24,72] (24x36 hexes):

$$H((6,3)A[24,72]\_i,x) = 1 + 3x + 6x^2 + 9x^3 + 12x^4 + 15x^5 + 18x^6 + 21x^7 + 24x^8 + \\ 27x^9 + 30x^{10} + 33x^{11} + 36x^{12} + 39x^{13} + 42x^{14} + 45x^{15} + 48x^{16} + 51x^{17} + \\ 54x^{18} + 57x^{19} + 60x^{20} + 63x^{21} + 66x^{22} + 69x^{23} + 59x^{24} + 48x^{25-35} + 47x^{36} + \\ 44x^{37} + 40x^{38} + 36x^{39} + +32x^{40} + 28x^{41} + 24x^{42} + 20x^{43} + 16x^{44} + 12x^{45} + \\ 8x^{46} + 4x^{47} + x^{48}$$

$$W((6,3)A[24,72]) = 3.6820224*10^7$$

## 3.2. Twisted (6,3)Hh*t*[*C*,*N*] Tori

By twisting a horizontal row of squares in the (4,4) net (*i.e.*, connecting each square with the next column of squares), the polyhex (6,3) net, resulted by a cutting procedure,[6],[7] will also be twisted. As a consequence, a chiral object is obtained (Figure 7). In the name of objects, *t* indicates the number of twisted rows (in the range 1to *c*).

General formulas for the Hosoya polynomial and Wiener index in tori of (6,3) AH*t*[*c*,*n*] series are:[11]

Case: $(n/2) > 2\,(c+1)$; $t < c$:

$$H(i,x) = 1 + 3kx^{k}{}_{,k=1,2,...,(c-1)} + [3c-1-(c-t)/2]x^{c} + \\ + (2c+t-k+1)x^{(c+k-1)}{}_{,k=2,3,...,t} + 2cx^{k}{}_{,k=c+t,c+t+1,...,(n/2-1)} + (2c-1)x^{n/2} + \\ + (2c-4k)x^{(n/2+k)}{}_{,k=1,2,...[(c-t)/2-1]} + (t+1)x^{[n/2+(c-t)/2]} \quad (63)$$

$$W_{((6,3)\mathrm{AH}t[c,n])} = \frac{cn}{2}\Bigg[\sum_{k=1}^{c-1} 3k^{2} + [3c-1-(c-t)/2]c + \sum_{k=2}^{t}(2c+t-k+1)(c+k-1) + \\ + \sum_{k=c+t}^{n/2-1} 2ck + (2c-1)n/2 + \sum_{k=1}^{(c-t)/2-1}(2c-4k)(n/2+k)\Bigg] \quad (64)$$

$$W_{((6,3)\mathrm{AH}t[c,n])} = \frac{cn}{24}\left[c^{3} + 3c^{2}n + 3cn^{2} + 3ct^{2} + 4t^{3} - 3nt^{2} - 4c - 4t\right] \quad (65)$$

Case: $(n/2) > 2(c+1)$; $t = c$.

$$H(i,x) = 1 + 3kx^{k}{}_{,k=1,2,...,(c-1)} + (3c-1)x^{c} + (3c-k)x^{(c+k)}{}_{,k=1,2,...,(c-1)} \\ + 2cx^{k}{}_{,k=2c,2c+1,...,(n/2-1)} + cx^{n/2} \quad (66)$$

$$W_{((6,3)AH,t=c)} = \frac{cn}{2}\left[\sum_{k=1}^{c-1} 3k^{2} + (3c-1)c + \sum_{k=1}^{c-1}(3c-k)(c+k) + \sum_{k=2c}^{n/2-1} 2ck + cn/2\right] \quad (67)$$

$$W_{((6,3)AH,t=c)} = \frac{c^{2}n}{24}(8c^{2} + 3n^{2} - 8) \quad (68)$$

(6,3) HH*2*[10,44] (detail)

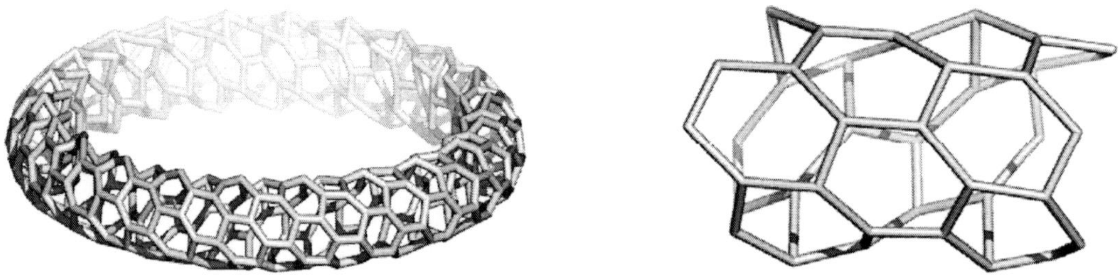

Figure 7: A twisted (chiral) (6,3)HH$t$ torus, in the HH ("horizontal twist", "horizontal cut") embedding.

# 4. WIENER INDEX OF ((4,8)3) NANOTUBES

A ((4,8)3) net is a trivalent decoration consisting of alternating squares and octagons. It can cover either a cylinder or a torus. Such a covering can be derived, from the (4,4) net, either by a cutting procedure[15] or by the leapfrog operation.[6] It also could appear by cascade Stone-Wales[16] edge flipping in a polyhex (6,3) net.[17] Optimized ((4,8)3) nanotubes are illustrated in Figure 8.

(a) TU((4,8)3)S[28,20] (b) TU((4,8)3)R[24,10]

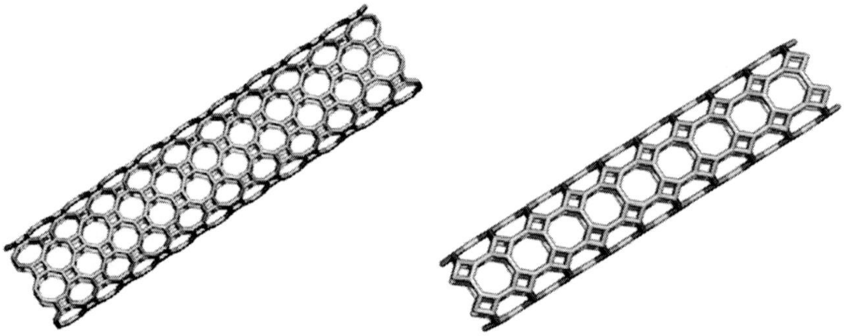

Figure 8: Nanotubes covered by ((4,8)3) nets, in (a) S (square) and (b) R (rhomb) embeddings, respectively.

## 4.1. Wiener Index of ((4,8)3)S (Square Tiled) Nanotubes

*Method.*[18] The method for deriving analytical formulas which calculate the Wiener index in ((4,8)3)S nanotubes is similar to that developed in refs.[9],[10]

Let us denote by $p$ the number of squares at level 1 in the tube and by $k$, $m$, $q$ the various levels (*i.e.*, the length) of the tube (Figure 9).

The sum from a vertex $v$ lying at level 1 to all other vertices on the same level 1 is given by:

$$s_1(p) = 4p^2 \tag{69}$$

There are two types of vertices: those not located on a square and vertices that lye on a square. For levels in the range $1 \le k \le p+1$, there are two different distance sums from vertex $v$ to all vertices lying at level $k$:

If $v$ does not belong to a square, then:

$$s_{0k}(p,k) = 4p^2 + 2(k-1)(2p+k) \tag{70}$$

otherwise,

$$s_{1k}(p,k) = 4p^2 + 2(k-1)(2p+k-2) \tag{71}$$

The distance from all vertices lying at level 1 to vertices at level $k$, $1 \le k \le p+1$ is given by:

$$s_k(p,k) = p[2 \cdot s_{0k}(p,k) + 2 \cdot s_{1k}(p,k)] = 8p \cdot [2p^2 + (k-1)(2p+k-1)] \tag{72}$$

The distance from level 1 to level $p+1$ is:

$$s_{p+1}(p) = 40p^3 \tag{73}$$

The total distance sum from level 1 to all levels up to $m$, $1 \le m \le p+1$ is:

$$st_m(p,m) = \sum_{k=1}^{m} s_k(p,k) = 4pm \cdot [4p^2 + \frac{1}{3}(m-1)(2m-1+6p)] \tag{74}$$

The total distance sum from level 1 to levels until $p+1$, is:

$$st_{p+1}(p) = 4p^2(p+1)[4p + \frac{1}{3}(8p+1)] \tag{75}$$

Calculate now the distance sums from all vertices lying on level 1, as follows:

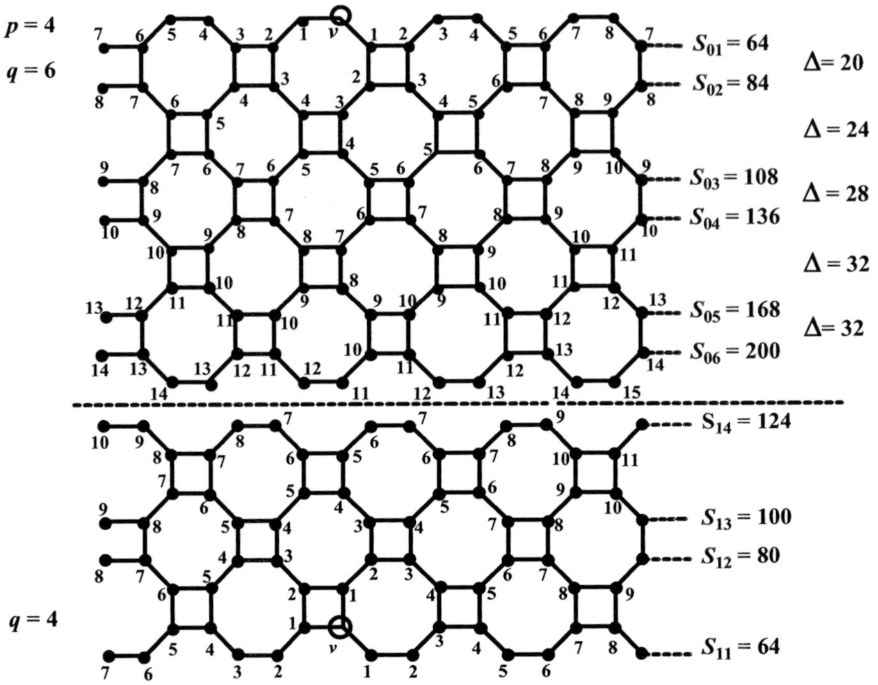


Figure 9: Distance sum from a vertex $v$ to vertices lying at levels $k$ =1..6 and 1..4.

The distance sums to vertices on levels $k > p+1$:

$$sl_k(p,k) = 40p^3 + 32p^2(k-p-1) \tag{76}$$

The total distance sums to all vertices in a tube having the length $m$, $m>p+1$:

$$stl_m(p,m) = \sum_{k=1}^{p+1} s_k(p,k) + \sum_{k=p+2}^{m} sl_k(p,k) \tag{77}$$

$$stl_m(p,m) = st_{p+1}(p) + 40p^3(m-p-1) + 16p^2(m-p)(m-p-1) \tag{78}$$

By substituting (75) in (78) it becomes:

$$stl_m(p,m) = 4/3 \cdot p^2(12 \cdot m^2 + 6 \cdot m \cdot p - 12m + 2p^2 + 3p + 1) \tag{79}$$

Now we are ready to calculate the Wiener index in a short tube, TU((4,8)3)S[4$p$,$q$] with the length $q \le p+1$, as:

$$W_s(p,q) = \frac{1}{2}[2 \cdot \sum_{m=1}^{q} st_m(p,m) - q \cdot 4p \cdot s_1(p)] \tag{80}$$

and after replacing $st_m(p,m)$ as in (74) and $s_1(p)$, the final formula is:

$$W_s(p,q) = \frac{2}{3} \cdot q \cdot p(q+1)(q^2 + 4 \cdot q \cdot p - q - 4p + 12 \cdot p^2) - 8 \cdot q \cdot p^3 \tag{81}$$

The subtraction of the last term in (79) is reasoned as follows: the reference vertex $v$ may be located at any level $m$, $1 \le m \le p+1$, each time looking at TU((4,8)3)S[4$p$,$q$] as being obtained by two smaller tubes sharing a common level, namely that containing the vertex $v$. It is obvious that the actual level of $v$ is counted twice.

If $q = p+1$, the formula (14) for calculating the Wiener index becomes:

$$W_{p+1}(p) = \frac{2}{3} \cdot p^2 (p+1)(17 \cdot p^2 + 23p + 2) \tag{82}$$

The Wiener index in a long TU((4,8)3)S[4$p$,$q$], $q > p+1$ is:

$$\begin{aligned} W(p,q) &= \frac{1}{2}[2\sum_{m=1}^{p+1} st_m(p,m) + 2\sum_{m=p+2}^{q} stl_m(p,m) - q \cdot 4p \cdot s_1(p)] \\ &= W_{p+1}(p) + \sum_{m=p+2}^{q} stl_m(p,m) - 8 \cdot p^3 \cdot (q-p-1) \end{aligned} \tag{83}$$

and, after replacing $W_{p+1}(p)$ as in (82) and $stl_m(p,m)$ as in (77), the final formula is:

$$W(p,q) = \frac{2}{3} \cdot p^2 (8 \cdot q^3 - p^3 + 4 \cdot p^2 \cdot q + 6 \cdot p \cdot q^2 - 6 \cdot q + p) \tag{84}$$

Numerical data for Wiener index in tubes TU((4,8)3)S[4$p$,$q$] of various dimensions are given in Tables 3 and 4 (see eqs. (81) and (84)).

**Table 3: Wiener index in short tubes, *s*TU((4,8)3)S[4*p*,*q*],** $q \leq p+1$

| *p* | *q* | *W* | *p* | *q* | *W* |
|---|---|---|---|---|---|
| 4 | 2 | 2336 | 5 | 2 | 4440 |
| 4 | 3 | 5824 | 5 | 4 | 20800 |
| 4 | 4 | 11392 | 5 | 5 | 35000 |
| 4 | 5 | 19520 | 5 | 6 | 54200 |
| 6 | 3 | 18144 | 7 | 3 | 28168 |
| 6 | 4 | 34368 | 7 | 4 | 52864 |
| 6 | 5 | 57120 | 7 | 6 | 132104 |
| 6 | 6 | 87408 | 7 | 7 | 189336 |
| 8 | 5 | 126080 | 7 | 8 | 260288 |
| 8 | 6 | 190016 | 3 | 2 | 1032 |
| 8 | 8 | 369664 | 3 | 3 | 2664 |
| 8 | 9 | 489216 | 3 | 4 | 5376 |

**Table 4: Wiener index in long tubes, TU((4,8)3)S[4p,q],** $q \geq p+1$**.**

| *p* | *q* | *W* | *p* | *q* | *W* |
|---|---|---|---|---|---|
| 3 | 6 | 15192 | 5 | 6 | 54200 |
| 4 | 5 | 19520 | 5 | 7 | 79200 |
| 4 | 6 | 30720 | 5 | 8 | 110800 |
| 4 | 7 | 45504 | 5 | 9 | 149800 |
| 7 | 8 | 260288 | 5 | 10 | 197000 |
| 7 | 9 | 346528 | 5 | 13 | 395800 |
| 8 | 10 | 631296 | 5 | 20 | 1296000 |

## 4.2. Wiener Index Of ((4,8)3)R (Rhomb Tiled) Nanotubes

**Method.[18]** We denote by *p* the number of rhombs on the level 1 and by *k, m, q* the various length of the tube (Figure 10).We thus obtain two formulas: one for $q \leq [p+1)/2]$ and one for $q \geq [p+1)/2]$ . There are three types of vertices at level 1: vertices lying on the outside part of the rhomb, $v_1$, vertices that are on the inside part of the rhomb, $v_2$, and vertices located on an octagon, $v_3$.

The sum of distances from $v_1$ and $v_2$ to all vertices lying at level $k = 1$ is given by:

$$s_{11}(p) = 3 \cdot p^2 + 1 + \frac{[1+(-1)^p]}{2} \tag{85}$$

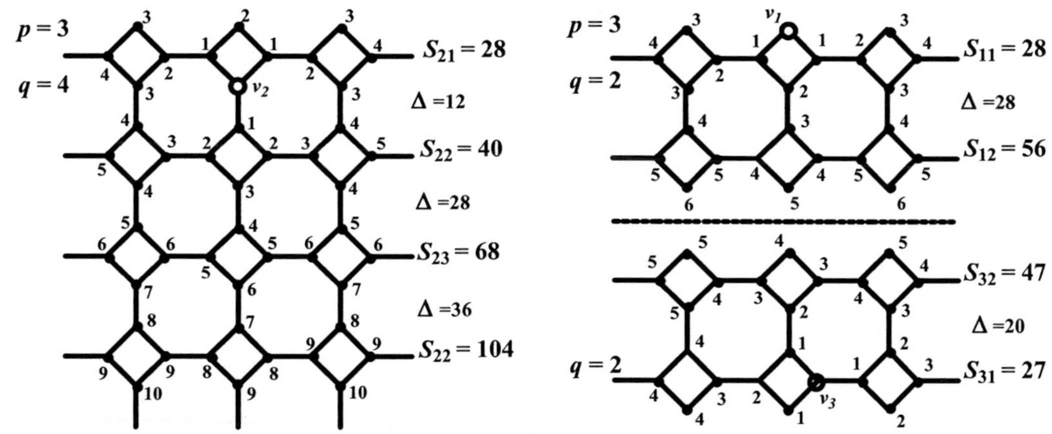


Figure 10: Distance sums from a vertex $v$ to vertices lying at levels $k$ = 1...4 and 1...2.

The sum from $v_3$ (which is on the octagon) to all vertices lying at level

$$k = 1 \text{ is: } s_{13}(p) = 3 \cdot p^2 \tag{86}$$

The sum from all vertices on the level 1 to all others vertices at the same level 1 is:

$$s_1(p) = 2 \cdot p(s_{11}(p) + s_{13}(p)) = 2 \cdot p(6 \cdot p^2 + 1 + \frac{[1+(-1)^p]}{2}) \tag{87}$$

For levels in the range $1 \le k \le \text{int}[(p+1)/2]$, there are three different distance sums from level 1 to level $k$:

From vertex $v_1$:

$$s_{1k}(p,k) = 3p^2 + 1 + \frac{[1+(-1)^p]}{2} + 4(k-1)(2k+p) \tag{88}$$

From vertex $v_2$:

$$s_{2k}(p,k) = 3p^2 + 1 + \frac{[1+(-1)^p]}{2} + 4(k-1)(2k+p-4) \tag{89}$$

From vertex $v_3$:

$$s_{3k}(p,k) = 3p^2 + 4(k-1)(2k+p-2) \tag{90}$$

The distance from all vertices lying at level 1 to vertices at level $k$, $1 \le k \le \mathrm{int}[(p+1)/2]$ is given by:

$$s_{ks}(p,k) = p[s_{1k}(p,k) + s_{2k}(p,k) + 2 \cdot s_{3k}(p,k)] \\ = p[12 \cdot p^2 + 3 + (-1)^p + 16(k-1)(2k+p-2)] \tag{91}$$

The distance sum from level 1 to all levels up to $m$, where $1 \le m \le \mathrm{int}[(p+1)/2]$ is:

$$st_{ms}(p,m) = \sum_{k=1}^{m} s_{ks}(p,k) = \frac{1}{3} p \cdot m \cdot [36p^2 + 24 \cdot p \cdot m + 32m^2 - 48m - 24p + \\ + 25 + 3 \cdot (-1)^p] \tag{92}$$

The Wiener index for short tubes (length $q$ in the range $1 \le q \le \mathrm{int}[(p+1)/2]$) is calculated as:

$$W_s(p,q) = \frac{1}{2}[2 \cdot \sum_{m=1}^{q} st_{ms}(p,m) - q \cdot s_1(p)] \tag{93}$$

and, after replacing $st_{ms}(p,m)$ as in (92) and $s_1(p)$ as in (87), the final formula is:

$$W_s(p,q) = \frac{1}{6} \cdot q \cdot p(16 \cdot q^3 + 16 \cdot p \cdot q^2 + 36 \cdot p^2 \cdot q + 3 \cdot q \cdot (-1)^p - 7 \cdot q - 16 \cdot p) \tag{94}$$

For calculating the Wiener index for long tubes, we derived formulas for three different tube length:

**Case (i):** $q = \mathrm{int}[(p+1)/2] + 1$.

The distance from all vertices lying at level 1 to the last level:

$$s_{klp1}(p) = p[28 \cdot p^2 + 12p[1 - (-1)^p] + 3 \cdot (-1)^p - 1] \tag{95}$$

The distance sum from level 1 to all levels up to $m = \mathrm{int}[(p+1)/2] + 1$ is

$$st_{mp1}(p) = p[\frac{28}{3} \cdot p^3 + 27p^2 - 7p^2(-1)^p + \frac{79}{6} p - \frac{17}{2} p \cdot (-1)^p - \frac{1}{2} + \frac{5}{2}(-1)^p] \tag{96}$$

The Wiener index of a tube of length $q = \text{int}[(p+1)/2]+1$ is

$$W_{p1}(p) = W_s(p, \text{int}[(p+1)/2]) + st_{mp1}(p) - \frac{s_1(p)}{2} \tag{97}$$

$$W_{p1}(p) = \frac{p}{24}[48p^4 + 280p^3 - 56p^3(-1)^p + 507p^2 - 207p^2(-1)^p + 302p - 190p(-1)^p - 51[1-(-1)^p]] \tag{98}$$

**Case (ii):** $q = \text{int}[(p+1)/2]+2$.

Analogously, the Wiener index of such tubes is calculated as:

$$W_{p2}(p) = \frac{p}{24}[48p^4 + 504p^3 - 56p^3(-1)^p + 1683p^2 - 375p^2(-1)^p + 2058p - 682p(-1)^p - 147[1-(-1)^p]] \tag{99}$$

**Case (iii):** $q > \text{int}[(p+1)/2]+2$.

For tubes of the third length, the distance from all vertices lying at level 1 to vertices at level $k \geq \text{int}[(p+1)/2]+2$ is:

$$s_{kl}(p,k) = p \cdot [28 \cdot p^2 - 12p(-1)^p + 60p - 2[1-(-1)^p] + 12 \cdot 4 \cdot p[k - \frac{2p+1-(-1)^p+8}{4}]] \tag{100}$$

The distance sum from level 1 to all levels until level *m*, where $m \geq \text{int}[(p+1)/2]+2$ is

$$st_{ml}(p,n) = p \cdot [\frac{4}{3}p^3 + 4p^2m + 24pm^2 + 4p^2 - 24pm - 2m[1-(-1)^p] + \frac{31}{6}p - \frac{1}{2}p(-1)^p - \frac{1}{2}(-1)^p + \frac{5}{2}] \tag{101}$$

If we take $l_2 = \text{int}[(p+1)/2]+2 = \dfrac{2p+1-(-1)^p+8}{4}$ then the Wiener index will be:

$$W_l(p,q) = W_{p2}(p,q) + \sum_{m=l_2+1}^{q} st_{ml}(p,m) - \frac{(q-l_2)\cdot s_1(p)}{2} \tag{102}$$

and, after replacing $W_{p2}(p,q)$ from (99), $st_{ml}(p,m)$ from (101) and $s_1(p)$ from (97), the final formula is:

$$W_l(p,q) = \frac{-p}{24} \cdot [4p^4 - 48p^2q^2 - 192pq^3 - 32p^3q + 24 \cdot q^2[1-(-1)^p] - p^2 - 3p^2(-1)^p + 68pq + 12pq(-1)^p - 3[1-(-1)^p]) \quad (103)$$

**Table 5: Wiener index in TU((4,8)3)R nanotubes**

| *p* | *q* | *W(p,q)* | *p* | *q* | *W(p,q)* |
|---|---|---|---|---|---|
| 4 | 2 | 1952 | 6 | 2 | 6000 |
| 4 | 3 | 5312 | 6 | 3 | 15228 |
| 4 | 4 | 11232 | 6 | 4 | 30516 |
| 4 | 5 | 20480 | 6 | 5 | 53580 |
| 4 | 8 | 75872 | 6 | 6 | 86148 |
| 4 | 10 | 143520 | 6 | 7 | 129948 |
| 5 | 1 | 755 | 7 | 2 | 9268 |
| 5 | 2 | 3580 | 7 | 3 | 23065 |
| 5 | 3 | 9355 | 7 | 4 | 45360 |
| 5 | 4 | 19210 | 7 | 5 | 78407 |
| 5 | 5 | 34345 | 7 | 6 | 124558 |
| 5 | 8 | 123430 | 7 | 7 | 186165 |
| 5 | 9 | 171685 | 7 | 10 | 487242 |
| 8 | 3 | 33424 | 9 | 3 | 46359 |
| 8 | 4 | 64768 | 9 | 4 | 88848 |
| 8 | 5 | 110464 | 9 | 5 | 149895 |
| 8 | 6 | 173568 | 9 | 7 | 342837 |
| 8 | 7 | 257152 | 9 | 9 | 656163 |
| 8 | 8 | 364288 | 9 | 10 | 867690 |
| 8 | 10 | 661504 | | | |

Note that the final formula (103) includes both formula (99) (which calculates $W_{p2}(p)$ for the tube length $q = l_2 = \dfrac{2p+1-(-1)^p+8}{4}$) and (98) (in the calculation of $W_{p1}(p)$, for the tube length $q = l_1 = [2p+1-(-1)^p+4]/4$). Examples are given in Table 5 (see eqs. (94) and (103)).

## 5. Distance Counting in ((4,8)3)S (Square Tiled) Tori

The ((4,8)3)S (square tiled) toroidal objects show the same embedding types as the corresponding open tubes (Figure 11).

According to the construction by cutting procedures,[15] the ((4,8)3)HS[*c*,*n*] isomer contains, on dimension "*c*", half of the number of (4,8) pairs in ((4,8)3)VS[*c*,*n*]. Conversely, the number of (4,8) pairs in ((4,8)3)VS[$c$,$n$], on dimension "$n$", is half of that in ((4,8)3)HS[$c$,$n$]. In other words, the "H" embedding isomer will expand, after optimization, in a horizontal plane (*i.e.*, perpendicularly to the z axis) while the "V" isomer, in a vertical plane. The above symbols also remind the location of the cut edges in the (4,4) net, by the cutting procedure. The same is true for the other lattices below.

(a) ((4,8)3)HS[20,100]

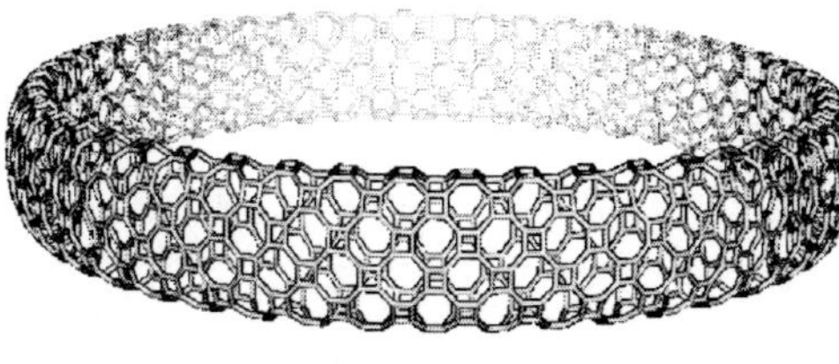

(b) ((4,8)3)VS[20,100]

Figure 11: Toroidal ((4,8)3)S structures, in HS ("horizontal square") and VS ("vertical square") embeddings, respectively.

General formulas for the Hosoya polynomial and Wiener index in tori ((4,8)3)S[$c$,$n$] of both H- and V-series are given in the following.[15] Several cases can be delimited:

## 5.1. Tori ((4,8)3)HS[$C$,$N$]; 0 Mod($C$,4)

(1) **Case**: $c < n$; **true** ((4,8)3)HS[$c$,$n$] **torus**; $c = 4p$; $n = 2r$

$$\begin{aligned} H(i,x) = 1 + m_k x^k{}_{,k=1,\dots,(2p-1)} + (m_{2p}-1)x^{2p} + (m_{(2p+k)} - 4k)x^{(2p+k)}{}_{,k=1,\dots,(p-1)} \\ + 4px^k{}_{,k=3p,\dots,(2r-1)} + (4p-1)x^{2r} + (4p-4k)x^{(2r+k)}{}_{,k=1,\dots,(p-1)} + \\ + x^{(2r+p)} \end{aligned} \tag{104}$$

$$\begin{aligned} W_{((4,8)3)HS[4p,2r]} = 4pr\Bigg[ \sum_{k=1}^{2p-1} m_k k + (m_{2p}-1)2p + \sum_{k=1}^{p-1}(m_{(2p+k)} - 4k)(2p+k) + \\ + \sum_{k=3p}^{2r-1} 4pk + (4p-1)2r + \sum_{k=1}^{p-1}(4p-4k)(2r+k) + (p+2r) \Bigg] \end{aligned} \tag{105}$$

where $m_k$ is defined by the following recursion:

$$\begin{aligned} & m_0 = 0;\ \ m_1 = 3;\ \ m_2 = 5, \ \ and \\ & m_k = m_{\left(\frac{k-\text{mod}(k,2)}{2}+1\right)} + m_{\left(\frac{k+\text{mod}(k,2)}{2}-1\right)} = m_{\left\lfloor \frac{k+2}{2} \right\rfloor} + m_{\left\lfloor \frac{k-1}{2} \right\rfloor} \end{aligned} \tag{106}$$

with $\lfloor x \rfloor$ being the greatest integer part of a real number $x$;
or, for $k = 3s + t$, and $t = 0, 1, 2$ ,

$$m_k = \frac{2 \cdot 3^{1/2} \sin\left(2k\frac{\pi}{3}\right)}{9} + 8k/3 = 8s + \begin{cases} 0, \text{ if } t = 0 \\ 3, \text{ if } t = 1 \\ 5, \text{ if } t = 2 \end{cases} \tag{107}$$

With (107), recursion (106) becomes:

$$\begin{aligned} m_k &= 8[(k - \text{mod}(k,3))/3] + 3\,\text{mod}(k,3)(2 - \text{mod}(k,3)) - \\ &- [(1 - \text{mod}(k,3))(5\,\text{mod}(k,3)/2] \end{aligned} \tag{108}$$

Expansion of relation (105) leads to the formula:

$$W_{((4,8)3)\text{HS}[4p,2r]} = \frac{16}{3} p^2 r(p^2 + 6r^2 + 3pr - 1) \tag{109}$$

which can be translated to $c$, $n$ dimensions as:

$$W_{((4,8)3)\text{H }S[c,n]} = c^2 n(c^2 + 24n^2 + 6cn - 16)/96 \tag{109'}$$

(2) **Case**: $c = n$; ((4,8)3)HS[$c$,$c$] = V((4,8)3)S[$c$,$c$], or ((4,8)3)HS[4$p$,4$p$] = ((4,8)3)VS[4$p$,4$p$].

$$\begin{aligned} H(i,x) = 1 + m_k x^k{}_{,k=1,\ldots,(2p-1)} + (m_{2p} - 1)x^{2p} + (m_{(2p+k)} - 4k)x^{(2p+k)}{}_{,k=1,\ldots,(p-1)} \\ + 4px^k{}_{,k=3p,\ldots,(4p-1)} + (4p-1)x^{4p} + (4p - 4k)x^{(4p+k)}{}_{,k=1,\ldots,(p-1)} + x^{5p} \end{aligned} \tag{110}$$

The summation of the corresponding terms, after appropriate handling, leads to the following simple relation for the Wiener index:

$$W(p) = \frac{32}{3} p^3 (31p^2 - 1) \tag{111}$$

Or

$$W(c) = c^3(31c^2 - 16)/96 \tag{111'}$$

(3) **Case**: $c > n$; ((4,8)3)HS[$c$,$n$]; this case turns to ((4,8)3)VS[$n$,$c$].

## 5.2. Tori ((4,8)3)VS[*C*,*N*]; 0 Mod(*N*,4); *C* = 2*p*; *N* = 4*r*

(1) **Case**: $2p < 4r < 4p$. Two additional parameters are needed:

$$d = 3r - 2p \tag{112}$$

$$s = \left| 2p - 2r \right| \tag{113}$$

Case: $d < 0$

$$\begin{aligned} H(i,x) = 1 + m_k x^k{}_{,k=1,\dots,(2r-1)} + (m_{2r} - 1)x^{2r} + (m_{(2r+k)} - 4k)x^{(2r+k)}{}_{,k=1,\dots,(r-1)} \\ + n \cdot x^k{}_{,k=3r,\dots,(2p-1)} + (4r-1)x^{2p} + (4r-4k)x^{(2p+k)}{}_{,k=1,\dots,(r-1)} + \\ + x^{(2p+r)} \end{aligned} \tag{114}$$

**Case**: $d > 0$

$$\begin{aligned} H(i,x) = 1 + m_k \cdot x^k{}_{,k=1,..,(2r-1)} + (m_{2r} - 1) \cdot x^{2r} + (m_{(2r+k)} - 4k) \cdot x^{(2r+k)}{}_{,k=1,..,(2p-2r-1)} \\ + (m_{2p} - (4s+1)) \cdot x^{2p} + (m_{(2p+k)} - 4(s+2k)) \cdot x^{(2p+k)}{}_{,k=1,..,d} \\ + (4r-4k) \cdot x^{(2p+k)}{}_{,k=(d+1),..,(r-1)} + x^{(2p+r)} \end{aligned} \tag{115}$$

(2) **Case**: $n = 2c$

$$\begin{aligned} H(i,x) = 1 + m_k x^k{}_{,k=1,\dots,(2p-1)} + (m_{2p} - 2)x^{2p} + \\ + (m_{(2p+k)} - 8k)x^{(2p+k)}{}_{,k=1,\dots,(p-1)} + x^{3p} \end{aligned} \tag{116}$$

The Wiener index corresponding to relation (116) is:

$$W(p) = \frac{p^3}{6}(436p^2 - 19) \tag{117}$$

$$W(c) = 2c^3(109c^2 - 16)/96 \tag{117'}$$

(3) **Case**: $4p < 4r < 6p$. The additional $d$ parameter is of the form:

$$d = 3p - 2r \tag{118}$$

while $s$ is the same as in (113).

$$\begin{aligned} H(i,x) &= 1 + m_k \cdot x^k{}_{,k=1,..,(2p-1)} + (m_{2p} - 1)\cdot x^{2p} + \\ &+ (m_{(2p+k)} - 4k)\cdot x^{(2p+k)}{}_{,k=1,..,(2r-2p-1)} \\ &+ (m_{2r} - (4s+1))\cdot x^{2r} + (m_{(2r+k)} - 4(s+2k))\cdot x^{(2r+k)}{}_{,k=1,..,d} \\ &+ (4r - 4k)\cdot x^{(3p+k)}{}_{,k=1,..,(r-1)} + x^{(p+2r)} \end{aligned} \tag{119}$$

(4) **Case**: $6p \le 4r$; true ((4,8)3)VS[$2p$,$4r$] torus.

$$\begin{aligned} H(i,x) = 1 &+ m_k x^k{}_{,k=1,...,(2p-1)} + (m_{2p} - 1)x^{2p} + (m_{(2p+k)} - 4k)x^{(2p+k)}{}_{,k=1,...,(p-1)} \\ &+ 4x^k{}_{,k=3p,...,(2r-1)} + (4p-1)x^{2r} + (4p-4k)x^{(2r+k)}{}_{,k=1,...,(p-1)} + x^{(p+2r)} \end{aligned} \tag{120}$$

$$\begin{aligned} W_{((4,8)3)VS[2p,4r]} = 4pr\Bigg[ &\sum_{k=1}^{2p-1} m_k k + (m_{2p} - 1)2p + \sum_{k=1}^{p-1}(m_{(2p+k)} - 4k)(2p+k) + \\ &+ \sum_{k=3p}^{2r-1} 4pk + (4p-1)2r + \sum_{k=1}^{p-1}(4p-4k)(2r+k) + (p+2r)\Bigg] \end{aligned} \tag{121}$$

By developing (121) results in the formulas:

$$W_{((4,8)3)VS[2p,4r]} = \frac{16}{3}p^2 r(p^2 + 6r^2 + 3pr - 1) \tag{122}$$

$$W_{((4,8)3)VS[c,n]} = c^2 n(2c^2 + 3n^2 + 3cn - 8)/24 \tag{122'}$$

Observe that relations [(105), (121)] and [(109), (122)] are identical. This is a consequence of the isotropy of the ((4,6)3)S net. In the above, *true torus* means the torus having the number of (4,8) pairs around the tube smaller than around the torus. Formulas (109), (122), (109'), and (122') are applicable particularly in true tori. On domains where no close formula is given, the following relations (accounting for the involved embedding isomers) are useful:[15]

$$W_{((4,8)3)HS[c,n]} = W_{((4,8)3)VS[c/2,2n]} \tag{123}$$

$$W_{((4,8)3)VS[c,n]} = W_{((4,8)3)HS[2c,n/2]} = W_{((4,8)3)VS[n/2,2c]} \tag{124}$$

Always an equivalent torus will be found, to verify the true result. Table 6 gives examples of Wiener index calculation, along with some of the equivalent tori.

**Table 6: Wiener index in ((4,8)3)S tori**

| Torus ((4,8)3)S | W | [*p*,*r*] | Equivalent torus |
|---|---|---|---|
| H[12,12] | 80,064 | [3,6] | V[12,12]* V[6,24] |
| H[12,14] | 122,640 | [3,7] | V[6,28] |
| H[12,16] | 178,176 | [3,8] | V[6,32] |
| H[12,24] | 564,480 | [3,12] | V[6,48] |
| H[24,6]* | 80,064 | [3,6] | V[6,24] |
| H[24,8]* | 155,136 | [4,6] | V[8,24] |
| H[24,10]* | 264,000 | [5,6] | V[10,24] |
| H[24,12]* | 413,568 | [6,6] | V[12,24] |
| V[12,12]* | 80,064 | [3,6] | V[6,24] |
| V[12,20] | 264,000 | [6,5] | V[10,24] |
| V[16,20]* | 563,200 | [5,8] | V[10,32] |
| V[16,24]* | 864,768 | [6,8] | V[12,32] |
| V[16,28] | 1,254,400 | [7,8] | V[14,32] |
| V[16,32] | 1,744,896 | [8,8] | V[16,32] |
| V[8,20] | 96,000 | [4,5] | V[10,16] |
| V[8,24] | 155,136 | [4,6] | V[12,16]* |
| V[8,28] | 234,752 | [4,7] | V[14,16]* |
| V[12,36] | 1,180,224 | [6,9] | V[18,24]* |
| V[12,40] | 1,564,800 | [6,10] | V[20,24]* |

* values non-calculable by eqs (109), (122), (109’), and (122’)

# 6. Wiener Index in ((5,7)3)SP (Spiral) Tori

A ((5,7)3)SP “spiral” covering can be derived from a square (4,4) net by switching and deleting appropriate edges or by Stone-Wales transformation of a hexagonal (6,3) net.[17],[19],[20] (Figure 12). The “SP” specification differentiates this type of covering (with the (5,7) pairs disposed in a spiral, in optimized lattice) from other, non-spiral, ((5,7)3) nets.

H/V embedding isomers are again possible. In ((5,7)3)HSP[*c*,*n*] tori, each (5,7) pair takes exactly four squares in the parent (4,4) net, so that $c/4$ such pairs lie around the tube. The *n*-dimension is, in this case, preserved. Conversely, in ((5,7)3)VSP[*c*,*n*] tori, the constant dimension is *c*, while the number of pairs around the torus equals $n/4$.

The total number of atoms is $c \times n$. After a Molecular Mechanics optimization, a ((5,7)3)VSP[*c*,*n*] torus looks as in Figure 12. We limit here to the ((5,7)3)HSP[*c*,*n*] tori.

((5,7)3)VSP[6,56] (detail)

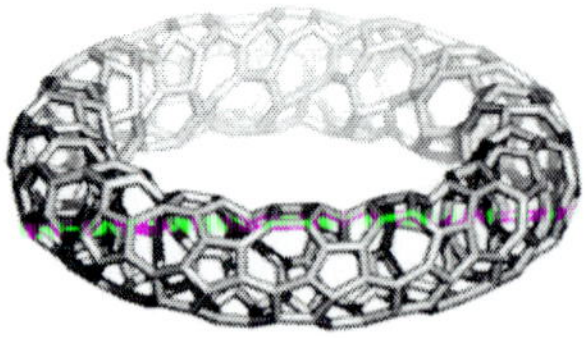

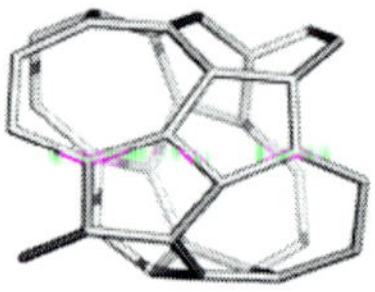

Figure 12: An optimized (5,7)3) net, in VSP ("vertical spiral") toroidal embedding.

We were not able to find neither a closed formula nor a recursion for calculating the Wiener index in ((5,7)3)SP[*c*,*n*] tori. We have succeeded, however, to find an approximate solution, only for the series ((5,7)3)HSP[*c*,*n*]:

$$W_{((5,7)3)HS[c,n]} = (cn/8)(\Delta_W + 2cn^2) \quad (125)$$

with the increment $\Delta_W$:

$$\Delta_W = 0.3369 \cdot c^{3.0593} \quad (126)$$

Relation (126) was obtained in a regression equation, with $R^2 = 0.9999$. Table 7 lists some *W* values for the ((5,7)3)HSP[8,*n*] tori.

## 7. Distance Counting in Naphthylenic Tori

Diudea has recently proposed a naphthylenic pattern[21], with the sequence: $C_6$, $C_6$, $C_4$, ..., $C_6$, $C_6$, $C_4$. The naphthylenic nets, embedded in the torus, are denoted: ((4,6,8)3)HN, ((4,6,8)3)VN, ((4,6,8)3)HNX and ((4,6,8)3)VNX (Figure 13). The local signature for ((4,6,8)3)N is: $t_{4j}(0, 2, 2)$; $t_{6j}(1, 3, 2)$; and $t_{8j}(2, 4, 0)$ while for ((4,6,8)3)NX, $t_{4j}(0, 4, 0)$; $t_{6j}(1, 3, 2)$; and $t_{8j}(0, 8, 0)$, with $j = 4, 6, 8$.

The X net originates in a rhomboidal ((4,8)3)R "bathroom floor" lattice, from which it can be obtained by the leapfrog operation.[21]

The Hosoya polynomial can be expressed either by general, closed formulae or by recursive formulae. The last one is the case of naphthylenic nets.

Let us introduce the difference Hosoya polynomial for a homologous series of graphs (*i.e.*, tori in which the tube girth is consecutively increased by one repeating unit) as:[11],[21]

**Table 7: Wiener index *W* in ((5,7)3)HSP[*c*,*n*] tori.**

| *c* | *n* | *W* | *c* | *n* | *W* |
|---|---|---|---|---|---|
| 8 | 22 | 174614 | 8 | 62 | 3825214 |
| 8 | 24 | 225816 | 8 | 64 | 4206656 |
| 8 | 26 | 286234 | 8 | 66 | 4612674 |
| 8 | 28 | 356636 | 8 | 68 | 5044036 |
| 8 | 30 | 437790 | 8 | 70 | 5501510 |
| 8 | 32 | 530464 | 8 | 72 | 5985864 |
| 8 | 34 | 635426 | 8 | 74 | 6497866 |
| 8 | 36 | 753444 | 8 | 76 | 7038284 |
| 8 | 38 | 885286 | 8 | 78 | 7607886 |
| 8 | 40 | 1031720 | 8 | 80 | 8207440 |
| 8 | 42 | 1193514 | 8 | 82 | 8837714 |
| 8 | 44 | 1371436 | 8 | 84 | 9499476 |
| 8 | 46 | 1566254 | 8 | 86 | 10193494 |
| 8 | 48 | 1778736 | 8 | 88 | 10920536 |
| 8 | 50 | 2009650 | 8 | 90 | 11681370 |
| 8 | 52 | 2259764 | 8 | 92 | 12476764 |
| 8 | 54 | 2529846 | 8 | 94 | 13307486 |
| 8 | 56 | 2820664 | 8 | 96 | 14174304 |
| 8 | 58 | 3132986 | 8 | 98 | 15077986 |
| 8 | 60 | 3467580 | 8 | 100 | 16019300 |
| 12 | 20 | 308520 | 16 | 20 | 574640 |
| 12 | 24 | 522288 | 16 | 24 | 962400 |
| 12 | 32 | 1212480 | 16 | 32 | 2201984 |

Figure 13: Naphthylenic ((4,6,8)3) tori, in (a) H ("horizontal") and VX ("vertical") embeddings, respectively.

$$\Delta H(G_1, G_2, x) = H(G_2, x) - H(G_1, x) \tag{127}$$

The polynomial coefficients are factorized so that the polynomial can be written as:

$$H(G,x) = A\mathbf{TX} \tag{128}$$

with A being a factor and T being the vector (of dimension d(G)) of coefficients. For two subsequent graphs relation (5) can be rewritten as:

$$\Delta H(G_1, G_2, x) = A_2\mathbf{T}_2\mathbf{X} - A_1\mathbf{T}_1\mathbf{X} \tag{129}$$

The vector $\mathbf{T}_2$ can be obtained by adding a (recursive) difference **R** to the vector $\mathbf{T}_1$.

## 7.1. The ((4,6,8)3)Hn Series

In series ((4,6,8)3)HN[*c*,*n*] with $c < n$, $A = cn/20$. The two dimensions, *c* and *n*, are expressed to account for the net type: $c = 5(2m); 5(2m+1)$ and $n = 4p; 4p+2$.

Thus:

I. In case $c_1$-even, $c = 5(2m)$: for $n = 4p$, the factor $A_1 = 2mp$ while for $n = 4p+2$ the result is $A_1 = m(2p+1)$.
II. In case $c_1$-odd, $c = 5(2m+1)$: for *n* as above, one obtains $A_1 = p(2m+1)$ and
III. $A_1 = 2mp + m + p + 1/2$. $A_2$ can be derived from $A_1$, *via* a difference $\Delta A$; in ((4,6,8)3)HN series, always $\Delta A = n/4$. Thus, the difference polynomial can be obtained by relation:

$$\Delta H(G_1, G_2, x) = \left[\mathbf{R}(A_1 + \Delta A) + \Delta A\mathbf{T}_1\right]\mathbf{X} \tag{130}$$

Keeping in mind the above values of $A_1$ and $\Delta A$, the difference polynomials for the next odd and even terms, respectively, are:

$$\Delta H(G_1, G_2, x)_{odd} = (n/4)\left[(2m+1)\mathbf{R}_{odd} + \mathbf{T}_1\right]\mathbf{X} \tag{131}$$

$$\Delta H(G_1, G_2, x)_{even} = (n/4)\left[(2m+2)\mathbf{R}_{even} + \mathbf{T}_1\right]\mathbf{X} \tag{132}$$

In naphthylenic tori ((4,6,8)3)HN[*c,n*] with $c < n$, the expressions for both $\mathbf{R}_{odd}$ and $\mathbf{R}_{even}$ are as follows:

$$\begin{aligned}
\mathbf{R}_{odd} &= 10x^{5m} + 40x^{5m+1} + 80x^{5m+2} + 100x^{5m+3} + 100x^{5m+4} + \\
&\left[100x^{5m+4k+5} + 100x^{5m+4k+6} + 100x^{5m+4k+7} + 100x^{5m+4k+8}\right]_{k=0,..,(m-2)} \\
&90x^{9m+1} + 76x^{9m+2} + 66x^{9m+3} + 56x^{9m+4} + 54x^{9m+5} + 52x^{9m+6} + \\
&50x^{9m+7} + \begin{bmatrix} 52x^{9m+4k+8} + 50x^{9m+4k+9} + \\ + 48x^{9m+4k+10} + 50x^{9m+4k+11} \end{bmatrix} k = 0,..,(y-1) + Z \\
&y = [n - 8m - 7 - (1 + n(\text{mod } 4))]/4 \\
&Z = 26x^{n+m} \text{ (for } n = 0(\text{mod } 4)); \\
&Z = 52x^{n+m-2} + 50x^{n+m-1} + 24x^{n+m} \text{ (for } n = 2(\text{mod } 4))
\end{aligned} \tag{133}$$

$$\begin{aligned}
\mathbf{R}_{even} &= 20x^{5m+3} + 60x^{5m+4} + 90x^{5m+5} + 100x^{5m+6} + 100x^{5m+7} + 100x^{5m+8} + \\
&+ \begin{bmatrix} 100x^{5m+4k+9} + 100x^{5m+4k+10} + \\ + 100x^{5m+4k+11} + 100x^{5m+4k+12} \end{bmatrix}_{k=0,..,(m-2)} + \\
&+ 92x^{9m+5} + 80x^{9m+6} + 72x^{9m+7} + 60x^{9m+8} + \\
&+ \begin{bmatrix} 52x^{9m+4k+9} + 50x^{9m+4k+10} + \\ + 48x^{9m+4k+11} + 50x^{9m+4k+12} \end{bmatrix}_{k=0,..,(y-1)} + Z \\
&y = [n - 8m - 7 - (1 + n(\text{mod} 4))]/4 \\
&Z = 26x^{n+m+1} \text{ (for } n = 0(\text{mod} 4)); \\
&Z = 52x^{n+m-1} + 50x^{n+m} + 24x^{n+m+1} \text{ (for } n = 2(\text{mod} 4))
\end{aligned} \tag{134}$$

The first two terms: ((4,6,8)3)HN[5,*n*] and ((4,6,8)3)HN[10,*n*], needed for starting the recursion, are given in Appendix.

The Wiener index $W$ is calculated as the first derivative of the Hosoya polynomial, under $x = 1$. From relation (5) it is immediate:

$$H'(G_2,1) = H'(G_1,1) + \Delta H'(G_1,G_2,1) \tag{135}$$

or, with the Wiener index symbol:

$$W(G_2) = W(G_1) + \Delta W(G_1,G_2) \tag{136}$$

where $\Delta W(G_1,G_2)$ is written (by keeping in mind (130)) as:

$$\Delta W(G_1,G_2) = (A_1 + \Delta A)\cdot W(\mathbf{R}) + (\Delta A / A_1)\cdot W(G_1) \tag{137}$$

Formulae for the contributions of recursions $W(\mathbf{R})$ to $\Delta W(G_1,G_2)$ are given in the following:

$$W(\mathbf{R}_{odd\_a}) = 2 \cdot (104 + 480m + 200mp + 800m^2 + 400p^2) \tag{138}$$

$$W(\mathbf{R}_{odd\_b}) = 2 \cdot (202 + 580m + 400p + 200mp + 800m^2 + 400p^2) \tag{139}$$

$$W(\mathbf{R}_{even\_a}) = 2 \cdot (408 + 1120m + 200p + 200mp + 800m^2 + 400p^2) \tag{140}$$

$$W(\mathbf{R}_{even\_b}) = 2 \cdot (606 + 1220m + 600p + 200mp + 800m^2 + 400p^2) \tag{141}$$

Finally, $W(G_2)$ can be written as:

$$W(G_2) = (1 + \Delta A / A_1) \cdot W(G_1) + (A_1 + \Delta A) \cdot W(R) \tag{142}$$

Appropriate substitutions lead to the following recursive relations for the Wiener index:

$$\begin{aligned} W(G_{2odd\_a}) = {} & ((2m+1)/2m)(W(G_1) + 416mp + 1920m^2p + 3200m^3p + \\ & + 1600mp^3 + 800m^2p^2) \end{aligned} \tag{143}$$

$$\begin{aligned} W(G_{2odd\_b}) = {} & ((2m+1)/2m)(W(G_1) + 404m + 1160m^2 + 1600m^3 + 1608mp + \\ & 2720m^2p + 3200m^3p + 2400mp^2 + 800m^2p^2 + 1600mp^3) \end{aligned} \tag{144}$$

$$\begin{aligned} W(G_{2even\_a}) = {} & (2(m+1)/(2m+1))(W(G_1) + 3872mp + 6080m^2p + 3200m^3p + \\ & 1200mp^2 + 1600mp^3 + 800m^2p^2 + 816p + 400p^2 + 800p^3) \end{aligned} \tag{145}$$

$$\begin{aligned} W(G_{2even\_b}) = {} & (2(m+1)/(2m+1))(W(G_1) + 606 + 2432m + 3240m^2 + 1600m^3 \\ & + 6264mp + 6880m^2p + 3200m^3p + 3600mp^2 + \\ & + 800m^2p^2 + 1600p^3m + 1812p + 1600p^2 + 800p^3) \end{aligned} \tag{146}$$

In the above relations, the subscripts $a$ and $b$ refer to the $n$ dimension of tori: $a$ for $n = 0$ (mod 4) and $b$ for $n = 2$ (mod 4), respectively. Relations (142) to (146) hold for $m \geq 1$ and $p \geq 3$.

The distance degree sequence DDS can be plotted as a "distance degree fingerprint"; for the series ((4,6,8)3)HN[15,$n$] (with only five repeating quadruplets and $d(G,k)$ being twice the values in the graph) is given in Figure 14.

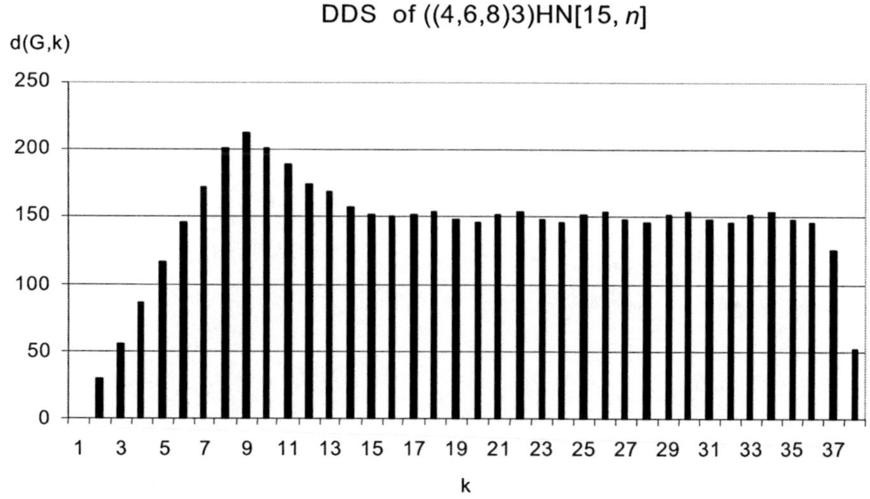


Figure 14: The distance degree sequence of ((4,6,8)3)HN[15,*n*] tori (normalized by 15*n*/20).

## 7.2. The ((4,6,8)3)Vnx Series

In series ((4,6,8)3)VNX[*c*,*n*], with $2c < n$, $c = div\,3$; $n = div\,8$, the factor $A$ is $cn/60$ while $\Delta A = n/20$. The difference polynomial is written here as:[21]

$$\Delta H(G_1, G_2, x) = (n/20)\left[(1 + c/3)\mathbf{R} + \mathbf{T}_1\right]\mathbf{X} \tag{147}$$

with **R** being $R_{odd}$, if *c* is even (and *c* = 0 (mod 6)) and $R_{even}$, if *c* is odd (and *c* = 3 (mod 6)). The expressions for both $R_{odd}$ and $R_{even}$ are as follows:

$$\begin{aligned} R_{odd} &= (cn/60)[75x^{5k} + 295x^{5k+1} + 580x^{5k+2} + 720x^{5k+j+3} :_{j=0,..(k-5)} + \\ &710x^{6k-1} + 625x^{6k} + 475x^{6k+1} + 310x^{6k+2} + 220x^{6k+3} + 205x^{6k+4} + \\ &160x^{6k+5} + 155x^{6k+6} + 150x^{6k+7} + 140x^{6k+8} + 145x^{6k+9} + 150x^{6k+10} + \\ &(145x^{6k+5m+11} + 140x^{6k+5m+12} + 140x^{6k+5m+13} + 145x^{6k+5m+14} \\ &+ 150x^{6k+5m+15}) :_{m=0,...,(y-1)} + 145x^{c/6+5n/8-4} + 140x^{c/6+5n/8-3} \\ &+ 140x^{c/6+5n/8-2} + 145x^{c/6+5n/8-1} + 75x^{c/6+5n/8}] \end{aligned} \tag{148}$$

$$k \geq 5; \quad k = c/6; \quad c = 0(\text{mod } 6); \quad y_{odd} = n/8 - c/6 - 3$$

$$R_{even} = (cn/60)[140x^{5k-2} + 425x^{5k-1} + 645x^{5k} + 720x^{5k+j+1} :_{j=0,..(k-5)} +$$
$$650x^{6k-3} + 480x^{6k-2} + 330x^{6k-1} + 215x^{6k}$$
$$175x^{6k+1} + 170x^{6k+2} + 150x^{6k+3} + 145x^{6k+4} + 150x^{6k+5} +$$
$$(145x^{6k+5m+6} + 140x^{6k+5m+7} + 140x^{6k+5m+8} + 145x^{6k+5m+9}$$
$$+150x^{6k+5m+10}) :_{m=0,..,(y-1)} +145x^{(c-3)/6+5n/8-3} \quad (149)$$
$$+140x^{(c-3)/6+5n/8-2} + 140x^{(c-3)/6+5n/8-1} + 145x^{(c-3)/6+5n/8} +$$
$$75x^{(c-3)/6+5n/8+1}]$$

$$k \geq 5; \quad k = (c+3)/6; \quad c = 3(\text{mod } 6); \quad y_{even} = n/8 - c/6 - 5/2$$

In drawing the difference polynomial (135), the following relation is involved:

$$W(T_1) = \left(\frac{60}{c \cdot n}\right) \cdot W(G_1) \quad (150)$$

and the difference in Wiener index (137) is here:

$$\Delta W(G_1, G_2) = \frac{n}{20} \cdot \left(1 + \frac{c}{3}\right) \cdot W(\mathbf{R}) + \left(\frac{3}{c}\right) \cdot W(G_1)$$
(151)

with $R_{odd}$ and $R_{even}$ given below:

$$W(R_{odd}) = 635 + \frac{425}{3} \cdot c + 15 \cdot c \cdot n + 40 \cdot c^2 + \frac{225}{8} \cdot n^2 \quad (152)$$

$$W(R_{even}) = 430 + \frac{295}{3} \cdot c + 45 \cdot n + 15 \cdot c \cdot n + 40 \cdot c^2 + \frac{225}{8} \cdot n^2 \quad (153)$$

The value for the Wiener index of $G_2$ (see (142)) is given by:

$$W(G_2) = \left(1 + \frac{3}{c}\right) \cdot W(G_1) + \frac{n}{20} \cdot \left(1 + \frac{c}{3}\right) \cdot W(R) \quad (154)$$

Appropriate substitutions lead to the following recursive relation for calculating the Wiener index in ((6,6,4),3)VX tori:

$$W(G_{2odd}) = ((c+3)/288 \cdot c)(288 \cdot W(G_1) + 3048 \cdot cn + 680 \cdot c^2 n + \\ + 192 \cdot c^3 n + 72 \cdot c^2 n^2 + 135 \cdot cn^3) \tag{155}$$

$$W(G_{2even}) = (1/288 \cdot c)(864 \cdot W(G_1) + 288 \cdot cW(G_1) + 6192 \cdot cn + \\ 3480 \cdot c^2 n + 1048 \cdot c^3 n + 192 \cdot c^4 n + 648 \cdot cn^2 \\ + 432 \cdot c^2 n^2 + 72 \cdot c^3 n^2 + 405 \cdot cn^3 + 135 \cdot c^2 n^3 \tag{156}$$

The recursive relations (151) to (156) hold for $c \geq 27$ (as deduced from (148) and (149)). Relations for $G_1$, needed for starting the recursion, are given in the Appendix.

Since ((4,6,8)3)VNX is an isotropic net, it follows that the embedding of ((4,6,8)3)HNX on the (4,4) net results in the same pattern. The difference is only in the net dimensions:

$$((4,6,8)3)\text{ VNX}\left[\frac{c}{3},\frac{2n}{8}\right] = ((4,6,8)3)\text{ HNX}\left[\frac{2c}{8},\frac{n}{3}\right] \tag{157}$$

Relations (1a) and (5a) in Appendix give the first term of each series. It indicates the origins of the repeating unit of the corresponding recursive formulae $R$ proposed herein. Tables 1a and 2a offer some examples of applying the recursive formulae herein proposed.

# 8. APPENDIX

## Hosoya Polynomial

## 8.1. The ((4,6,8)3)HN Series

$$H((((4,6,8)3)\text{HN}[5,n]),x) = \left(\frac{cn}{20}\right)\begin{bmatrix} 20 + 30x + 56x^2 + 66x^3 + 56x^4 + 54x^5 + 52x^6 + 50x^7 + \\ (52x^{4k+8} + 50x^{4k+9} + 48x^{4k+10} + 50x^{4k+11})_{k=0,..,(y-1)} + Z \end{bmatrix} \tag{1a}$$

$n > 2c; \quad y = [n - 7 - (1 + n(\text{mod}4))]/4$

$Z = 26x^n$ (for $n = 0(\text{mod}4)$);

$Z = 52x^{n-2} + 50x^{n-1} + 24x^n$ (for $n = 2(\text{mod}4)$)

$$H((((4,6,8)3)\text{HN}[10,n]),x) = \left(\frac{cn}{20}\right)\begin{bmatrix} 20 + 30x + 56x^2 + 86x^3 + 116x^4 + 136x^5 + 132x^6 + 122x^7 + \\ 112x^8 + 102x^9 + (98x^{4k+10} + 98x^{4k+11} + 102x^{4k+12} + \\ 102x^{4k+13})_{k=0,..,(y-1)} + Z \end{bmatrix} \tag{2a}$$

$n > 2c; \quad y = [n - 8 - (4 - n(\text{mod}4))]/4$

$Z = 98x^{n-2} + 98x^{n-1} + 76x^n + 26x^{n+1}$ (for $n = 0(\text{mod}4)$);

$Z = 74x^n + 24x^{n+1}$; (for $n = 2(\text{mod}4)$)

$$W_{((4,6,8)3)\text{ HN}[10,4p]} = 2p(512 + 200p + 800p^2) \tag{3a}$$

$$W_{((4,6,8)3)\mathrm{HN}[10,4p+2]} = (2p+1)(808+1000p+800p^2) \quad (4a)$$

**Table 1a: Examples of calculating *W* index in ((4,6,8),3)HN tori**

| | $H_2$ | $H_1$ | $W_2$ | $W_1$ | $W(R_{odd})$ | $W(R_{even})$ | *m* | *p* | *n* |
|---|---|---|---|---|---|---|---|---|---|
| 1 | [30,70] | [25,70] | 163973040 | 112346150 | | 138846 | 2 | 17 | *b* |
| 2 | [35,70] | [30,70] | 226028670 | 163973040 | 141742 | | 3 | 17 | *b* |
| 3 | [40,80] | [35,80] | 440975360 | 333502400 | | 186968 | 3 | 20 | *a* |

## 8.2. The Series ((4,6,8),3)VNX

$$H(((4,6,8)3)\mathrm{VNX}\,[3,n]),x) = (cn/60)[60+90x+170x^2+170x^3+155x^4+160x^5+ 155x^6+150x^7+140x^8+145x^9+150x^{10}+ (145x^{10+1+5k}+140x^{10+2+5k}+140x^{10+3+5k}+145x^{10+4+5k}+150x^{10+5+5k})_{k=0,\ldots,y-1}+ 145x^{5n/8-4}+140x^{5n/8-3}+140x^{5n/8-2}+145x^{5n/8-1}+75x^{5n/8}]$$

with

$$y=(5n/8-15)/5; \quad n\ge 4\cdot 8=32 \quad (5a)$$

**Table 2a: Examples of calculating *W* index in ((4,6,8)3)VNX tori**

| | $VX_2$ | $VX_1$ | $W_2$ | $W_1$ | $WR_{odd}$ | $WR_{even}$ |
|---|---|---|---|---|---|---|
| 1 | [30,64] | 27,64 | 45421440 | 35803440 | | 176245 |
| 2 | [33,72] | 30,72 | 77715198 | 62763120 | 219085 | |
| 3 | [36,72] | 33,72 | 94798944 | 77715199 | | 231915 |
| 4 | [39,80] | 36,80 | 151398780 | 126275520 | 280775 | |
| 5 | [42,80] | 39,80 | 179593120 | 151398780 | | 295505 |
| 6 | [45,88] | 42,88 | 272044410 | 232324400 | 350385 | |

## REFERENCES

[1] H. Wiener, Structural determination of paraffin boiling points, *J. Amer. Chem. Soc.*, 1947, 69, 17-20.

[2] H. Hosoya, Topological index. A newly proposed quantity characterizing the topological nature of structural isomers of saturated hydrocarbons, *Bull. Chem. Soc. Japan*, 1971, 44, 2332-2339.

[3] H. Hosoya, On some counting polynomials in chemistry, *Discr. Appl. Math.*, 1988, 19, 239-257.

[4] I Gutman, S. Klavžar, M. Petkovšek, and P. Žigert, On Hosoya polynomials of benzenoid graphs, *MATCH, Commun. Math. Chem.*, 2001, 43, 49-66.

[5] A L. Ivanovskii, Simulation of nanotubular forms of matter, *Russ. Chem. Rev.*, 1999, 68, 103-118.

[6] M. V. Diudea and P. E. John, Covering polyhedral tori, *MATCH, Commun. Math. Comput. Chem.*, 2001, 44, 103-116.

[7] M. V. Diudea, O. Ursu, and B. Parv, Hex tubes from square tubes, *Studia Univ. "Babes-Bolyai"*, 2003, 48, 11-20.

[8] M. V. Diudea, Graphenes from 4-valent tori. *Bull. Chem. Soc. Japan*, 2002, 75, 487-492.

[9] M. V. Diudea, M. Stefu, B. Parv, and P. E. John, Wiener index of armchair polyhex nanotubes, *Croat. Chem. Acta*, 2004, 77, 111-115.

[10] E. P. John and M. V. Diudea, Wiener index of zig-zag polyhex nanotubes, *Croat. Chem. Acta,* 2004, 77, 127-132.

[11] M. V. Diudea, Hosoya polynomial in tori, *MATCH, Commun. Math. Comput. Chem.*, 2002, 45, 109-122.

[12] M. V. Diudea, Layer Matrices in Molecular Graphs, *J. Chem. Inf. Comput. Sci.*, 1994, 34, 1064-1071.

[13] M. V. Diudea, I. Gutman, and L. Jäntschi, *Molecular Topology*. Nova Science, New York, 2002.

[14] E. V. Konstantinova and M. V. Diudea, The Wiener polynomial derivatives and other topological indices in chemical research, *Croat. Chem. Acta*, 2000, 73, 383-403.

[15] M. V. Diudea, B. Parv, P. E. John, O. Ursu, and A. Graovac, Distance counting in tori, *MATCH, Commun. Math. Comput. Chem.*, 2003, 49, 23-36.

[16] A J. Stone and D. J. Wales, Theoretical studies of icosahedral $C_{60}$ and some related species, *Chem. Phys. Lett.*, 1986, 128, 501-506.

[17] M. V. Diudea, Nanotube covering modification, *Studia Univ. "Babes-Bolyai"*, 2003, 48 (2), 17-26.

[18] M. Stefu and M. V. Diudea, Wiener index of $C_4C_8$ nanotubes. *MATCH, Commun. Math. Comput. Chem.*, 2004, 50, 133-144.

[19] M. V. Diudea, B. Parv, and E. C. Kirby, Azulenic tori, *MATCH, Commun. Math. Comput. Chem.*, 2003, 47, 53-70.

[20] M. Deza, P. W. Fowler, M. Shtogrin, and K. Vietze, Pentaheptite modifications of the graphite sheet, *J. Chem. Inf. Comput. Sci.*, 2000, 40, 1325-1332

[21] M. V. Diudea, Topology of naphthylenic tori, *Phys. Chem. Chem. Phys.*, 2002, 4, 4740-4746.

In: Nanostructures: Novel Architecture
Editor: Mircea V. Diudea, pp. 167-174
ISBN: 1-59454-499-9


*Chapter 8*

# COUNTING KEKULÉ NUMBERS IN TORENES A CONJECTURE

***Peter E. John**[a1] **and Mircea V. Diudea**[b]*

[a] Institute of Mathematics,Technical University Ilmenau PSF 100565, D-98684 Ilmenau, Germany

[b]Faculty of Chemistry and Chemical Engineering, Babes-Bolyai University, 400028 Cluj, Romania

## ABSTRACT

An algorithm for calculating the number of Kekulé structures of a polyhex torus (in polynomial time), based on a *conjecture*, is presented. The algorithm, called B, is given in the very short, therefore the interesting reader find more information (the algorithm A) in ref. 2.

## 1. INTRODUCTION

Crop circles fullerenes[1] recently discovered in the soot are presumably torus-shaped „graphitoids". Such structures may be called *toroidal polyhexes* or *torenes*, if all their polygonal faces are hexagons and *toroidal fullerenes*, if their faces are of mixed ring sizes.[2,3] Note that the torus is a closed surface *S* that can carry graphs *G* such that all its vertices have degree 3 and all faces of the embedding of *G* in *S* are hexagons (Figure 1). Pioneering papers on this topic are due to Kirby,[4] Randić *et al*,[5] and John.[6]

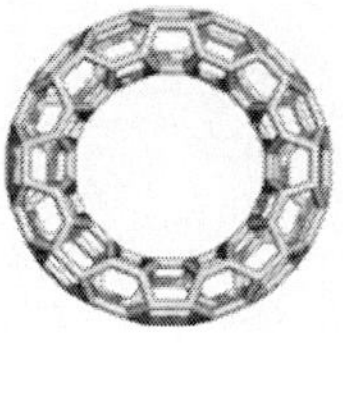

Figure 1: A polyhex torus.

[1] E-mail address: peter.john@mathematik.TU-Ilmenau.de

In this paper, an algorithm B for calculatig the number of Kekulé structures of a polyhex torus (in polynomial time) **based on a conjecture** is presented. The algorithm given here is very short, therefore the interesting reader find more informations (especially another algorithm A) in ref. 2.

Let **T** denote the class of all toroidal polyhexes.[2] The graph of $T \in \mathbf{T}$ can be drawn in the plane (equipped with the regular hexagonal lattice $L$) using the representation of the torus $T$ by a parallelogram $P$ with the usual boundary identification (see Figure 2, *e.g.*, parallelogram $A = A_1, A_1', B_1'', B_1'''$; identify first line $A_1 B_1'''$ with line $A_1'B_1''$ and second „circle" $A_1 A_1'$ with „circle" $B_1''' B_1''$).

For every $T \in \mathbf{T}$, three (not necessarily incongruent) representations $T_1$, $T_2$, and $T_3$ of its graph are obtained (Figure 2). For $i = 1, 2, 3$ there is $T_i = (p_i, q_i, t_i)$ , with $p_i$ being the number of hexagons met by $s_i$, $q_i$ - the number of layers of hexagons parallel to $s_i$ that are covered by $P$, and $t_i$ (the *torsion*) - the number of edges of $L$ intersected by $s_i$ between $A$ and $B_i$.

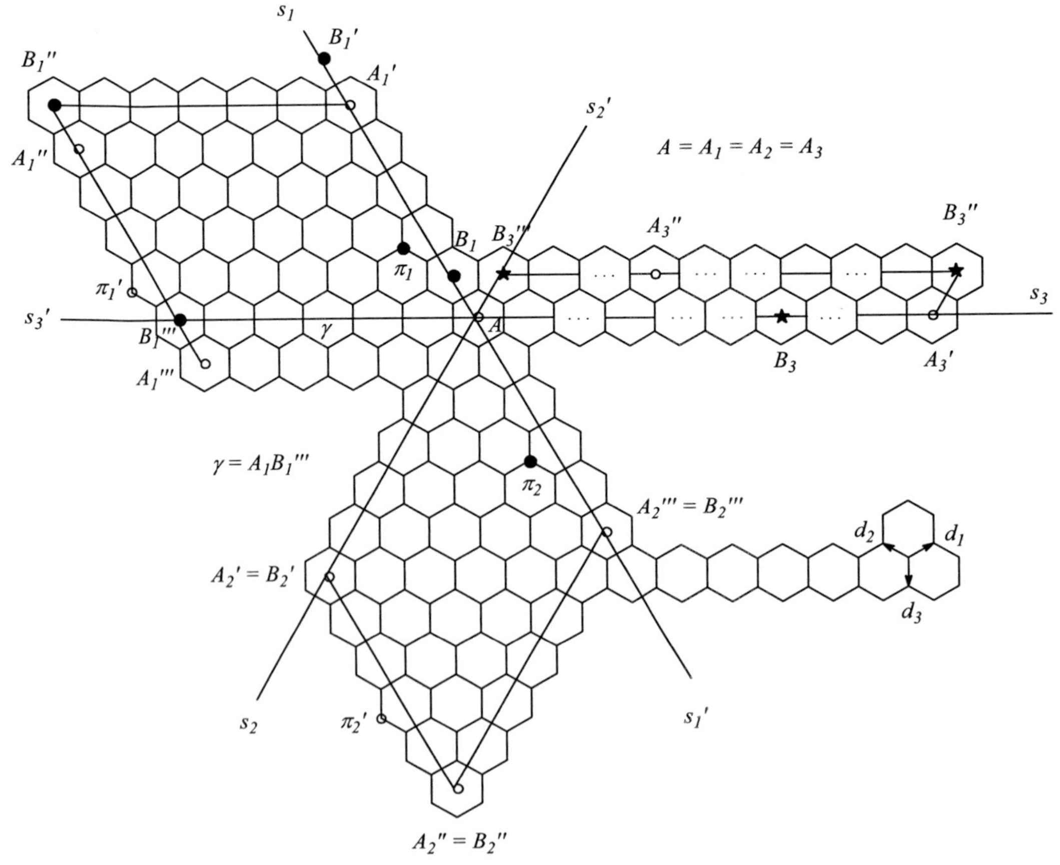


Figure 2: Three representations of a torus $T$ in the regular hexagonal lattice $L$.

We may assume $p_1 \leq p_2 \leq p_3$. Let $p = p_1$, $q = q_1$, $t = t_1$, $s = s_1$, $T^* = T_1$: then $T^* = T^*(p,q,t)$ is the canonical representation of the graph $T \in \mathbf{T}$. For our example, given in Figure 2, clearly $p = 5$, $q = 6$ and $t = 1$.The parameters $p_i$, $q_i$, $t_i$, $i = 2, 3$, can easily be found using, *e.g.*, the method described in ref. [7].

## 2. DEFINITIONS

Let $G = (V,E)$ be a connected graph with vertex set $V = V(G)$ and edge set $E = E(G)$. Graph $G$ is called *bipartite* if and only if its vertices can be coloured black and white such that every edge connects a white vertex with a black one.

A *matching* $M$ of $G$ is a set of pairwise disjoint edges of $G$. Matching $M$ is called *perfect* iff $M$ covers all vertices of $G$ . Let $\mathbf{M} = \mathbf{M}(G)$ denote the set of all perfect matchings of $G$; set $m = m\{G\} = |\mathbf{M}|$. Note that , for $i = 1, 2, 3$, the set of all edges of direction $d_i$ (see Figure 2) is a perfect matching of $T$.

## 3. THE ALGORITHM

Assume that $T^*$ has been drawn such that $s$ lies horizontally in the plane (see Figure 3). Graph $T^*$ being bipartite, its vertices can be coloured such that every vertical edge connects a black top vertex with a white bottom vertex. Let $\mathbf{P} = \mathbf{P}(T^*)$ denote the set of all vertical edges of $T^*$ intersected by s ; clearly, $|\mathbf{P}| = p$.

Graph $G(0)$ is obtained from $T^*$ by omitting all edges of $\mathbf{P}$. Clearly, the graph $G(0)$ is the embedding of a planar graph in a cylinder. Note that $m = m\{G(0)\} = 2^q$, see, *e.g.*, ref.[2,8]

For $T \in \mathbf{T}$ we shall now determine the number $m = m\{T\} = h(p,q,t)$.

First we define a $p$ x $p$ *circulant matrix*[9] $\mathbf{S}_1 = circ(0,1,...,0,0)$ and a $p$ x $p$ *skew circulant* matrix $\mathbf{S}_2 = scirc(0,1,...,0,0)$. Matrix $\mathbf{I} = circ(1,0,...0,0) = scirc(1,0,...0,0)$ denote the $p$ x $p$ *unit matrix*.

### 3.1. Algorithm B

Let $\lambda = 1, 2$.

(B.1) Calculate matrices:

$$\mathbf{W}_\lambda = \mathbf{W}_\lambda\{G(0)\} = \mathbf{W}_\lambda(p,q,0) = (\mathbf{I} + \mathbf{S}_\lambda)^q \tag{1}$$

(B.2) Define polynomials:

$$f_\lambda(T;x) := \det\left(x(\mathbf{S}_\lambda)^t + \mathbf{W}_\lambda\right) \tag{2}$$

(B.3) For $\mu = 1,2$ calculate numbers:

$$g(T;\lambda,\mu) = 1/2\left|f_\lambda(T;1) - (-1)^\mu f_\lambda(T;-1)\right| \tag{3}$$

(B.4) Conjecture :

$$m(T) = h(p,q,t) = \begin{cases} g(T;1,1) + g(T;2,2) & \text{if } p \text{ is } odd \\ g(T;1,2) + g(T;2,1) & \text{if } p \text{ is } even \end{cases} \quad (4)$$

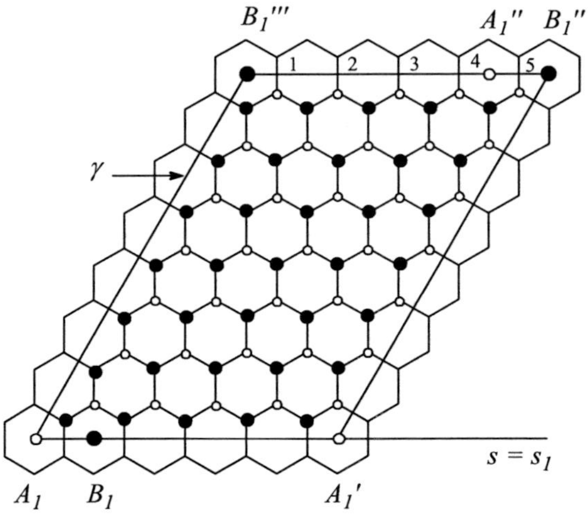


Figure 3: Representation of $T^* = T_1(5,6,1)$, s = $s_1$ lies horizontally.

## 3.2. Example

Torene graph $T$, given in Figure 2, has three different representations $T^*= T_1$, $T_2$, and $T_3$. We calculate the number of Kekule‘ structures for $T^*= T_1= T(5,6,1)$ ($p$ is *odd*) and for $T_2=T(6,5,0)$ ($p$ is *even*).

(a) For $T = T(5,6,1)$ are

$$\mathbf{W}_1 = \mathbf{W}_1(5,6,1) = circ(\mathbf{I} + \mathbf{S}_1)^6 = circ(7,7,15,20,15) \quad (5)$$

and

$$\mathbf{W}_2 = \mathbf{W}_2(5,6,1) = scirc(\mathbf{I} + \mathbf{S}_2)^6 = scirc(-5,5,15,20,15) \quad (6)$$

The polynomials of (**B**.2) are for $T = T(5,6,1)$:

$$f_1(T;x) = x^5 + 35x^4 - 1535x^3 + 20555x^2 + 705x + 64 \quad (7)$$

and

$$f_2(T;x) = -x^5 - 25x^4 - 2125x^3 + 9375x^2 - 15625x \quad (8)$$

Because of $p$ is *odd*, calculate by (**B**.3):

$$g(T;1,1) = 35 + 20555 + 64 = 20654 \tag{9}$$

$$g(T;2,2) = 1 + 2125 + 15625 = 17751 \tag{10}$$

and finally by (**B**.4) is

$$m\{T\} = m\{T(5,6,1)\} = h(5,6,1) = 38405 \tag{11}$$

(b) For $T = T(6,5,0)$ are]

$$\mathbf{W}_1 = \mathbf{W}_1(5,6,0) = circ(\mathbf{I} + \mathbf{S}_1)^5 = circ(1,5,10,10,5,1) \tag{12}$$

$$\mathbf{W}_2 = \mathbf{W}_2(6,5,0) = circ(\mathbf{I} + \mathbf{S}_2)^5 = scirc(1,5,10,10,5,1) \tag{13}$$

The polynomials of (**B**.2) are for $T = T(6,5,0)$:

$$f_1(T;x) = x^6 + 6x^5 - 615x^4 + 7160x^3 + 7155x^2 + 7776x \tag{14}$$

and

$$f_2(T;x) = x^6 + 6x^5 + 645x^4 - 5300x^3 + 22785x^2 + 1656x + 32 \tag{15}$$

Because of $p$ is *even*, calculate by (**B**.3):

$$g(T;1,2) = 6 + 7160 + 7776 = 14942 \tag{16}$$

$$g(T;2,1) = 1 + 645 + 22785 + 32 = 23463 \tag{17}$$

and finally by (**B**.4) is

$$m\{T\} = m\{T(6,5,0)\} = h(6,5,0) = 38405 = m\{T(5,6,1)\} \tag{18}$$

The polynomials and values for torene T(5,6,*t*), $t = 0, 2, 3, 4$ , are given in the Appendix of this paper.

## 3.3. Remarks

1) Note that matrix $\mathbf{W}_\lambda$ is identical with matrix $\mathbf{W}_\lambda$ (0) given in ref.[2] The reader can therein find an example for $T = T(5,6,1)$, the calculation of matrix $\mathbf{W}_\lambda = \mathbf{W}_\lambda(0)$ being made by using of Algoritm **A** (see Figure 4), described in ref. 2.Explicite formulas for elements of $\mathbf{W}_\lambda$ are given in Remark (3) of that paper.
2) It is simple to see that for integer $j$:

$$h(p,q,t) = h(p,q,q-t) = h(p,q,t+jp) \tag{19}$$

3) For natural numbers $n$, $s \geq 0$ and integer $r$

define numbers:

4)

$$\begin{bmatrix} n \\ r \end{bmatrix}_s := \sum_k \binom{n}{r+ks}, \quad \begin{Bmatrix} n \\ r \end{Bmatrix}_s := \sum_k (-1)^k \binom{n}{r+ks}$$

where integer $k$ satisfies the condition: $-\frac{r}{s} \leq k \leq \frac{n-r}{s}$

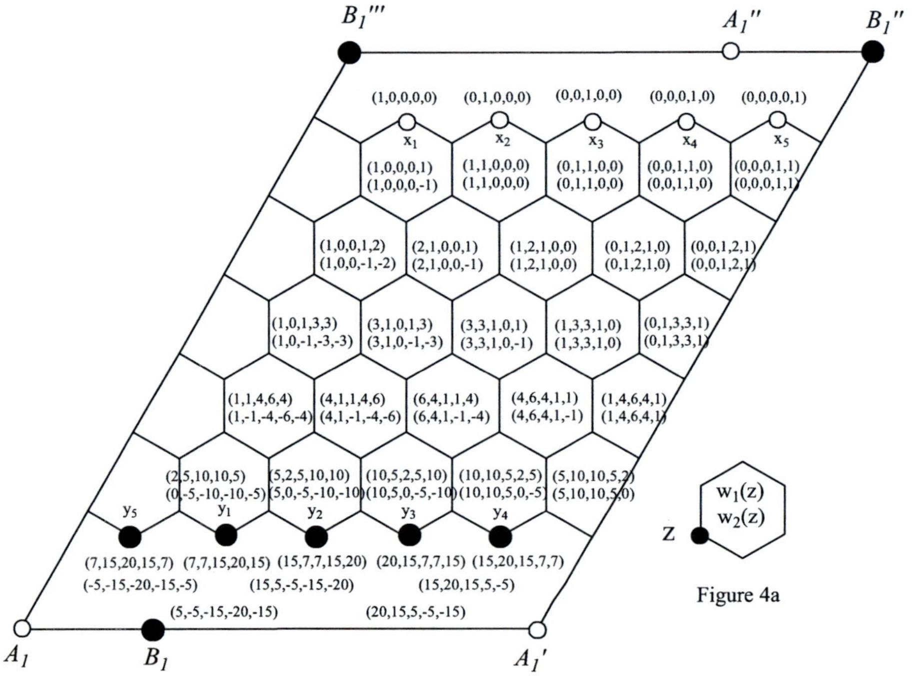


Figure 4: The entries are arranged as indicated in Fig. 4a.

For $\lambda$ = 1,2 the matrices $W_\lambda$ we can represent as $W_1 = circ(c_0, c_1,...,c_{p-1})$ and $W_2 = scirc(s_0,s_1,...,s_{p-1})$. It is simple to prove by induction over $q$ , that for $i$ = 0, 1, 2, ..., $p$-1,

$$c_i = c_i(p,q,0) = \begin{bmatrix} q \\ i \end{bmatrix}_p \qquad \text{and } s_i = s_i(p,q,0) = \begin{Bmatrix} q \\ i \end{Bmatrix}_p$$

Some interesting properties of these *cyclic binomial numbers* will be discussed elsewhere.

## 4. Appendix

For the remaining $t$ = 0,2,3,4 are the coefficients of $f_\lambda(T;x)$ polynomials:

| t | λ = 1 | λ = 2 |
|---|---|---|
| 0 | 1; -35; -1535; -20555; 705; -64 | 1; 35; 2125; - 9375; 15625; 0 |
| 2, 4 | 1; 75; 1025; 20515; -1855; 64 | 1; 75; 2125; - 3125; 15625; 0 |
| 3 | 1; 100; 2630; 20900; 2305; 64 | -1; -100; -2750; -12500; -15625; 0 |

The $g(T;\lambda,\mu)$ and $m(T) = h(5,6,t)$ values are:

| t | g(T; 1, 1) | g(T; 2, 2) |
|---|---|---|
| 0; 2; 4 | 20654 | 17751 |
| 3 | 21064 | 18376 |

## Acknowledgements

This paper is dedicated to Professor Klaus Möckel (Mühlhausen/Germany) on his 75th birthday.

## References

[1] J. Liu, H. Dai, J. H. Hafner, D. T. Colbert, R. E. Smalley, S. J. Tans, C. Dekker, Fullerene „crop circles“, *Nature*, 1997, 385, 780-781.

[2] P. E. John, Kekule' Count in Toroidal Hexagonal Carbon Cages, *Croat.Chem.Acta*, 1998, 71, 435 - 447.

[3] E. C. Kirby, T. Pisanski, Aspects of Topology, Genus and Isomerism in Closed 3-Valent Networks, *J. Math. Chem.*, 1998, 23, 151-167.

[4] E. C. Kirby, Cylindrical and toroidal polyhex structures, *Croat. Chem. Acta*, 1993, 66, 13-26.

[5] M. Randić, Y. Tsukano, H. Hosoya, Kekule' Structures for Benzenoid Tori *Natural Science Report, Ochanumizu University*, 1994, 45, 101-119.

[6] P. E. John, B. Walther, Sketch About Possible Structures and Numbers of Perfect Matchings of Toroidal Hexagonal Systems, *Preprint No. M* 08/94 ( *TU Ilmenau*)

[7] E. C. Kirby, R. B. Mallion, P. Pollak, Toroidal Polyhexes, *J. Chem. Soc. Faraday Trans.*, 1993, 89, 1945-1953.

[8] Sachs, H., Hansen, P.; Zheng, M.: Kekule' Count in Tubular Hydrocarbons, *Commun. Math. Comput. Chem.* (*MATCH*), 1996, 33, 169-241.

[9] P. J. Davis, *Circulant Matrices*, John Wiley et Sons, New York, Chichester, Brisbane, Toronto, 1979.

In: Nanostructures: Novel Architecture
Editor: Mircea V. Diudea, pp. 175-191
ISBN: 1-59454-499-9


*Chapter 9*

# FULLY-RESONANT-AZULENOID NETWORKS: SOME ASPECTS OF CONSTRUCTION AND TAXONOMY

***Edward. C. Kirby***[1]
The Resource Use Institute 14 Lower Oakfield,
Pitlochry, Perthshire PH16 5DS, Scotland UK

## ABSTRACT

Fully-resonant-azulenoid networks - those that in chemical terms can be described as a set of essentially disconnected azulene nuclei - are defined a little more carefully than in earlier work, and several examples are illustrated with a view to the generation of tubes and tori. A method of generating certain members of this set by conversion of toroidal benzenoids is described, and its relevance for a partial characterization pointed out. The speculative possibility of novel properties is briefly discussed.

## 1. INTRODUCTION

Azulene[1-4] is the 5,7 bicyclic hydrocarbon $C_{10}H_8$ (see Figure 1). It is closely related to, and can undergo thermal rearrangement to, naphthalene, but it is blue in colour, and its physico-chemical properties differ from those of naphthalene.

Figure 1: (a) Azulene and (b) naphthalene, each represented by one Kekulé structure.

[1] E-mail address: ekirby@cix.co.uk

In this paper a fully-resonant-azulenoid network is defined as an as-yet-hypothetical carbon network with at least one Kekulé structure equivalent to a set of azulene units all interconnected by essentially single bonds. In part this is equivalent to saying, in graph-theoretical language, that it is the realisation of a graph G that everywhere is locally planar, has only vertices of degree-3 or degree-2 and that has a 2-factor consisting of a set of disjoint 10-cycles that accounts for every vertex. This is a necessary but not a sufficient description, for it is true of both azulenoid and naphthalenoid systems that are fully resonant, as well as for different fully-resonant-azulenoids of the same size.

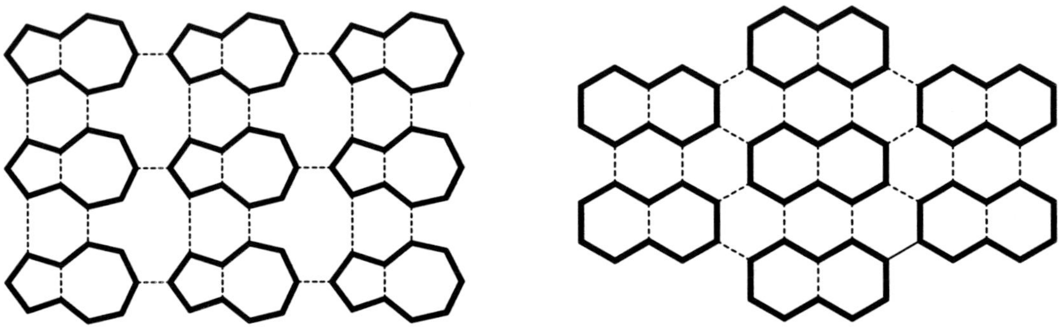

Figure 2: Examples of a fully-resonant-azulenoid and of a fully- resonant-naphthalenoid finite planar system.

For a full characterisation, the position of other connecting edges (see Figure 2) must be made clear. Such networks can in principle cover the closed surfaces of tori and Klein bottles as well as cylinders, and a finite or infinite plane. Not quite so obvious is the fact, discussed later, that certain sphere coverings can be fully resonant azulenoid too.[5]

The concept of being fully-resonant can be said to have its roots in the work of Armit and Robinson[6] on the aromatic sextet, and more specifically of Clar and Zander[7-9] who showed that the most stable benzenoids are those that can be drawn as a set of benzene rings connected by single bonds. The simplest example, apart from the rather artificial one of benzene itself, is triphenylene (Figure 3), and a number of studies have been made of this class.[10-21]

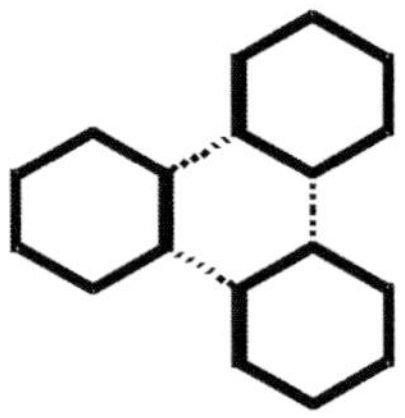

Figure 3: Triphenylene, a fully benzenoid structure.

More recently, studies of other fully-arenoid hydrocarbons (although mostly of benzenoid type) have also been made,[22-28,19,29] and here arenoid species related to the non-classical/non-benzenoid aromatic system azulene are examined and compared with similar benzenoid systems.

## 1.1. A Note on Terminology

### *1.1.1. Fully-Resonant Networks*

The use of this particular arenoid system in this context was first suggested in an earlier paper,[28] where a few examples were shown. They were described as being *fully-azulenoid*, this being a logical and specific term analogous to, for example, the often-used *fully-benzenoid* or *fully-naphthalenoid*. However, this name has proved to be inadequate, as it is liable to be misunderstood. See for example reference[30] where, with equally valid logic, it was taken to mean a surface that merely is completely tiled with contiguous azulene units only. For this latter network, I have in the past used simply the unqualified term *azulenoid*, as for example in *toroidal azulenoid*,[31] although I now think that *pentaheptite* as used by, for example, Crespi[32] and Deza[30] and their respective co-workers is probably better and less ambiguous.

So, in the sense used here, the networks being discussed will be referred to as being *fully-resonant-azulenoid*. Dias also has used the term resonant, in the field of benzenoids.[14,15]

In part this minor confusion arises from an important and interesting difference between azulenoid/pentaheptite and benzenoid systems. The infinite benzenoid sheet is also *fully-benzenoid*, but an analogous azulenoid one is not and cannot be *fully-resonant-azulenoid*; it is in fact *fully-pentalenoid* (Figure 4), and pentalene is of course a '4n' rather than a '4n+2' system and therefore potentially is destabilising, using Hückel's arguments, although all such reasoning, developed for relatively small molecules, should always be treated with caution in the context of large networks.

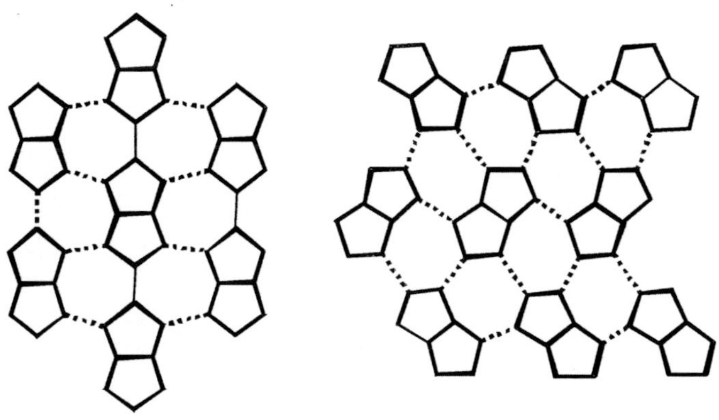

Figure 4: Pentaheptite (azulenoid) networks: they are fully-pentalenoid.

### *1.1.2.*

A *polyhex* is an assembly of hexagons, regular in the case of the plane, that can cover a surface, *i.e.*, the well known 'honeycomb' structure, while a *benzenoid* is a polyhex that has a Kekulé structure (a 1-factor), but here the distinction is unimportant, and such terms are used interchangeably. Neither of these, strictly, refer to a graph-theoretical object, because they both embody geometric information. However they do also have graph-theoretical characteristics – such as being composed of points and connections between pairs of points. The term *benzenoid* itself is of course a shorthand term for *benzenoid graph* or *benzenoid hydrocarbon* in, respectively, a graph theoretical or a chemical context.

*1.1.3.*

*Polarity* is used here to indicate the direction of alignment of azulenes in relation to other azulenes within the network, the angular relationship typically being 0° or 180°. This differs from chemical usage, where the term can have meaning (a different one) within a single substructure - such as azulene itself, which has a dipole moment.

## 1.2. Why Study Fully-Resonant-Azulenoid Systems?

A reason for being interested in carbon systems of this particular type lies in the chemistry of azulene itself.[33,34] It differs in a number of ways from its benzenoid counterpart naphthalene. For example, it shows the typical aromatic property of reaction by substitution, but reacts under milder conditions. Under reacting conditions it can easily polarize by electron drift towards the five membered ring ('trying' to form two aromatic sextets), thus enabling stable tropylium salts to be formed, or protonation to take place, at position-1 or -3 of the five-membered ring (see Figure 5). Azulene itself, and many of its derivatives, have an absorption spectrum that extends well into the visible region as intense and pleasingly pure colours. Whilst it is not clear how, or indeed if, such electrical, optical and other chemical properties would manifest themselves within the kind of networks discussed here, the possibilities do seem sufficiently intriguing as to be worthy of further study.

# 2. The Generation of Fully-Resonant-Azulenoid Patterns

## 2.1 Patterns Deduced From The Possible Combinations Of Possible Positions For Edge Addition

Figure 5 shows the azulene graph (c.f. Figure 1a) with its degree-2 vertices numbered in sequence round the outer circuit. New edges can be formed by connecting a numbered vertex with any of the other numbered vertices. Its adjacent neighbour(s) were excluded in order to consider only 0° or 180° polarities. By definition, to be a valid tiling pattern of the type required, every degree-2 vertex of every azulene must be connected to another degree-2 vertex of an azulene. There are 22 opportunities for an extra edge to be formed between a pair of the eight numbered vertices, namely, 1-3..8; 2-4..8; 3-4..8; 4-6..8; 5-7..8 and 6-8, and from these, four edges are needed for each potential tiling pattern.

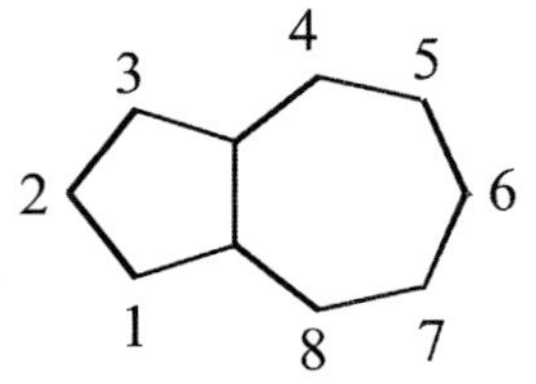


Figure 5: The azulene graph, labelled according to the usual chemical convention, where reactive sites correspond to degree-2 vertices.

Forty two distinct combinations were counted using a simple algorithm. Eliminating symmetry-equivalent repeats of those already found left eight distinct patterns, which are shown in Figure 6.

This number, eight, is thus a lower bound on the number of tiling patterns of this type, *i.e.*, fully-resonant-azulenoids with a repeating pattern unit that contains one azulene. It is stated as a lower bound because, since part of the algorithm involved simple inspection, it is always possible that one or more were missed.

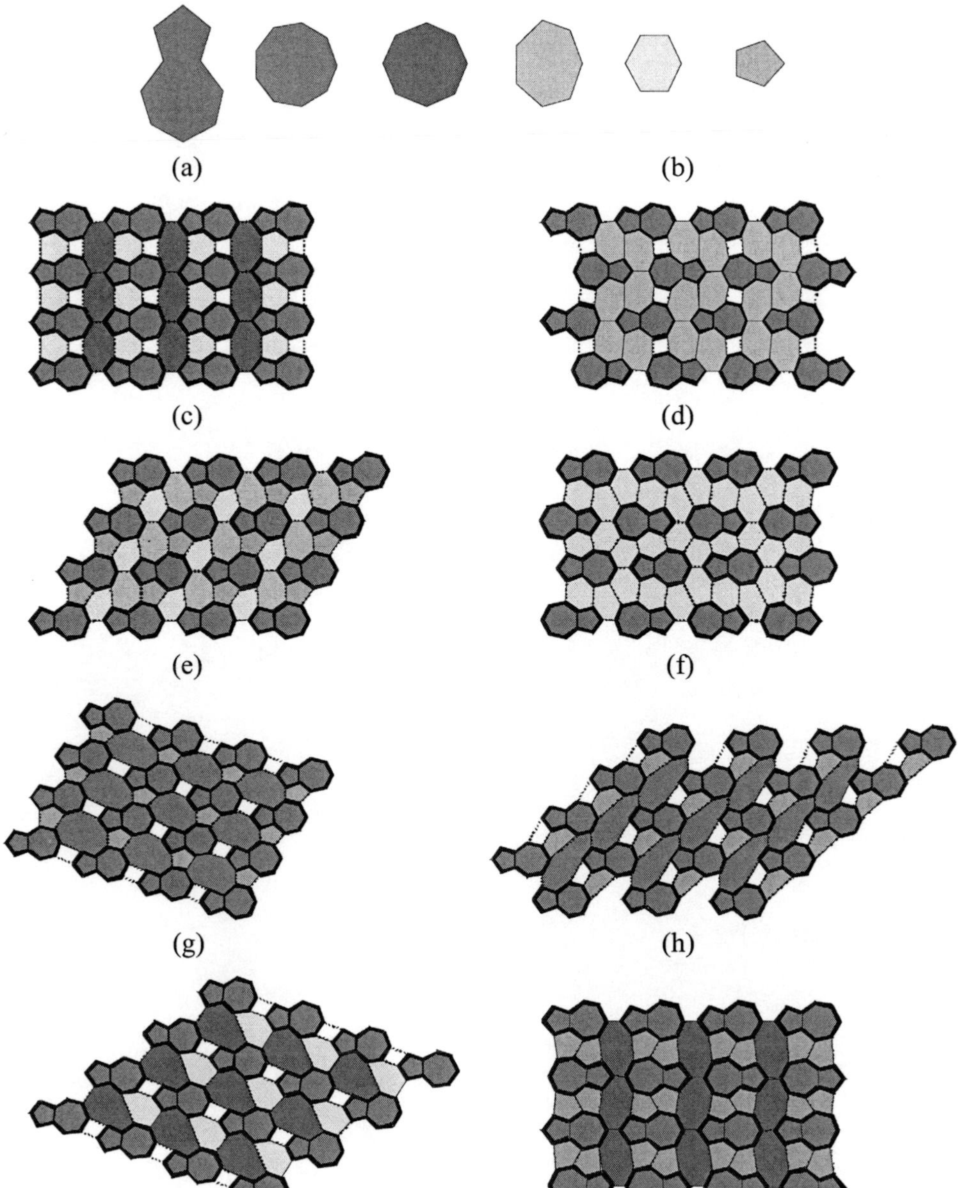


Figure 6: Eight patterns found by considering combinations of azulene-azulene connections. For ease of recognition, azulene units are drawn to a constant size and shape, in parallel, while variable distortion of size and shape is allowed for the empty set of rings.

## 2.2. Fully-Resonant-Azulenoid Patterns Deduced From Parallel Strings of Azulenes

This is a simple procedure, used earlier,[28] whereby completing the connections between strings of 2-6 coupled azulenes generates two empty-ring patterns depending upon the initial choice of connection. (See Figure 7.) It is used more generally here to include consideration of both vertical and diagonal rows of connections as well as horizontal ones. These terms, and the description 'parallel', are used in an informal sense, to refer to the stylised representations drawn for convenience, without necessarily implying anything about the exact geometry of a real structure. They are, however, quite distinct from other, for example radial, arrangements that are referred to later.

For example, within each of the two groups shown in Figure 6, *i.e.*, (i) with single azulene row polarity (**a**,**c**,**e**,**f** and **g**), and (ii) with alternate rows that oppose (**b**,**d** and **h**), the members can all be interchanged by this process. Azulene rows are made to 'slide past' one another, as it were, either horizontally, vertically or diagonally, depending on the pattern. That is, each connection of each azulene row to its two neighbouring rows is displaced one or more connections backwards or forwards. Each pair of rows thus transformed is independent of the others, so although only homogeneous patterns are shown in Figure 6, many more isomer possibilities exist if displacements vary from row to row.

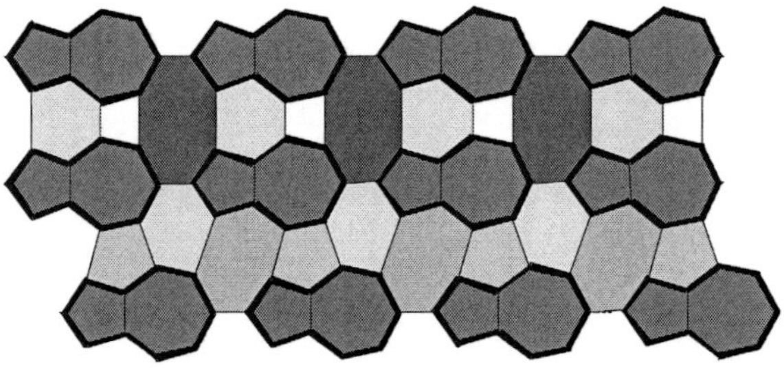

Figure 7: Isomerism induced by 'sliding' rows of azulenes. In this example the empty ring set of Figure 6a is converted to a form symmetrically equivalent to that of Figure 6c by displacing the third row of azulenes one place to the right. A second displacement reproduces 6c exactly, and a third one restores the original 6a (ignoring of course, the shape, if, as here, a periphery is present).

## 2.3. Construction by Tiling an Empty Surface

### *2.3.1. The Smallest Tiles*

For regular patterns of like polarity, such as Figure 6**a**,**c**,**e**,**f** and **g**, deduced by the methods of 2.1 and/or 2.2, it is a simple matter - if one already knows the pattern - to construct a tile (consisting of one azulene and three other rings) to reproduce it. It is not always easy to find possible tilings directly by trial and error.

### *2.3.2. Some More Complex Tiling Constructions*

No systematic attention has yet been given to patterns where the repeating unit is larger, and contains more than one azulene (see Figure 8). Examples of such patterns are Figures 6**b**,**d** and **h** and 9**a**-**d**, where the first one (Figure 9**a**) is closely related to Figure 6**a**. These three patterns of Figure 6 (**b**,**d** and **h**) that have alternate rows of opposing polarity, were deduced systematically as discussed (2.2.1), but they are not so amenable to the method of generating from a benzenoid (2.2.2) because the periphery of the necessary tile does not relate so easily and obviously to the periphery of a benzenoid. The tile of Figure 8 is one that can be used to create the pattern of Figure 6**d**. Again, it is laborious to derive this without having the overall pattern already to hand.

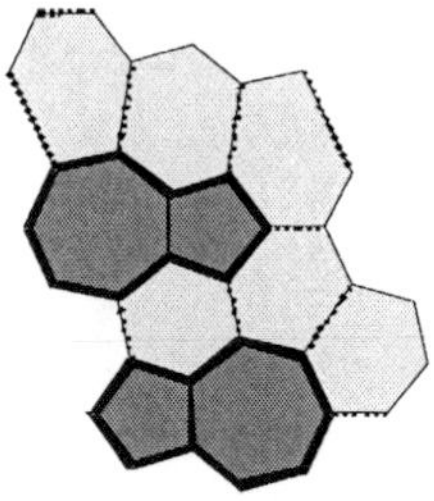

Figure 8: An example of a tile used to construct the pattern of Figure 6d. Its periphery does not conform to an obvious benzenoid.

For each of the 'homogeneous' patterns in Figure 6, many variant 'heterogeneous' ones may be drawn in which either the repeating tile contains more than one azulene, or even where connection sets between pairs of rows and/or pairs of columns vary throughout the structure, sometimes due to varying polarities A few examples are shown in Figure 9.

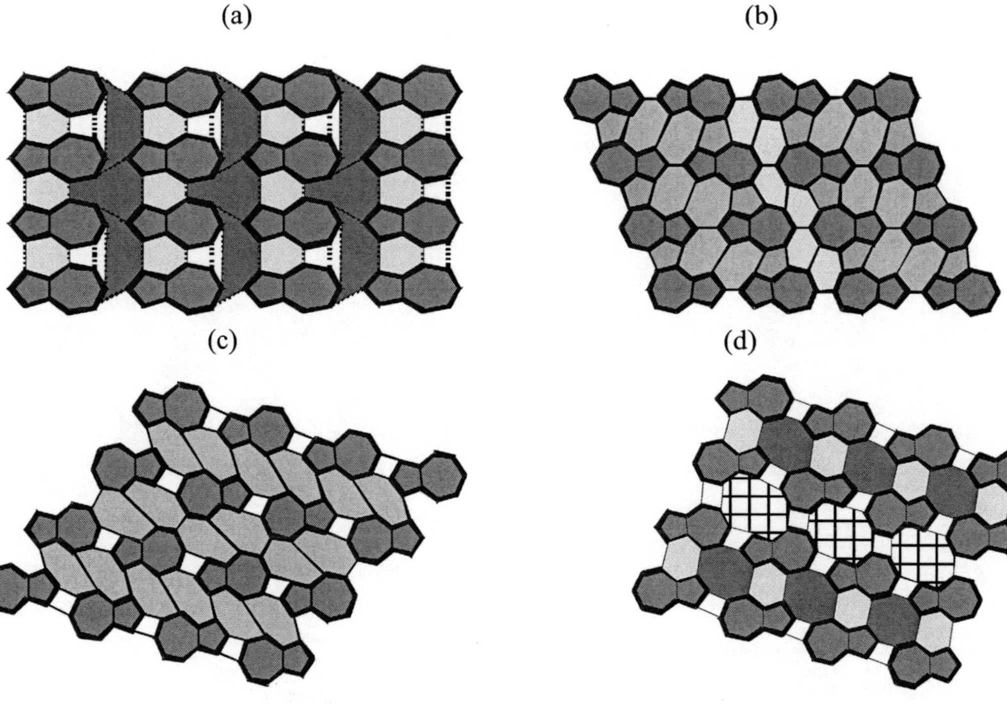


Figure 9: Four examples of patterns where, if constructed by tiling, more than one azulene must be included in the tile. Hatched rings are ten-sided.

## 2.4. Patterns Construction by a Re-Tiling Conversion Of Benzenoid Networks

### *2.4.1. The Transformation*

A *general* relationship between finite planar benzenoids and fully-resonant-azulenoids has not been found, but nevertheless at least some of the networks mentioned here can be constructed for the infinite plane or a boundless closed surface such as the torus, by conversion of a suitable benzenoid. The well known Stone-Wales transformation (Figure 10) - effectively rotating the internal bond of a pyrene unit - can be generalised,[35-42] and if this technique is applied to the most compact possible five-hexagon benzenoid,[43] tiles are generated (see Figure 11) that can be used to construct two of the patterns of Figure 6.

Figure 10: The Stone-Wales transformation.

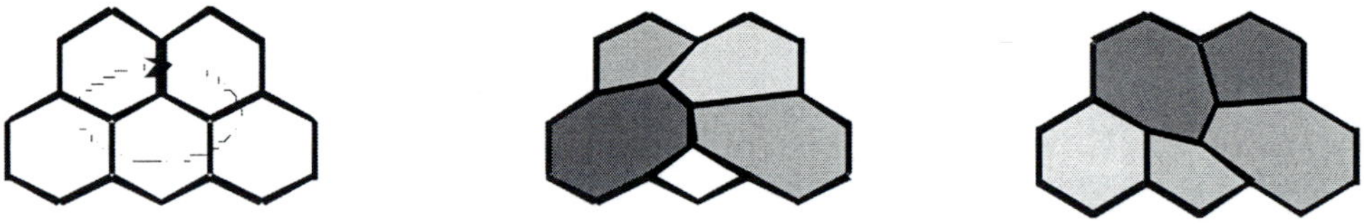

Figure 11: Conversion of a benzenoid to a fully-resonant azulenoid by a generalized Stone-Wales procedure. Clockwise rotation of the inner three-vertex subgraph in (a) relative to the periphery, by one connection position, gives the tile (b), or its mirror image, by anticlockwise movement, and this generates a pattern equivalent to Figure 6g. Rotation by two positions gives (c), which generates tiling patterns equivalent to Figure 6c. The polarity of the latter can vary depending on which set of pentagon-heptagon pairs is defined as azulenoid and full. In each case the result will be a fully-resonant-azulenoid provided that it is boundless (for example, toroidal). Note that in these drawings it is the geometry of the five-hexagon benzenoid tiles, rather than of the azulene units, that, in the interests of clarity, has remained undistorted.

The single polarity patterns (Figure 6 **a**, **c**, **e**, **f** and **g**) that can be constructed by conversion of certain benzenoids as above (followed in some cases by lateral sliding), does not in general provide a characterisation of simple planar azulenoids with a periphery, because some pruning of faces at the edges is necessary, but it does if the surface has no bound. Toroidal polyhexes are well characterised,[44] and up to isomorphism they can always be defined by three parameters (see Figure 12). It follows that if an azulenoid pattern can be related to a known toroidal polyhex then this provides valuable (although not complete) encoding information.

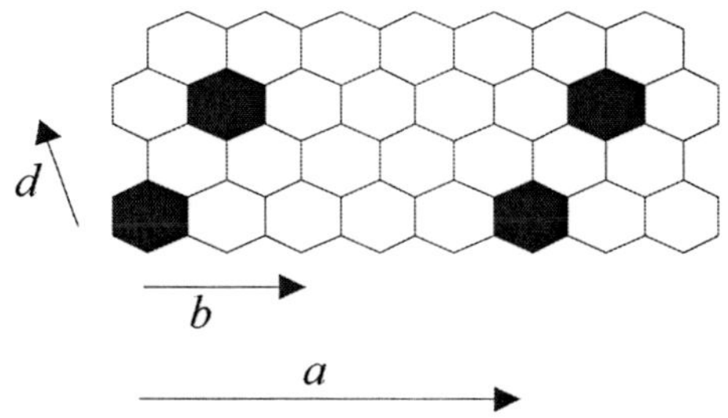


Figure 12: The notation used to define a toroidal polyhex in terms of its face-labelled planar biperiodic pattern. The visual algorithm applied is (i) select an arbitrary hexagon (here shown in black at the lower left) as hexagon-zero for the origin; (ii) count rightwards along the horizontal row of adjacent hexagons until a repeat is found at the $a^{th}$ one; (iii) return to the starting point and count along the same row, this time until, at the $b^{th}$ one, a third hexagon-zero is seen along the upward acute-angled diagonal; (iv) count $d$ hexagons up this diagonal to reach it. The torus is then described as $a,b,d$ (in this example 5-2,2). There may be several equivalent $a,b,d$ representations (up to six). To find them from the origin-hexagon, vary the row of hexagons to be defined as horizontal, by rotation of the grid, and also examine the mirror images. A canonical form, if required, is based on selecting the one with minimum $b$ from those of minimum $d$.[44,45] For example, were the above pattern extended, it would be found that the standardised code value is 10,5,1.

### *2.4.2. Which Toroidal Benzenoids Are Convertible to Fully-Resonant-Azulenoids?*

There are some interesting questions here, and little attention has yet been given to the matter. An elementary and obvious condition for a toroidal benzenoid to be convertible to a fully-resonant toroidal azulenoid, by the process shown in Figure 11, is that the number of vertices ($2.a.d$) must be a multiple of 10 and (equivalently) the number of hexagons ($a.d$) a multiple of 5. This condition is necessary but we do not know for sure whether or not it is sufficient.

It is shown above (Figure 11) that if a toroidal benzenoid can be tiled with (a) then it can be converted, but is this the only benzenoid tile that can be so transformed? So far other possibilities, notably the two preceding 5-hexagon benzenoids in the list used,[43] which can, likewise, be regarded as pyrene derivatives, have not been examined closely, but appear less easy to tile with, and to offer no advantage.

### *2.4.3. Which Toroidal Benzenoids Can Be Tiled with Figure 11a?*

Using the notation of Figure 12, we know that if, for at least one of the equivalent forms of the toroidal benzenoid, $a$ mod 5 =0 and $d$ is even, or *vice versa*, then that benzenoid is convertible. This is a sufficient condition, but that it is not necessary can be seen from the toroidal benzenoid with ten vertices. There is only one isomer, which can be described by any of three equivalent forms (5,2,1; 5,3,1 or 5,4,1), none of which has an even value for $d$, yet the tile 11a exactly covers 5,3,1 and accounts for single azulene tori (2.6.2). Another example is 15,2,1. It has an equivalent form 15,8,1 onto which three of these tiles can be superimposed. The reason is immediately apparent from examination of a labelled biperiodic hexagon grid, for tiles of the right shape accounting for every hexagon can be outlined.

The answer to the question posed is unknown at present, but tentatively it is conjectured that *any* toroidal benzenoid with a multiple of ten vertices can be converted to a fully-resonant-azulenoid.

## 2.5. Non-Parallel, Circular Patterns

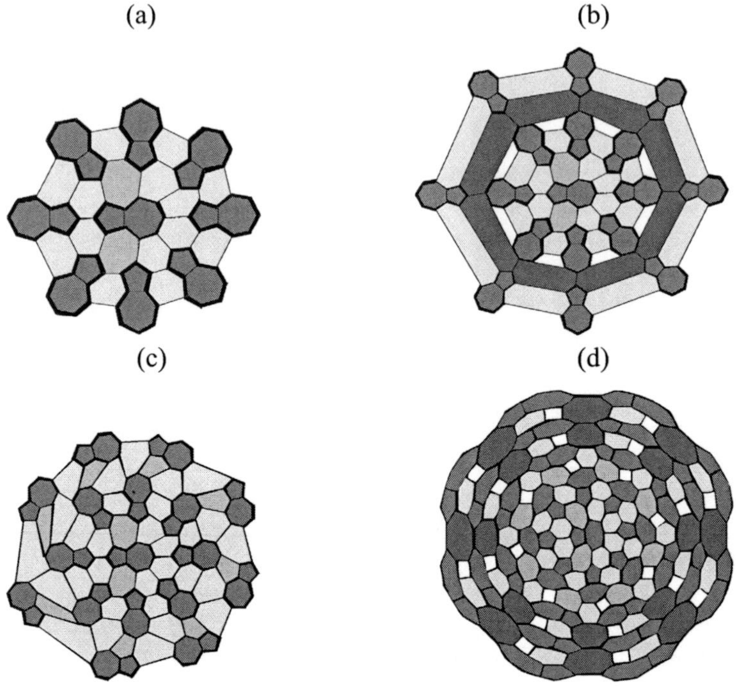


Figure 13: (a) A nine-fully-resonant-azulene cap, and (b-d) the beginnings of three types of half-capped cylinder. In (b) the first cylinder ring has a circumference of eight azulenes parallel to the axis; in (c) they are perpendicular to the axis and in (d) perpendicular but to a different pattern. There is, inevitably, increasing distortion with distance from the centre when drawing these in the plane.

A further set of patterns, where the azulenes have widely varying polarities, not merely opposing ones, form another group. The simplest and most straightforward has a central azulene from which others radiate outwards. However, only the smaller azulene caps (Figure 13) embed easily and smoothly into a roughly spherical surface, although others, such as that shown in Figure 14, may do so if constrained by being connected, as a cap, to a cylinder end.

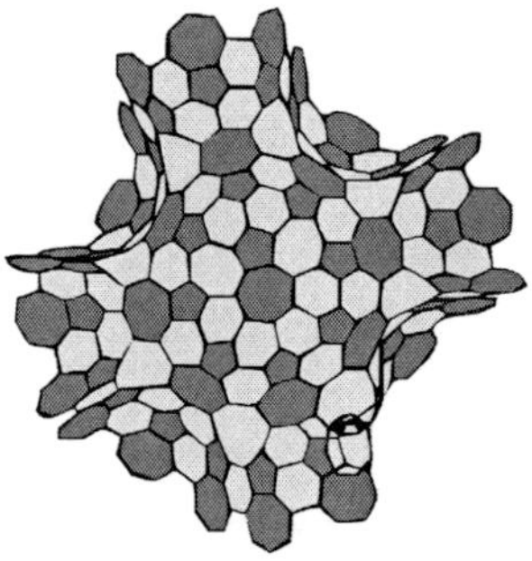

Figure 14: The nine azulene cap (Figure 13a) extended with a second band of radially oriented azulenes to give a 33 azulene structure with 330 vertices. Under a molecular modelling algorithm (Hyperchem[46]) it adopted this fluted shape. (Note that here, different sizes of non-azulenoid ring are not differentiated by colour).

(a) (b)

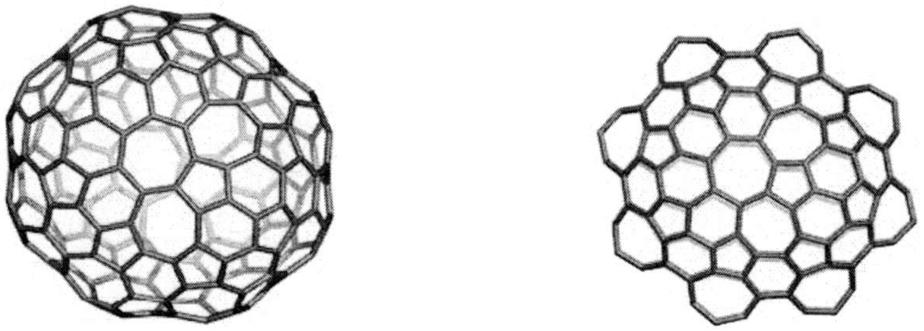

Figure 15: Two fully-resonant-azulenoid spheres made by joining a pair of the 9-azulene caps (Figure 13a) with (a) a 4-6-6 and (b) a 4-4-8 ring pattern where the caps are joined. (In (b) the other cap fits invisibly underneath).

## 2.6. Some HMO Features of Toroidal Fully-Resonant-Azulenoids

### *2.6.1. Selected Eigenvalues of Tori Related to Figure 6*

For comparison, small (6x6=36 azulene units; 360 vertices), simple and untwisted tori covered with each of the patterns of Figure 6 were examined (Table1). Tori can sometimes be more revealing than either geometrically planar systems or cylinders, about the differences between different patterns of covering, because they avoid the complications of a periphery effect. Preliminary testing of the effect of size, in terms of both of the principal dimensions of the torus, on the number and position of zero eigenvalues is shown in Table 2.

**Table 1. Some features of the eigenvalue spectra of 360 vertex untwisted tori constructed by "gluing" parallelograms containing 6 × 6 azulene units cut from patterns a-h shown in Figure 6. Rows A-G show the presence of certain particular eigenvalues, as their degeneracy, while rows H-M give the eigenvalues at particular positions ($\lambda_2$, $\lambda_3$, $\lambda_{n/2}$, $\lambda_{n/2+1}$, $\lambda_{n-1}$ and $\lambda_n$; n = 360).**

| | | Pattern type; see Figure 6 | | | | | | | |
|---|---|---|---|---|---|---|---|---|---|
| | E* | a | b | c | d | e | f | g | h |
| A | 2 | | | | 1 | | | | |
| B | √2 | | | 6 | 4 | | | | |
| C | 1 | 5 | 1 | 13 | 2 | 19 | 1 | 8 | 20 |
| D | 0 | 6 | 9 | 4 | 1 | | | 3 | 2 |
| E | -1 | 1 | 1 | 3 | | 5 | | 2 | 2 |
| F | -√2 | | | 6 | 4 | | | | |
| G | -2 | 4 | 2 | | 2 | | 1 | 2 | 2 |
| H | 2 | 2.969 | 2.967 | 2.952 | 2.963 | 2.952 | 2.982 | 2.944 | 2.968 |
| I | 3 | 2.969 | 2.967 | 2.952 | 2.963 | 2.952 | 2.982 | 2.944 | 2.968 |
| J | 180=n/2 | 0 | 0 | 0.201 | 0.212 | 0.026 | 0.296 | 0 | 0.224 |
| K | 181=n/2+1 | 0 | 0 | 0 | 0.039 | -0.026 | -0.060 | 0 | 0.216 |
| L | 359=n-1 | -2.771 | -2.656 | -2.618 | -2.767 | -2.631 | -2.622 | -2.758 | -2.615 |
| M | 360=n | -2.796 | -2.667 | -2.618 | -2.796 | -2.631 | -2.625 | -2.785 | -2.618 |

* Eigenvalue or eigenvalue position in an ordered list

**Table 2. The occurrence of zero eigenvalues in some toroidal structures. The letters a-h denote the patterns shown in Figure 6. Results show the position of zeros relative to n/2, where n is the number of vertices, within an ordered set of eigenvalues, with their degeneracy in parentheses. A dash indicates no zero; a blank that the structure was not examined. An odd number of rows is not permissible for b, d and h - where rows have opposing polarity**

| R* | a | b | c | d | e | f | g | h |
|---|---|---|---|---|---|---|---|---|
| 6 x 6 | -2..+3 (6) | -5..+3 (9) | +1..+4 (4) | +3 (1) | - | - | 0..+2 (3) | +4..+5 (2) |
| 7 x 6 | -1..0 (2) | -6..0 (7) | - | - | | | | |
| 8 x 6 | 0..+1 (2) | -7..+3 (11) | +3..+4 (2) | +3 (1) | | | | |
| 6 x 7 | - | n/a | - | n/a | | | | n/a |
| 6 x 8 | - | -3..+3 (7) | - | +3 (1) | | | | |
| 7 x 8 | | | | - | | | | |
| 8 x 8 | | | | +3 (1) | | | | |

* 'Rectangle' of azulenes used to make the torus: Columns of Rows

### *2.6.2. The Smallest Possible Fully-Resonant-Toroidal-Azulenoids*

Clearly, these have one azulene and ten vertices. Table 3 shows the eigenvalues of the 10-vertex cubic toroidal graphs (Figure 16) obtained from a one-azulene section excised from the patterns of Figure 6 in the same way as described above. These are simply for formal consideration because, even if they can be made, such small strained structures would be highly unlikely to assume a toroidal geometry.

**Table 3. Eigenvalues of the structures shown in Figure 12; c and e are isospectral**

| Eigenvalue position in ordered list. | a | c | e | f | g |
|---|---|---|---|---|---|
| 1 | 3 | 3 | 3 | 3 | 3 |
| 2 | 2.1149 | 1.6180 | 1.6180 | 1.8794 | 1.5616 |
| 3 | 1.6180 | 1.3028 | 1.3028 | 1.2631 | 1 |
| 4 | 0.6180 | 1 | 1 | 1 | 1 |
| 5 | -0.2541 | 0.6180 | 0.6180 | 0.5157 | 1 |
| 6 | -0.3820 | -0.3820 | -0.3820 | -0.3473 | 0 |
| 7 | -0.6180 | -0.6180 | -0.6180 | -1.1826 | -1 |
| 8 | -1.6180 | -1.6180 | -1.6180 | -1.5321 | -2 |
| 9 | -1.8608 | -2.3028 | -2.3028 | -2 | -2 |
| 10 | -2.6180 | -2.6180 | -2.6180 | -2.5962 | -2.5616 |

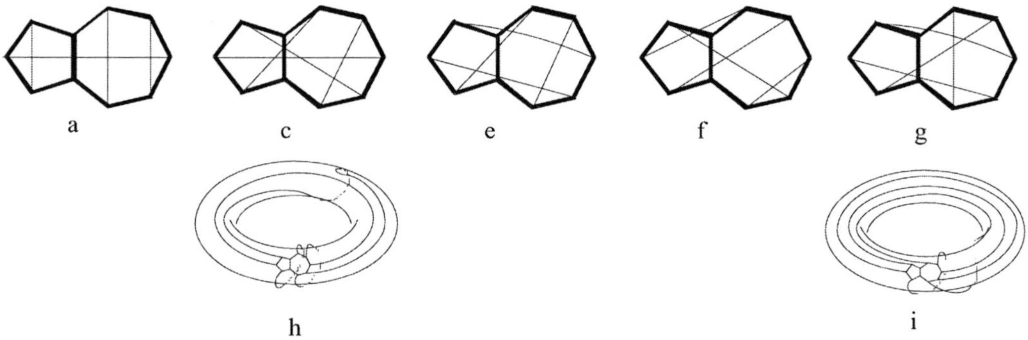


Figure 16: Formal diagrams of some one-azulene tori in accordance with the patterns of Figure 6. It can easily be seen how a,e and f can be embedded in a toroidal surface. This is perhaps not quite so obvious for the twisted structures c and g, and h and i show how this can be done.

### *2.6.3. Selected Eigenvalues of Fully-Resonant-Azulenoid Spheres*

**Table 4. Selected eigenvalues of two 18-azulene, 180-vertex fully-resonant-azulenoid spheres**

| R* | a | b | c | d | e | f | g | h |
|---|---|---|---|---|---|---|---|---|
| 6 × 6 | -2..+3 (6) | -5..+3 (9) | +1..+4 (4) | +3 (1) | - | - | 0..+2 (3) | +4..+5 (2) |
| 7 × 6 | -1..0 (2) | -6..0 (7) | - | - | | | | |
| 8 × 6 | 0..+1 (2) | -7..+3 (11) | +3..+4 (2) | +3 (1) | | | | |
| 6 × 7 | - | n/a | - | n/a | | | | n/a |
| 6 × 8 | - | -3..+3 (7) | - | +3 (1) | | | | |
| 7 × 8 | | | | - | | | | |
| 8 × 8 | | | | +3 (1) | | | | |

*'Rectangle' of azulenes used to make the torus: Columns of Rows

## 3. Discussion

For a surface in which is embedded a fully-benzenoid network, the pattern visible on any reasonably sized sample, i.e. one of an area large enough to contain several full hexagons, will be exactly the same. As has been shown, however, fully-resonant-azulenoids are more complicated than this. Several distinct patterns are possible and may co-exist, so that the surface does not have to be homogeneous in the sense of being tiled with a single uniform pattern. This complication arises of course from the (necessary) fact that empty rings may have a variety of sizes. As was mentioned earlier, benzenoids can be fully-benzenoid, but azulenoids (pentaheptites) cannot be fully-resonant-azulenoid.

Figure 16 and Table 3 show this in relation to the smallest possible members of this class, containing just one azulene. This can be embedded on a torus in several ways, albeit that some are severely twisted, giving rise to the different patterns of three variable empty rings. The smallest possible toroidal benzenoid,[44] has just three hexagons, one full and two empty.

Another difference may be observed for some surfaces. A simple planar fully-benzenoid structure has a unique set of full rings, because, following from the definition, any edge between two degree-2 vertices on the periphery must be part of a full ring, and once one ring is defined, all the rest are too. For a boundless surface such as the torus, this restriction does not apply, and there is a choice of ring sets that could be defined as full. For a fully-resonant-azulenoid however, the full ring set is nearly always unique even on a boundless surface. Thus, for the eight patterns shown in Figure 6, **a**,**b**,**d** and **g**, have only one set of pentaheptite pairs to fill. Of the remainder, **e**, **f** and **h** have only one set of 7-rings, but an apparent choice of 5-rings. The choice is illusory, however, because the resulting set of azulenes cannot be made disjoint. This leaves **c**, the one pattern where there is a clear alternative set of azulenes that could be defined as the full set, although the pattern of the empty set is the same but 'turned upside down'. Whether such considerations have any relevance to the likely properties of such systems is unknown.

Another question one may speculate about is how the polarity of azulenes might manifest itself in the properties of these different patterns, that is, what might the differences be between the uniform row patterns of Figures 6**a**,**c**,**e**, **f** and **g**, and those, such as Figures 6**b**, **d** and **h**, with alternating rows that oppose, or Figure 9**b**, where columns oppose? It is known that only equal numbers of opposing azulenes can be embedded within an otherwise all-polyhex sheet without distorting its 'rectangularity' in terms of rings,[37] (which complicates the construction of tubes and tori, but can also be used to advantage where bending or narrowing of a tube is required[47]). Figure 6**d**, intuitively the 'simplest' system of Figure 6, since members of the empty set are all hexagons, is an example of such pairing. This, possibly more stable, configuration might be more attractive for chemical purposes, although for any application as an electrical conductor or semiconductor, one would expect that the polarity of azulenes should all be same.

If we move away (Figures 13-15) from considering azulenes in straight rows we see how these systems can also be embedded on spheres or capped cylinders. This is another way in which they differ from fully-benzenoids but again, this difference arises through the necessary relaxation in requirements for the empty ring set – it is easy to cover a sphere with disjoint hexagons if any size of ring induced by the interconnections of full rings is acceptable. If the stipulation is for hexagons only, then it cannot be done.

Table 1 shows only selected features of the eigenvalue spectra of a standard torus constructed from the eight patterns of Figure 6. Considering integer values, only Figure **d**, the one with opposing azulenes with a polyhex background, has a (single) eigenvalue of two, although more have its negative –2. The most 'popular' eigenvalue (one) is found for every pattern, with degeneracies as high as 19 and 20 for **e** and **h** respectively. For the zero eigenvalues (non-bonding orbitals), **d** has just one, at $\lambda_{n/2+3}$. Neither **e** nor **f** have two- or zero-eigenvalues despite an initial naïve expectation that they might, since they have four-membered rings. At this simple Hückel level, only Figures 6**a** and **b** have sets of multiple zero eigenvalues – degeneracy 6 and 9 respectively – that straddle the HOMO-LUMO energy level boundary (see table 2); *a priori* a potential indicator of possible metallic properties.[32,48,49] It is known that among benzenoid cylinders and tori, the occurrence of these is influenced by the dimensions of the object.[48] The structures considered here have not been examined systematically, but a brief comparison of a few structures of different sizes is shown in Table 2, and this does show that both dimensions (parallel or perpendicular to the azulene axis) have an effect. In so far as periodicity occurs, this seems to be two, rather than the commonly observed three, among benzenoids, but it must be born in mind that calculations have been done on only a very small sample. Table 4 shows selected eigenvalues of the two spheres of Figure 16. Neither of them have zero (non bonding) levels.

## 4. Conclusions

Fully-resonant-azulenoids form an interesting and intriguing sub group of generalized fullerene structures such as spheres, nanotubes, and tori, although there is as yet no available evidence one way or the other as to whether they would in fact show any unusual properties. As a group they are much more complicated to characterise than their benzenoid counterparts, and only pointers and some partial results are given here. Possible network patterns meeting the fully-resonant-azulenoid criterion may repeat regularly or, within some constraints, irregularly over the surface, giving rise to different isomers, and this complexity is in addition to the possible connectional twisting that can occur in a cylinder or torus. Some of the patterns can be derived by conversion of suitable benzenoids, with or without additional displacement. Where this is possible, its relationship to the already well-characterised cylindrical and toroidal benzenoids provides the means for at least a simple, even if only partial, characterisation.

## References


[1] A S. Pfau and P. A. Plattner, *Helv. Chim. Acta*, 1936, 19, 858.

[2] D Piesse, *Chem. News*, 1863, 8, 245.

[3] D Piesse, *Compt. rend.*, 1864, 57, 1016.

[4] P A. Plattner and A. S. Pfau, *Helv. Chim. Acta*, 1937, 20, 224.

[5] E C. Kirby, *From the Football Field to the Beach: Toward Toroidal and Other Nonplanar Fullerenes. The Chemical Intelligencer*, 1999, April, 46-48.

[6] J. W. Armit and R. Robinson, *J. Chem. Soc.*, 1925, 1604.

[7] E. Clar, *Polycyclic Hydrocarbons*, Academic Press: London, 1964.

[8] E. Clar, *The Aromatic Sextet*, John Wiley: London, 1972.

[9] E. Clar and M. Zander, *J. Chem. Soc.*, 1958, 1861.

[10] J-Is. Aihara and T. Tamaribuchi, Aromatic Character of Graphite Intercalation Compounds, *J. Chem. Soc., Farady Trans.*, 1994, 90, 3513-3516.

[11] S J. Cyvin, Note on the Series of Fully Benzenoid Hydrocarbons with a Constant Number of Isomers, *Chem. Phys. Lett.*, 1991, 181, 431-434.

[12] S J. Cyvin, B. N. Cyvin and J. Brunvoll, All-Benzenoid Systems: Additional Classes of Pericondensed Conjugated Hydrocarbons, *MATCH*, 1990, 25, 105-113.

[13] S J. Cyvin, B. N. Cyvin and J. Brunvoll, All-Benzenoid Systems: An Algebra of Invariants, *Croat. Chem. Acta*, 1993, 66, 101-112.

[14] J R. Dias, Total Resonant Sextet Benzenoid Hydrocarbon Isomers and Their Molecular Orbital and Thermodynamic Characteristics, *Thermochimica Acta*, 1987, 122, 313-337.

[15] J R. Dias, Benzenoid Series Having a Constant Number of Isomers. 3. Total Resonant Sextet Benzenoids and Their Topological Characteristics, *J. Chem. Inf. Comput. Sci.*, 1991, 31, 89-96.

[16] E. C. Kirby, Why Can So Few Benzenoids Be Completely Drawn with Clar's Resonant Sextets? An Analysis Using 'Branching Graphs' and a 'Coiled-Hexagon Code', *J. Chem. Soc., Faraday Trans.*, 1990, 86, 447-452.

[17] J. V. Knop, W. R. Müller, K. Szymanski, S. Nikolić and N. Trinajstić, *Math. Chem.*, 1993, 29, 81.

[18] J. V. Knop, W. R. Müller, K. Szymanski and N. Trinajstić, On the Enumeration of 2-Factors of Polyhexes, *J. Comput. Chem.*, 1986, 7, 547-564.

[19] S. Nikolić, M. Randić, D. J. Klein, D. Plavšić and N. Trinajstić, The Conjugated-Circuit Model: Application to Benzenoid Hydrocarbons, *J. Mol. Structure*, 1989, 198, 223-237.

[20] M. Randić, Fully Benzenoid Systems Revisited, *THEOCHEM*, 1991, 75, 139-153.

[21] M. Randić, D. J. Klein, H. Zhu, N. Trinajstić and T. Živković, Aromatic Properties of Fully-Benzenoid Hydrocarbons. *FIZIKA A* 3, 1994, 2, 61-75.

[22] J. Brunvoll, B. N. Cyvin, S. J. Cyvin, E. C. Kirby and I. Gutman, Fully-Naphthalenoid Hydrocarbons: Enumeration and Classification Polycyclic Aromatic Compounds, 1995, 4, 219-229.

[23] I Gutman, J. Brunvoll, B. N. Cyvin, E. Brendsdal and S. J. Cyvin, Essentially Disconnected Fully-Acenoids, and Fully-Arenoid Hydrocarbons, *Chemical Physics Letters*, 1994, 219, 355-359.

[24] Gutman and S. Cyvin, Fully-Arenoid Hydrocarbons, *MATCH*, 1994, 30, 93-102.

[25] Gutman and S. J. Cyvin, *Commun. Math. Chem.*, 1994, 30, 93.

[26] Gutman, S. J. Cyvin, V. Petrovic and A. Teodorovic, Fully-Napththalenoid Hydrocarbons and Their Conjugation Modes Polycyclic Aromatic Compounds 1994, 4, 183-189.

[27] E. C. Kirby, Cylindrical and Toroidal Polyhex Structures, *Croat. Chem. Acta*, 1993, 66, 13-26.

[28] E. C. Kirby, Fully Arenoid Toroidal Fullerenes, Both Benzenoid and Non-Benzenoid, *MATCH*, 1996, 33, 147-156.

[29] S. Nikolić and N. Trinajstić, A Theoretical Study of Fully Arenoid Systems Gazz, *Chim. Ital.*, 1990, 120, 685-689.

[30] M. Deza, P. W. Fowler, M. Shtogrin and K. Vietze, Pentaheptite Modifications of the Graphite Sheet, *J. Chem. Inf. Comp. Sci.*, 2000, 40, 1325-1332.

[31] E. C. Kirby, On Toroidal Azulenoids and Other Shapes of Fullerene Cage Fullerene Science and Technology, 1994, 2, 395-404.

[32] V. H. Crespi, L. X. Benedict, M. L. Cohen and L. S. G., Prediction of a Pure-Carbon Covalent Metal, *Phys. Review* B, 1996, 53, R13303-R13305.

[33] D. Lloyd, *Non - Benzenoid Conjugated Carbocyclic Compounds*, Elsevier Science Publishers B.V.: New York and Amsterdam, 1984, Vol. 16.

[34] D. Lloyd, What is Aromaticity?, *J. Chem. Inf. Comp. Sci.*, 1996, 36, 442-447.

[35] D. Babić, S. Bassoli, M. Casartelli, F. Cataldo, A. Graovac, O. Ori and B. York, Generalized Stone-Wales Transformations Molecular Simulation, 1995, 14, 395-401.

[36] A T. Balaban, T. G. Schmalz, H. Y. Zhu and D. J. Klein, Generalizations of the Stone-Wales Rearrangement for Cage Compounds, Including Fullerenes, *Theochem - Journal of Molecular Structure*, 1996, 363, 291-301.

[37] E. C. Kirby, On the Partially Random Generation of Fullerenes, *Croat. Chem. Acta*, 2000, 73, 983-991.

[38] E. C. Kirby and T. Pisanski, Even Tori Can Be Odd Paper in preparation 2000.

[39] D. Mitchell, P. W. Fowler and F. Zerbetto, A Generalized Stone-Wales Map: Energetics and Isomerizations of C-40 Carbon Cages, *Journal of Physics* B - *Atomic Molecular and Optical Physics*, 1996, 29, 4895-4906.

[40] E. Osawa and K. Honda, Stone-Wales Rearrangements Involving Heptagonal Defects, *Fullerene Science and Technology* 1996, 4, 939-961.

[41] E. Osawa, Z. Slanina, K. Honda and X. Zhao, Stone-Wales Rearragement as the Double Olefin-Carbene 1, 2-CC Bond Shift. Denied Role of Triplet Biradicals, Fullerene Science and Technology 1998, 6, 259-270.

[42] E. Osawa, H. Ueno, M. Yoshida, Z. Slanina, X. Zhao, M. Nishiyama and H. Saito, Combined Topological and Energy Analysis of the Annealing Process in Fullerene Formation.Stone-Wales Interconversion Pathways among Ipr Isomers of Higher Fullerenes, *J. Chem. Soc., Perkin Trans.* 2, 1998, 2, 943-950.

[43] J. V. Knop, W. R. Müller, K. Szymanski and N. Trinajstić, Computer Generation of Certain Classes of Molecules; SKTH/Kemija u industriji: Zagreb, 1985.

[44] E. C. Kirby, R. B. Mallion and P. Pollak, Toroidal Polyhexes, *J. Chem. Soc. Faraday Trans.*, 1993, 89, 1945-1953.

[45] E. C. Kirby and P. Pollak, How to Enumerate the Connectional Isomers of a Toroidal Polyhex Fullerene, *J. Chem. Inf. Comp. Sci.*, 1998, 38, 66-70 and 1256.

[46] Hyperchem, Version 7; Hypercube, Inc., Gainsville, Florida, USA.

[47] B. Borštnik and D. Lukman, Molecular Mechanics of Toroidal Carbon Molecules, *Chem. Phys. Lett.*, 1994, 228, 312-316.

[48] M. V. Diudea, T. S. Balaban, E. C. Kirby and A. Graovac, Energetics and π-Electronic Structure of Polyhex Nanotubes, *Phys. Chem. Chem. Phys.*, 2003, 19, 4210.

[49] M. Yoshida, M. Fujita, P. W. Fowler and E. C. Kirby, Non-Bonding Orbitals in Graphite, Carbon Tubules, Toroids and Fullerenes, *J. Chem. Soc. Faraday Trans*, 1997, 93, 1037-1043.

In: Nanostructures: Novel Architecture
Editor: Mircea V. Diudea, pp. 193-202
ISBN: 1-59454-499-9


*Chapter 10*

# TOPOLOGICAL COORDINATES FOR SCHLEGEL DIAGRAMS OF FULLERENES AND OTHER PLANAR GRAPHS

***István László***[1]

Department of Theoretical Physics, Institute of Physics and Center for Applied Mathematics Budapest University of Technology and Economics, H-1521 Budapest, Hungary

## ABSTRACT

We extended the topological coordinate method for 3-connected planar graphs. The two-dimensional planar representation of these graphs was obtained only relying on the topological information contained in the connectivity structure. The results were presented by drawing Schlegel diagrams for fullerenes.

## 1. INTRODUCTION

There are complicated nanostructures, where only the connectivity network of the atoms is given but one wants to know the Cartesian coordinates as well. A natural solution to this problem is the application of an inter-atomic potential in order to find the minimal total energy which is calculated from the three dimensional arrangement of the atoms. Such kind of procedures need, however, good initial coordinates which can be constructed only in a complicated way using inter-atomic distances and angles. With the help of the topological coordinate method good initial coordinates can be constructed and they can be used also for the visualization of the molecule or cluster as well. There are already developed topological coordinates for fullerenes,[1],[2] tori,[3],[4] nanotubes [5],[6] and two-dimensional periodic planar structures.[7] Here we present the application of topological coordinate method for non-periodic planar structures.

---

[1] E-mail address: laszlo@feynman.phy.bme.hu

## 2. The Topological Coordinates

Suppose that the atomic arrangement of carbon atoms is given by an $N$ - vertex graph $G=(V,E)$ where $V$ is the set of vertices and $E$ is the set of edges. The vertices characterize the atoms and the edges the first neighbor bonds. The $\mathbf{A}$ adjacency matrix of this graph has matrix elements $A_{ij}=1$ if $(i,j)\in E$ and $A_{ij}=0$ otherwise. Here $(i,j)$ is the edge between vertices $i$ and $j$. The eigenvalues $a_k$ are arranged in descending order, that is

$$a_1 > a_2 \geq a_3 \geq \ldots \geq a_n \tag{1}$$

and $\mathbf{c}^{\mathrm{k}}$ is the eigenvector with the eigenvalue $a_k$.

For any subset $U$ of $V$ let $G|U$ denote the sub-graph of $G$ induced by $U$. That is $G|U=(U,E')$, where $E'=$ all the edges of $G=(V,E)$ that join two vertices in $U$. Let further $\mathrm{supp}(\mathbf{c}):=\{i|c_i \neq 0\}$ denote the support of eigenvector $\mathbf{c}$. The notations $\mathrm{supp}^{+}(\mathbf{c}):=\{i|c_i > 0\}$ and $\mathrm{supp}^{-}(\mathbf{c}):=\{i|c_i < 0\}$ are used for the positive and negative support of $\mathbf{c}$. The topological coordinates are constructed from the so-called bilobal eigenvectors. The eigenvector $\mathbf{c}$ is bilobal if both $G|\mathrm{supp}^{+}(\mathbf{c})$ and $G|\mathrm{supp}^{-}(\mathbf{c})$ are connected.[1-5]

### 2.1. Two-Dimensional Spherical Structures, the Fullerenes

If we know only the number of atoms and the connectivity network in a two-dimensional spherical molecule, all of our structural information can be described with the help of a graph $G=(V,E)$. The first three bilobal eigenfunctions $\mathbf{c}^{\mathrm{k}_1}$, $\mathbf{c}^{\mathrm{k}_2}$ and $\mathbf{c}^{\mathrm{k}_3}$ of the corresponding adjacency matrix $\mathbf{A}$ give the topological coordinates as:[1],[2]

$$x_i = S_1 c_i^{k_1} \tag{2}$$

$$y_i = S_2 c_i^{k_2} \tag{3}$$

and

$$z_i = S_3 c_i^{k_3} \tag{4}$$

where $S_\alpha = 1$ or $S_\alpha = \dfrac{1}{\sqrt{a_1 - a_{k_\alpha}}}$ or any other appropriate scaling factor. It was found that the most realistic coordinates can be found by the scaling factors $S_\alpha = \dfrac{1}{\sqrt{a_1 - a_{k_\alpha}}}$.[6]

## 2.2. Toroidal Structures

The topological coordinates of a torus are calculated from four bilobal eigenvectors of the corresponding adjacency matrix **A**. That is, if $\mathbf{c}^{k_1}$, $\mathbf{c}^{k_2}$, $\mathbf{c}^{k_3}$ and $\mathbf{c}^{k_4}$ are the four bilobal eigenvectors then the $x_i$, $y_i$ and $z_i$ topological coordinates are:[4]

$$x_i = S_1 c_i^{k_1}(1 + S_4 c_i^{k_4}) \tag{5}$$

$$y_i = S_2 c_i^{k_2}(1 + S_4 c_i^{k_4}) \tag{6}$$

$$z_i = S_3 c_i^{k_3} \tag{7}$$

## 2.3. The Nanotubes

An ideal nanotube can be thought of as a hexagonal network of carbon atoms that has been rolled up to make a cylinder. The hexagonal network (the graphene sheet) is usually generated with the help of two unit vectors $\mathbf{a}_1 = a(\sqrt{3}/2,\ 1/2)$ and $\mathbf{a}_2 = a(\sqrt{3}/2,\ -1/2)$ where the unit cell contains two atoms at positions $(0,0)$ and $(a,0)$. In this hexagonal network a parallelogram is constructed from the super cell vectors $\mathbf{b}_1 = \mathrm{m}\ \mathbf{a}_1 + \mathrm{n}\,\mathbf{a}_2$ and $\mathbf{b}_2 = \mathrm{p}\,\mathbf{a}_1 + \mathrm{q}\,\mathbf{a}_2$ where m, n, p and q are integers. The nanotube is obtained by rolling up this super cell, which is carried out by identifying the two opposite sides that are parallel with $\mathbf{b}_2$. In the case of a hexagonal structure this is a simple geometric transformation. If the unit cell contains more than two atoms, that is the network is not hexagonal, then one can not find the position of the atoms in the unit cell and the rolling up technique does not work easily.[8-22] This problem can be solved with the help of the topological coordinate method. Namely the connectivity network is determined by the number of unit cell atoms and by their neighbors. A neighbor is in the same unit cell or in one of the 8 neighboring unit cells. Although an affine transformation does not change the neighboring structure of the connectivity network, it can produce, however, orthogonalized and normalized unit cell vectors. Thus we suppose further that the unit cell is described by two orthonormal vectors $\mathbf{a}_1$ and $\mathbf{a}_2$. Let k, l and r be three integers then the vector $\mathbf{ka}_1 + \mathbf{la}_2$ points to the unit cell $(\mathrm{k},\ \mathrm{l})$, and $(\mathrm{k}, \mathrm{l}, \mathrm{r})$ means the $r$-th atom in this unit cell.

The neighbors of the atom $(k, l, r)$ can be in the unit cells $(k, l)$, $(k+1, l)$, $(k-1, l)$, $(k, l+1)$, $(k, l-1)$, $(k+, l+1)$, $(k-1, l-1)$, $(k+1, l-1)$ and $(k-1, l+1)$ where the corresponding types of neighbors in order are of type $t = 0$, 1, 2, 3, 4, 5, 6, 7, 8 and 9. Let $\mathbf{a}_3 = \mathbf{a}_1 \times \mathbf{a}_2$ be the vectorial product of the vectors $\mathbf{a}_1$ and $\mathbf{a}_2$, and thus we obtain the relation $\mathbf{b}_1 \times \mathbf{b}_2 = (\mathrm{mq} - \mathrm{np})\mathbf{a}_3$. For the scalar product of vectors $\mathbf{a}_1$ and $\mathbf{a}_2$ we have $\mathbf{a}_\mathrm{i} \cdot \mathbf{a}_\mathrm{j} = \delta_{ij}$. Here we remark once more that in the topological coordinate method we do not need the coordinates of these unit cell vectors $\mathbf{a}_1$ and $\mathbf{a}_2$ .

We say that an atom of coordinates $(k, l, r)$ is in the super cell of vectors $\mathbf{b}_1 = \mathrm{m}\ \mathbf{a}_1 + \mathrm{n}\,\mathbf{a}_2$ and $\mathbf{b}_2 = \mathrm{p}\,\mathbf{a}_1 + \mathrm{q}\,\mathbf{a}_2$ if with $\mathbf{c} = \mathrm{k}\mathbf{a}_1 + \mathrm{l}\mathbf{a}_2$ the following relations are valid:

$$(\mathbf{b}_1 \times \mathbf{b}_2) \cdot \mathbf{a}_3 = \mathrm{mq} - \mathrm{np} > 0 \tag{8}$$

$$(\mathbf{b}_1 \times \mathbf{c}) \cdot \mathbf{a}_3 = \mathrm{ml} - \mathrm{nk} \geq 0 \tag{9}$$

$$(\mathbf{c} \times \mathbf{b}_2) \cdot \mathbf{a}_3 = \mathrm{kq} - \mathrm{lp} \geq 0 \tag{10}$$

$$\{(\mathbf{c} - \mathbf{b}_2) \times \mathbf{b}_1\} \cdot \mathbf{a}_3 = (\mathrm{k} - \mathrm{p})\mathrm{n} - (\mathrm{l} - \mathrm{q})\mathrm{m} > 0 \tag{11}$$

$$\{\mathbf{b}_2 \times (\mathbf{c} - \mathbf{b}_1)\} \cdot \mathbf{a}_3 = \mathrm{p}(\mathrm{l} - \mathrm{n}) - \mathrm{q}(\mathrm{k} - \mathrm{m}) > 0 \tag{12}$$

or

$$(\mathbf{b}_1 \times \mathbf{b}_2) \cdot \mathbf{a}_3 = \mathrm{mq} - \mathrm{np} < 0 \tag{13}$$

$$(\mathbf{b}_1 \times \mathbf{c}) \cdot \mathbf{a}_3 = \mathrm{ml} - \mathrm{nk} \leq 0 \tag{14}$$

$$(\mathbf{c} \times \mathbf{b}_2) \cdot \mathbf{a}_3 = \mathrm{kq} - \mathrm{lp} \leq 0 \tag{15}$$

$$\{(\mathbf{c} - \mathbf{b}_2) \times \mathbf{b}_1\} \cdot \mathbf{a}_3 = (\mathrm{k} - \mathrm{p})\mathrm{n} - (\mathrm{l} - \mathrm{q})\mathrm{m} < 0 \tag{16}$$

$$\{\mathbf{b}_2 \times (\mathbf{c} - \mathbf{b}_1)\} \cdot \mathbf{a}_3 = \mathrm{p}(\mathrm{l} - \mathrm{n}) - \mathrm{q}(\mathrm{k} - \mathrm{m}) < 0 \tag{17}$$

If the network of a nanotube is constructed by identifying the two $\mathbf{b}_2$ sides (they are parallel with the vector $\mathbf{b}_2$) then the atoms $(k_1, l_1, r_1)$ and $(k_2, l_2, r_2)$ are identified if $r_1 = r_2$ and for the vectors $\mathbf{c}_1 = k_1\mathbf{a}_1 + l_1\mathbf{a}_2$ and $\mathbf{c}_2 = k_2\mathbf{a}_1 + l_2\mathbf{a}_2$ the following relation is valid:

$$\mathbf{c}_2 - \mathbf{c}_1 = \mathrm{u}\,\mathbf{b}_1 \tag{18}$$

where u is an integer in the range $-1 \le u \le 1$.

In the case of toroidal structure each opposite sides of type $\mathbf{b}_1$ and $\mathbf{b}_2$ are identified and the condition for identifying the two atoms $(k_1, l_1, r_1)$ and $(k_2, l_2, r_2)$ is $r_1 = r_2$ and

$$\mathbf{c}_2 - \mathbf{c}_1 = \mathrm{u}\,\mathbf{b}_1 + \mathrm{v}\,\mathbf{b}_2 \tag{19}$$

with integers u and v in the ranges $-1 \le u \le 1$ and $-1 \le v \le 1$.

From the Equations 8-19 follows that we do not need the special Descartes coordinates of unit cell atoms in a plane for the determination of the connected graph $G = (V, E)$. The initial data contain only the integers m, n, p, q and the list of neighbors for each unit cell atoms. A neighboring atom is described by two numbers the serial number r and the type number t.

Let us turn back to the special case of nanotube topological coordinates. From the initial data first we construct the graph $G = (V, E)$ of the corresponding torus and we calculate its topological coordinates. Next we transform it with a simple geometric transformation into a tube. Applying these two transformations one after the other, we obtain the topological coordinates of the nanotube as:[5]

$$x_i = S_3 c_i^{k_3} \tag{20}$$

$$y_i = S_4 c_i^{k_4} \tag{21}$$

$$z_i = R \arccos\left(S_1 c_i^{k_1} / R\right) \text{ if } c_i^{k_2} \ge 0 \tag{22}$$

$$z_i = R\left( 2\pi - \arccos\left(S_1 c_i^{k_1} / R\right) \right) \text{if } c_i^{k_2} < 0 \tag{23}$$

Here the radius R governs the size of the auxiliary torus. The vectors $\mathbf{c}^{k_1}$, $\mathbf{c}^{k_2}$, $\mathbf{c}^{k_3}$ and $\mathbf{c}^{k_4}$ are the four bilobal eigenvectors of the adjacency matrix $G = (V, E)$.

## 3. Two-Dimensional Periodic Planar Structures

The idea is very similar to that applied for the nanotubes. First we construct a super cell in the periodic two-dimensional planar structure, then by identifying the opposite edges we obtain the graph $G = (V, E)$ of the auxiliary torus and its topological coordinates are transformed back to the original planar super cell with the formulas:[7]

$$x_i = r \arccos\left(S_4 c_i^{k_4} / r\right) \text{ if } c_i^{k_3} \ge 0 \tag{24}$$

$$x_i = - r \arccos\left(S_4 c_i^{k_4} / r\right) \text{ if } c_i^{k_3} < 0 \quad (25)$$

$$y_i = 0, \quad (26)$$

$$z_i = R \arccos\left(S_1 c_i^{k_{14}} / R\right) \text{ if } c_i^{k_2} \geq 0 \quad (27)$$

$$z_i = R\left( 2\pi - \arccos\left(S_1 c_i^{k_1} / R\right) \right) \text{ if } c_i^{k_2} < 0 \quad (28)$$

The radii $R$ and $r$ determine the size and aspect ration of the auxiliary torus,[7] and $\mathbf{c}^{k_1}$, $\mathbf{c}^{k_2}$, $\mathbf{c}^{k_3}$ , $\mathbf{c}^{k_4}$ are four bilobal eigenvectors of its adjacency matrix.

## 4. Topological Coordinates for Planar Graphs

Our examples for planar graphs will be fullerene structures as they can be generated easily with the help of the spiral algorithm.[2] One generates a spiral sequence of pentagons and hexagons consistent with a given fullerene of $N$ carbon atoms and winds it up in order to obtain the connectivity structure. Since all $C_N$ fullerene polyhedra have $\frac{N}{2}+2$ faces of which 12 are pentagonal the spiral sequence is described by the position of the 12 pentagons.

Let us suppose that the graph $G = (V, E)$ is planar and using the Equations (2-4) it has a spherical representation with the coordinates $(x_i, y_i \ z_i)$. The usual stereographic coordinates give the Schlegel-diagram which is a planar representation of the graph $G = (V, E)$.[23]

Here, another method for two-dimensional representation of planar graphs is presented. It is based on the eqs (24-28). First we construct a two-dimensional periodic network containing polygons which has nearly as many vertices as the outer polygon of the planar graph. We put the planar graph in the unit cell of this periodic structure by connecting it to the vertices of the polygon. The final Schlegel diagram is obtained by printing out the topological coordinates of eqs. (24-28) from a chosen unit cell. Figure 1 is obtained using two atoms in the unit cell and $(m, n, p, q) = (7, 7, -7, 7)$.

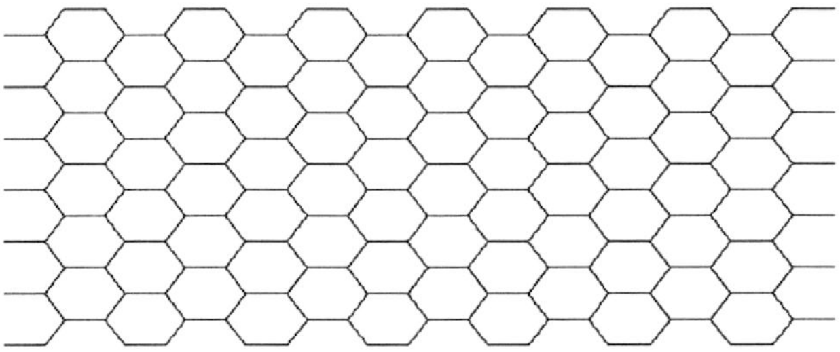

Figure 1: A hexagonal network built up by using topological coordinates.

In Figure 2 we put another hexagon into each hexagon of Figure 1. It was supposed that the vertices of an interior hexagon are in a unit cell and the vertices of its neighboring hexagon are in the same unit cell or in a neighboring unit cell.

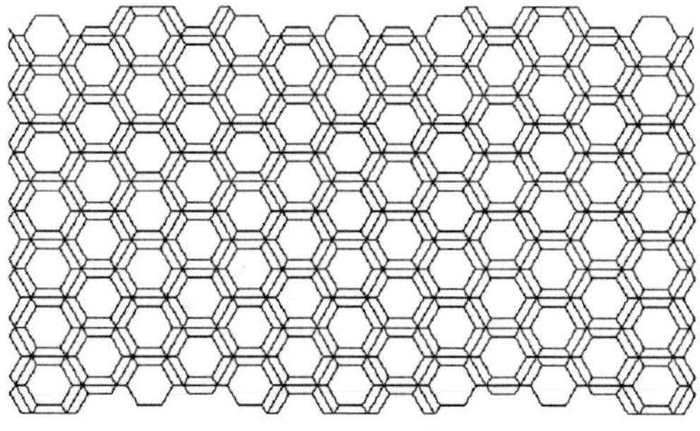

Figure 2: Hexagons in the hexagonal network, by topological coordinates.

Replacing the interior hexagons by pentagons we obtained the structure in Figure 3.

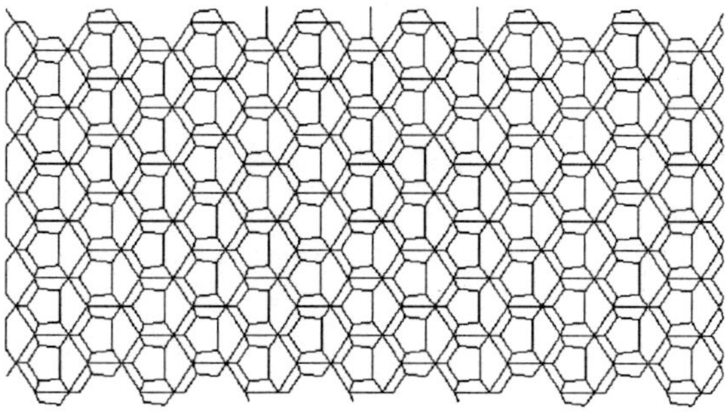

Figure 3: Pentagons in the hexagonal network by topological coordinates.

If the question is the Schlegel diagram of a fullerene first we chose a hexagon or a pentagon as the outer polygon. We put this outer polygon into a unit cell in the same way as in Figures 2 and 3 by keeping the connectivity structure of the fullerene.

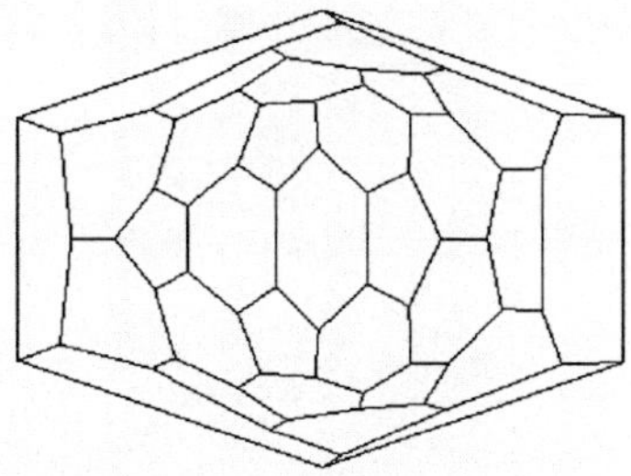

Figure 4: Topological Schlegel diagram of the $C_{60}$ fullerene molecule. The outer face is a hexagon.

In Figures 4 and 5 we present two Schlegel diagrams of the $C_{60}$ fullerene of $I_h$ symmetry and spiral structure $(1, 7, 9, 11, 13, 15, 18, 20, 22, 24, 26, 32)$.

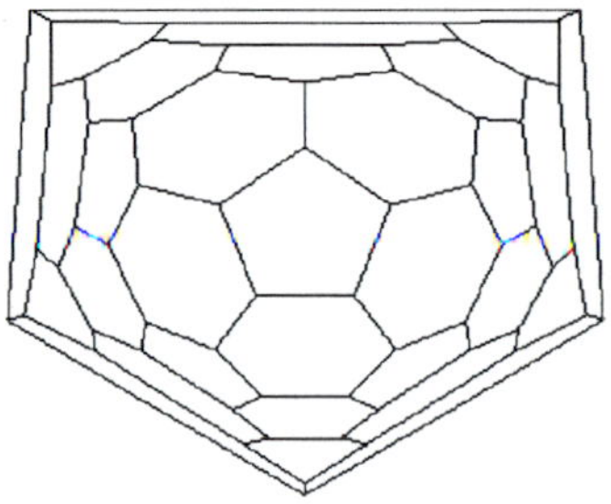

Figure 5: Topological Schlegel diagram of the $C_{60}$ fullerene molecule. The outer face is a pentagon.

Two topological Schlegel diagrams of the fullerene $C_{100}$ of $C_1$ symmetry and spiral structure $(1, 7, 10, 13, 19, 26, 34, 40, 42, 44, 47, 49)$ are given in Figures 6 and 7. In view to obtain a smaller number of vertices in the auxiliary torus we used $(m, n, p, q) = (2, 2, -2, 2)$ or other small coordinates for the super cell. The number vertices can be even smaller if we put the fullerene or the other planar graph only in one unit cell of the super cell.

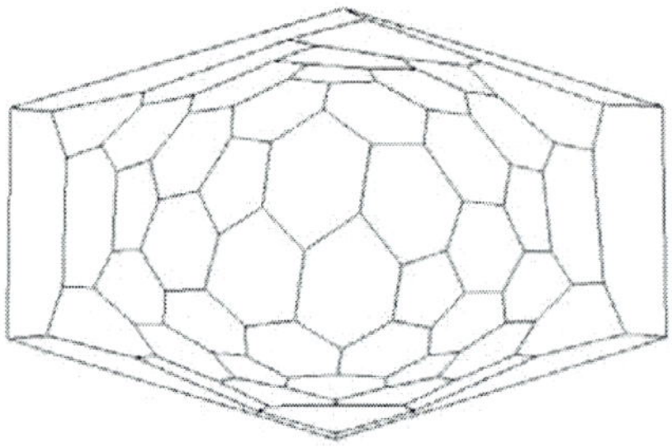

Figure 6: Topological Schlegel diagram of a $C_{100}$ fullerene molecule. The outer face is a hexagon.

Figure 7 shows that the topological coordinate method does not always give a symmetric outer polygon. The shape of the graph can be modified by the parameters of the super cell.

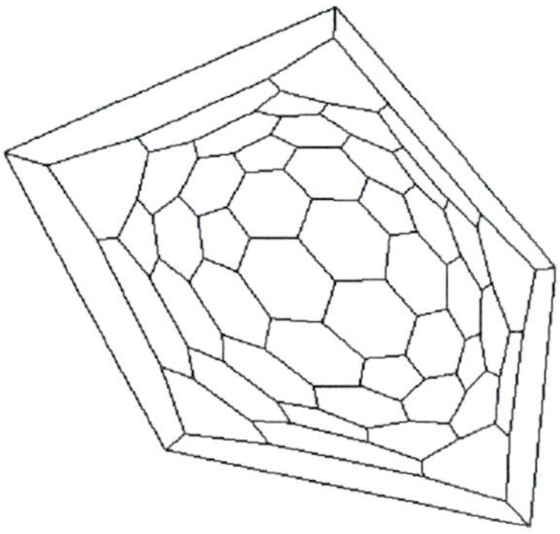

Figure 7: Topological Schlegel diagram of a $C_{100}$ fullerene molecule. The outer face is a pentagon.

## 5. CONCLUSIONS

Although most of the previous structures were three-coordinated carbon networks, the topological coordinate method is not restricted to them. Taking into account Lovász and Schrijver's theorem about the null space embedding of planar graphs,[24] it can be supposed that the topological coordinate method is applicable to any 3-connected planar graph. The consequences and problems if the graph is 2- or 1-connected have been examined in ref. 6.

Finally, we have found that the topological coordinate method can be used even for non symmetric 3-connected planar graphs.

## ACKNOWLEDGEMENT

This work was supported by grants from OTKA (T 038191, T 043231). The author thanks Prof. André Rassat for the fruitful discussions.

## REFERENCES

[1] D. E. Manolopoulos, P. W. Fowler, *Molecular Graphs, Point Groups, and Fullerenes, J. Chem. Phys.*, 1992, 96, 7603-7614.

[2] P. W. Fowler, D. E. Manolopoulos, *An Atlas of Fullerenes*, Clarendon Press, Oxford, 1995.

[3] A Graovac, D. Plavšić, M. Kaufman, T. Pisanski, E. C. Kirby, Application of the Adjacecy Matrix Eigenvectors Method to Geometry Determination of Toroidal Carbon Molecules, *J.* Chem. Phys., 2000, 113, 1925-1931.

[4] László, A. Rassat, P. W. Fowler, A. Graovac, Topological Coordinates for Toroidal Structures, *Chem. Phys. Letters*, 2001, 342, 369-374.

[5] László, A. Rassat, The Geometric Structure of Deformed Nanotubes and the Topological Coordinates, *J. Chem. Inf. Comput. Sci.*, 2003, 43, 519-524.

[6] László, Topological Coordinates for Nanotubes, *Carbon*, **xx,** xxx-xxx, (2004).

[7] László, The Electronic Structure of Nonpolyhex Carbon Nanotubes, *J. Chem. Inf. Comput. Sci.*, **xx,** xxx-xxx, (2004).

[8] L. Mackay, H. Terrones, Diamond from Graphite, *Nature*, 1991, 352, 762-762.

[9] S. Ihara, S. Itoh, J.-i. Kitakami, Toroidal Forms of Graphitic Carbon, *Phys. Rev.*, 1993, *B*47, 12908-12911.

[10] S. Ihara, S. Itoh, J.-i. Kitakami, Helically Coiled Cage Forms of Graphitic Carbon, *Phys. Rev.*, 1993, B48, 5643-5647.

[11] S. Itoh, S. Ihara, Toroidal Forms of Graphitic Carbon. II. Elongated Tori, *Phys. Rev.*, 1993, B48, 8323-8328.

[12] E. C. Kirby, On Toroidal Azulenoids and Other Shapes of Fullerene Cage, *Full. Sci. Technol.*, 1994, 2, 395-404.

[13] S. Amelinckx, X. B. Zhang, D. Bernaerts, X. F. Zhang, V. Ivanov, J. B. Nagy, A Formation Mechanism for Catalytically Grown Helix-shaped Graphite Nanotubes, *Science*, 1994, 265, 635-639.

[14] E. Avron, J.Berger, Tiling Rules for Toroidal Molecules, *Phys. Rev.*, 1995, *A*51, 1146-1149.

[15] E. Avron, J.Berger, Classification Scheme for Toroidal Molecules, *J. Chem. Soc., Faraday Trans.*, 1995, 91, 4037-4045.

[16] László, A. Rassat, Toroidal and Spherical Fullerene-like Molecules with only Pentagonal and Heptagonal Faces, *Int. J. Quantum. Chem.*, 2001, 84, 16-139.

[17] V. Diudea, Toroidal Graphenes from 4-valent Tori, *Bull. Chem. Soc. Japan*, 2002, 75, 487-492.

[18] V. Diudea, Topology of Naphthylenic Tori, *Phys. Chem. Chem. Phys.*, 2002, 4, 4740 -4746.

[19] V. Diudea, P. E. John, A. Graovac, M. Primorac, T. Pisanski, Leapfrog and Related Operations on Toroidal Fullerenes, *Croat. Chem. Acta*, 2003, 76, 153-159.

[20] L. Bíró, G. I. Márk, A. A. Koós, J. B. Nagy, Ph. Lambin, Coiled Carbon Nanotube Structures with Supraunitary Nonhexagonal Ring Ratio, *Phys. Rev.*, 2002, *B*66, 165405-1 –165405-6.

[21] Y. Xia, H. Hosoya, Analysis of the Relationship among the Graphs Isomorphic to Multilayered Cyclic Fence Graphs (MLCFG), *J. Chem. Inf. Comput. Sci.*, 2002, 42, 1004-1010.

[22] T. Réti, K. Böröczky, Topological Characterization of 2D Finite Cellular Systems. *Mater. Sci. Forum*, 2003, 414, 471-481.

[23] M. Nakahara, Geometry Topology and Physics. Adam Hilger, Bristol, Philadelphia and New York, 1991.

[24] Lovász, A. Schrijver: On the Null Space of a Colin de Verdiere Matrix. Annales de l'Institute Fourier (Grenole), 1999, 49, 1017-1026.

In: Nanostructures: Novel Architecture
Editor: Mircea V. Diudea, pp. 203-242
ISBN: 1-59454-499-9


*Chapter 11*

# COVERING NANOSTRUCTURES

***Mircea V. Diudea****
Faculty of Chemistry and Chemical Engineering "Babeş-Bolyai" University,
400084 Cluj, ROMANIA

## ABSTRACT

Tiling of a tubular nanostructure can be modified, from the tetragonal (4,4) to hexagonal (6,3) and other coverings of interest, by following various routes: (i) cutting procedures; (ii) Stone-Wales SW edge rotations and (iii) operations on maps. Cutting procedures, as implemented in the Torus software program, provide tubular/toroidal objects of various size, embedding and covering. Stone-Wales edge rotations, by CageVersatile program, enable covering changes in both spherical and tubular objects. Several new coverings and routes of possible isomerization reactions were identified by this procedure. The most versatile and deeply involved in the electronic structure of their transforms are some geometric operations on maps, including the newly proposed Capra operation. Examples are given for all the three classes of transformations.

A corannulenic Cor covering is proposed, by analogy to the well-known perfect Clar structure, to account for some supra-organized fully-resonant $\pi$-electronic aromatic systems of fullerenes. The constructive demands for such DCor coverings are given.

Supramolecular graphitoids, as infinite periodic surfaces of negative curvature and genus higher than zero, is shown to be constructible by self-assembling of various repeat units, some of them appearing as 3D junctions of nanotubes. In this respect, new representations of the Dyck graph are proposed.

## 1. INTRODUCTION

Covering a local planar surface by various polygonal or curved regions is an ancient human activity. It occurred in house building, particularly in floor, windows and ceiling decoration. There were well known three regular Platonic tessellations: (4,4), (6,3) and (3,6). The Greek and Roman mosaics were very appreciated in this respect.

* E-mail address: diudea@chem.ubbcluj.ro

Covering is nowadays a mathematically founded science.[1] It makes use of the Graph and Set Theory and often inspires from the Arts and Architecture.

Covering transformation is one of the ways in understanding chemical reactions occurring in nanostructures.[2-4]

The article is organised as follows. The second section introduces to some cutting procedures performed on the chessboard tessellation embedded in torus and cylinder. Next, some isomerization pathways are defined on tubular nanostructures, in terms of cascade Stone-Wales edge flipping. A forth section deals with the main Graph-Theoretical operations on a map while the fifth section gives examples of their molecular realization. The last two sections provide pertinent conclusions and references.

## 2. Cutting Procedures on (4,4) Tessellation

Embedding[5] (*i.e.*, drawing of a graph on a (closed) surface with no crossing lines) of the hexagonal trivalent (6,3) net (cf. the Schäfli notation) in the torus or the cylinder is achieved mainly by the well-known graphite zone-folding.[6-11]

The method finds an equivalent planar parallelogram, tiled with the graphitic polyhex lattice. When two opposite edges are identified, a tube is formed, which is completely defined by two integer parameter (*k*,*l*), according to Hamada *et al.*[12] The two parameters quantify (in terms of the primitive lattice vectors of graphene) the *chiral* vector *Ch* on which the tube is rolled up. In case: *armchair* A, (*k*,*k*) and *zigzag* Z, (*k*,0), the tube is achiral.

When the two remaining edges are further identified (*cf.*, the *translation* vector *T*) the torus thus obtained is completely defined by four integers (*e.g.*, (*k*,*l*,*p*,*q*)), reducible to three parameters, according to Kirby *et a*l.[7],[10]

An alternative to the parallelogram procedure uses the topological coordinates, extracted from the adjacency matrix eigenvectors.[13-16]

### 2.1. Basic Polyhex Constructions

Our construction starts with embedding of the tetragonal (4,4) net on either the torus or cylinder.[17-22] The tetragonal objects thus obtained are named by a string specifying the tiling and dimensions of the net: (4,4)[*c*,*n*] (for tubes, Tu is added in the front of), with the (integer) parameters in the tetragonal brackets being the number of atoms in the tube cross-section and the number of cross-sections around the torus (or along the tube), respectively. The objects consist of $c \times n$ vertices, $c \times n$ squares and $4 \times c \times n/2$ edges, 4 being the vertex degree of the net (which is a regular graph). This covering appears in some metal clusters and silicon nanotubes.[23],[24]

Next, the tetragonal (4,4) net is modified to give, most often, a hexagonal (6,3) net or other patterns, of chemical interest. Several cutting procedures we have developed in this respect.[18-22],[25],[26]

A cutting operation consists of deleting appropriate edges in a tetragonal (4,4) lattice in order to produce some larger polygonal faces. By deleting each second *horizontal* edge and alternating edges and cuts in each second row it results in a standard (6,3)H/Z pattern (Figure 1,a). A *vertical* action of the above algorithm leads to a standard (6,3)V/A pattern (Figure

1,b). After optimizing, *e.g.*, by a Molecular Mechanics procedure, (6,3) lattices appear like in Figure 2. The specifications Z (zigzag) and A (armchair) come from the shape of the tube cross-section.

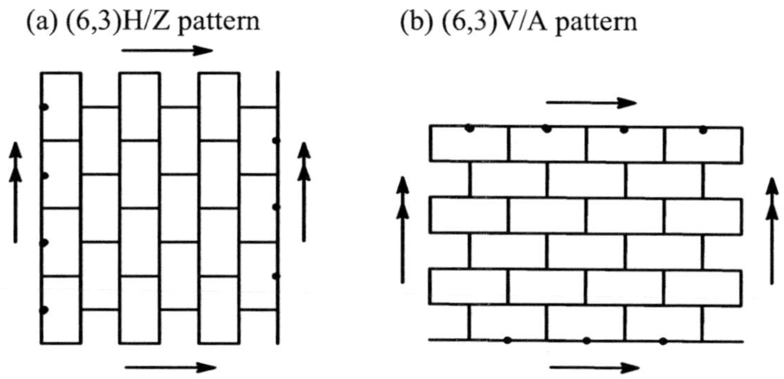


Figure 1: The (6,3) covering by H- (a) and V- (b) cutting of the (4,4) net.

Note that each hexagon consumes exactly two squares of the (4,4) lattice.

By construction, the number of hexagons in the H/Z-objects is half the number of squares on dimension $c$ while, in V/A-objects, the reduced number of hexagons appears on dimension $n$. Thus, (Tu)(6,3)Z[2$c$,$n$] has the same number of hexes as its embedding isomer (Tu)(6,3)A[$c$,2$n$]. Recall that, the above cutting procedure leaves unchanged the number of vertices in the original (4,4) lattice.

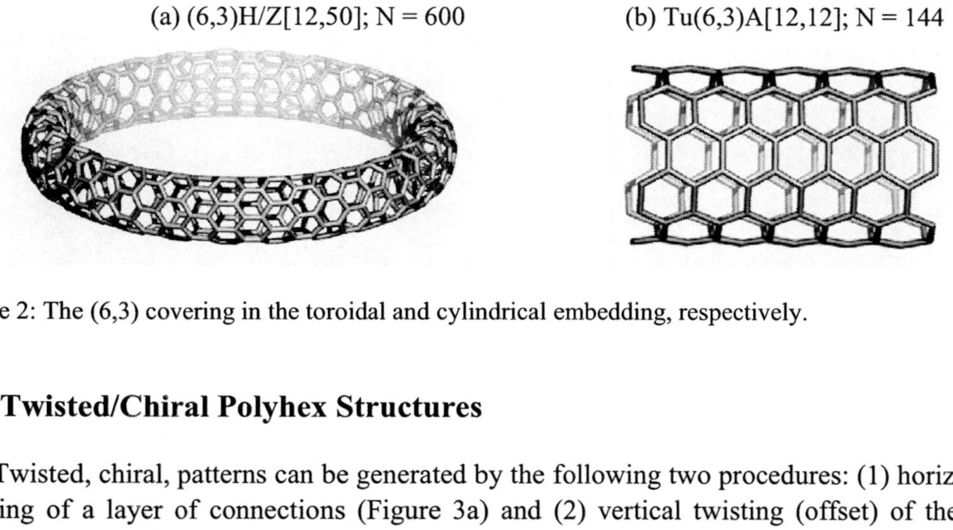


Figure 2: The (6,3) covering in the toroidal and cylindrical embedding, respectively.

## 2.2. Twisted/Chiral Polyhex Structures

Twisted, chiral, patterns can be generated by the following two procedures: (1) horizontal twisting of a layer of connections (Figure 3a) and (2) vertical twisting (offset) of the end connections (not shown). In cylindrical embedding, only one twisting type (namely the one parallel to the cylinder generator) will work, because of the open ends. Edge cutting is further needed to change squares into hexagons (Figure 3b). Twisting an even number of layers is needed to obtain a hexagonal net.

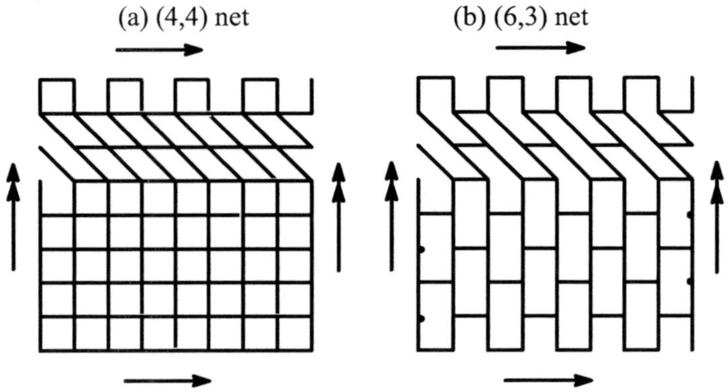


Figure 3: Twisted (4,4) pattern (a) and its (6,3) derivative (b).

Each of the above twistings superimpose on the two basic cuttings, thus resulting four classes of twisted tori: (i) H-twist, H-cut HH*t*[*c*,*n*]; (ii) V-twist, H-cut, VH*t*[*c*,*n*]; (iii) H-twist, V-cut, HV*t*[*c*,*n*] and (iv) V-twist, V-cut, VV*t*[*c*,*n*].[27] The twisting parameter *t* takes integer values in the range [0,*c*]. It is just the deviation (in number of hexagons) of the chiral (*i.e.*, rolling-up) vector to the zigzag line, in the graphite sheet representation.[6-11] Correspondence between the name of tori in our approach X*t*[*c*,*n*] and the four integer parameter (*k*,*l*,*p*,*q*) notation was given elsewhere.[27] Note that the HV lattice is essentially a HH one, the objects differing, however, as embedding and electronic structure.

The type of cutting prevails in embedding realization and the shape of objects. Figures 4 and 5 illustrate some optimized twisted polyhex tori.

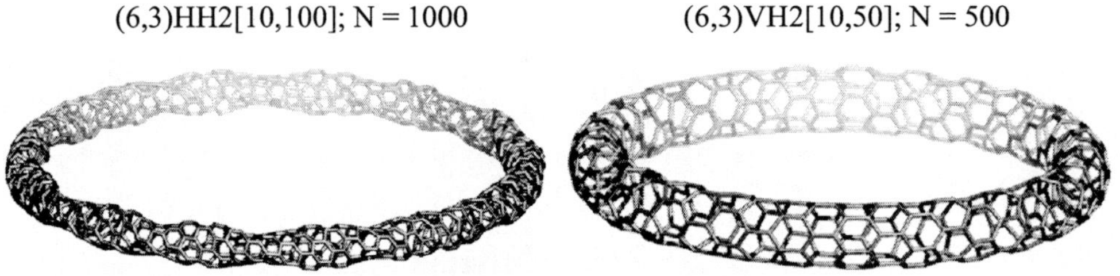


Figure 4: A twisted (6,3) covering in horizontal H-cutting (optimized) toroids.

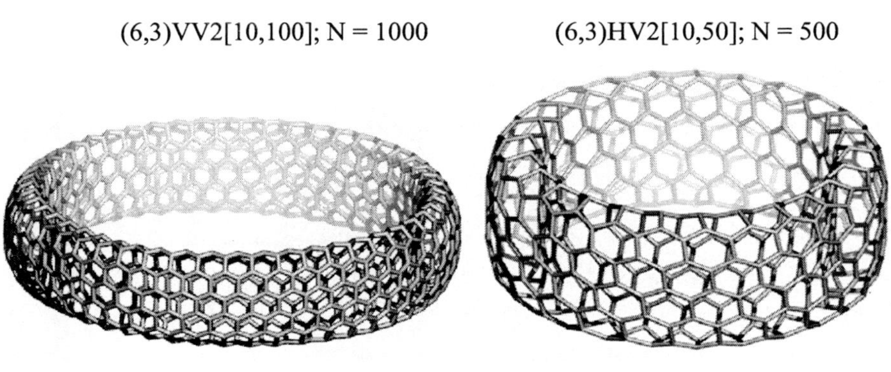


Figure 5: A twisted (6,3) covering in vertical V-cutting (optimized) toroids.

Tu(6,3)Z[12,12]

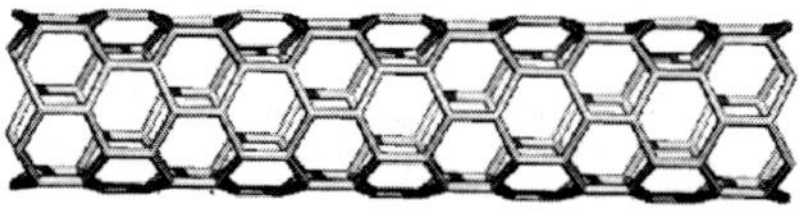

Tu(6,3)A[12,12]

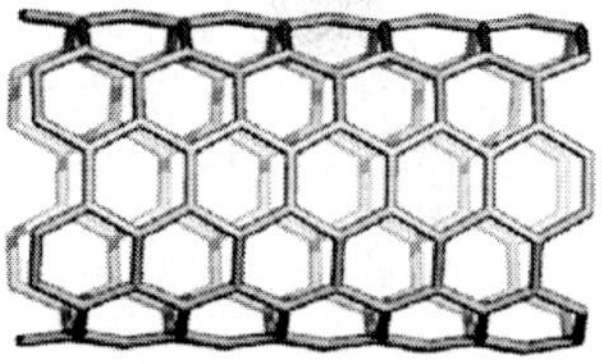

Tu(6,3)HH6[12,12]

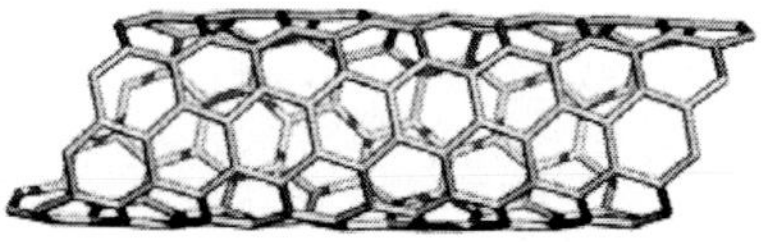

Tu(6,3)HV8[12,12]

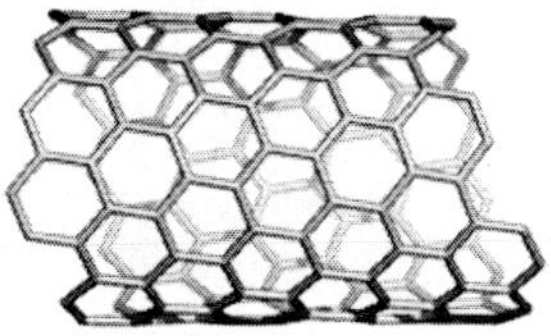

Tu(6,3)HH10[12,12]

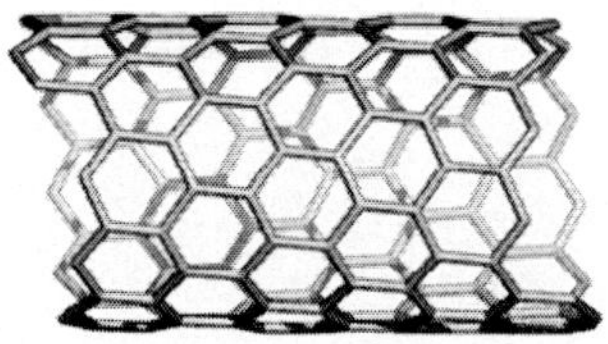

Tu(6,3)HH12[12,12]; Tu(6,3)HV12[12,12]
= Tu(6,3)Z[24,6]

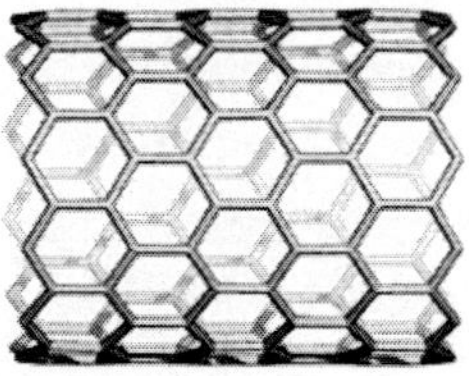

Figure 6: Nanotube twisting; diameter doubling at $t = c$ is evident by comparing the top left corner tube with the bottom right one.

In case of tubes, Tu(6,3)HH*t*[*c*,*n*], the number of atoms on the *c*-dimension increases as *t* increases, concomitant with the shortening of *n*-dimension (in general, non-integer values, having a statistical meaning).[28] The twisting preserves the type of net (Z, in this case) and the total number of hexagons as well (*i.e.*, $(c/2) \times (n\text{-}1)$), the same as in the non-twisted tube). Correspondingly, the number of "zigzags" (*i.e.*, the number of end-hexagons) increases, from $c/2$ up to twice the initial value. The final object will be Tu(6,3)HH*c*[*c*,*n*] = Tu(6,3)Z[2*c*,*n*/2]. Note that diameter doubling of single walled nanotubes has been observed experimentally[29],[30] and termed "tube coalescence".

In case of Tu(6,3)HV*t*[*c*,*n*], the twisting keeps the *c*, *n* dimensions but stepwise adds the Z-character (*i.e.*, adds zigzag hexagons), finishing in a pure Z-net, particularly Tu(6,3)Z[2*c*,*n*/2]. The crenel and zigzag tips coexist, as can be seen from Figure 6 (the right column).

Clearly, both local covering and embedding (*i.e.*, global covering) contribute to the structural (and electronic) properties of the tubular nanostructures.

Correspondence between the name of polyhex tubes in our approach and the two integer parameter (*k*,*l*) notation is given below:

$$\text{TUH/Z}[c,n]:\quad c = 2k;\quad t = 0;\quad \text{Z}(c/2,0) \tag{1}$$

$$\text{TUHH}t[c,n]:\quad c = 2k + l;\quad t = |c - 2k|;\quad k = (c - t)/2;\quad l = t \tag{2}$$

$$\text{TUHV}t[c,n]:\quad c = k + l;\quad t = |k - l|;\quad k = (c + t)/2;\quad l = (c - t)/2 \tag{3}$$

$$\text{TUV/A}[c,n]:\quad c = 2k;\quad t = 0;\quad \text{A}(c/2, c/2) \tag{4}$$

$$t_{hh} = \frac{|c_{hv} - t_{hv}|}{2};\quad c_{hh} = \frac{3c_{hv} + t_{hv}}{2} \tag{5}$$

$$t_{hv} = \frac{|c_{hh} - 3t_{hh}|}{2};\quad c_{hv} = \frac{c_{hh} + t_{hh}}{2} \tag{6}$$

with $n$ remaining indefinite.

## 2.3. Other Coverings

A ((4,8)3)S covering can be derived from the tetragonal (4,4) covering by deleting appropriate edges (Figure 7). The cut edge lay either horizontally or vertically, which results in two embedding isomers; however, the net is isotropic.[31] Letter "S" comes from the "tetragonal" shape of the tetragon (Section 3).

According to the building procedure, the following relations account for the involved toroidal embedding isomers:

$$((4,8)3)\text{HS}[c,n] = ((4,8)3)\text{VS}[c/2, 2n] \tag{7}$$

$$((4,8)3)\text{VS}[c,n] = ((4,8)3)\text{HS}[2c, n/2] = ((4,8)3)\text{VS}[n/2, 2c] \tag{8}$$

Construction of a ((5,7)3)SP pattern ("SP" comes from "spiral" – Section 3) needs, in our procedure, four vertex rows (*i.e.*, a 0 mod($c$,4) lattice).[32] It is called the pentaheptite tessellation and shows a local ($t_5$, $t_7$) signature of (1,3) type. Lattices ((5,7)3) with various local signatures have been proposed[4],[32-35] (Section 3).

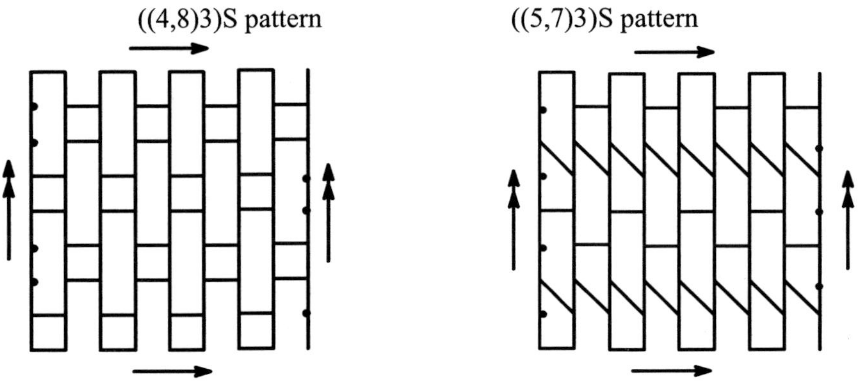


Figure 7: Non-graphitic coverings in toroidal nanostructures.

Rings of size 5 and 7 were also used for curvature matching and strain relief in otherwise all-hexagon tori.[36-40] Figure 8 shows examples of optimized objects with the two above coverings. They can also be drawn by the Stone-Wales transformation of a polyhex net[4],[41] (Section 3).

((4,8)3)SH[20,64]; N = 1280 ((5,7)3)SPH[20,64]; N = 1280

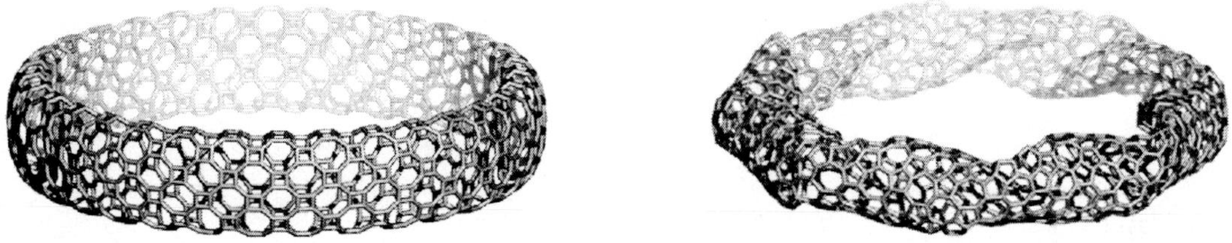

Figure 8: Non-graphitic optimized tori.

Two phenylenic patterns: ((4,6,8)3)H/V and ((4,6,8)3)HX/VX we designed on the tetragonal (4,4) net (Figure 9).[25] Their local signature is: $t_{4j}(0, 2, 2)$; $t_{6j}(2, 0, 4)$; and $t_{8j}(2, 4, 2)$, $j = 4, 6, 8$.

The resulting toroidal objects differ in orientation and number of the (4,6,8) repeat units. Figure 10 illustrates some phenylenic tori, as optimized objects.

((4,6,8)3)H-net ((4,6,8)3)HX-net

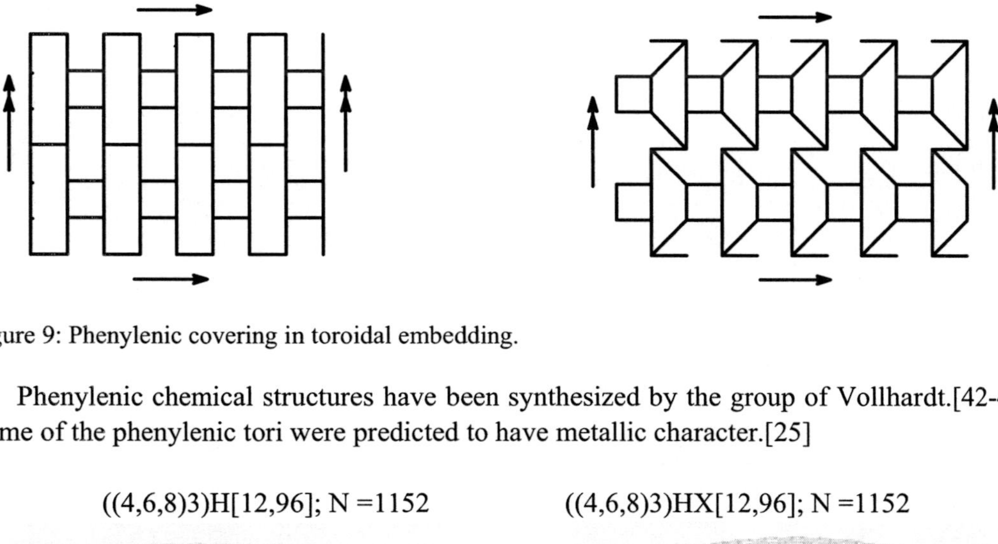

Figure 9: Phenylenic covering in toroidal embedding.

Phenylenic chemical structures have been synthesized by the group of Vollhardt.[42-44] Some of the phenylenic tori were predicted to have metallic character.[25]

((4,6,8)3)H[12,96]; N =1152 ((4,6,8)3)HX[12,96]; N =1152

Figure 10: Optimized phenylenic tori.

In close analogy to the phenylenic net, we proposed[25] the naphthylenic net, with the sequence: $C_6$, $C_6$, $C_4$,..., $C_6$, $C_6$, $C_4$. The HN/VN lattice has the local signature: ((4,6,8)3)HN, $t_{4j}(0, 2, 2)$; $t_{6j}(1, 3, 2)$; and $t_{8j}(2, 4, 0)$, while the net HNX/VNX has the signature: $t_{4j}(0, 4, 0)$; $t_{6j}(1, 3, 2)$; and $t_{8j}(0, 8, 0)$, with $j = 4, 6, 8$. This last net can be obtained from the square (4,4) lattice by a double leapfrog operation (Section 3). Figure 11 gives the above patterns and Figure 12 presents some optimized toroidal objects.

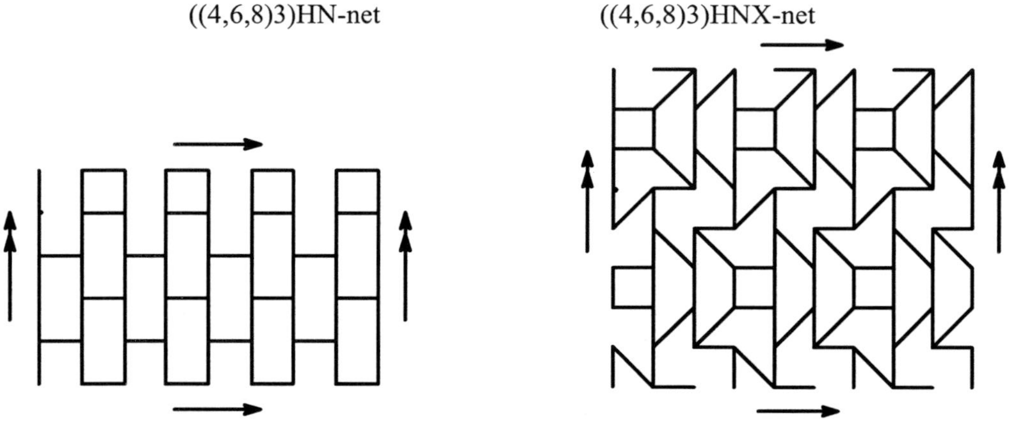


Figure 11: Naphthylenic covering in toroidal embedding.

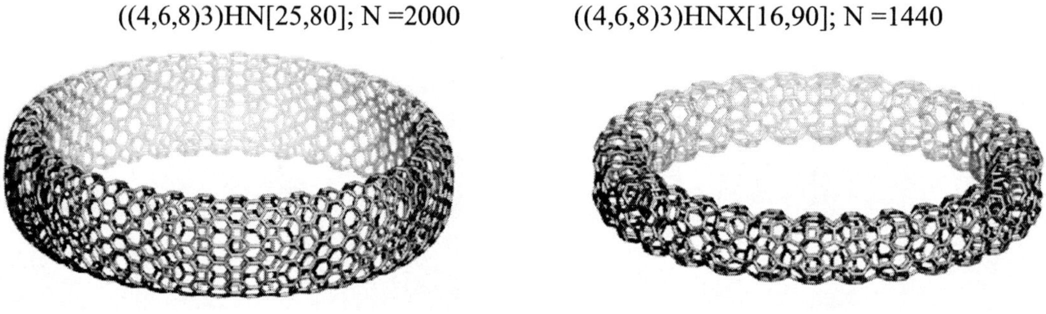


Figure 12: Optimized naphthylenic tori.

Insulating is the major trend of the π-electronic shell of naphthylenic toroidal nanostructures, in the case of even-dimensional objects. On the contrary, odd-dimensional objects show open-shell structure. A particular position is occupied by the X-net: the toroidal objects are all properly closed-shell, due to its double leapfrog character.[26]

## 3. Isomerization Routes on Tubular Nanostructures

A hexagonal tessellation, embedded either in torus or cylinder, can be modified such that the number of points (*i.e.*, atoms) is preserved; the only allowed changing, in the following, is its connectivity.

There is a well-known edge rotation procedure, patterned by Stone and Wales.[45] The bold edge shares two cycles of size ($s_m$, $s_n$) to be reduced, after rotation, to ($s_{m-1}$, $s_{n-1}$).

Correspondingly, the two cycles joined by this edge will increase their size from ($s_p$, $s_r$) to ($s_{p+1}$, $s_{r+1}$) in the Stone-Wales SW isomerization,[41] (Figure 13a, b). Most often, the edge flipping runs as a cascade SW transformation (Figure 13c).

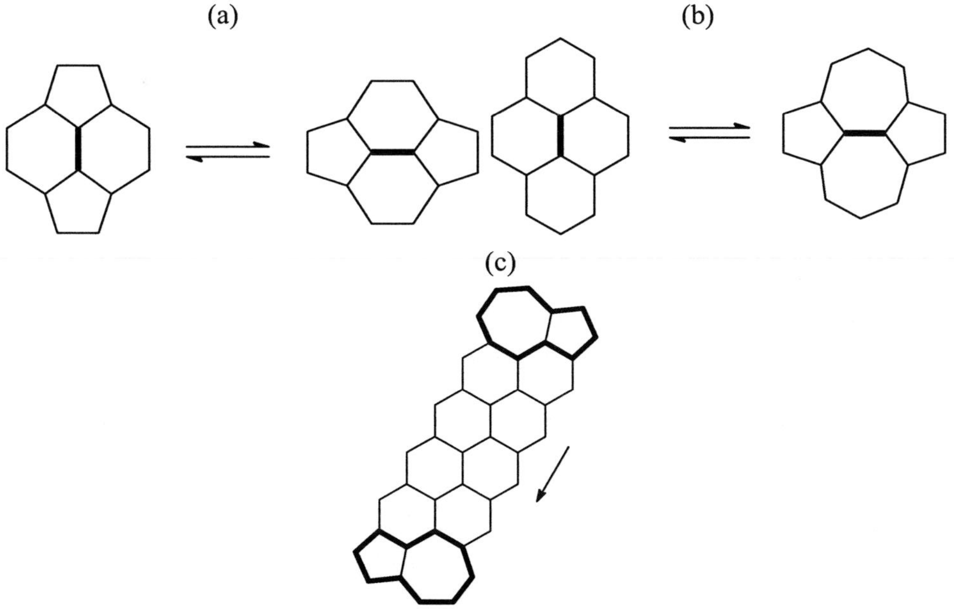


Figure 13: The Stone-Wales rearrangement, as local (a and b) or long-range transformation (c - the glide direction indicated by the arrow).

This section presents several ways of modifying the (6,3) tubular lattice, leading to various coverings, possible isomerization routes in the real experiments.

## 3.1. Isomerizations on (6,3) Net

Let (6,3) net be embedded in the cylinder, as Tu(6,3)Z[$c$,$n$], or, in the ($k$,$l$) notation,[12] "zigzag" ($c$/2,0) nanotube (Figure 14a). The embedding Tu(6,3)A[$c$,$n$] is an "armchair" nanotube and denoted ($c$/2,$c$/2) ((Figure 15a).

Let denote by $H_{(i,j),(p,r)}$ the edges lying parallel to the horizontally oriented tube generator, in the schematic lattice representation; the first subscript bracket encodes the relative location of the start-point of rotating edges along the tube while the second one the location of edges around the tube. Mark $V_{(i,j),(p,r)}$ the edges lying perpendicular to the tube generator. The marked edges will be rotated in the following isomerization and the above symbols play the role of a true *rotational operator*, as shown in the following.[41]

The hexagonal (6,3) covering is transformed into the "rhomboidal-bathroom-floor" tiling[46] R((4, 8)3), by operations:

$$H_{(1,3),(1,3)}((6,3)\mathrm{Z}) = ((4,\ 8)3)\mathrm{R} = V_{(1,3),(1,3)}((6,3)\mathrm{A}) \qquad (9)$$

and the corresponding objects are illustrated in Figures 14 and 15. The starting objects in these figures represent pair isomers of different embeddings in the cylinder of the same (6,3) net having 16 x 8 = 128 atoms.

(a) $H_{(1,3),(1,3)}$ ((6,3)Z) (b) ((4,8)3)R

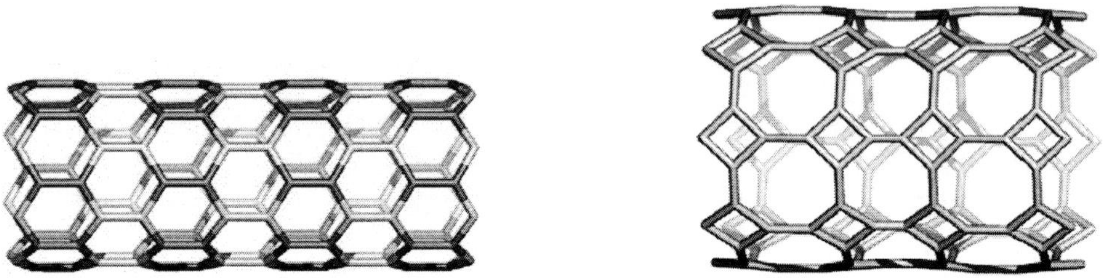

Figure 14: A (6,3)Z nanotube and its SW isomer resulted by rotating the marked bonds.

When edge rotation follows a spiral path, like that illustrated in Figure 16, the covering is, after optimization by an MM procedure, a "tetragonal-bathroom-floor" ((4,8)3)S lattice.[41]

(a) $V_{(1,3),(1,3)}$((6,3)A) (b) ((4, 8)3)R

Figure 15: A (6,3)A nanotube and its SW isomer.

$V_{(1,5),(1,5)}$((6,3)Z) ((4,8)3)S

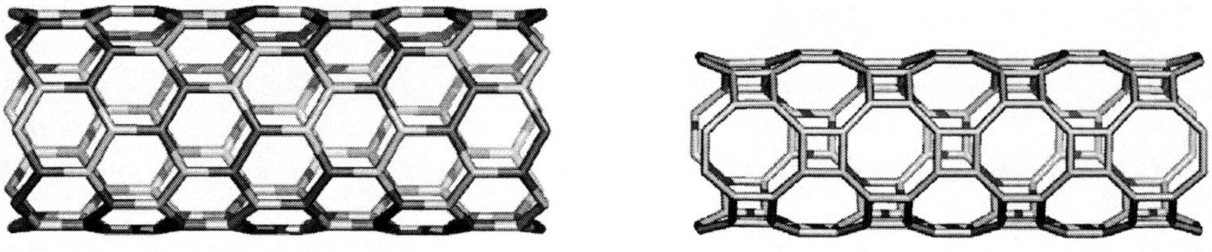

Figure 16: A spiral path of SW edge rotation and its transform.

The operation can be written as:

$$V_{(1,5),(1,5)}((6,3)Z) = ((4,\ 8)3)S \qquad (10)$$

Similarly, the operation:

$$V_{(1,5),(1,5),1a}\ ((6,3)Z) = ((5,\ 7)3)SP \qquad (11)$$

leads to a spiral ((5, 7)3)SP net,[32] Figure 17. Note the combination *V* and Z, for describing a spiral path and the subscript *1a* for a "leave one row/column out" way in getting an "alternating" spiral net.

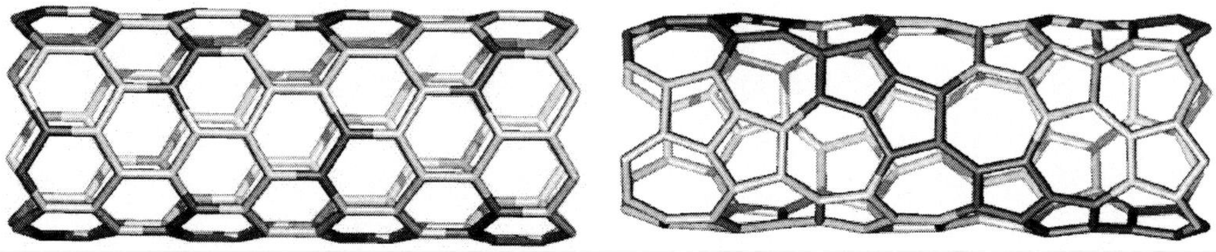


Figure 17: A spiral path of SW edge rotation and its spiral net product.

The same operations can be done on the (6,3)A net thus resulting the corresponding pair embedding isomers.

Different pentaheptite[4] ((5,7)3) lattices can be obtained by the following operations:

$$H_{(1,5),(1,5)}((6,3)\mathrm{Z}) = ((5,7)3)\mathrm{V} \tag{12}$$

$$V_{(1,5),(1,5)}((6,3)\mathrm{A}) = ((5,7)3)\mathrm{H} \tag{13}$$

These coverings will be illustrated in the below.

## 3.2. Isomerizations on ((4,8)3) Net

The ((4,8)3) covering, particularly ((4,8)3)R, transforms to either (6,3)A or (6,3)Z net by operations:

$$H_{(1,4),(1,4)}(((4,8)3)\mathrm{R}) = (6,3)\mathrm{A} \tag{9'}$$

$$V_{(1,4),(1,4)}(((4,8)3)\mathrm{R}) = (6,3)\mathrm{Z} \tag{9''}$$

as a unique intermediate of the (6,3) net isomerization.

Other isomerizations of this covering are:

$$H_{(1,7),(1,7)}\,(((4,8)3)\mathrm{R}) = ((5,7)3)\mathrm{H} \tag{14}$$

$$V_{(1,7),(1,7)}\,(((4,8)3)\mathrm{R}) = ((5,7)3)\mathrm{V} \tag{15}$$

the corresponding objects being illustrated in Figures 18 and 19.

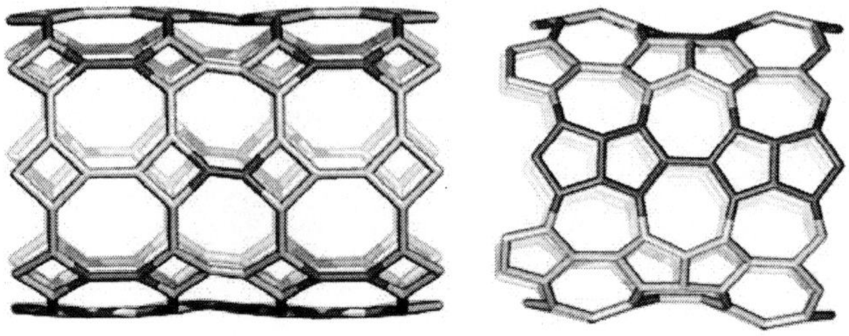


Figure 18: SW isomerization of ((4,8)3)R to ((5,7)3)H lattice.

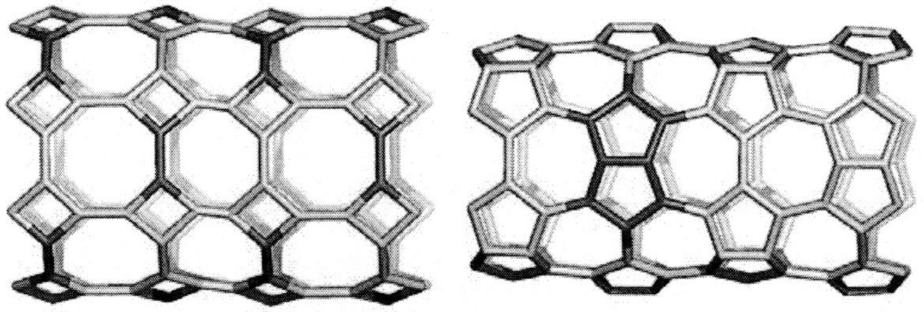


Figure 19: SW isomerization of ((4,8)3)R to ((5,7)3)V lattice.

Note that the pentaheptite H/V((5, 7)3) lattice is encountered in the chemical net of $ThMoB_4$.[47] It is a *2-isohedral* tiling,[48] (*i.e.*, it has only two face orbits), with the local signature $(t_5, t_7) = (1, 3)$. Crespi *et al.*[35] stated that such nanotubes would show metallic behavior. This statement appears quite strange, since these pentaheptite nets can be termed *fully-pentalenoid* (well-known as an anti-aromatic unit) rather than fully-azulenoid, according to Kirby.[49]

Other combinations of SW rotation, on ((4,8)3)R, are as follows:

$$V_{(1,3),(1,6)}(H_{(1,4),(1,7)}(((4,8)3)R) = ((5,7)3)HA \quad (16)$$

$$H_{(1,6),(1,3)}(V_{(1,7),(1,4)}(((4,8)3)R) = ((5,7)3)VA \quad (17)$$

The resulting net is a *2-isohedral* tiling with a (2, 4) signature. It is a periodic (*i.e.*, face-regular) covering, described as a capped tubulene elsewhere.[50] The objects are presented in Figures 20 and 21, respectively, both as optimized tubes and geodesic projections.

Other isomerizations are derivable form the two above ones:

$$H_{(1,6),(1,5)}(V_{(1,7),(1,4)}(((4,8)3)R) = ((5, 6, 7)3)HA \quad (18)$$

$$V_{(1,5),(1,6)}(H_{(1,4),(1,7)}(((4,8)3)R) = ((5, 6, 7)3)VA \quad (19)$$

and the objects are presented in Figures 22 and 23. This novel lattice has the local signature: $t_{5j}(0, 4, 1)$; $t_{6j}(2, 2, 2)$; and $t_{7j}(1, 4, 2)$, $j = 5, 6, 7$.

$H_{(1,4),(1,7)}(((4,8)3)R)V_{(1,3),(1,6)}(H_{(1,4),(1,7)}(((4,8)3)R))$

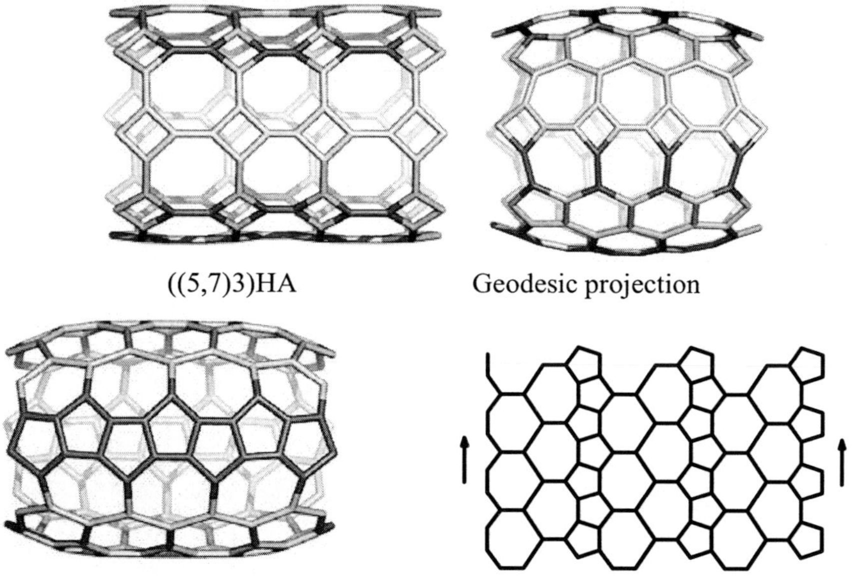


Figure 20: The pathway to ((5,7)3)HA lattice.

$V_{(1,7),(1,4)}(((4,8)3)R)H_{(1,6,),(1,3)}(V_{(1,7),(1,4)}(((4,8)3)R))$

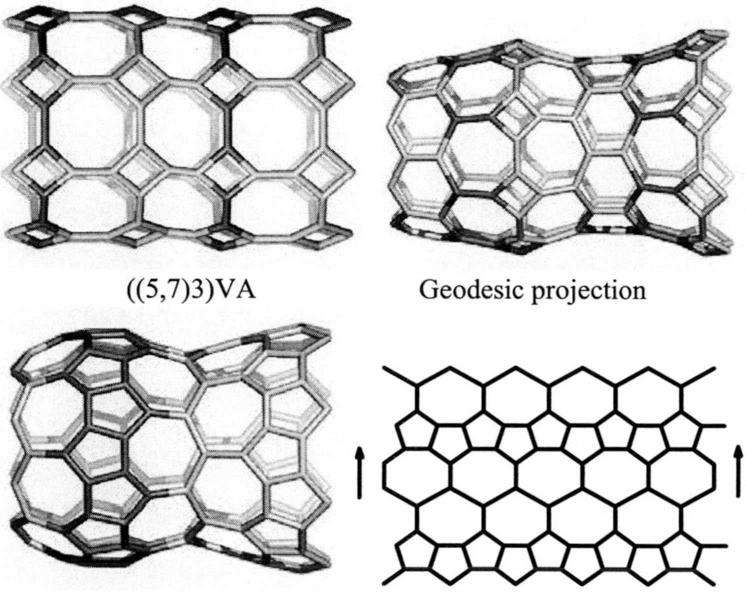


Figure 21: The pathway to ((5,7)3)VA lattice.

In the VA embedding, ((5, 6, 7)3) covering is deducible from $C_{60}$ by cutting off two hexagonal parallel rings. It was described as a capped tubulene elsewhere.[50]

Note that the ((4,8)3) lattices are deductible from the square (4,4) covering, by some basic operations on maps[51] (Section 4).

$H_{(1,4),(1,7)}(((4,8)3)R)V_{(1,5),(1,6)}(H_{(1,4),(1,7)}(((4,8)3)R))$

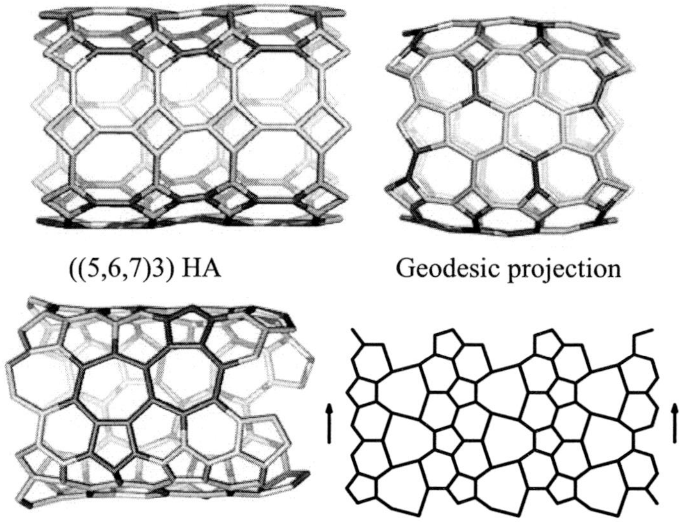


Figure 22: The pathway to ((5,6,7)3)HA lattice.

At this stage, a short comment on the energetic of Stone-Wales isomerization is needed. At high temperature, the nanotube wall yields strain by formation of a dislocation dipole (5-7), a defect in the hexagonal lattice.[52, 53] The primary SW defect is most likely followed by other bond rotations, thus forming some (5-7-7-5) pairs. Relaxation of the nanotube, either by brittle fracture or by plastic glide of the separated dislocations (also termed "ductile" behavior) is achieved, depending on the helicity of the tube and the imposed conditions. Ductile behavior is usually encountered in (*k*,*k*) armchair tubes, and involves the separation of the (5-7-7-5) pairs, by cascade bond flipping, into two individual (5-7) units.

$V_{(1,7),(1,4)}(((4,8)3)R)H_{(1,6),(1,5)}(V_{(1,7),(1,4)}(((4,8)3)R))$

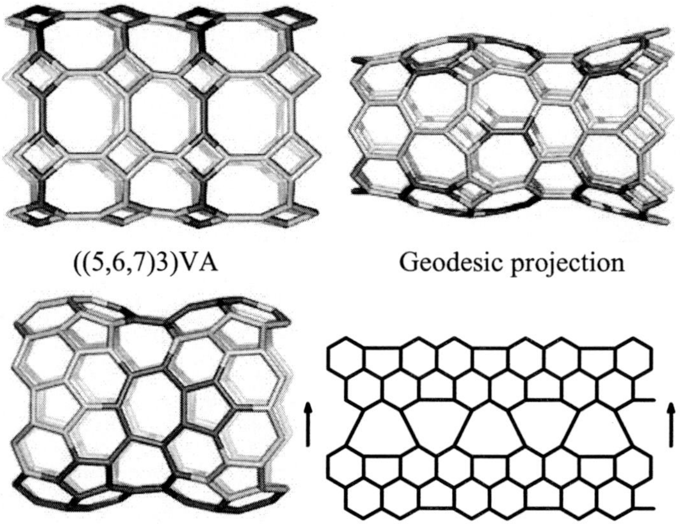


Figure 23: The route to ((5,6,7)3)VA lattice.

The initial (5-7-7-5) defect undergoes an interesting sequence of transformations when the tube is placed under tension in the 5–10% range.[54] Bonds emanating from the vertices of the pentagons rotate to form a hexagon that is separated from the rest of the tube by (5-7) pairs. Further bond rotations increase the number of rotated hexagons in this extended defect. Segments of a tube, with an altered helicity, are thus formed. This is interesting because, depending on the helicity of the initial tubes, such defects naturally form different metal/metal, semiconductor/semiconductor, and/or semiconductor/metal hetero-junctions, possibly leading to all-nanotube-based quantum dot structures.[54] Various addimer-based defect structures were constructed, placed under strain, and annealed at high temperature (2500–3000$^0$ K). The defect structures on armchair tubes were not observed to be stable. More promising, in this respect, appeared the ($k$,0) zigzag tubes. Addimers was found to induce particular sequences of transformations giving rise to different classes of extended defects.

The activation energy for defect formation depends on the applied strain[52] magnitude and direction, as well as on the lattice curvature. A relatively high barrier energy, in the range of 5-7 eV (or about 115-160 kcal/mol), at a strain above 6%, was estimated for the SW edge rotation by several quantum chemical calculations (see also the cited refs. in 52, 53).

The mechanical response of nanotubes subject to tensile strain was theoretically simulated and defects examined experimentally by scanning tunneling microscopy STM.[54],[55]

The above SW edge rotations have been performed by the CageVersatile 1.1 original program.[56]

## 4. Operations on Maps

A map $M$ is a combinatorial representation of a closed surface.[57],[58] Several transformations (*i.e.*, operations) on maps are known and used for various purposes.

Let us denote in a map: $v$ - number of vertices, $e$ - number of edges, $f$ - number of faces and $d$ - vertex degree. A subscript "0" will mark the corresponding parameters in the parent map.

Recall some basic relations in a map:

$$\sum d\,v_d = 2e \qquad (20)$$

$$\sum s\,f_s = 2e \qquad (21)$$

where $v_d$ and $f_s$ are the number of vertices of degree $d$ and number of $s$-gonal faces, respectively. The two relations are joined in the famous Euler[59] formula:

$$v - e + f = \chi(M) = 2(1-g) \qquad (22)$$

with $\chi$ being the Euler *characteristic* and $g$ the genus[5] of a graph (*i.e.*, the number of handles attached to the sphere to make it homeomorphic to the surface on which the given graph is embedded; $g = 0$ for a planar graph and 1 for a toroidal graph). Positive/negative $\chi$ values indicate positive/negative curvature of a lattice. This formula is useful for checking the consistency of an assumed structure.

In this section the attention is focused on operations allowing the transformation of both 4- and 3-valent lattices into coverings of chemical interest.

## 4.1. Simple Operations on Maps

***Dualization Du*** of a map starts by locating a point in the center of each face. Next, two such points are joined if their corresponding faces share a common edge. It is the (Poincaré) *dual Du*(*M*). The vertices of *Du*(*M*) represent the faces of *M* and *vice-versa*.[57] Thus the following relations exist:

$$Du(M): \tag{23}$$

Dual of the dual returns the original map: *Du*(*Du*(*M*)) = *M*. Tetrahedron is self dual while the other Platonic polyhedra form pairs: *Du*(Cube) = Octahedron; *Du*(Dodecahedron) = Icosahedron (see Figure 24 for symbols hereafter used). It is also known the Petrie dual.[57]

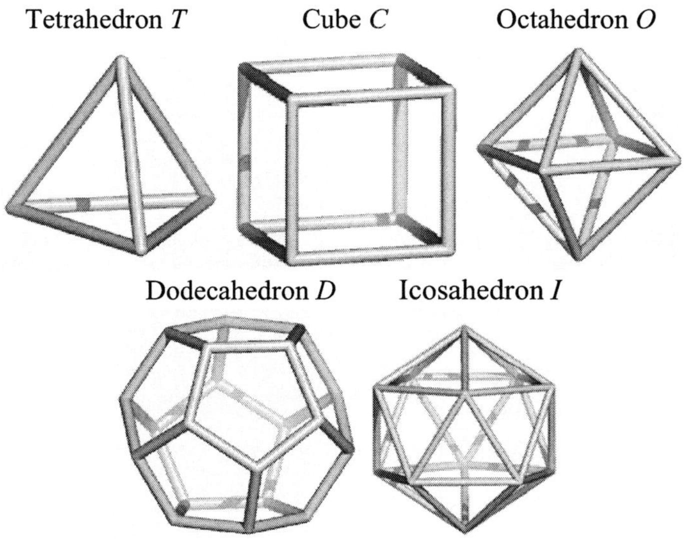


Figure 24: The five Platonic polyhedra.

*Medial Me* is achieved as follows:[57] put the new vertices as the midpoints of th e original edges. Join two vertices if the original edges span an angle (and are consecutive within a rotation path around their common vertex in *M*). Medial is a 4-valent graph and *Me*(*M*) = *Me*(*Du*(*M*)), as illustrated in Figure 25. The transformed parameters are:

$$Me(M): \tag{24}$$

Medial operation rotates parent $s$-gonal faces by $\pi/s$. Points in the medial represent original edges, thus this property can be used for topological analysis of edges in the parent polyhedron. Similarly, the points in dual give information on the topology of parent faces.

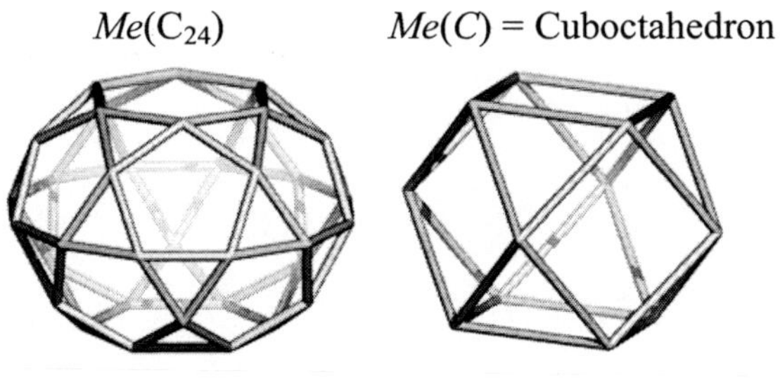


Figure 25: Medial.

***Truncation Tr*** is achieved by cutting off the neighborhood of each vertex by a plane close to the vertex, such that it intersects each edge incident to the vertex.[57] Truncation is similar to the medial, the transformed parameters being:

$$Tr(M): \tag{25}$$

This was the main operation used by Archimedes in building up the well-known 13 solids.[3] Figure 26 illustrates this operation.

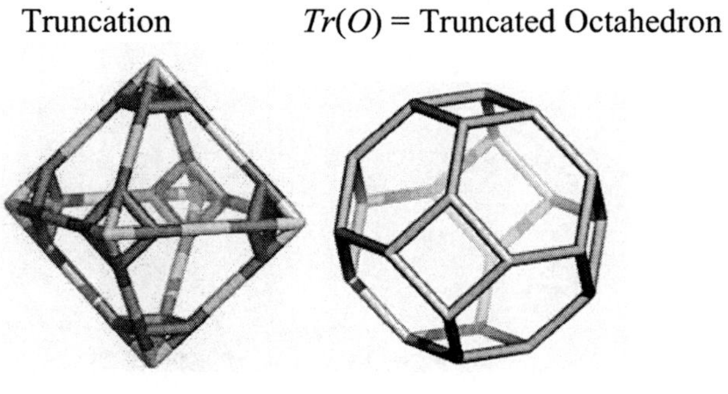


Figure 26: Truncation.

***Polygonal $P_s$ capping*** ($s$ = 3, 4, 5) of a face is achieved as follows:[51] add a new vertex in the center of the face. Put $s$-3 points on the boundary edges. Connect the central point with one vertex (the end points included) on each edge. In this way the parent face is covered by trigons ($s$ = 3), tetragons ($s$ = 4) and pentagons ($s$ = 5). The $P_3$ operation is also called *stellation* or (centered) *triangulation*. When all the faces of a map are thus operated, it is referred to as an omnicapping $P_s$ operation. The resulting map shows the relations:

$$P_s(M): \tag{26}$$

$$e = se_0$$

$$f = s_0 f_0$$

so that the Euler's relation holds.

Maps transformed by the above operations form dual pairs:

$$Du(P_3(M)) = Le(M) \quad (27)$$

$$Du(P_4(M)) = Me(Me(M)) \quad (28)$$

$$Du(P_5(M)) = Sn(M) \quad (29)$$

Vertex multiplication ratio by this dualization is always:

$$v(Du)/v_0 = d_0 \quad (30)$$

Relations (27) to (29) enable the construction of Archimedes/Catalan objects, when applied on the Platonic solids.[3],[57] Relations (27) and (29) will be detailed below. Eq 30 comes out from: $v(Du) = f(P_s(M)) = s_0 f_0 = d_0 v_0$ (eqs 23 and 26). Note that all the operation parameters herein presented refer to regular maps (*i.e.*, having all vertices and faces of the same valence/size).

Figures in the following give details and examples of the operations realization.

$P_3(C)$ $P_3(D)$

Figure 27: $P_3$ operation = stellation, (centered) triangulation, etc.

$P_4(C) = (C4)^a$ $P_4(D) = (C10)^a$

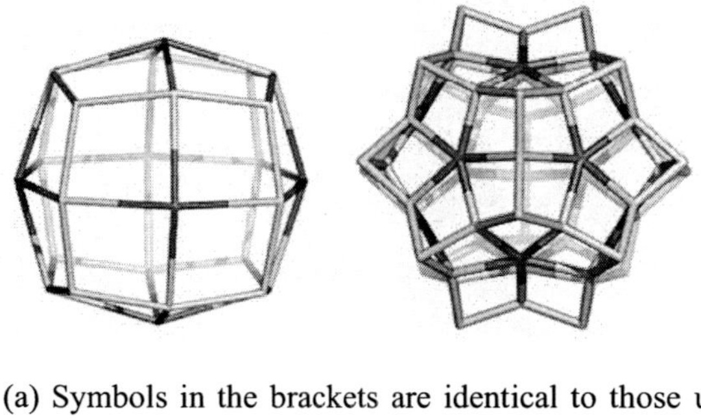

Figure 28: $P_4$ operation; (a) Symbols in the brackets are identical to those used in ref. 3 for Catalan objects (*i.e.*, duals of the Archimedean solids)

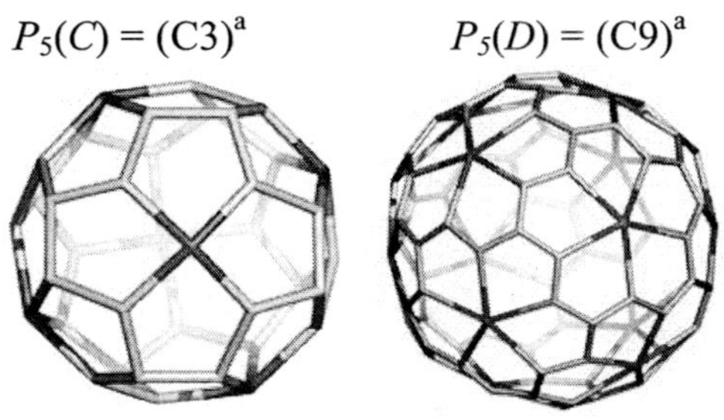


Figure 29: $P_5$ operation = pentangulation; (for other operation names the reader may consult www.georgehart.com\virtual-polyhedra\conway_notation.html) (a) see Figure 28.

## 4.2. Composite Operations

**Leapfrog** *Le* is a composite operation,[60-65] which can be written as:

$$Le(M) = Du(P_3(M)) = Tr(Du(M)) \tag{31}$$

A sequence of stellation-dualization rotates the parent s-gonal faces by π/s. Leapfrog operation is illustrated, for a tetragonal face, in Figure 30.

A bounding polygon, of size $2d_0$, is formed around each original vertex. In the most frequent cases of 4- and 3-valent maps, the bounding polygon is an octagon and a hexagon, respectively (Figure 31).

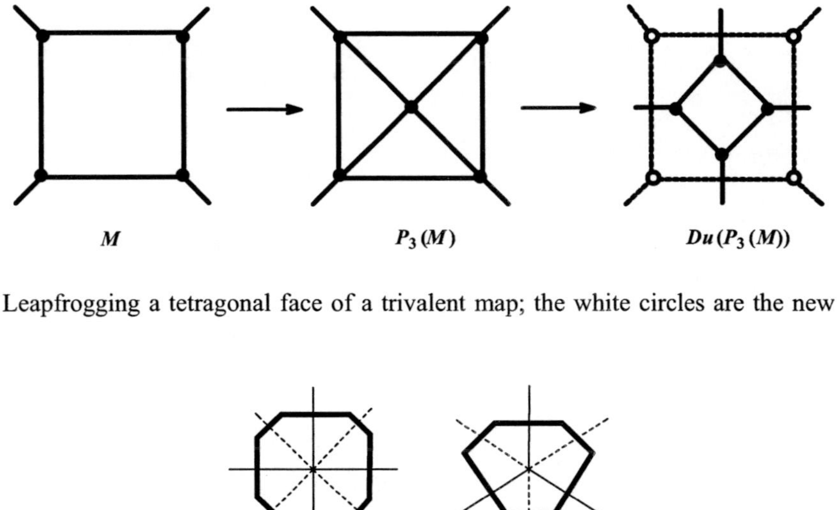


Figure 30: Leapfrogging a tetragonal face of a trivalent map; the white circles are the new vertices of *Le*(*M*).

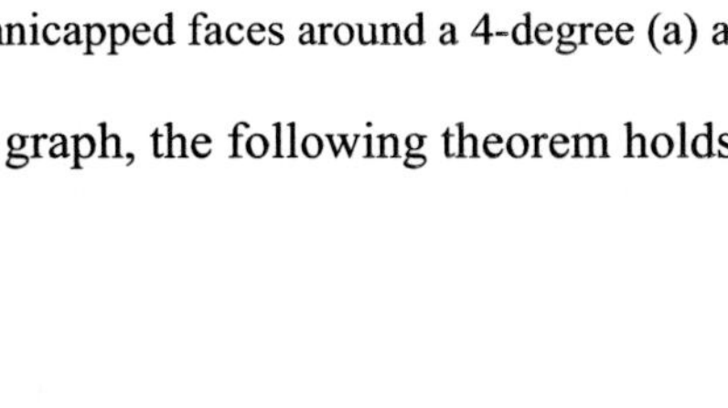


Figure 31: Dualization of the omnicapped faces around a 4-degree (a) and 3-degree (b) vertex.

If the map is a $d_0$ regular graph, the following theorem holds:[64],[65]

***Theorem 1***

The number of vertices in Le(M) is $d_0$ times larger than in the original map M, irrespective of the tessellation type.

The demonstration follows from the observation (Figure 30) that, for each vertex of $M$, $d_0$ new vertices result in $Le(M)$:

$$v / v_0 = d_0 v_0 / v_0 = d_0 \tag{32}$$

The same result can be derived from eqs 23, 26 and 27 (with $s = 3$) or directly from eqs 27 and 30. In trivalent maps, $Le(M)$ is the *tripling* operation. The complete transformed parameters are:

$$Le(M): \tag{33}$$

$$e = 3e_0$$

$$f = v_0 + f_0$$

being the same as for $Tr(M)$, eq 25.

Note that the vertex degree in $Le(M)$ is *always* 3, as a consequence of the involved triangulation $P_3$. In other words, dual of a triangulation is a *cubic net*.[57] It is also true that truncation always provides a trivalent net.

A nice example of using *Le* operation is: *Le*(Dodecahedron) = Fullerene $C_{60}$. The leapfrog operation can be used to insulate the parent faces by surrounding bounding polygons (see below).

***Quadrupling*** *Q* is another composite operation, achieved by the sequence:

$$Q(M) = E\_(Tr_{P_3}(P_3(M))) \tag{34}$$

where $E\_$ means the (old) edge deletion (dashed lines, in Figure 32) of the truncation $Tr_{P3}$ of each vertex where $P_3$ capping faces of a parent face are incident.

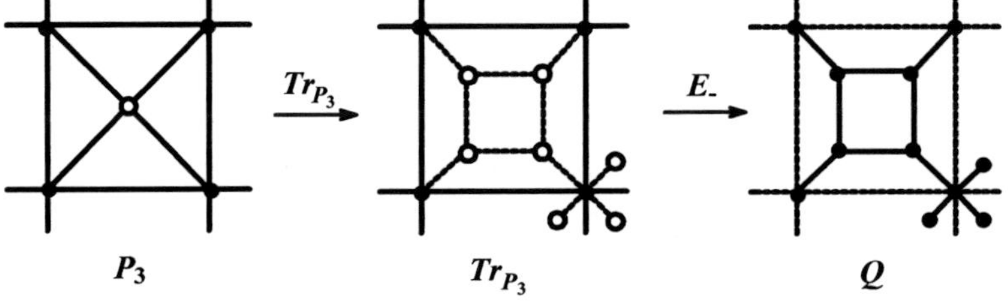


Figure 32: Quadrupling a tetragonal face of a 4-valent map.

***Theorem 2***

The vertex multiplication ratio in a Q transformation is $d_0 + 1$ irrespective of the original map tessellation.

With the observation that, for each vertex of *M*, $d_0$ new vertices result in *Q*(*M*) and the old vertex is preserved (Figure 32), the demonstration is immediate:[64],[65]

$$v = d_0 v_0 + v_0 \tag{35}$$

$$v / v_0 = d_0 + 1 \tag{36}$$

The complete transformed parameters are:

$$Q(M): \tag{37}$$

*Q* operation leaves unchanged the initial orientation of the polygonal faces. Note that, the quadrupling transform of a 4-valent map is not a regular graph anymore (because of mixing the new trivalent vertices with the parent 4-valent ones). Only *Q*(*M*) of a 3-valent map is a 3-regular graph.

*Q* insulates the parent faces always by hexagons. An example of this operation is: *Q* (Dodecahedron) = Fullerene $C_{80}$. It is also called "chamfering" (edge chamfering being equivalent to vertex truncation).[66]

Figure 33 illustrates the two above operations.

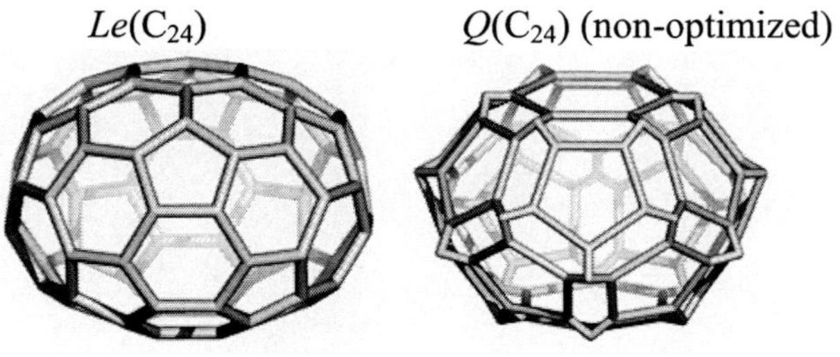


Figure 33: Two composite operations, leapfrog *Le* and quadrupling/chamfering *Q* applied on $C_{24}$

***Snub Sn*** is a composite operation that can be written as:

$$Sn(M) = Dg(Me(Me(M))) = Du(P_5(M)) \tag{38}$$

where *Dg* means the inscribing diagonals in the quadrilaterals resulting by *Me*(*Me*(*M*)).[57]

The true dual of the snub is the $P_5(M)$ transform:[51] $Du(Sn(M)) = P_5(M)$. Similar to the medial operation, *Sn*(*M*) = *Sn*(*Du*(*M*)). In case of *M* = *T*, the snub *Sn*(*M*) = *I*.

Of chemical interest is the easy transformation of the snub (*i.e.*, regular pentavalent graph) into the leapfrog transform (*i.e.*, regular trivalent graph - see below), by deleting the edges of the triangle joining any three parent faces (Figure 34, in black). This is particularly true in the snub of trivalent parent maps.

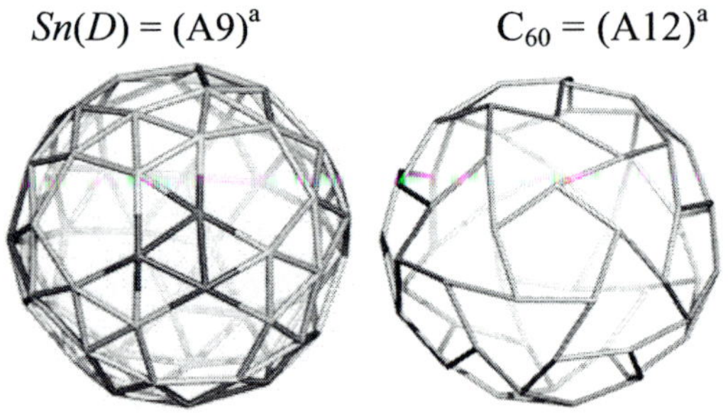


Figure 34: Snub; (a) Symbols in the brackets are identical to those used in ref. 3 for the Archimedean solids. Note the insulated pentagons in $C_{60}$.

The transformed parameters are derivable from eqs 23, 26 and 29 (with $s = 5$):

$$Sn(M): \tag{39}$$

$$e = 5e_0$$

$$f = v_0 + 2e_0 + f_0$$

The multiplication ratio is $d_0$ (eqs 29 and 30), the same as for $Le(M)$, both of them involving the dualization.

***Capra*** *Ca* - the goat, is the Romanian corresponding of the *leapfrog* English children game. It is a composite operation,[51],[67] necessarily coming from the Goldberg's[66] multiplying factor $m$:

$$m = (a^2 + ab + b^2); a \geq b; a + b > 0 \tag{40}$$

that predicts $m$ (in a 3-valent map) as shown in Table 1:

**Table 1: Goldberg factors m by operations on a 3-valent map**

| Operation | Symbol | *m* |
|---|---|---|
| identity | *I*: (1,0) | 1 |
| leapfrog | *Le*: (1, 1) | 3 |
| quadrupling (chamfering) | *Q*: (2, 0) | 4 |
| capra | *Ca*: (2, 1) | 7 |

The $m$ factor is also related to the formula giving the volume of truncated pyramid, of height $h$: $V = mh/3$, coming from the ancient Egypt. According to the $m$-value in trivalent maps, Capra is the *septupling* $S_1$ operation. A second septupling operation $S_2$ was defined in ref. 51.

Capra is achieved by truncating the vertex located in the center of parent faces of a pentangulation $P_5$ transform (Figure 35). Note that, $P_5$ involves an $E_2$ (*i.e.*, edge trisection).

This operation results in a map that preserves the original vertices while the parent $s$-gonal faces are rotated by $\pi/(3/2)s$.

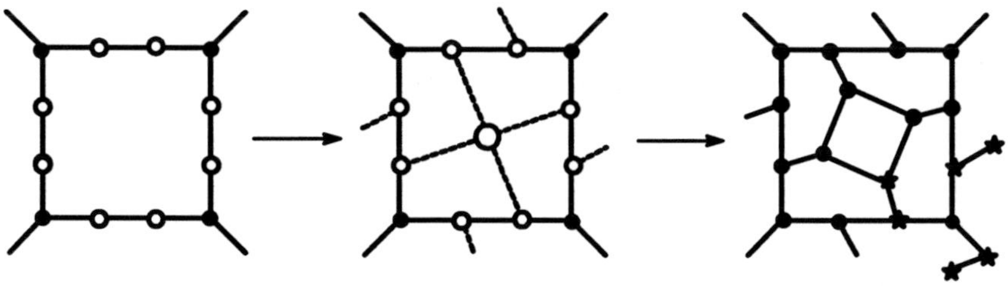


Figure 35: Septupling a tetragonal face of a 3-valent map.

The transformation can be written as:[67]

$$Ca(M) = Tr_{P_5}(P_5(M)) \tag{41}$$

with $Tr_{P5}$ meaning the truncation of each vertex where $P_5$ capping faces of a parent face are incident.

*Ca* insulates any face of *M* by its own hexagons, which are not shared with any old face (in contrast to *Le* or *Q*).

***Theorem 3***

The vertex multiplication ratio in a Ca transformation is $2d_0 + 1$ irrespective of the original map tiling.

Observe that, for each vertex of *M*, $2d_0$ new vertices (Figure 35, the starred ones) result in *Ca*(*M*) and the old vertex is preserved, thus demonstrating the theorem:[51]

$$v = 2d_0 v_0 + v_0 \tag{42}$$

$$v / v_0 = 2d_0 + 1 \tag{43}$$

*Ca* operation can continue by an $E_n$ homeomorphic transformation of the edges bounding the parent-like faces (*i.e.*, those resulted by $Tr_5$) thus resulting open maps with all polygons of the same (6+$n$) size. This last simple operation is hereafter denoted[51] by $Op_n$. The sequence leading to open objects within this operation can be written as:

$$Op_n(Ca(M)) = Op_n(Tr_{P_5}(P_5(M))) \tag{44}$$

When $n = 1$, this specification is omitted. The $Op_1$ objects (*i.e.*, molecules) have open-shell $\pi$-electronic structure, eventually including non-bonding orbitals.

In case of a 3-valent regular map the vertex multiplication ratio by Capra is 7. Clearly, maps/graphs of degree greater than 3 will not be regular anymore. Table 2 lists the net parameters in Capra transforms,[67] either in simple or iterative application on finite or infinite (*i.e.*, open) objects.

**Table 2: The transformed parameters by the iterative *Ca* operation**

| Operation | | Parameter | eq |
|---|---|---|---|
| 1 | $Ca(M)$ | $v_1 = v_0 + 2e_0 + s_0 f_0 = (2d_0 + 1)v_0$<br>$e_1 = 7e_0$<br>$f_1 = s_0 f_0 + f_0$ | (45) |
| 2 | $Op(Ca(M))$ | $v_{1Op} = v_1 + s_0 f_0 = (3d_0 + 1)v_0$<br>$e_{1Op} = 9e_0$<br>$f_{1Op} = f_1 - f_0 = s_0 f_0$ | (46) |
| 3 | $Ca_2(M)$ | $v_2 = v_1 + 2e_1 + s_1 f_1 = v_1 + 2e_1 + 7s_0 f_0 = v_0 + 32e_0$<br>$e_2 = 7^2 e_0$<br>$f_2 = s_1 f_1 + f_1 = 8s_0 f_0 + f_0$ | (47) |
| 4 | $Op(Ca_2(M))$ | $v_{2Op} = v_{1Op} + 2e_{1Op} + 7f_{1Op} = v_0 + 38e_0$<br>$e_{2Op} = 3v_0 + 108e_0$<br>$f_{2Op} = (7+1)f_{1Op} = 16e_0$ | (48) |
| 5 | $Ca_n(M)$ | case $d_0 > 3$<br>$v_n = 8v_{n-1} - 7v_{n-2}\ ; n \geq 2$<br>case $d_0 = 3$<br>$v_n = 7^n \cdot v_0$<br>$e_n = 7^n \cdot e_0$<br>$f_n = (7^n - 1) \cdot (s_0 f_0)/6 + f_0$ | (49)<br><br>(49′)<br>(49″)<br>(49‴) |

Since pentangulation of a face can be done either clockwise or counter-clockwise, it results in an enantiomeric pair of objects: *CaS*(*M*) and *CaR*(*M*), in terms of the sinister/rectus stereochemical isomers.[67] Figure 36 illustrates Schlegel projections of such a pair derived from the Cube.

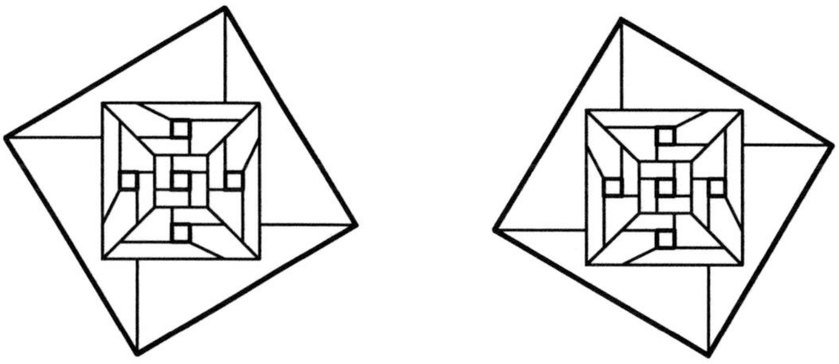

Figure 36: Enantiomeric pair of the Capra transforms of the Cube (Schlegel projection)

The sequence *CaS*(*CaS*(*M*)) results in a twisted transform, while *CaR*(*CaS*(*M*)) will straighten the (central part of) structure. The sequence *Du*(*Op*(*Ca*(*M*))) is just the snub *Sn*(*M*), being this transform a chiral one. Examples will be given in Section 5.

## 4.3. Supra-Coverings

According to the Clar theory, any polyhedral map may be looked for a perfect Clar PC structure[68],[69] (Figure 37), which is a disjoint set of faces (built up on all vertices in *M*) whose boundaries form a 2-factor. Recall that a *k*-factor is a regular *k*-valent spanning subgraph. A PC structure is associated with a Fries structure,[70] which is a Kekulé structure having the maximum possible number of benzenoid (alternating single-double edge) faces. A Kekulé structure is a set of pairwise disjoint edges/bonds of the molecule (over all its atoms) that coincides with a perfect matching and a 1-factor in Graph Theory.[71] A trivalent polyhedral graph, like that of fullerenes, has a PC structure if and only if it has a Fries structure.[58] Such structures represent *total resonant sextet* TRS benzenoid molecules and it is expected to be extremely stable, in the valence bond theory.[58],[72] Leapfrog *Le* is the only operation that provides PC transforms (see Figure37a, $M = C_{60}$).

Coverings larger than benzene, like naphthalene or azulene (*i.e.*, a pair of pentagonal-heptagonal carbon rings), appear only slightly less favorable, in the valence-bond based conjugated circuit theory.[2],[34],[72],[73]

By extension, corannulenic supra-faces[51] (eventually called *flowers* - Figure 37b,c) may be viewed as (*i*) intersected, (*ii*) joint or (*iii*) disjoint units. While the first case is a trivial one, the last two deserve more attention. A coronene-like flower is hereafter symbolized as [*n*]Fw[*x*,*y*], with *n*, *x*, *y* being the folding of the core and its surrounding polygonal faces, respectively. When $x = y = 6$ the last bracket can be omitted.

A joint corannulenic JCor structure is usually an Archimedean (([5]Fw,[6]Fw)3) supra-covering. Exceptional cases are the Platonic (([5]Fw)3) (a) and (([6]Fw)3) (b) tessellations. A tiling is called Platonic if it consists of a single type of faces (as the Platonic solids). Archimedean is that tiling consisting of more than one (usually two) type(s) of faces. Note the extension of Platonic and Archimedean tessellation, including here supra-faces (*e.g.*, [*n*]corannulenes) among simple polygonal faces.

The case (a) is unique: $Ca(D) = C_{140}$, having a single type of flowers, namely [5]Fw, as a supra-face, but is not a proper JCor, since the additional PC-Fries requirement is not obeyed.

This is achieved by the sequence (*Le*,*Ca*), the only way to the JCor covering. Partial joined flowers could appear by iterating *Le* operation (Figure 37a).

The case (b) is encountered in polyhex tori built up by (*Le*,*Ca*) sequence. Note the metallic and chiral character of such toroidal objects.

A disjoint corannulenic DCor structure is a *disjoint set* of flowers, covering all the vertices in the molecular graph; it represents the connected equivalent of the Fries structure. Only the (*Le*,*Q*) sequence provides a DCor covering. Such a transform is necessarily associated with a PC structure, but the reciprocal is not always true. This supra-covering may be an Archimedean one. By virtue of the involved *Le* operation, both JCor and DCor are associated to PC.

The supra-organized corannulenic units are expected to contribute to the stability of the whole molecule (see Table 3).

(a) *Le*($C_{60}$) = $C_{180}$, $PCC_{180}$, complementary structure

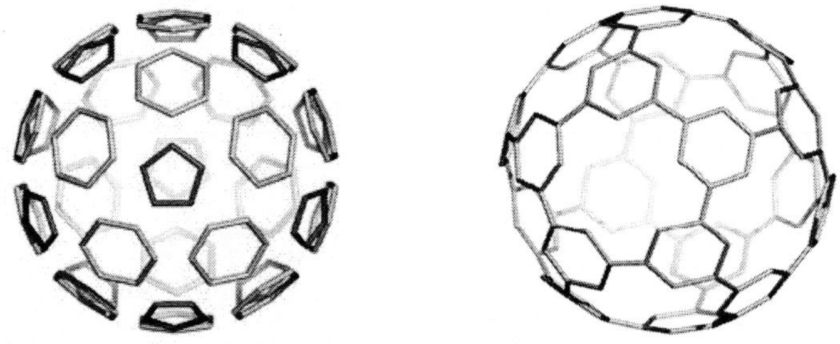

(b) *Q*($C_{60}$) = $C_{240}$, Dcor  $C_{240}$, complementary (Fries-like) structure

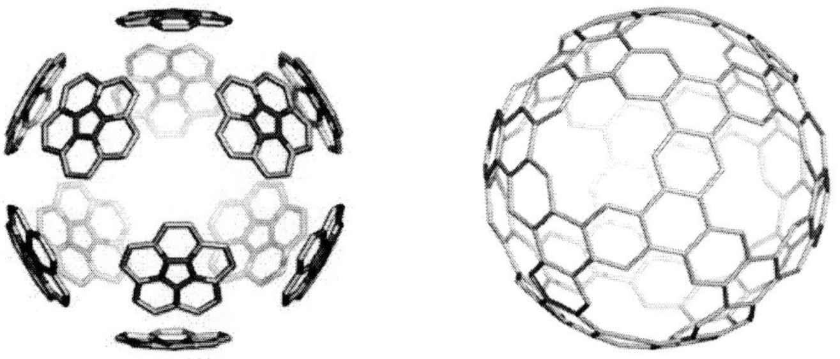

(c) *Ca*($C_{60}$) = $C_{420}$, Jcor$C_{420}$, complementary structure

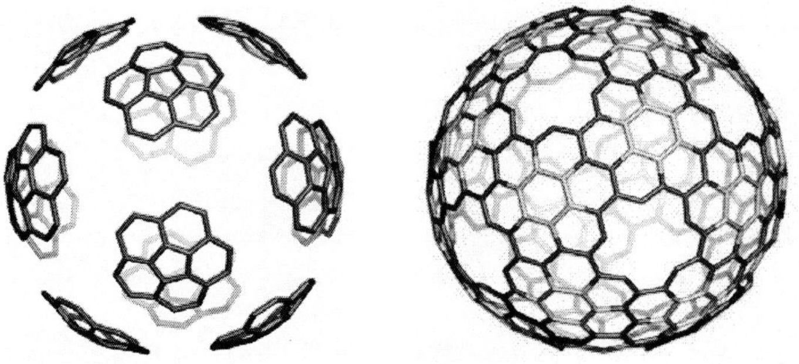

Figure 37: The three operations on maps: Leapfrog *Le*, Quadrupling *Q* and Capra *Ca*, and their transforms on $C_{60}$, along with their perfect Clar PC, corannulenic DCor/JCor and Fries-like structures (represented here as complementary, connected structures).

The need for a disjoint DCor system comes from the sense of the (large) ring current. Theoretical calculations by the ipsocentric CTOCD method[74] show diatropic current circulation, close to the corannulenic external boundary, which intensity could diminish due to counter-sense currents, in eventual adjacent (*i.e.*, edge sharing) Cor units.[75-78] This could be the case of JCor covering (Figure 37c).

**Table 3: PM3 data for families of cages, at least one member showing DCor covering.**

| | Name | Sym | HF /at (kcal/mol) | Gap (eV) | DIFF[d] (kcal/mol) | Type |
|---|---|---|---|---|---|---|
| | **$N = 96$** | | | | | |
| 1 | *Le*(*Ro*(32:5))[a,b] | $C_{3h}$ | 14.595 | 5.754 | -42.606 | Clar |
| 2 | *Le*(*Q* (Cube)) | $O_h$ | 14.152 | 5.970 | 0 | Clar/Cor |
| | **$N = 240$** | | | | | |
| 3 | *Le*(80)[c] | $D_{5d}$ | 8.500 | 3.697 | -459.815 | Clar |
| 4 | *Le*(*Ro*1(80:1)) | $C_{2h}$ | 7.055 | 4.686 | -113.064 | Clar |
| 5 | *Le*(*Ro*2(80:1)) | $C_{2h}$ | 7.021 | 4.732 | -104.831 | Clar |
| 6 | *Le*(*Ro*3(80:1)) | $C_{2h}$ | 7.067 | 4.696 | -115.997 | Clar |
| 7 | *Le*(80:1) | $D_{5d}$ | 6.756 | 4.773 | -41.298 | Clar |
| 8 | *Le*(80:6) | $D_{5h}$ | 6.664 | 5.060 | -19.157 | Clar |
| 9 | *Le*(*Q* (20:1)) = *Le*(80:7) | $I_h$ | 6.584 | 5.113 | 0 | Clar/Cor |
| | **$N = 288$** | | | | | |
| 10 | *Le*(*Ro*1(96:185)) | $S_4$ | 6.361 | 4.737 | -0.518 | Clar |
| 11 | *Le*(*Ro*2(96:185)) | $C_s$ | 6.349 | 4.730 | 2.813 | Clar |
| 12 | *Ro*(*Le*(*Ro*3(96:185))) | $C_s$ | 6.384 | 3.826 | -7.324 | - |
| 13 | *Le*(*Ro*3(96:185)) | $C_{2v}$ | 6.394 | 4.742 | -10.011 | Clar |
| 14 | *Le*(96:185) | $C_2$ | 6.425 | 4.764 | -18.988 | Clar |
| 15 | *Le*(96:184) | $C_{6h}$ | 6.387 | 4.752 | -8.150 | Clar |
| 16 | *Le*(*Q*(24:1)) = *Le*(96:187) | $D_{6d}$ | 6.359 | 4.696 | 0 | Clar/Cor |
| | **$N = 360$** | | | | | |
| 17 | *Q*(*Ro*1(90:1)) | $C_s$ | 5.846 | 3.638 | -56.983 | - |
| 18 | *Le*(*Le*(40:39)) | $D_{5d}$ | 6.099 | 4.472 | -147.992 | Clar |
| 19 | *Q*(*Le*(30:3)) | $C_{2v}$ | 5.942 | 4.435 | -91.433 | Clar /Cor |
| 20 | *Q*(*Le*(30:2)) | $C_{2v}$ | 5.867 | 4.307 | -64.481 | Clar /Cor |
| 21 | *Q*(*Le*(30:1)) = *Q*(90:1) | $D_{5h}$ | 5.688 | 4.064 | 0 | Clar /Cor |

*a* Isomer resulting by Stone-Wales edge rotation

*b* Numbering in *An Atlas of Fullerenes*, P. W. Fowler and D. E. Manolopolous, Oxford University Press, Oxford, U.K., 1995.

*c* $80(5\ 5^5\ (6^5)^6\ 5^5\ 5)$

*d* DIFF = difference in global HF energy between *Le*(*Q*(*M*)) and the other objects within a given family of this table

Calculations given in Table 3 have been performed by using the PM3 Hamiltonian (as provided by HyperChem package).[79]

The *Le*(*Q*(*M*)) transforms appear to have the lowest heat of formation HF per atom, in comparison to other members of these families, selected at *N* divisible by 3×4 (*i.e.*, the multiplicity corresponding to *Le* and *Q*, respectively). The unique *N* value within a family ensured no size dependence of the results.

Even though no complete search was made among thousands of possible topological isomers of such rather large cages, the trend is clear, and some close (or even lower) values (entries 10 to 12, Table 3) to each family *Le*(*Q*(*M*)) reference value (entries 2, 9, 16 and 21) can be eventually explained by some distortive effect of π-electrons.[80] If more than one PCorS structures are possible, those with higher symmetry would eventually have lower HF values (entries 19-21).

Note that *Le* operation, when operates alone, not necessarily leads to more stable molecules despite the PC character of the resulting transforms (compare entries 12, 17 with 13, 18). The other members of these families were generated by some Stone-Wales rotations leading, in general, to structures with lower symmetry.

As shown above, if a molecule has a DCor covering, an additional stabilization is expected. It is also possible that such molecules behave in magnetic field similarly to planar corannulenes, yet investigated by several authors.[75-78] We hope the hypothesis concerning DCor (see also[81-83]) will be further proved in the near future.

Apart of the energetic aspects, we conclude here that *Le*(*M*) provides disjoint PC covering while *Q*(*Le*(*M*)) covers a trivalent lattice by a disjoint corannulenic DCor pattern. *Ca*(*M*) paves a map by joint Cor units and a combination with the other two operations cannot be immediately interpreted.

## 5. Examples of *Le*, *Q*, and *Ca* Realization

This section illustrates the "molecular" realization or, in other words, the transformation of molecules (such as graphitoids) by the Section title mathematical operations.

It is well-established that *Le* operating on fullerenes provides properly closed shell, Clar fullerenes.[58],[84] In (6,3) tori, it is involved in criteria (divisibility by 6 of the hexagonal net dimensions[58],[60-63],[72],[84]) for metallic character.[27]

The two remaining operations, more related to each other, do not change, in general, the shell-type of the parent. Exceptions exist, the most notably being the transformed *Ca* of nanotubes, which all have properly closed shell, disregarding the character of their parent shell.[67]

### 5.1. Molecular Structures

Recall that *Le* and *Ca* rotate the parent bonds while *Q* preserves the original orientation. Among the three operations, only *Ca* provides chiral transforms.

A sequence of *CaS, CaR* provides non-twisted/non-chiral objects (Figure 38). This is more clearly observed when spiral chirality is developed, as shown in Figure 39.

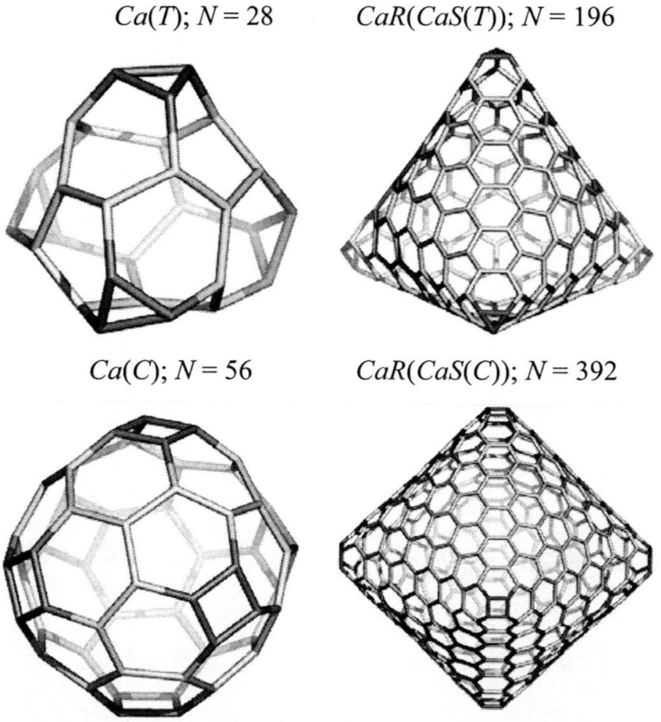


Figure 38: Capra transforms of Tetrahedron $T$ and Cube $C$.

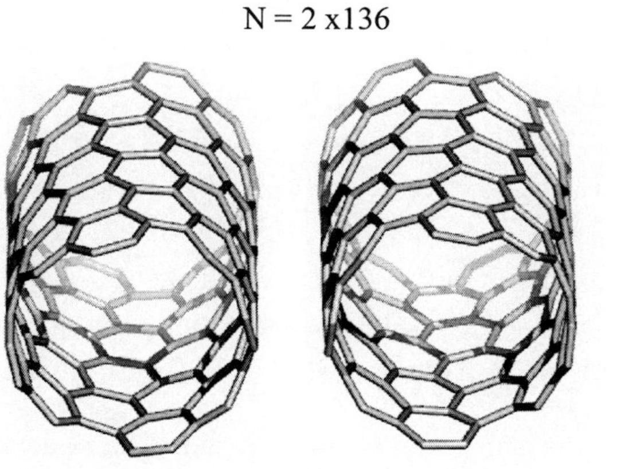


Figure 39: Spiral chirality in nanotubes by Capra operation

Observe that a map operation, performed on the Platonic solids, leads to a transformed lattice embedded in the dual $Du(M)$, as can be seen in Figure 38. Iterative operations will not further change the primary embedding, thus all derived objects belong to the symmetry class of $Du(M)$. This is exemplified by the sequence: $Tr(Du(D)) = Le(C_{20}) = C_{60}$, the transform belonging to the icosahedral $I_h$ symmetry.

The bond rotation has to be kept in mind when operating on orientable maps. Table 4 gives the multiplication of dimensions and re-orientation of the transform bonds, according to each embedding in polyhex tori.

**Table 4: Dimension multiplication in polyhex tori by leapfrog *Le*, quadruple/chamfering *Q* and capra *Ca* operations (for the last one, objects also given in (*k*,*l*,*p*,*q*) notation)**

| | Start | Final | Start | Final |
|---|---|---|---|---|
| | ***Le*** | | | |
| 1 | H/Z[*c*,*n*] | V/A[*c*,3*n*]<br>$N = 3cn$ | V/A[*c*,*n*] | H/Z[3*c*,*n*]<br>$N = 3cn$ |
| | ***Q*** | | | |
| 2 | H/Z[*c*,*n*] | H/Z[2*c*,2*n*]<br>$N = 4cn$ | V/A[*c*,*n*] | V/A[2*c*,2*n*]<br>$N = 4cn$ |
| | ***Ca*** | | | |
| 3 | H/Z[*c*,*n*][a] | $HV(c/2)[(3/2)c,(14/3)n]$<br>$HH(c/2)[(5/2)c,(14/5)n]$<br>$HHc[2c,(7/2)n]$<br>$N = 7cn$ | V/A[*c*,*n*][a] | $HH(c/2)[(9/2)c,(14/9)n]$<br>$HH2c[3c,(7/3)n]$<br>$HV(3c/2)[(5/2)c,(14/5)n]$<br>$N = 7cn$ |
| | | $(c/2,-c,n/2,2n)^{b}$<br>$N = 2(2kq - 3lp)$<br>$= 2(-7lp) = 7cn$ | | $(c/2,2c,n/2,-n)^{b}$<br>$N = 2(2lp - 3kq)$<br>$= 2(-7kq) = 7cn$ |

*a* H[*c*,*n*] = V[*n*,*c*]; when, in H[*c*,*n*], *c* > *n* go to V series, and *vice versa*.

*b* Double twisted toroids, as different embeddings but topologically identical to the corresponding, single twisted, X*t*[*c*,*n*] objects.

## 5.2. Supramolecular Structures

Sequence of different operations enable the construction of various objects, with both positive and negative curvature.[67],[85-88]

Recall that, an orientable surface *S* (*i.e.*, having two sides) is characterized by the Euler parameter $\chi(M) = 2(1-g)$ (see also eq 22, Section 4). $\chi(M)$ is directly calculable by the genus[5] *g*, being the number of holes performed in a plastic sphere to make it homeomorphic to *S*. Positive/negative $\chi$ values indicate positive/negative curvature of a lattice embedded in *S*. An embedding is a representation of a graph on a closed surface such that no crossing lines appear.[5],[89]

The negative curvature can be induced in graphite (the reference, of zero curvature) by replacing hexagons by sevenfold, or larger sized rings. Such nets cannot form closed polyhedra, in the opposite of those objects derived from graphite by introducing pentagons or smaller rings, which bring positive curvature, causing the cage closure. Thus, the negatively curved structures belong to infinite periodic surfaces, with zero mean curvature. Our goal is to generate various repeat units useful for various purposes, particularly in building up some interesting infinite lattices.

The operation leading to open (repeat) units is $Op_n$, which is an $E_n$ homeomorphic transformation of the parent faces. Most often $n = 1$, thus being omitted - see also section 4.2 and eq 44. It can be associated with any composite operation. Figure 40 illustrates the *Op*(*Ca*(*M*)) sequence.

(a) $Op(Ca(T))$; $S = 10.432$ (b) $Op(Ca(C))$; $S = 2.171$

$N = 40$; $E = 54$; $F_7 = 12$; $g = 2N = 80$; $E = 108$; $F_7 = 24$; $g = 3$

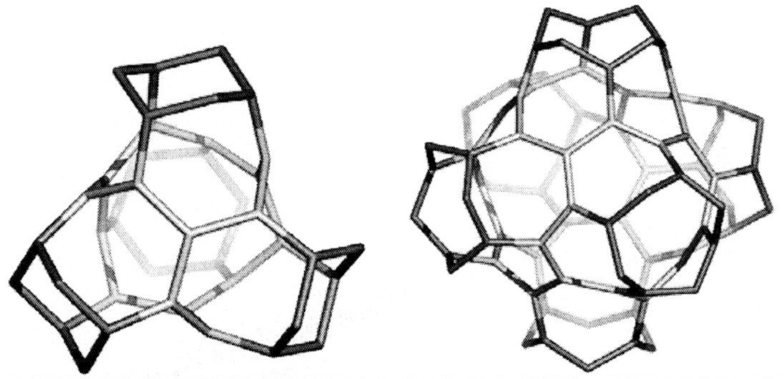

Figure 40: Opening *Op* operation leads to negative curvature lattices, embedded as repeat units in surfaces of genus 2 and 3, respectively.

The open (transformed) maps cannot longer be embedded in a surface of genus zero (*e.g.*, in the plane or the sphere). Thus, according to (22), negative $\chi$ values need $g > 1$ (see Figure 40).

The genus[5] of an open Capra object, $Op(Ca(M))$, is calculable, by using the results in Table 2 and eq 22, as:[67]

$$\chi(Op(Ca(M))) = v_{1Op} - e_{1Op} + f_{1Op} = v_0 - e_0 = 2 - 2g \quad (50)$$

$$g = (2 - v_0 + e_0)/2 = f_0/2 \quad (51)$$

Relation (51) comes out from the spherical character of the parent polyhedron, for which the Euler relation is written as: $v_0 - e_0 + f_0 = 2$. Since $f_0$ is just the number of open faces, (51) can be extended to any open object, derived by a composite operation from a spherical polyhedron.

Clearly, lattices with $g > 1$ will have negative $\chi(Op(M))$ and consequently negative curvature. For the five Platonic solids, the genus of the corresponding $Op(Ca(M))$ is: 2 (Tetrahedron); 3 (Cube); 4 (Octahedron); 6 (Dodecahedron) and 10 (Icosahedron).

Note that, *Op*(*Ca*(*M*)) leads to *all heptagon covering* of negatively curved units. An interesting example of using the above operation sequence is as follows: the *Ca* transform of Dodecahedron (Figure 41a) is opened by $Op_1$ and next capped by appropriate all pentagon decorated caps (Figure 41b and c). This is just the generalized way[51] to phantasmagoric fullerenes,[90] eventually covered by only pentagons and heptagons. Further capring/septupling of the Fowler's $C_{260}$ ($I_h$) leads to a more relaxed object, by additional hexagons (Figure 41d).

(a) $Ca(D)$ (b) $Op(Ca(D))$; $S = 1.574$

$N = 140$ $N = 200$; $E = 270$; $F_7 = 60$; $g = 6$

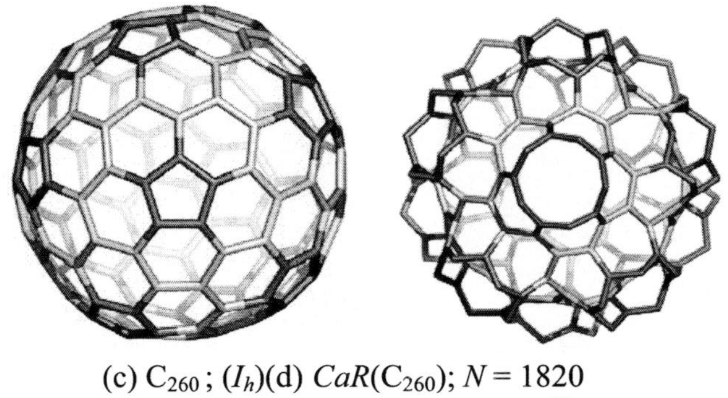

(c) $C_{260}$ ; $(I_h)$(d) $CaR(C_{260})$; $N = 1820$

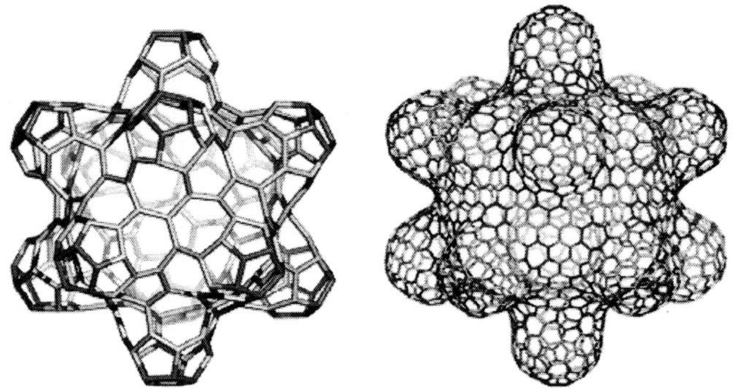

Figure 41: The $Op(Ca(D))$ can be capped by all pentagon decorated caps, leading to the Fowler's phantasmagoric fullerene $C_{260}$, covered only pentagons and heptagons. (to be continued).

Iterative *septupling* of open objects leads to more relaxed units, as appears by comparing data in Figures 40, 41 and 42, 43 (strain energy per $sp^2$ atom, in kcal/mol, in terms of the POAV1 theory).[91-94] Only the trivalent Platonic solids were transformed, by reason of obtaining cubic/trivalent transforms. Such units can be viewed as rational models for 3D junctions of nanotubes.[85],[95]

$CaR(Op(CaS(T)))$ $CaR(Op(CaS(C)))$

$N = 232$; $E = 330$; $F = 96$; $g = 2$ $N = 464$; $E = 660$; $F = 192$; $g = 3$

$S = 1.114$ $S = 0.428$

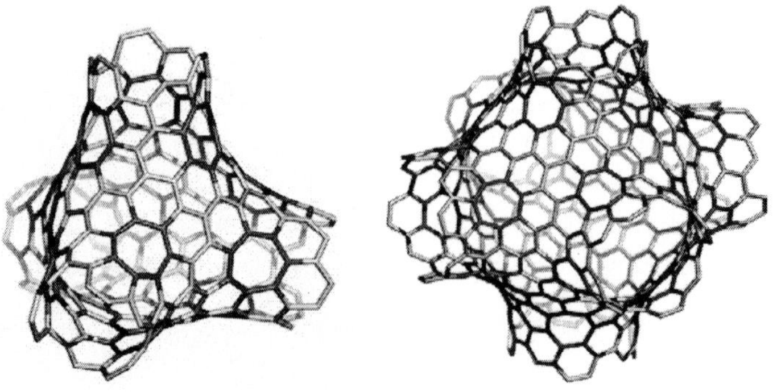

Figure 42: Units of negative curvature, derived by iterative septupling of Tetrahedron and Cube

*CaR*(*Op*(*CaS*(*D*)))
$N = 1160$; $E = 1650$; $F = 480$; $g = 6$; $S = 0.335$

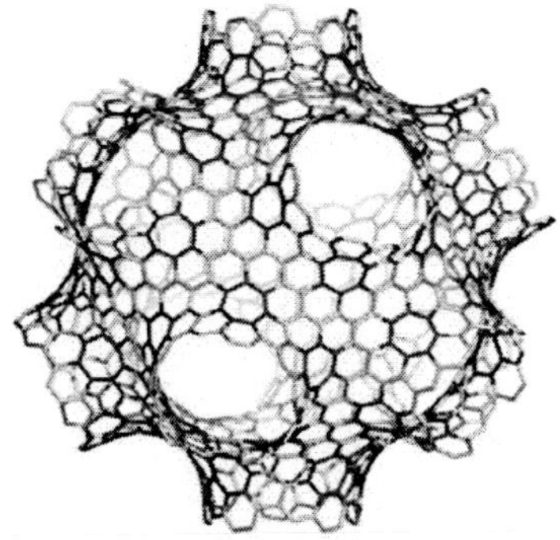

Figure 43: Unit of negative curvature, derived by iterative septupling of Dodecahedron.

The sequence *Le*(*Op*(*Ca*(*M*))) enables a covering by disjoint azulenic (5,7) units,[51] shown as infinite objects in Figure 44. By using *Le*(*Op*(*Ca*(*T*))), as a building block, we constructed a supra-molecular (open) dodecahedron, of genus 21, covered by azulenic disjoint units (separated by hexagons - Figure 44). They represent the first 3D-infinite fully-resonant-azulenoid objects (of $g > 0$), in terms of Kirby extension[34],[49] of the older Clar theory[69] on aromatic systems.

*Le*(*Op*(*Ca*(*T*)))
$N = 120$; $E = 174$; $F = 52$; $g = 2$

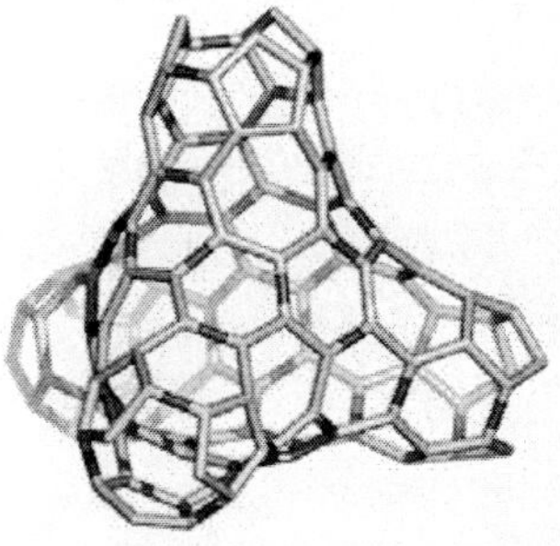

*Le*(*Op*(*Ca*(*C*)))
$N = 240$; $E = 348$; $F = 104$; $g = 3$

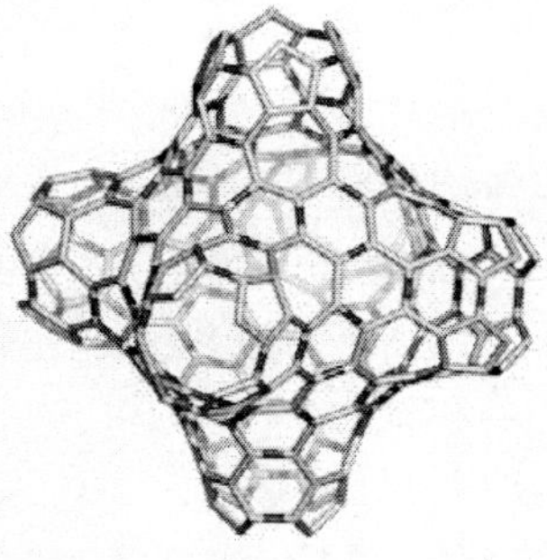

*Le*(*Op*(*Ca*(*D*)))
$N = 600$; $E = 870$; $F = 260$; $g = 6$

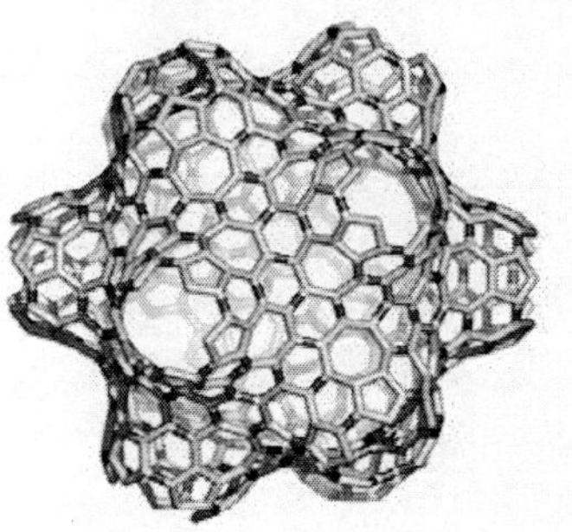

Supra-(azulenic)-*D*(*Le*(*Op*(*Ca*(*T*))))
$N = 2400$; $g = 21$

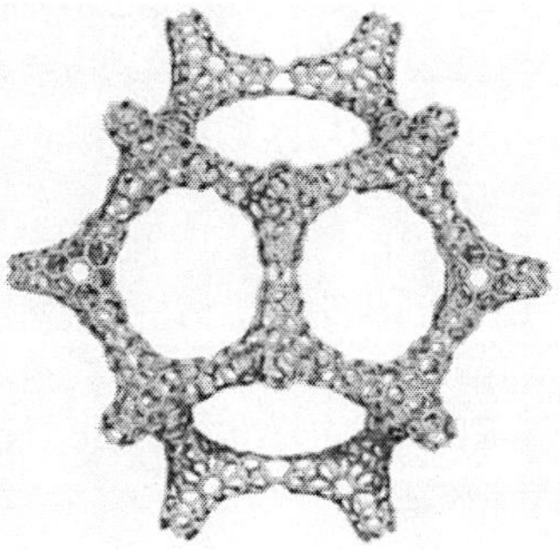

Figure 44: Units covered by disjoint azulenic pairs, representing fully-resonant-azulenoid, and aromatic structures.

The building block *Op*(*Ca*(*T*)), (Figure 40a), enabled construction of a supra-molecular (open) dodecahedron,[51] a multi torus of genus 21, entirely covered by heptagons $F_7$ (Figure 45a). A new capring leads to a more relaxed object (Figure 45b). Such supra-dodecahedra represent one of the twin labyrinths interlaced in construction of the FRD-type surface; they could appear by a self-assembling process, as shown in ref. 95.

(a) *Supra-D*(*Op*(*Ca*(*T*))) (b) *Ca*(*Supra-D*(*Op*(*Ca*(*T*))))

$N = 620$; $E = 900$; $F_7 = 240$; $g = 21$ $N = 4100$; $g = 21$

$S = 7.876$ $S = 1.075$

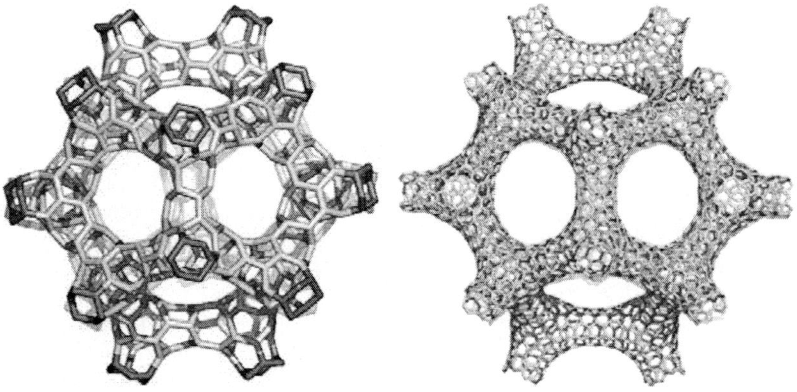

Figure 45: Supramolecular structures built-up by tetrahedral units.

## 5.3. Dyck Graph

The celebrate Dyck graph,[96-98] on 32 vertices of valence 3, can be represented in LCF notation (see also ref. 99) as $[5,-5,13,-13]^8$. It has 48 edges, 12 octagons, girth 6, diameter 5, and chromatic number 2. It is non-planar and has the minimal genus $g = 1$ (*i.e.*, there exists an embedding of the graph on the torus).[98] It is also Hamiltonian and has the graph spectrum: $3\left(\sqrt{5}\right)^6\ 1^9\ (-1)^9\ \left(-\sqrt{5}\right)^6\ -3$. Its unique symmetry has been discussed by several authors.[98],[100-104] Cycle counting revealed 12 octagons and 16 hexagons.

This amazing graph has been involved in the tessellation of the so-called "plumber nightmare",[105] appearing at 3D junction of carbon nanotubes. Such a genus 3 surface can be covered by infinite lattices having (repeat) units of 32 vertices/atoms, being representations[51] of the Dyck graph (Figure 46). These units, corresponding to zigzag (Figure 46a) and armchair (Figure 46b) tubes, are inserted in the lattice by identification of the opening faces (not counted in the Euler theorem). They can be obtained by the sequences: (a) *Op*(*Q*(*C*)) and (b) $Op_{2a}$(*Q*(*C*)), the subscript 2a meaning an alternating "open 2" operation. Infinite lattices, built up by units of Figure 46, were presented elsewhere.[95] The unit derived by the sequence *Op*(*Q*(*T*)) enabled the construction of a supra-dodecahedron, entirely covered by octagons (Figure 47a).

(a) $Op(Q(C))$ (b) $Op_{2a}(Q(C))$

$N = 56; E = 72; F_8 = 12; g = 3$ $N = 56; E = 72; F_8 = 12; g = 3$

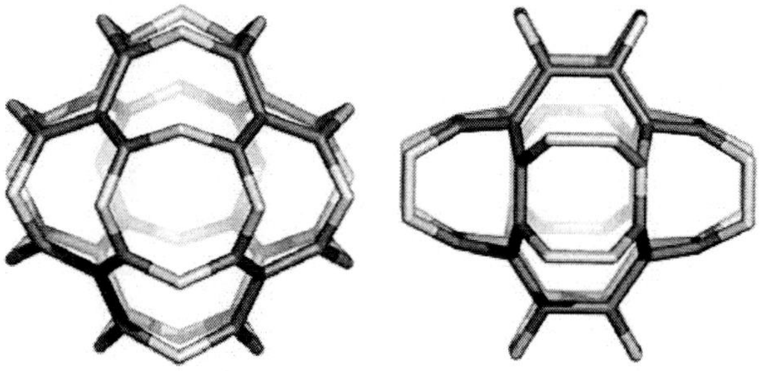

Figure 46: Repeat units of 32 atoms, representing the Dyck graph, within infinite lattices embedded in a genus 3 negative surface.

An "open 2" operation $Op_2(Ca(C))$ provides the $F_8$ unit (Figure 47b), proposed[104] for building octahedral lattices covered by octagons. Since the face identification of its opening faces is quite impossible, the insertion is achieved by cutting off these faces. It is equivalent to identify the opening faces in the $Op_{2a}(Q(C))$ unit (Figure 46b). The face identification was successful in case of the $F_7$ unit $Op(Ca(C))$, shown in Figure 40b, which is just the representation of the famous Klein graph,[105-107] within an infinite lattice, embedded in a genus 3 surface.[95]

(a) Supra-$D(Op(Q(T)))$; (b) $Op_2(Ca(C))$

$N = 380; E = 540; F_8 = 120; g = 21$ $N = 104; E = 132; F_8 = 24; g = 3$

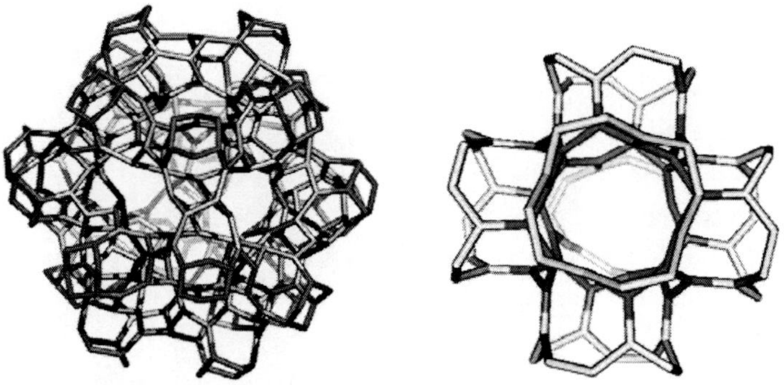

Figure 47: Structures covered by octagonal faces.

Let us resume to the finite, trivalent Dyck graph. The guiding idea was the number of octagonal faces, namely $F_8 = 12$. Take a look at the unit $Op(Q(T))$ (Figure 48a): it is $F_8 = 6$, *i.e.*, half of the Dyck graph, with only four missing atoms. Keeping in mind the identification procedure used in infinite lattice construction, a kind of interlaced dimer $Di(Op(Q(T)))$ (Figure 48b) can be built up by identifying all the four opening faces of the $Op(Q(T))$ unit. The question is how the four supplementary atoms (in deep black) have to be connected to give the true Dyck graph. Among the $6^4 = 1296$ permutations we found the object $Di(Op(Q(T)))_p$ illustrated in Figure 48c, having the all topological and spectral characteristics of the finite, trivalent Dyck graph.[51] A similar procedure, by $Ca$, led to $Di(Op(Ca(T)))$

(Figure 48d), which is not the Klein graph, despite some topological characteristics are identical. Up to now, no permutation on *Di*(*Op*(*Ca*(*T*))) was found to give the Klein graph.

(a) *Op*(*Q*(*T*)) (b) *Di*(*Op*(*Q*(*T*)))

$N = 28$; $E = 36$; $F_8 = 6$; $g = 2$ $N = 32$; $E = 48$; $F_8 = 12$ ; $g = 3$

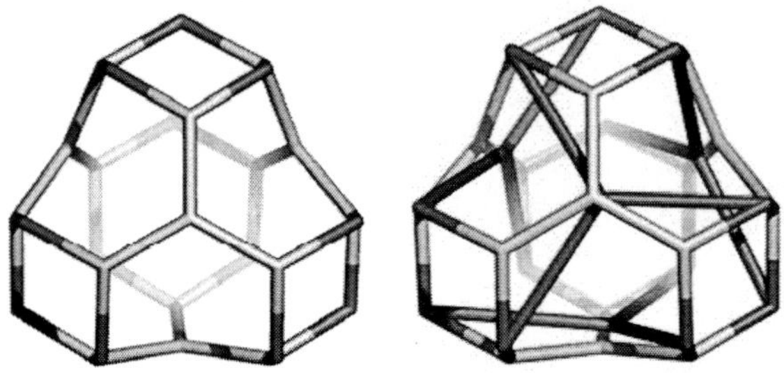

(c) *Di*(*Op*(*Q*(*T*)))$_p$ = Dyck graph (d) *Di*(*Op*(*Ca*(*T*))) = Klein graph?

$N = 32$; $E = 48$; $F_8 = 12$; $g = 3$ $N = 56$; $E = 84$; $F_7 = 24$; $g = 3$

(16 hexagons are internal rings)

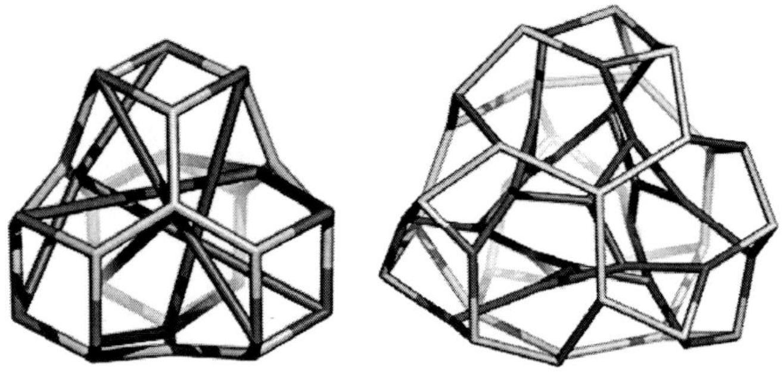

Figure 48: A way to the finite Dyck graph representation.

Interestingly, the object in Figure 48c appears of genus 3; despite the "open" Tetrahedron is of genus 2, over this the toroidal structure is superimposed. The Euler formula is obeyed for $g = 3$, when the 16 hexagonal rings are considered as "non-faces".

Conversely, in case of the toroidal embedding as HV4[4,8] or, in four integer parameters (4,0,4,0), the 12 octagons are not counted as faces (Figure 49). Thus, it is of genus 1, which is a normal result.[51]

HV4[4,8]; side HV4[4,8]; top

$N = 32$; $E = 48$ ; $F_6 = 16$ ; $g = 1$(12 octagons are external rings)

Figure 49: A toroidal embedding of the Dyck graph.

## 6. Conclusions

Three main ways for modifying the tiling of a nanostructure were presented: to, by following various routes: (i) cutting the tetragonal (4,4) lattice, embedded in the torus or cylinder, to give hexagonal (6,3) and other coverings of interest; (ii) Stone-Wales SW edge rotations enabling covering changes in both spherical and tubular objects and (iii) operations on maps, the last being the most versatile deeply involved in the electronic structure of their transforms. Our CageVersatile program enables both SW and map operations. Examples were given for all the three routes.

A corannulenic Cor covering was proposed, by analogy to the well-known perfect Clar structure, to give account for some supra-organized fully-resonant $\pi$-electronic aromatic systems of fullerenes. The constructive demands for DCor full covering were given.

Construction of graphitoids with trigonal $sp^2$ atoms can be extended to infinite periodic surfaces of negative curvature and genus higher than zero. In this respect, new representations of the Dyck graph were proposed. Supramolecular structures could eventually be obtained by self-assembling of such repeat units, some of them appearing as 3D junctions of nanotubes.

## Acknowledgements

Helpful discussions with Professors R. R. King (University of Georgia, Athens, Georgia, USA) and A. Ceulemans (Catholic University of Leuven, Belgium) are highly acknowledged.

## References

[1] B. Grünbaum and G. C. Shephard, *Tilings and Patterns*, Freeman, New York, 1985.
[2] D. J. Klein and H. Zhu, in: A. T. Balaban, (Ed.), *From Chemical Topology to Three - Dimensional Geometry*, Plenum Press, New York, 1997, pp. 297-341.
[3] B. de La Vaissière, P. W. Fowler, and M. Deza, *J. Chem. Inf. Comput. Sci.*, 2001, 41, 376-386.
[4] M. Deza, P. W. Fowler, M. Shtorgin, and K. Vietze, *J. Chem. Inf. Comput. Sci.*, 2000, 40, 1325-1332.
[5] F. Harary, *Graph Theory*, Addison-Wesley, Reading, MA, 1969.
[6] E. C. Kirby, *Croat. Chem. Acta*, 1993, 66, 13-26.
[7] E. C. Kirby, R. B. Mallion, and P. Pollak, *J. Chem. Soc. Faraday Trans.*, 1993, 89, 1945-1953.
[8] A Ceulemans, L. F. Chibotaru, S. A. Bovin, and P. W. Fowler, *J. Chem. Phys.*, 2000, 112, 4271-4278.
[9] D. J. Klein, *J. Chem. Inf. Comput. Sci.*, 1994, 34, 453-459.
[10] E. C. Kirby and P. Pollak, *J. Chem. Inf. Comput. Sci.*, 1998, 38, 66-70.
[11] A L. Ivanovskii, *Russ. Chem. Rev.*,1999, 68, 103-118.
[12] N. Hamada, S. Sawada, and A. Oshiyama, *Phys. Rev. Lett.*, 1992, 68, 1579-1581.
[13] T. Pisanski and J. Shawe-Taylor, *J. Chem. Inf. Comput. Sci.*, 2000, 40, 567-571.
[14] A Graovac, D. Plavšić, M. Kaufman, T. Pisanski, and E. C. Kirby, *J. Chem. Phys.*, 2000, 113, 1925-1931.

[15] I. Laszlo, A. Rassat, P. W. Fowler, and A. Graovac, *Chem. Phys. Lett.*, 2001, 342, 369-374.
[16] Laszlo and A. Rassat, *J. Chem. Inf. Comput. Sci.*, 2003, 43, 519-524.
[17] M. V. Diudea, B. Parv, and O. Ursu, *Torus* 1.1 software program, Babes-Bolyai University, 2001.
[18] M. V. Diudea, I. Silaghi-Dumitrescu, and B. Parv, *MATCH, Commun. Math. Comput. Chem.*, 2001, 44, 117-133.
[19] M. V. Diudea and E. C. Kirby, *Fullerene Sci. Technol.*, 2001, 9, 445-465.
[20] M. V. Diudea, *Bull. Chem. Soc. Japan*, 2002, 75, 487-492.
[21] M. V. Diudea, B. Parv, and O. Ursu, O., *Studia Univ. "Babes-Bolyai"*, 2003, 48, 3-10.
[22] M. V. Diudea, O. Ursu, and B. Parv, *Studia Univ. "Babes-Bolyai"*, 2003, 48, 11-20.
[23] A K. Singh, T. M. Briere, V. Kumar, and Y. Kawazoe, *Phys. Rev. Lett.*, 2003, 91, 146802-1 - 146802-4.
[24] K. Singh, V. Kumar, and Y. Kawazoe, *J. Mater. Chem.*, 2004, 14, 555-563.
[25] M. V. Diudea, *Fullerenes, Nanotubes Carbon Nanostruct.*, 2002, 10, 273-292.
[26] M. V. Diudea, *Phys. Chem. Chem. Phys.*, 2002, 4, 4740-4746.
[27] M. V. Diudea, in: M. V. Diudea, Ed., *Nanostructures-Novel Architecture*, NOVA, New York, 2005 (in press).
[28] M. V. Diudea, T. S. Balaban, E. C. Kirby, and A. Graovac, *Phys. Chem.,Chem. Phys.*, 2003, 5, 4210 – 4214.
[29] S. L. Fang, A. M. Rao, P. C. Eklund, P. Nikolaev, A. G. Rinzler and R. E. Smalley, *J. Mater. Res.*, 1998, 13, 2405-2411.
[30] M. Terrones, H. Terrones, F. Banhart, J-C. Charlier, and P. M. Ajayan, *Science*, 2000, 288, 1226-1229.
[31] M. V. Diudea, B. Parv, P. E. John, O. Ursu, and A. Graovac, *MATCH Commun. Math. Comput. Chem.*, 2003, 49, 23-36.
[32] M. V. Diudea, B. Parv, and E. C. Kirby, *MATCH, Commun. Math. Comput. Chem.*, 2003, 47, 53-70.
[33] E. C. Kirby, *Fullerene Sci. Technol.* 1994, 2, 395-404.
[34] E. C. Kirby, Fully Arenoid Toroidal Fullerenes, Both Benzenoid and Non-Benzenoid, *MATCH, Commun. Math. Comput. Chem.*, 1996, 33, 147-156.
[35] V. H. Crespi, L. X. Benedict, M. L. Cohen, and S. G. Louie, *Phys. Rev. B*, 1996, 53, 13303-13305.
[36] S. Itoh, S. Ihara, and J-i. Kitakami, *Phys. Rev., B* 1993, 47, 1703-1704.
[37] S. Ihara, S. Itoh and J-i. Kitakami, *Phys. Rev., B* 1993, 47, 12908-12911.
[38] D. Babić, D. J. Klein, and T. G. Schmalz, *J. Mol. Graphics Modell.* 2001, 19, 222-231.
[39] S. Iijima, T. Ichihashi, and Y. Ando, *Nature* 1992, 356, 776-778.
[40] S. Itoh and S. Ihara, *Phys. Rev., B*, 1993, 48, 8323-8328.
[41] M. V. Diudea, *Studia Univ. "Babes-Bolyai"*, 2003, 48( 2), 17-26.
[42] K. P. C. Vollhardt and D. L. Mohler, in: B. Halton (Ed.), *Advances in Strain in Organic Chemistry*, JAI Press, London, 1966, pp. 121-160.
[43] B. C. Berris, G. H. Hovakeemian, Y.-H. Lai, H. Mestdagh, and K. P. C. Vollhardt, *J. Am. Chem. Soc.*, 1985, 107, 5670-5687.
[44] K. P. C. Vollhardt, *Pure Appl. Chem.*, 1993, 65, 153-156.
[45] A J. Stone and D. J. Wales, *Chem. Phys. Lett.* 1986, 128, 501-503.

[46] M. Stefu and M. V. Diudea, *MATCH Commun. Math. Comput. Chem.*, 2004, 50, 133-144.
[47] M. O'Keeffe and B. G. Hyde, *Crystal Structures*, Mineralogical Society of America, Washington DC, 1996.
[48] B. Grünbaum, H. D. Löckenhoff, G. C. Shephard, and A. Temesvari, *Geom. Dedicata*, 1985, 19, 109-174.
[49] E. C. Kirby, in: M. V. Diudea, Ed., *Nanostructures-Novel Architecture*, NOVA, New York, 2005 (in press).
[50] M. V. Diudea, Cs. L. Nagy, I. Silaghi-Dumitrescu, A. Graovac, D. Janežić, and D. Vikic-Topić, *J. Chem. Inf. Model.,* 2005, 45, 293-299.
[51] M. V. Diudea, *Forma*, 2004, 19 (3), 131-163.
[52] H. F. Bettinger, B. I. Yakobson, and G. E. Scuseria, *J. Amer. Chem. Soc.*, 2003, 125, 5572-5580.
[53] Ge. G. Samsonidze, G. G. Samsonidze, and B. I. Yakobson, Comput. Mater. Sci., 2002, 23, 67-72.
[54] D. Orlikowski, M. B. Nardelli, J. Bernholc, and Ch. Ronald, *Phys. Rev. B*, 2000, 61, 14194-14203.
[55] Q. Zhao, M. B. Nardelli, and J. Bernholc, *Phys. Rev. B*, 2000, 65, 144105-1 - 144105-6.
[56] M. V. Diudea and M. Stefu, CageVersatile CV 1.1, „Babes-Bolyai" University, Cluj 2003.
[57] T. Pisanski and M. Randić, in *Geometry at Work*, M. A. A. Notes, 2000, 53, 174-194.
[58] P. W. Fowler and T. Pisanski, *J. Chem. Soc. Faraday Trans.*, 1994, 90, 2865-2871.
[59] L. Euler, *Novi Comment. Acad. Sci. I. Petropolitanae*, 1758, 4, 109-160.
[60] P. W. Fowler, *Phys. Lett.*, 1986, 131, 444-450.
[61] P. W. Fowler and J. I. Steer, *J. Chem. Soc.*, *Chem. Commun.*, 1987, 1403-1405.
[62] P. W. Fowler and K.M.Rogers, *J. Chem. Soc.*, *Faraday Trans.*, 1998, 94, 1019-1027.
[63] P. W. Fowler and K.M.Rogers, *J. Chem. Soc.*, *Faraday Trans.*, 1998, 94, 2509-2514.
[64] M. V. Diudea and P. E. John, *MATCH, Commun. Math. Comput. Chem.*, 2001, 44 103-116.
[65] M. V. Diudea, P. E. John, A. Graovac, M. Primorac, and T. Pisanski, *Croat. Chem. Acta*, 2003, 76, 153-159.
[66] M. Goldberg, *Tohoku Math. J.*, 1937, 43, 104-108.
[67] M. V. Diudea, *Studia Univ. "Babes-Bolyai"*, 2003, 48 (2), 3-16.
[68] E. Clar, *Polycyclic Hydrocarbons*, Acad. Press, London, 1964.
[69] E. Clar, *The Aromatic Sextet*, Wiley, New York, 1972.
[70] K. Fries, *J. Liebigs Ann. Chem.*, 1927, 454, 121-324.
[71] W-Ch. Shiu, P. Ch. B. Lam, and H. Zhang, *Theochem*, 2000, 4, 0210.
[72] J. R. Dias, *J. Chem. Inf. Comput. Sci.*, 1999, 39, 144-150.
[73] J. V. Knop, W. R. Műller, K. Szymanski, S. Nikolić and N. Trinajstić, *MATCH, Commun. Math. Comput. Chem.*, 1993, 29, 81-106.
[74] E. C. Kirby, *MATCH, Commun. Math. Comput. Chem.*, 1996, 33, 147-156.
[75] T. A. Keith and R. F. W. Bader, *Chem. Phys. Lett.*, 1993, 210, 223-231.
[76] A Acocella, R. W. A. Havenith, E. Steiner, P. W. Fowler, and L. W. Jenneskens, *Chem. Phys. Lett.*, 2002, 363, 64-72.
[77] P. W. Fowler and E. Steiner, *Chem. Phys. Lett.*, 2002, 363, 259-266.

[78] P. Lazzeretti, in: J. W. Emsley, J. Feeney and L. H. Sutcliffe (Eds.) *Progress in Nuclear Magnetic Resonance Specroscopy*, Elsevier, Amsterdam, 2000, 36, 1-88.
[79] HyperChem$^{TM}$, Release 4.5 for SGI, © 1991-1995, HyperCube, Inc.
[80] P. W. Fowler and A. Rassat, *Phys. Chem. Chem. Phys.*, 2002, **4**, 1105-1113.
[81] H. Hosoya, Y. Tsukano, K. Nakada, S. Iwata, and U. Nagashima, *Croat. Chem. Acta*, 2004, 77, 89-95.
[82] H. Hosoya, Y. Tsukano, M. Ohuchi, and K. Nakada, in: M. Doyama, J. Kihara, M. Tanaka, and R. Yamamoto (Eds.), *Computer Aided Innovation of New Materials II*, 1993, Elsevier, pp. 155-158.
[83] J.-I. Aihara, T. Yamabe and H. Hosoya, *Synth. Metals*, 1994, 64, 309-313.
[84] M. Yoshida, M. Fujita, P. W. Fowler, and E. C. Kirby, *J. Chem. Soc., Faraday Trans.*, 1997, 93, 1037-1043.
[85] Cs. L. Nagy and M. V. Diudea, *Studia Univ. "Babes-Bolyai"*, 2003, 48 (2), 37-46.
[86] A L. Mackay and H. Terrones, *Nature*, 1991, 352, 762-762.
[87] D. Vanderbilt and J. Tersoff, *Phys. Rev. Lett.*, 1992, 68, 511-513.
[88] T. Lenosky, X. Gonze, M. Teter, and V. Elser, *Nature*, 1992, 355, 333-335.
[89] J. L. Gross and T. W. Tucker, *Topological graph theory*, Dover Pubs, Dover, 2001.
[90] A Dress and G. Brinkmann, *MATCH, Commun. Math. Comput. Chem.*, **33** (1996) 87-100.
[91] R.C. Haddon, *J. Am. Chem. Soc.*, **112** (1990) 3385- 3389.
[92] R.C. Haddon and K. Raghavachari, in: W.E. Billups and M. A. Ciufolini, *Buckminsterfullerenes*, VCH, 1993, p. 185-215.
[93] R.C. Haddon and S.-Y. Chow, *J. Am. Chem. Soc.*, **120** (1998) 10494-10496.
[94] R.C. Haddon, *J. Phys. Chem. A*, **105** (2001) 4164-4165.
[95] Cs. L. Nagy and M. V. Diudea, in: M. V. Diudea, Ed., *Nanostructures-Novel Architecture*, NOVA, New York, 2005 (in press).
[96] W. Dyck, *Math. Ann.*, 1880, 17, 510-516.
[97] G. W. Hart, 2004, www.georgehart.com.
[98] A Ceulemans, E. Lijnen, L. J. Ceulemans, and P. W. Fowler, *Mol. Phys.*, 2004, 102, 1149.
[99] M. V. Diudea, I. Gutman, and L. Jäntschi, *Molecular Topology*, Nova, New York, 2002, pp. 295-306.
[100] R. B. King, *Croat. Chem. Acta*, 2002, 75, 447-473.
[101] R. B. King, *Discrete Mathematics*, 2002, 244, 203-210.
[102] R. B. King, in: M. V. Diudea, Ed., *Nanostructures-Novel Architecture*, NOVA, New York, 2005 (in press).
[103] A Ceulemans, R. B. King, S. A. Bovin, K. M. Rogers, A. Troisi, and P. W. Fowler, *J. Math. Chem.*, 1999, 26, 101-123.
[104] E. Lijnen and A. Ceulemans, in: M. V. Diudea, Ed., *Nanostructures-Novel Architecture*, NOVA, New York, 2005 (in press).
[105] R. B. King, *J. Phys. Chem.*, 1996, 100, 15096-15104.
[106] F. Klein, *Math. Ann.*, 1879, 14, 428-471.
[107] F. Klein, *Gesammelten Mathematischen Abhandlungen*, Springer Verlag, Berlin, 1923, Vol. 3, pp. 90-136.

In: Nanostructures: Novel Architecture
Editor: Mircea V. Diudea, pp. 243-267

ISBN: 1-59454-499-9


*Chapter 12*

# ALGORITHMS FOR BASIC OPERATIONS ON MAPS

***Monica Ştefu[a1], Daniela Butyka[a], Mircea V. Diudea[a], Lorentz Jäntschi[b] and Bazil Pârv[c]***

[a]Faculty of Chemistry and Chemical Engineering
"Babeş-Bolyai" University, 400084, Cluj, Romania
[b]Technical University, Cluj, Romania
[c]Department of Computer Science, Faculty of Mathematics and Computer Science"Babeş-Bolyai" University, 400084, Cluj, Romania

## ABSTRACT

Covering operations in nanostructure modeling are presented as algorithms implemented in the original software program CageVersatile 1.1. The basic operations described are: Dual, Medial, Stellation, Truncation, Leapfrog, Quadruple and Capra. Output file objects are illustrated.

## 1. INTRODUCTION

Covering a local planar surface by various polygonal or curved regions is nowadays a mathematically founded science.[1] The Geometry of (regular) polygons and polyhedra, Graph Theory and Set Theory concentrate in the ground statement of an interdisciplinary science, often inspired from the Arts and Architecture, and implemented in Computer Science.

Covering transformation is one of the ways in understanding chemical reactions occurring in nanostructures.[2-4]

A map $M$ is a combinatorial representation of a closed surface.[5],[6] Several transformations (*i.e.*, operations) on maps are known and used for various purposes.

Recall some basic relations in a map:

[1] E-mail address: mstefu02@yahoo.com

$$\sum d\,v_d = 2e \quad (1)$$

$$\sum s\,f_s = 2e \quad (2)$$

where $v_d$ and $f_s$ are the number of vertices of degree $d$ and number of $s$-gonal faces, respectively. The two relations are joined in the famous Euler formula:

$$v - e + f = \chi(M) = 2(1-g) \quad (3)$$

with $\chi$ being the Euler *characteristic* and $g$ the genus[7] of a graph (*i.e.*, the number of handles attached to the sphere to make it homeomorphic to the surface on which the given graph is embedded; $g = 0$ for a planar graph and 1 for a toroidal graph). Positive/negative $\chi$ values indicate positive/negative curvature of a lattice.

The above relations, along with some rules for checking the connectivity and cycle and/or polyhedral face identification were kept in mind when the CageVersatile CV1.1 software program was written.

## 2. Input, Output, and Helper Functions

The molecules are considered from the Graph Theoretical point of view, with atoms as vertices and bonds between atoms as edges. The map operations, theoretically founded in ref. [8], are presented in the order: Dual, Medial, Stellation, Truncation, Leapfrog, Quadruple and Capra. Within these operations, new vertices are put and linked, by the map operations rules, while the old vertices and edges are, in general, removed. For details see ref. 5.

The program, written in the PHP programming language, has as input a .hin (Hyperchem) file and writes the results as output .txt and/or .hin files. It can be executed online at the internet address http://193.226.7.140/~monica/graph/cv1.1/

### 1.1. Input Data

In the first part of the program, the .hin file is tacken into array ***a***. Then, the cycles with chosen maximal size are searched; the distinct faces are put into the array ***c***. The basic condition is: an edge is shared to maximum two cycles/rings/faces.

The array ***a*** contains the list of vertices of the initial hydrogen depleted molecular graph. The element $a[i]$ ($i$=1,$n$; $n$ is the number of vertices in the initial molecular graph) is an array containing the coordinates, valence and the vertices linked to the vertex $i$.

In the array ***c***, $c[i,0]$ is the number of faces contained in the vector $c[i]$, which begins at the vertex $i$; $nv = c[i,j,0]$ is the number of vertices of the face $c[i,j]$, while $c[i,j,1]$ ... $c[i,j,nv]$ is the list of vertices of the face $c[i,j]$.

The array ***m*** stores the list of edges of the initial molecular graph and some additional new vertices, used within operations. It has a structure similar to the array ***c***; $m[i,j]$ represents an edge with the end-points in $m[i,j,1]$ and $m[i,j2]$. In some cases, $m[i,j,3]$ .. $m[i,j,6]$ are used to memory new vertices in connection with the edge $m[i,j]$.

The algorithms given below are based on the arrays ***a***, ***c*** and ***m***, derived from the input .hin file.

## 1.2. Output Files

All basic operations on maps described here provide .hin files as the output. The general procedure for producing an output file is: (1) generate an operation-specific array, (2) concatenate the array elements in a string and (3) write the resulting strings in the .hin format file.

The operation-specific arrays have the same structure for all operations. For example, in the MEDIAL operation, the operation-specific array ***med*** has for every atom/ vertex *i* an element in the form of array *med*[i] with the .hin file structure:

*med*[*i*,1] = "atom ".*i*." - ".*tipa*." CA - 0 ".*x*." ".*y*." ".*z*." ";
where *i* is the atom number, *tipa* is the atom type and *x*, *y*, *z* are the coordinates;
*med*[*i*,2] contains the atom valence (denoted by *v*);
*med*[*i*,3] ... *med*[*i*,*v*+2] contain the atoms adjacent to atom *i*.

Before producing the content of the .hin file, a header must also be provided for compatibility with the Hyperchem program. This header has the general form:

;*OpName* + newLine + "forcefield mm+" + newLine + "sys 0" + *texti* + "mol 1" + newLine

where + means string concatenation, *OpName* is the name of the operation, hard-coded in every basic map operation, newLine is the line terminator, while *texti* is an explanatory text to be included in the header; mol1 marks the beginning of the first molecular graph.

## 1.3. Helper Operations

This section presents the helper functions, invoked in the basic map operations as follows:

- for Dual: Center, OnSphere, Much_D, ad2l, WriteF
- for Medial: AdMe, ad2l, WriteF
- for Stellation: Center_D, Ini, WriteF
- for Truncation: Ad_CoordT, Ad_P, AdTr, WriteF
- for LeapFrog: Ad_Coord, Center_D, P_New, ad2l, Much_D, WriteF
- for Quadruple: IniQ, AdCycleInt, AdCoordQ, AdLegQ, MeanQ, WriteF
- for Capra: AdPi, Center_D, Ad_P_Middle, AdLeg_C, P_New_Capra, AlreadyVisited, Nrc, Verif_M2, Verif_M, Inv_56, WriteF.

Function Grow increases the size of the initial molecular graph by the factor mar = 2 to 6 times. It is used before the calling of the procedures for the operations.

```
function Grow(a,n,mar){
// in: n, mar
// in-out: a
// purpose: increases the coordinates
 for i := 1 to n
 a[i,2]:=a[i,2]*mar;
 a[i,3]:=a[i,3]*mar;
 a[i,4]:=a[i,4]*mar;
 endfor i
end function Grow
```

Function Center computes the center of the molecular graph defined by the coordinates (xc, yc, zc) and the radius r of the sphere.

```
function Center(n, a, xc, yc, zc, r)
// in: n, a
// out: xc, yc, zc, r
// purpose: computes the center of the molecular graph by the coordinates
// xc,yc, zc and the radius of the sphere.
 // the center coordinates, as arithmetic mean
 xc := 0; yc := 0; zc := 0;
 for i := 1 to n
 xc := xc + a[i,2];
 yc := yc + a[i,3];
 zc := zc + a[i,4];
 endfor i
 xc := xc/n; yc := yc/n; zc := zc/n;
 // the sphere radius is the arithmetic mean of the distances
 // between the center and the vertices.
 r := 0;
 for i := 1 to n
 r := r + sqrt((xc-a[i,2])^2 + (yc-a[i,3])^2 +(zc-a[i,4])^2);
 endfor i
 r := r/n;
end function Center
```

Function OnSphere returns the adapted coordinates (x, y, z) for the new vertices lying on a sphere with the center of coordinates xc, yc, zc and radius r.

```
function OnSphere(xc, yc, zc, r, x, y, z)
// in: xc, yc, zc, r
// in-out: x, y, z
// purpose: computes the adapted coordinates for a vertex lying on a
// sphere of center (xc, yc, zc) and radius r.
 // rp is the distance between the old vertex and the sphere center.
 rp := sqrt((xc-x)^2 + (yc-y)^2 + (zc-z)^2);
```

```
x := x*r/rp; y := y*r/rp; z := z*r/rp;
end function OnSphere
```

Function Much_D visits all the edges and puts the new vertex v on a side of the edge (m1,m2), in the array elements m[i,j,5] or m[i,j,6]. The function is used in the Dual and the Leapfrog operations.

```
function Much_D(v, m1, m2, n, m)
// in: v, n, m1, m2
// in-out: m
// purpose: puts the new vertex v on a side of the edge (m1, m2).
 for i := 1 to n
 for j := 1 to m[i,0]
 if ((m[i,j,1] = m1) and m[i,j,2] = m2)) or
 ((m[i,j,2] = m1) and m[i,j,1] = m2)) then
 if m[i,j,5] = 0 then {
 m[i,j,5] := v; m[i,j,6] := 0; }
 else
 m[i,j,6] := v;
 endif
 endfor j
 endfor i
end function Much_D
```

The function AdLeg puts the link (from, to) in the array med and concatenates the corresponding strings.

```
function AdLeg(from, to, med)
// in: from, to
// in-out: med
// purpose: puts in the array med the link from 'from' to 'to' vertices.
 med[from, 2] := med[from, 2] + 1;
 loc := med[from, 2];
 med[from, loc+2] := to and " a";
end function AdLeg
```

The function ad2l links in both directions the vertices l1 and l2.

```
function ad2l(l1,l2,med)
// in: l1, l2
// in-out: med
// purpose: links the given vertices in both directions
 adleg(l1,l2,med);
 adleg(l2,l1,med);
end function ad2l
```

The function AdMe links the vertices lying in the middle of the edges (m1, m2) and (m2, m3). It is invoked in the Medial operation.

```
function AdMe(m1, m2, m3, n, m, med)
// in: m1, m2, m3, n, m
// in-out: med
// purpose: links the vertices stored in the m[k,kk,5] array elements of // the edges
(m1, m2) and (m2, m3).
 for k := 1 to n
 for kk := 1 to m[k,0]
 if ((m1 = m[k,kk,1]) and (m2 = m[k,kk,2])) or
 ((m1 = m[k,kk,2]) and (m2 = m[k,kk,1]))
 leg1 := m[k,kk,5];
 if ((m2 = m[k,kk,1]) and (m3 = m[k,kk,2])) or
 ((m2 = m[k,kk,2]) and (m3 = m[k,kk,1]))
 leg2 := m[k,kk,5];
 endfor kk
 ad2l(leg1, leg2, med); // links vertices leg1 and leg2
 endfor k // in both directions.
end function AdMe
```

Function Center_D returns the coordinates x, y, z of the new vertex nn, the center of the face c[i,j]. The function is called in the Leapfrog and Capra operations.

```
function Center_D(i, j, nn, a, tipa, x, y, z, c, st)
// in: i, j, a, tipa
// out: x, y, z
// in-out: nn, c, st
// purpose: computes the coordinates of the new vertex which is the
// center of the face c[i,j] and stores the new vertex in
// c[i,j,c[i,j,0]+1].
 x := 0; y := 0; z := 0;
 nn := nn + 1;
 st[nn,1] := "atom " + nn + " - " + tipa + " CA - 0 ";
 for k := 1 to c[i, j, 0]
 x := x + a[c[i,j,k]],2]; // the coordinates of the center of the
 y := y + a[c[i,j,k]],3]; // face c[i,j] is the average of the
 z := z + a[c[i,j,k]],4]; // coordinates of the boundary vertices.
 endfor k
 x := x/c[i,j,0]; y := y/c[i,j,0]; z := z/c[i,j,0];
 st[nn,1] := st[nn, 1] + x + " " + y + " " + z + " ";
 c[i,j,c[i,j,0]+1] := nn;
end function Center_D
```

Function Ini concatenates the information about valences, old coordinates and old adjacent vertices to the string st, in order to preserve the old vertices in the output file. It is called in the Stellation operation.

```
function Ini(a, n, tipa, st)
// in: a, n, tipa
// in-out: st
// purpose: updates the string st with the valences, old coordinates and
// old adjacent vertices info.
 for i := 1 to n
 st[i, 2] := a[i,8];// old valence
 // old coords
 st[i, 1] := "atom " + i + " - " + tipa + " CA - 0 " +
 a[i, 2] + " " + a[i, 3] + " " + a[i, 4] + " ";
 k := 0;
 for j := 9 to a[i,8]*2 + 8 step 2
 k := k + 1; // old adjacent vertices of the vertex i
 st[i, k+2] := a[i,j] + " a";
 endfor j
 endfor i
end function Ini
```

Function Ad_CoordT computes the coordinates (x, y, z) of a new vertex lying on the edge m[i,j] and puts the new vertex with the valency 0 in the final array tr. The parameter k is either 2 or 0.5, defining the position of the new vertex with respect to the ends of the edge. This function is used in the Truncation and the Capra operations.

```
function Ad_CoordT(m, i, j, k, x, y, z, nn, tr, tipa, a)
// in: m, i, j, k, tipa, a
// out: x, y, z
// in-out: nn, tr
// purpose: computes the coordinates (x,y,z) of a new vertex lying on the
// edge m[i,j]. There are two new vertices trisecting the edge.
// The parameter k defines the position of the new vertex with
// respect to the edge bounds. Valid values are 2 or 0.5.
 nn := nn + 1; // the number of vertices in the new graph
 tr[nn,2] := 0;
 tr[nn,1] := "atom " + nn + " - " + tipa + " CA - 0 ";
 x := (a[m[i,j,1],2] + k * a[m[i,j,2],2])/(1+k);
 y := (a[m[i,j,1],3] + k * a[m[i,j,2],3])/(1+k);
 z := (a[m[i,j,1],4] + k * a[m[i,j,2],4])/(1+k);
 tr[nn,1] := tr[nn,1] + x + " " + y + " " + z + " ";
end function Ad_CoordT
```

Function `Ad_P` puts two new vertices on the edge m[i,j] by calling Ad_CoordT twice with k = 2 and k = 0.5, and stores them in m[i,j,5] and m[i,j,6]. Both Ad_P and AdTr functions are invoked in the Truncation operation.

```
function Ad_P(tr, m, i, j, nn, tipa, a)
// in: i, j, tipa, a
// in-out: tr, m, nn
// purpose: puts two new vertices on the edge m[i,j] by calling Ad_CoordT
 Ad_CoordT(m,i,j,2,x,y,z,nn,tr,tipa,a); // the vertex closer to m[i,j,2]
 m[i,j,5] := nn;
 Ad_CoordT(m,i,j,0.5,x,y,z,nn,tr,tipa,a)// the vertex closer to m[i,j,1]
 m[i,j,6] := nn;
 ad2l(nn-1,nn,tr); // links the new vertices related to the old edge
end function Ad_P // m[i,j], in the final array tr.
```

Function AdTr adds a link between the two new vertices located on the edges (m1, m2) and (m2, m3), nearer to the vertex m2.

```
function AdTr(m1, m2, m3, n, m, tr)
// in: m1, m2, m3, n, m
// in-out: tr
// purpose: links the two new vertices located on the edges (m1, m2) and // (m2, m3),
that is closer to the vertex m2.
 for k := 1 to n
 for kk := 1 to m[k,0]
 if ((m1 = m[k,kk,1]) and (m2 = m[k,kk,2])) leg1 := m[k,kk,5];
 if ((m1 = m[k,kk,2]) and (m2 = m[k,kk,1])) leg1 := m[k,kk,6];
 if ((m2 = m[k,kk,1]) and (m3 = m[k,kk,2])) leg2 := m[k,kk,6];
 if ((m2 = m[k,kk,2]) and (m3 = m[k,kk,1])) leg2 := m[k,kk,5];
 endfor kk
 ad2l(leg1, leg2, tr); // links the vertices leg1 and leg2
 endfor k// in both directions.
end function AdTr
```

The function Ad_Coord sums up the coordinates of the vertex vc in the corresponding sum variables x, y, and z. The function is called by the P_New function.

```
function Ad_Coord(x, y, z, vc, a)
// in: vc, a
// in-out: x, y, z
// purpose: sums the coordinates of a[vc] into x, y, and z.
 x := x + a[vc,2];
 y := y + a[vc,3];
 z := z + a[vc,4];
end function Ad_Coord
```

Function P_New puts a new vertex inside the face c[i,j], in the final array leaf, near the edge (c[i,j,k], c[i,j,kk]). It is used in the Leapfrog and Capra operations.

```
function P_New(i, j, k, kk, a, c, tipa, xf, yf, zf, nn, leaf, c2)
// in: i, j, k, kk, a, c, tipa, xf, yf, zf,
// in-out: nn, leaf, c2
// purpose: adds a new vertex inside the face c[i,j] in the array leaf,
// near the edge (c[i,j,k], c[i,j,kk]).
// (xf, yf, zf) define the center of the face c[i,j].
 nn := nn + 1; // new vertex
 leaf[nn, 1] := "atom " + nn + " - " + tipa + " CA - 0 ";
 x := 0; y := 0; z := 0;
 Ad_Coord(x, y, z, c[i,j,k], a);
 Ad_Coord(x, y, z, c[i,j,kk], a);
 x = (x + xf)/3; // Computes the coordinates x,y and z as the mean of
 y = (y + yf)/3; // the coordinates of the two ends of the edge
 z = (z + zf)/3; // (c[i,j,k), c[i,j,kk)) and the face center.
 leaf[nn,1] := leaf[nn,1] + x + " " + y + " " + z + " ";
 c2[i,j,k] := nn;
 leaf[nn,2] := 0;
end function P_New
```

Function IniQ puts the coordinates of old vertices into the qa string. It is invoked in the Quadrupling and Capra operations.

```
function IniQ(a, n, tipa, qa)
// in: a, n, tipa
// in-out: qa
// purpose: initializes the array qa with the old vertices.
 for i := 1 to n
 qa[i,1] := "atom " + i + " - " + tipa + " CA - 0 ";
 qa[i,1] := qa[i,1] + a[i,2] + " " + a[i,3] + " " + a[i,4] + " ";
 qa[i,2] := 0;
 endfor i
end function IniQ
```

Function WriteF writes the final version of the array qa to the output file referred by the file pointer fp.

```
function WriteF(fp, n, qa)
// in: fp, n, qa
// purpose: writes the array qa to the text file referred by fp.
 for i := 1 to n // + means string concatenation.
 print fp, qa[i,1] + qa[i,2]; // writes the coordinates and
 for j := 1 to qa[i,2] // valence of the vertex i.
 print fp, " " + qa[i,j+2]; // Writes the linked vertices.
```

```
 endfor j
 print fp, newLine; // Writes the line terminator(s).
 endfor i
 print fp, "endmol 1";
end function WriteF
```

Function `MeanQ` returns the coordinates of the new vertex corresponding to `p2` as the weighted average of three consecutive vertices p1, p2 and p3 in the final string strc. This function is used in the AdCoordQ function.

```
function MeanQ(a, p1, p2, p3, strc)
// in: a, p1, p2, p3
// out: strc
// purpose: computes the coordinates of a new vertex as weighted average
// of the coordinates of vertices p1, p2, p3 and returns them
// in the string strc.
 x := (a[p1,2] + a[p2,2]*2 + a[p3,2])/4;
 y := (a[p1,3] + a[p2,3]*2 + a[p3,3])/4;
 z := (a[p1,4] + a[p2,4]*2 + a[p3,4])/4;
 strc := x + " " + y + " " + z + " ";
end function MeanQ
```

Function AdCoordQ puts the vertex coordinates of the new cycle inside the old cycle c[i,j]. AdCoordQ and AdLegQ functions are called by the AdCycleInt function, in the Quadrupling operation. They update the final string qa.

```
function AdCoordQ(a, c, no, i, j, qa)
// in: a, c, no, i, j
// in-out: qa
// purpose: puts the coordinates of the new cycle inside the old cycle
// c[i,j].
 lc := c[i,j,0];
 n := no - lc;
 MeanQ(a, c[i,j,lc], c[i,j,1], c[i,j,2], strc);
 qa[n+1,1] := qa[n+1,1] + strc; // the coords of the first vertex
 MeanQ(a, c[i,j,lc-1], c[i,j,lc], c[i,j,1], strc);
 qa[no,1] := qa[no,1] + strc; // the coords of the last vertex
 for k := 2 to c[i,j,0]-1
 MeanQ(a, c[i,j,k-1], c[i,j,k], c[i,j,k+1], strc);
 qa[n+k,1] := qa[n+1,1] + strc; // the coords of the other vertices
 endfor k
end function AdCoordQ
```

Function AdLegQ adds the edges of the new cycle located inside the old cycle c[i,j].

```
function AdLegQ(a, c, no, i, j, qa)
// in: a, c, no, i, j
// in-out: qa
// purpose: adds the edges of the new cycle located inside the old
// cycle c[i,j].
 lc := c[i,j,0]; // the size of the cycle
 p1 := no - lc + 1;// the first vertex in the cycle
 // Adds the links between the vertices of the new cycle.
 ad2l(p1, no, qa);
 for k := 1 to c[i,j,0]-1
 ad2l(p1+k-1, p1+k, qa);
 endfor k
 // Adds the links between the vertices of the new cycle and the
 // corresponding vertices of the old cycle.
 for k := 1 to c[i,j,0]
 Ad2l(p1+k-1, c[i,j,k], qa);
 endfor k
end function AdLegQ
```

Function `AdCycleInt` adds a similar inner cycle for every old cycle and puts the edges for the new cycle between the new and corresponding old vertices. It implements an important step in the `Quadrupling` operation.

```
function AdCycleInt(a, n, c, tipa, qa, no)
// in: a, n, c, tipa
// out: no
// in-out: qa
// purpose: adds a similar inner cycle for every old cycle and updates
 the links between the new and the corresponding old vertices.
 no := n; // The old vertices are kept.
 for i := 1 to n
 for j := 1 to c[i,0] // for every cycle c[i,j]
 for k := 1 to c[i,j,0] // for every vertex c[i,j,k] of the cycle
 no := no + 1; // a new vertex to be added
 qa[no, 1] := "atom " + no + " - " + tipa + " CA - 0 ";
 qa[no, 2] := 0;
 endfor k
 AdCoordQ(a, c, no, i, j, qa);
 AdLegQ(a, c, no, i, j, qa);
 endfor j
 endfor i
end function AdCycleInt
```

Function `P_New_Capra` puts a new vertex inside the face c[i,j], in the array leaf, close to the edge (c[i,j,k], c[i,j,kk]). Unlike the function P_New, the new vertex will be selected. This function is called by the Ad_P_Middle function, described below.

```
function P_New_Capra(i, j, k, kk, a, c, tipa, xf, yf, zf, nn, cip, c2)
// in: i, j, k, kk, a, c, tipa, xf, yf, zf
// in-out: nn, cip, c2
// purpose: adds a new vertex inside the face c[i,j] in the array cip,
// close to the edge (c[i,j,k], c[i,j,kk]).
// (xf, yf, zf) define the center of the face c[i,j]
 P_New(i, j, k, kk, a, c, tipa, xf, yf, zf, nn, cip, c2);
 // the vertices within the new cycle are marked by 's'.
 cip[nn,1] := replace("CA - 0", "CA s 0", cip[nn,1]); // string repl.
end function P_New_Capra
```

Function Ad_P_Middle adds a new cycle, of the same folding, in the center of old cycle c[i,j]. This function is used by the Capra operation.

```
function Ad_P_Middle(cip, m, i, j, nn, c, x, y, z, a, tipa, c2)
// in: m, i, j, x, y, z, a, tipa
// in-out: cip, nn, c, c2
// purpose: adds a new cycle in the center of the old cycle c[i,j].
 nn := nn - 1; // Removes the vertex from the center.
 nv = c[i,j,0];
 c[i,j,nv+2] := "x"; // Marks the visited cycle.
 // Adds the vertices of the new cycle inside the old cycle.
 for k := 1 to nv-1
 P_New_Capra(i, j, k, k+1, a, c, tipa, x, y, z, nn, cip, c2);
 endfor k
 P_New_Capra(i, j, nv, 1, a, c, tipa, x, y, z, nn, cip, c2);
 // Adds edges of the new cycle.
 for k := 1 to nv-1
 Ad2l(c2[i,j,k], c2[i,j,k+1], cip);
 endfor k
 Ad2l(c2[i,j,1], c2[i,j,nv], cip);
end function Ad_P_Middle
```

All the functions described below in this section are invoked by the Capra operation. The function Inv56 returns the value 5 if receives the value 6, and 6 otherwise.

```
function Inv56(i)
// in: i
// purpose: if the argument has the value 6, returns 5;
// otherwise returns 6.
 if i = 6
 then return 5
```

```
  else return 6
  endif
end function Inv56
```

The function Verif_M2 returns the position (of the added points on the edges) put by the last link.

```
function Verif_M2(m, i, j, c)
// in: m, i, j, c
// purpose: returns the position of vertex put by the last link
// if no link is found, returns 0.
 for k := 1 to c[i,j,0] // Verifies if the links exist
 ii := c[i,j,k]; // from the central cycle at the first or
 if k < c[i,j,0] // second vertex on the old edge.
 then iiu := c[i,j,k+1];
 else iiu := c[i,j,1];
 endif
 for jj := 1 to m[ii,0]
 if m[ii,jj,2] = iiu then // for the edge ii, iiu
 if m[ii,jj,7] > 0 then // If the link was found
 // returns the position of the last link on the edge m[ii,jj].
return m[ii,jj,7];
 endif
 endfor jj
 for jj := 1 to m[iiu, 0]
 if m[iiu,jj,2] = ii then // for the edge iiu, ii
 if m[iiu,jj,7] > 0 then// If the link was found
 // returns the position of the last link on the edge m[ii,jj].
return Inv56(m[iiu,jj,7]);
 endif
 endfor jj
 endfor k
 return 0;
end function Verif_M2
```

Function `Verif_M` puts a link between the central cycle and a new vertex on the edges.

```
function Verif_M(cip, m, ii, iiu, m56, pc)
// in: ii, iiu, m56, pc
// in-out: cip, m
// purpose: puts the link from the vertex pc of the central cycle to
// a vertex from the edge ii - iiu, in the final array cip.
 for jj := 1 to m[ii, 0]
 if m[ii,jj,2] = iiu then // for the edge ii, iiu
 ad2l(m[ii,jj,m56], pc, cip); // links the vertices
 m[ii,jj,7] := m56;
```

```
endif
endfor jj
end function Verif_M
```

Function AlreadyVisited checks if the cycle was already visited or it shares an edge with a visited cycle.

```
function AlreadyVisited(c, i, j, m)
// in: c, i, j, m
// purpose: returns true (0) if
// a) the cycle was already visited
// b) the cycle shares an edge with an already visited cycle.
 nv = c[i,j,0];
 if c[i,j,nv+2] = "x" then return true; // The cycle was visited.
 if Verif_M2(m, i, j, c) = 0 then return true;
 return false;
end function AlreadyVisited
```

The function Nrc counts the number of cycles.

```
function Nrc(c, n, nci)
// in: c, n
// out: nci
// purpose: returns the number of cycles.
 nci := 0;
 for i := 1 to n
 for j := 1 to c[i,0]
 nci := nci + 1;
 endfor j
 endfor i
end function Nrc
```

Function AdLeg_C adds edges connecting the vertices of the new cycle lying in the center of the old cycle, to the old cycle, c[i,j].

```
function AdLeg_C(cip, m, i, j, nn, c, c2, n, p12)
// in: i, j, nn, c2, n, p12
// in-out: cip, m, c
// purpose: puts the links between the central cycle and the points added
// on the edges of the old cycle c[i,j].
 nv := c[i,j,0];
 poz1 := Verif_M2(m, i, j, c);
// the position of the last link on the edges
 if poz1 = 0 then poz1 := 4 + p12;
 poz2 := inv56(poz1);
 for k := 1 to c[i,j,0] // for every vertex c[i,j,k] from c[i,j]
```

```
ii := c[i,j,k];
if k < c[i,j,0]
then iiu := c[i,j,k+1]
else iiu := c[i,j,1]
endif
Verif_M(cip, m, ii, iiu, poz2, c2[i,j,k]); // Puts a link to the new
Verif_M(cip, m, iiu, ii, poz1, c2[i,j,k]); // vertex added on the old
endfor k // edge ii, iiu.
end function AdLeg_C
```

The function Ad_Pi adds two vertices on each link and then links to each other and to the ends of the edge.

```
function Ad_Pi(cip, m, i, j, nn, tipa, a)
// in: i, j, tipa, a
// in-out: cip, m, nn
// purpose: adds two vertices on each link and then links them with
// the ends of the edge.
Ad_CoordT(m, i, j, 2, x, y, z, nn, cip, tipa, a);
ad2l(m[i,j,2], nn, cip);
m[i,j,6] := nn;
Ad_CoordT(m, i, j, 0.5, x, y, z, nn, cip, tipa, a);
m[i,j,5] := nn;
// Links the new vertices to each other.
ad2l(nn-1, nn, cip);
// Links the new vertices with the ends of the edge.
ad2l(m[i,j,1], nn, cip);
end function Ad_Pi
```

## 3. BASIC MAP OPERATIONS

### 3.1. Dual

Function Dual operates all the faces of the array c by putting a new point in the center of each face. Every two such points are then joined if their corresponding faces share a common edge (Figure 1).

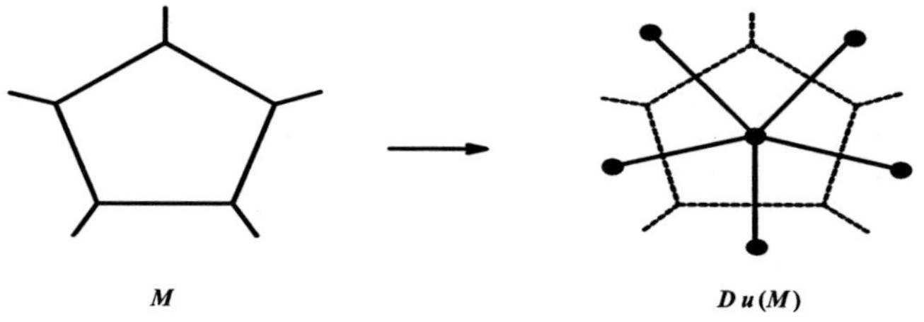


Figure 1: Dual operation.

```
function Dual(a, n, c, fp, m, tipa, sfer, texti)
// in: a, n, c, fp, tipa, sfer, texti
// in-out: m
// purpose: puts a new point in the center of each face
// joins every two such new points if their corresponding faces
// share a common edge.
// textd: the .hin file header
// newLine is the line terminator.
textd := "Dual " + newLine + "forcefield mm+" + newLine +
"sys 0" + texti + "mol 1" + newLine;
print fp, textd;
Center(n, a, xc, yc, zc, r);
nn := 0; // the number of initial faces = number of the new vertices
for i := 1 to n // Visits the faces to compute their center coords.
for j := 1 to c[i,0];
nn := nn + 1;
du[nn,2] := 0; // Initializes the valences.
du[nn,1] := "atom " + nn + " - " + tipa + " CA - 0 ";
x := 0; y := 0; z := 0;
for k := 1 to c[i,j,0]
x := x + a[c[i,j,k],2]; // Coordinates x, y, z of the center
y := y + a[c[i,j,k],3]; // of the face c[i,j]
z := z + a[c[i,j,k],4]; // are averages of vertex coords.
endfor k
x := x / c[i,j,0]; y := y / c[i,j,0]; z := z / c]i,j,0];
if sfer then // if the new vertices lye on the sphere
OnSphere(xc, yc, zc, r, x, y, z);
endif
du[nn,1] := du[nn,1] + x + " " + y + " " + z + " ";
c[i,j, c[i,j,0]+1] := nn; // the new vertex for the face c[i,j]
endfor j
endfor i
// adds the links in the new graph
for i := 1 to n
for j := 1 to m[i,0]
m[i,j,5] := 0;
endfor j
endfor i
for i := 1 to n
for j := 1 to c[i,0]
v := c[i,j,c[i,j,0]+1];
for k := 1 to c[i,j,0] - 1
m1 := c[i,j,k];
m2 := c[i,j,k+1];
Much_D(v, m1, m2, n, m); // Puts the new vertex v in the array m.
endfor k
```

```
  Much_D(v, c[i,j,1], c[i,j,c[i,j,0]], n, m);
  endfor j
  endfor i
  for i := 1 to n
  for j := 1 to m[i,0]// Visits the edges and links the
  if m[i,j,6] > 0 then// vertices across the shared edge.
  ad2l(m[i,j,5], m[i,j,6], du);
  endif
  endfor j
  endfor i
  // Write the strings from du in the output file.
  WriteF(fp, nn, du);
end function Dual
```

### 3.2. Medial

By this operation, a new vertex is put in the middle of every old edge and the new vertices are linked if they belong to consecutive edges, within a rotational path around their common vertex. Only the new vertices are retained. Optionally, the new vertices can be embedded on the sphere.

In the medial transform, the number of vertices equals the number of edges in the parent map. Function Medial visits the edges and puts a new vertex in the middle of every edge. The local array med will contain the new structure/map.

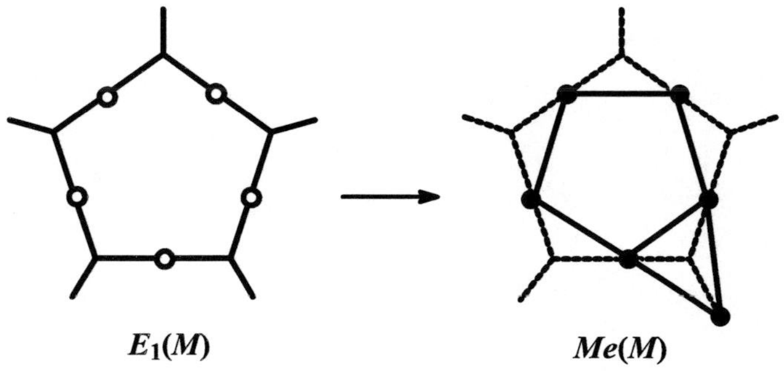


Figure 2: Medial operation

```
function Medial(n, a, c, m, fp, tipa, sfer, texti)
// in: n, a, c, fp, tipa, sfer, texti
// out: m
// purpose: puts a new point in the middle of every old edge
// the new vertices are linked if they belong to consecutive
// edges within a rotational path around their common vertex.
 // .hin file header
 // newLine is the line terminator.
 textm := "Medial " + newLine + "forcefield mm+" + newLine +
```

```
"sys 0" + texti + "mol 1" + newLine;
print fp, textm;
Center(n, a, xc, yc, zc, r);
nn := 0; // the number of resulting vertices
for i := 1 to n
for j := 1 to m[i,0]// For every edge
nn := nn + 1;// adds a new vertex.
med[nn,2] := 0;
med[nn,1] := "atom " + nn + " - " + tipa + " CA - 0 ";
// the coords of the edge middle
x := (a[m[i,j,1],2] + a[m[i,j,2],2])/2;
y := (a[m[i,j,1],3] + a[m[i,j,2],3])/2;
z := (a[m[i,j,1],4] + a[m[i,j,2],4])/2;
if sfer then // if the new points will lye on the sphere
OnSphere(xc, yc, zc, r, x, y, z);
endif
med[nn,1] := med[nn,1] + x + " " + y + " " + z + " ";
m[i,j,5] := nn;
endfor j
endfor i
// Add the links to the other new vertices.
for ii := 1 to n
for jj := 1 to c[ii,0] // for each face
for kk := 1 to c[ii,jj,0] - 2
// Connects the new vertices located at every consecutive edges.
// (m1, m2) and (m2, m3)
m1 := c[ii, jj, kk];
m2 := c[ii, jj, kk+1];
m3 := c[ii, jj, kk+2];
AdMe(m1, m2, m3, n, m, med);
endfor kk
// Connects the start and end vertices, closing the ring.
AdMe(c[ii,jj,c[ii,jj,0]-1],c[ii,jj,c[ii,jj,0]],c[ii,jj,1],n,m,med);
AdMe(c[ii,jj,c[ii,jj,0]],c[ii,jj,1],c[ii,jj,2],n,m,med);
endfor jj
endfor ii
// Writes the strings from med in the output file.
WriteF(fp, nn, med);
end function Medial
```

## 3.3. Stellation

This operation adds a new vertex in the center of every face and connect it with the boundary vertices. The old vertices and edges are kept, and all the faces of the resulting graph are triangles. Function Stellation builds the array `st` and writes it in the output file.

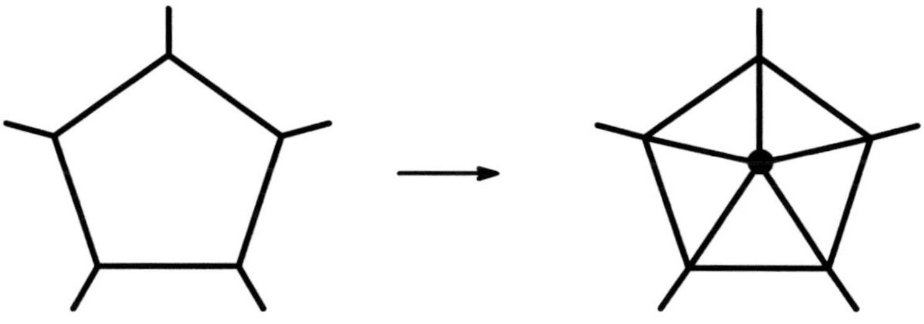


Figure 3: Stellation operation

```
function Stellation(n, a, c, fp, m, tipa, texti)
// in: n, a, c, fp, tipa, texti
// in-out: m
// purpose: adds a new vertex in the center of every face
// and connects it with the boundary vertices. The old vertices
// and edges are kept.
 // .hin file header
 // newLine is the line terminator.
 texts := "Stellation " + newLine + "forcefield mm+" + newLine +
 "sys 0" + texti + "mol 1" + newLine;
 print fp, texts;
 Ini(a, n, tipa, st); // puts the initial vertices and edges in st.
 nn := n; // nn stores the number of final vertices.
 for i := 1 to n
 for j := 1 to c[i,0] // visits all parent faces;
 Center_D(i,j,nn,a,tipa,x,y,z,c,st); // the nn-th vertex is the
 st[nn,2] := 0; // center of the c[i,j] face.
 for k := 1 to c[i,j,0]
 ad2l(nn,c[i,j,k],st); // joins all boundary vertices to the
 endfor k // center of the face.
 endfor j
 endfor i
 // writes the strings from st in the output file.
 WriteF(fp, nn, st);
end function Stellation
```

## 3.4. Truncation

This operation puts a new vertex on each edge incident in an old vertex, and join them around the old vertex, which is finally cut off. In the function `Truncation`, the final vertices are put in the local array `tr`, which is then written in the .hin file.

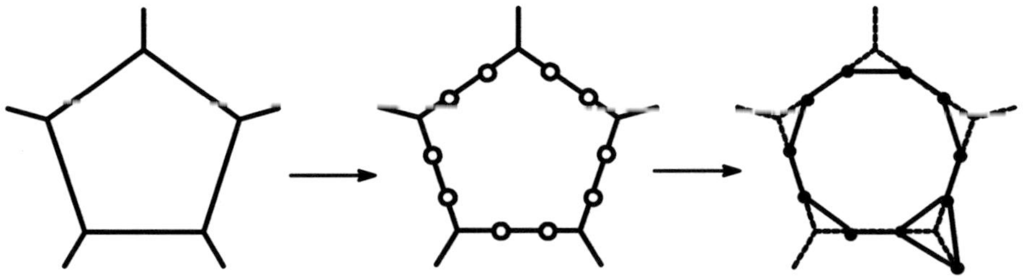


Figure 4: Truncation operation

```
function Truncation (n, a, c, m, fp, tipa, texti)
// in: n, a, c, m, fp, tipa, texti
// purpose: adds a new vertex on each edge around an old vertex
// and connects them around the old vertex which is cut off;
//connects the two vertices on the same edge.
 // .hin file header
 // newLine is the line terminator.
 textt := "Truncation " + newLine + "forcefield mm+" + newLine +
 "sys 0" + texti + "mol 1" + newLine;
 print fp, textt;
 nn := 0; // the number of final vertices
 for i := 1 to n
 for j := 1 to m[i,0] // On each edge two vertices
 Ad_P(tr, m, i, j, nn, tipa, a); // are added and linked.
 endfor j
 endfor i
 for ii := 1 to n
 for jj := 1 to c[ii,0]
 for kk := 1 to c[ii,jj,0]-2 // for each cycle c[ii,jj]
 m1 := c[ii,jj,kk]; // Connects the new vertices
 m2 := c[ii,jj,kk+1]; // lying on the edges (m1,m2) and (m2,m3)
 m3 := c[ii,jj,kk+2]; // around the old vertex m2.
 AdTr(m1, m2, m3, n, m, tr);
 endfor kk
 AdTr(c[ii,jj,c[ii,jj,0]-1],c[ii,jj,c[ii,jj,0]],c[ii,jj,1],n,m,tr);
 AdTr(c[ii,jj,c[ii,jj,0]],c[ii,jj,1],c[ii,jj,2],n,m,tr);
 endfor jj
 endfor ii
 // Writes the strings from array tr in the output file.
 WriteF(fp, nn, tr);
end function Truncation
```

### 3.5. Leapfrog

This operation puts vertices on both sides of the old edges. Next, it links these new vertices inside of each face if they correspond to consecutive edges and then links every two vertices from different faces if they correspond to the same edge. Only the new vertices and edges remain in the resulting graph. During the construction, the local array leaf contains information about new vertices, which is written in the .hin file.

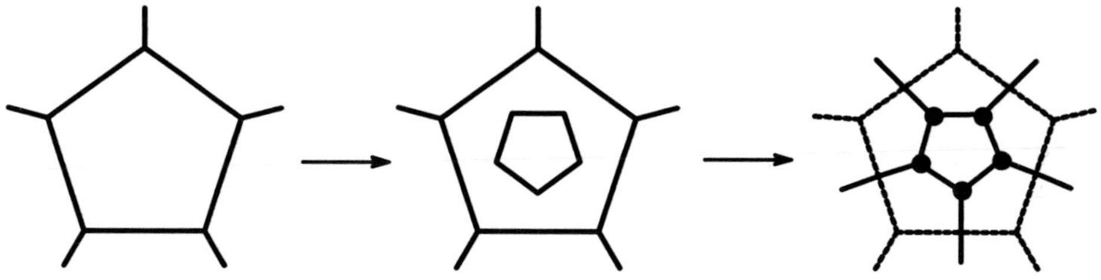


Figure 5: Leapfrog operation.

```
function Leapfrog(a, n, c, fp, m, tipa, texti)
// in: a, n, c, fp, tipa, texti
// in-out: m
// purpose: puts vertices on both sides of the old edges;
// links the new vertices inside the faces if they correspond to // consecutive edges
and links every two vertices from different
// faces if they correspond to the same edge;
// only the new vertices remain.
// .hin file header
// newLine is the line terminator.
textl := "Leap Frog " + newLine + "forcefield mm+" + newLine +
"sys 0" + texti + "mol 1" + newLine;
print fp, textl;
nn := 0; // the number of new vertices
nf := 0; // the number of new faces
for i := 1 to n
for j := 1 to c[i,0] // For every face c[i,j]
// computes the coordinates (xf,yf,zf) of the face center.
Center_D(i,j,nf,a,tipa,xf,yf,zf,c,st);
nv := c[i,j,0]; // the number of vertices in the face c[i,j]
v := c[i,j,nv+1];
for k := 1 to nv-1
// For each boundary edge, put in c2 a new vertex inside the face.
P_New(i,j,k,k+1,a,c,tipa,xf,yf,zf,nn,leaf,c2);
endfor k
P_New(i,j,nv,1,a,c,tipa,xf,yf,zf,nn,leaf,c2);
for k := 1 to nv-1
```

```
ad2l(c2[i,j,k],c2[i,j,k+1],leaf); // Adds the links between the
m1 := c[i,j,k]; // new vertices c2[i,j,k],
m2 := c[i,j,k+1]; // c2[i,j,k+1] inside the
Much_D(c2[i,j,k],m1,m2,n,m); // face c[i,j].
endfor k
ad2l(c2[i,j,1],c2[i,j,nv],leaf); // Links vertices at the end
Much_D(c2[i,j,nv],c[i,j,1],c[i,j,nv],n,m); // of the cycle c[i,j].
endfor j
endfor i
for i := 1 to n
for j := 1 to m[i,0]
if m[i,j,6] > 0 then
ad2l(m[i,j,5],m[i,j,6], leaf);
endif
endfor j
endfor i
// Writes the strings from leaf in the output file.
 WriteF(fp, nn, leaf);
end function Leapfrog
```

### 3.6. Quadrupling

This operation adds a new cycle inside of each cycle/face and connects the two cycles vertex by vertex. Next, the old edges are deleted. The transformation preserves the initial orientation of all parent faces in the map as well as the initial vertices (Figure 6).

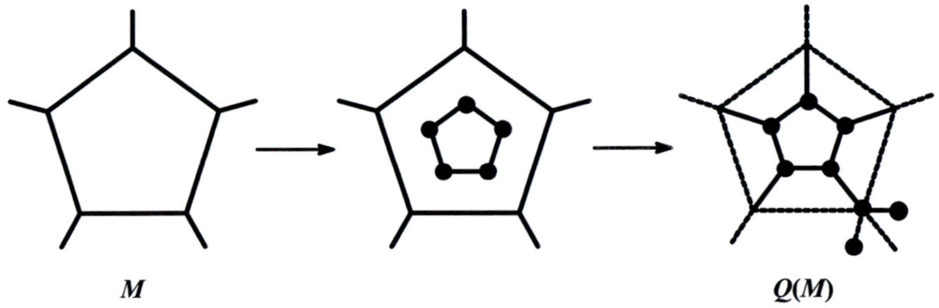


Figure 6: Quadrupling operation

```
function Quadrupling (a, n, c, fp, tipa, texti)
// in: a, n, c, fp, tipa, texti
// purpose: adds a new cycle inside of each cycle/face and connects the
//two cycles vertex by vertex.
// The transformation preserves the initial orientation of all
// parent faces. The initial vertices are preserved while the
// old edges are removed.
 // .hin file header
```

```
// newLine is the line terminator.
textq := "Quadrupling " + newLine + "forcefield mm+" + newLine +
"sys 0" + texti + "mol 1" + newLine;
print fp, textq;
IniQ(a,n,tipa, qa); // Puts in qa the initial vertices without edges.
AdCycleInt(a,n,c,tipa,qa,no);
// Writes the strings from qa in the output file.
WriteF(fp, no, qa);
end function Quadrupling
```

### 3.7. Capra

This transformation puts two new points on each edge of the map. Next, puts a new cycle of the same folding in the center of each face and makes (1, 4) connections to the boundary points, starting with a new vertex[9],[10].

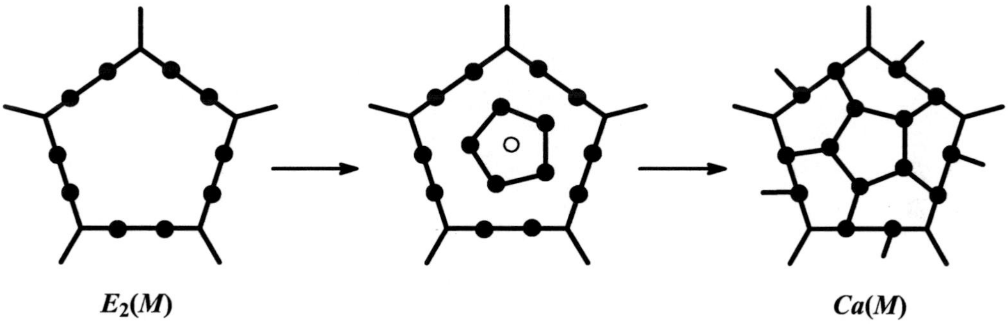


Figure 7: Capra operation

```
function Capra(a, n, c, m, fp, tipa, texti, p12)
// in: a, n, c, m, fp, tipa, texti, p12
// purpose: puts two new points on each edge of the map.
// Puts a new cycle of the same size in the center of each face
// and makes (1,4) connections to the boundary points,
// starting with a new vertex.
// .hin file header
// newLine is the line terminator.
textc := "Capra " + newLine + "forcefield mm+" + newLine +
"sys 0" + texti + "mol 1" + newLine;
print fp, textc;
Nrc(c, n, nci);
IniQ(a,n,tipa,cip); // Puts in the final array cip the initial
nn := n;// vertices without edges.
for i := 1 to n// Initialization
for j := 1 to m[i,0]
m[i,j,7] := ""; // for vertices related to the edges
```

```
endfor j
for j := 1 to c[i,0] // and for vertices related to the cycles.
nv := c[i,j,0];
c[i,j,nv+2] := "";
endfor j
endfor i
for i := 1 to n
for j := 1 to m[i,0]
Ad_Pi(cip,m,i,j,nn,tipa,a); // Adds two new linked vertices on
endfor j // every old edge.
endfor i
nc := 0;
do while nc < nci// do while exists a cycle not visited
for ix := 1 to n
for jx := 1 to c[ix,0]
i := ix; j := jx;
if (nc = 0) or (AlreadyVisited(c,i,j,m)) then
// if it is the first cycle or the current cycle shares an edge
// with an already visited cycle
nc := nc + 1;
v := c[i,j,0];
// Adds a vertex in the face center; it will be deleted later.
Center_D(i,j,nn,a,tipa,x,y,z,c,c3);
c[i,j,0] := v;
// Adds a rotated new cycle in the middle of the old cycle.
Ad_P_Middle(cip,m,i,j,nn,c,x,y,z,a,tipa,c2);
// Adds the links between the new cycle and the vertices
// added on the old edges.
AdLeg_C(cip,m,i,j,nn,c,c2,n,p12);
endif
endfor jx
endfor ix
loop
// Writes the strings from cip in the output file.
WriteF(fp, no, cip);
end function Capra
```

## 4. CONCLUSIONS

Based on the Graph Theory, a self-consistent software program was developed. Its user interface lists all the faces of an input map/polyhedron, and allows the user to select the above-described map operations in order to obtain transformed maps. The program was tested on objects having up to thousands points, both closed and open.

## REFERENCES


[1] B. Grünbaum and G. C. Shephard, *Tilings and Patterns*, Freeman, New York, 1985.

[2] D. J. Klein and H. Zhu, in: A. T. Balaban, (Ed.), *From Chemical Topology to Three - Dimensional Geometry*, Plenum Press, New York, 1997, pp. 297-341.

[3] B. de La Vaissière, P. W. Fowler, and M. Deza, *J. Chem. Inf. Comput. Sci.*, 2001, 41, 376-386.

[4] M. Deza, P. W. Fowler, M. Shtorgin, and K. Vietze, *J. Chem. Inf. Comput. Sci.*, 2000, 40, 1325-1332.

[5] T. Pisanski, and M. Randić, in *Geometry at Work*, M. A. A. Notes, 2000, 53, 174-194.

[6] P. W. Fowler and T. Pisanski, *J. Chem. Soc. Faraday Trans*. 1994, 90, 2865-2871.

[7] F. Harary, *Graph Theory*. Addison-Wesley, Reading, MA, 1969.

[8] Diudea, M. V.; John, P. E.; Graovac,A.; Primorac, M.; Pisanski, T. Leapfrog and Related Operations on Toroidal Fullerenes. *Croat. Chem. Acta*, 2003, 76, 153-159.

[9] M. V. Diudea, Covering Nanostructures, in: M. V. Diudea, Ed., *Nanostructures-Novel Architecture*, NOVA, New York, 2004 (in press).

[10] M. V. Diudea, *Studia Univ. "Babes-Bolyai"*, 2003, 48 (2), 3-16.

In: Nanostructures: Novel Architecture
Editor: Mircea V. Diudea, pp. 269-297
ISBN: 1-59454-499-9


*Chapter 13*

# ZEOLITE-LIKE POSSIBLE CARBON AND BORON NITRIDE ALLOTROPES

***R. Bruce King***[1]
Department of Chemistry, University of Georgia
Athens, Georgia 30602 USA

## ABSTRACT

Structures for low-density zeolite-like possible carbon and boron nitride allotropes can be constructed from networks of trigonal ($sp^2$) carbon atoms containing hexagons, heptagons, and/or octagons. This article surveys the geometry, topology, and symmetry of these structures. The most symmetrical such structures are obtained by leapfrog transformations of either the Dyck tessellation of 12 octagons or the Klein tessellation of 24 heptagons. Such tessellations can be embedded into genus 3 infinite periodic minimal surfaces of negative curvature. The automorphism group, $^4O$, of the Dyck tessellation is a solvable group of order 96, the structure of which relates to the dual of the Dyck tessellation as the regular tripartite graph $K_{4,4,4}$. However, the automorphism group, $^7O$, of the Klein tessellation is a simple group of order 168, which can be generated from the prime number 7 in a way analogous to the generation of the icosahedral rotation group $I$ ($\approx$ the alternating group $A_5$) from the prime number 5.

## 1. INTRODUCTION

The experimentally known allotropes of carbon (Figure 1) can be classified into the following four categories:

(a) *Diamond.* An infinite three-dimensional lattice of $sp^3$ (tetrahedral) carbon atoms, each of which is bonded to four other carbon atoms to form interlocking $C_6$ rings similar to those in cyclohexane or adamantane;
(b) *Graphite.* Infinite planar layers of hexagons of $sp^2$ (trigonal) carbon atoms.

---

[1] E-mail address: rbking@chem.uga.edu

(c) *Fullerenes.* Finite molecular cages of $sp^2$ carbon atoms containing only $C_6$ and $C_5$ rings of which the truncated icosahedral $C_{60}$ is the prototypical example.
(d) *Nanotubes.* Tubes constructed from hexagons of $sp^2$ carbon atoms with curved ends containing some $C_5$ pentagons.

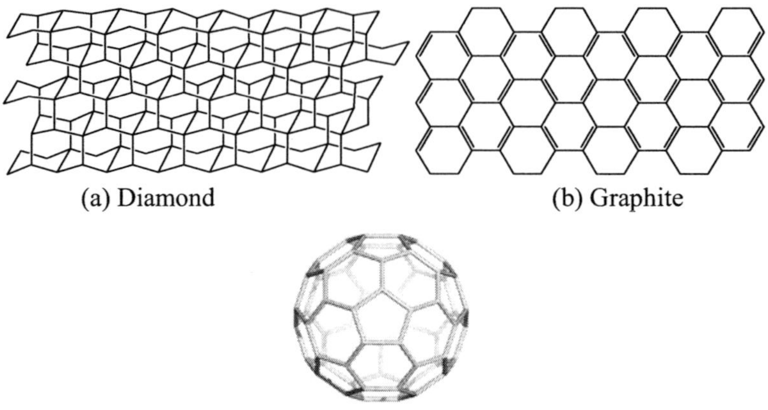

(a) Diamond (b) Graphite

(c) $C_{60}$ Fullerene (Truncated Icosahedron)

Figure 1: (a) A portion of the diamond lattice; (b) A portion of a planar hexagonal sheet of the graphite structure; (c) A front view of the truncated icosahedral structure of the $C_{60}$ fullerene.

Among these carbon allotropes diamond and graphite have been known since antiquity, the fullerenes were discovered in the 1980s, and the nanotubes in the 1990s. The building blocks of all of the known carbon allotropes other than diamond are trigonal $sp^2$ carbon atoms, which are favored to form hexagons. Introduction of pentagons of $sp^2$ carbon atoms as in the fullerenes and nanotubes leads to sites of positive curvature.

A question of interest is whether carbon allotropes can be obtained having structures based on networks of trigonal carbon vertices with only heptagonal and hexagonal carbon rings. The presence of the heptagons leads to sites of negative curvature and precludes the formation of closed polyhedra. The resulting infinite structures resemble those of zeolites and provide the possibility of a new porous low-density form of carbon having cavities suitable for trapping small molecules. Such carbon allotropes would be non-oxidic non-metallic analogues of zeolites.

The possibility of negative curvature allotropes of carbon was apparently first recognized by Mackay and Terrones[1] in 1991. In 1992 Vanderbilt and Tersoff[2] first postulated the so-called D168 carbon structure with a unit cell of genus three containing 24 heptagons and 80 hexagons and a total of 168 carbon atoms. Such negative curvature carbon allotropes are conveniently called schwarzites, in honor of the 19$^{th}$ century mathematician H. A. Schwarz, who first studied some of the mathematics relevant to understanding these structures; these structures have not yet been obtained experimentally. A major obstacle in the synthesis of schwarzites is the selective formation of carbon rings of the desired sizes (*e.g.*, hexagons and heptagons) without forming carbon rings of undesired sizes (*e.g.*, pentagons).

Boron nitride analogues of diamond and graphite are known with the stoichiometry $(BN)_x$ in which pairs of adjacent carbon atoms are replaced by isoelectronic BN pairs thereby suggesting the possible existence of negative curvature boron nitrides analogous to the schwarzites. In this connection, boron nitride structures $(BN)_x$ are most favorable energetically if they contain only B–N bonds without any B–B or N–N bonds. However, the presence of carbon rings with an odd number of atoms (*e.g.*, pentagons in fullerenes and heptagons in schwarzites) makes impossible analogous $(BN)_x$ structures if only the most energetically favorable B–N bonds are allowed. Thus in order to obtain $(BN)_x$ structures having only B–N bonds, only $(BN)_r$ rings with an even number of atoms having alternating boron and nitrogen atoms are allowed. Therefore, the most favorable positive curvature BN cage analogous to the fullerenes contains only $(BN)_2$ squares and $(BN)_3$ hexagons and the most favorable negative curvature IPMS boron nitride structure contains only $(BN)_4$ octagons and $(BN)_3$ hexagons. In this connection the truncated octahedral $B_{12}N_{12}$ structure has been shown computationally[3],[4] to be a favorable structure for a cage $(BN)_x$ boron nitride.

The renaissance of elemental carbon chemistry arising from the discoveries first of the fullerenes and then of the nanotubes has raised new and interesting mathematical questions relating to their topologies and symmetries. Such mathematical questions have provided new opportunities for mathematical chemists. The possibility of creating even more complicated zeolite-like carbon or boron nitride allotrope structures based on negative curvature networks containing hexagons, heptagons, and/or octagons opens up even richer areas of mathematics. This article provides an overview of some of the areas of mathematics relevant to understanding these structures including pertinent aspects of geometry, topology, and group theory.

## 2. Geometry: Platonic Tessellations on Riemann Surfaces

### 2.1. Euler's Theorem and the Leapfrog Transformation

The standard version of Euler's theorem relating the numbers of vertices ($v$), edges ($e$), and faces ($f$) is

$$v - e + f = 2 \tag{1}$$

This theorem applies to polyhedra which can be embedded in a sphere or a surface homeomorphic to a sphere and thus have genus zero. For example, this theorem thus applies to all five regular polyhedra depicted in Figure 2 (see Table 1).

First apply equation (1) to a regular polyhedron with all degree 3 vertices so that $2e = 3v$ since each edge connects exactly two vertices and each vertex is an endpoint of exactly three edges. In addition each edge is shared by exactly two faces leading to the relationship

Now consider regular polyhedra containing only triangular, square, or pentagonal faces so that

$$2e = 3v = 3f_3 + 4f_4 + 5f_5 \tag{3}$$

Substituting this into Euler's equation (equation 1) gives

$$\sum (6-k) f_k = 12 \tag{4}$$

$$\sum f_n = 2e \tag{2}$$

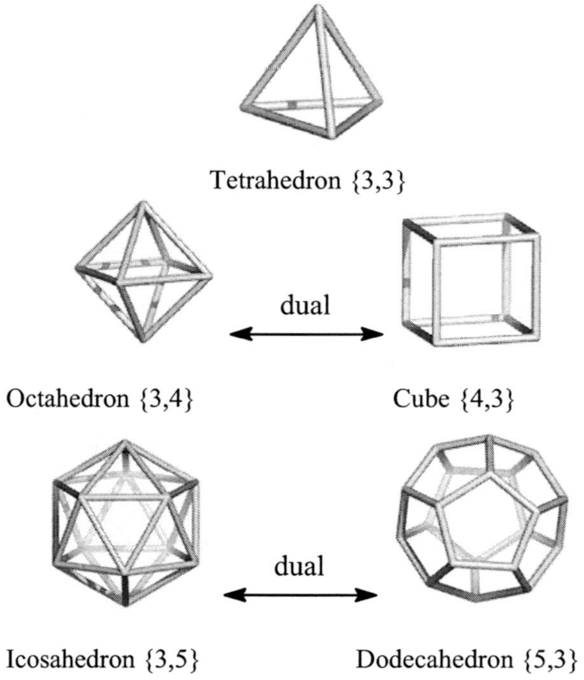


Figure 2: The five regular polyhedra showing the dual relationships.

Setting in turn $f_4 = f_5 = f_{>5} = 0$, $f_3 = f_5 = f_{>5} = 0$ and $f_3 = f_4 = f_{>5} = 0$ gives the solutions $f_3 = 4$, $f_4 = 6$, and $f_5 = 12$ for the regular tetrahedron, the cube, and the regular dodecahedron, respectively (Figure 2). Using a similar approach for regular polyhedra with all degree 4 vertices gives the equation

$$2e = 3v = 3f_3 + 4f_4 + 5f_5 \tag{5}$$

which, when substituted into Euler's equation (equation 1) gives

$$\sum (4-k) f_k = 8 \tag{6}$$

Setting $f_4 = f_5 = f_{>5} = 0$ gives $f_3 = 8$ corresponding to the regular octahedron; no other regular polyhedra can be generated from equation (6). Finally, a similar approach for regular polyhedra with all degree 5 vertices leads to

$$2e = 5v = 3f_3 + 4f_4 + 5f_5 \Rightarrow (10-3k) f_k = 20 \tag{7}$$

Setting $f_4 = f_5 = f_{>5} = 0$ gives $f_3 = 20$ corresponding to the regular icosahedron; no other regular polyhedra can be generated from equation (7).

**Table 1: Properties of the Regular Polyhedra (Figure 2)**

| Polyhedron | Face Type | Vertex Degrees | Number of Edges | Number of Faces | Number of Vertices |
|---|---|---|---|---|---|
| Tetrahedron | Triangle | 3 | 6 | 4 | 4 |
| Octahedron | Triangle | 4 | 12 | 8 | 6 |
| Cube | Square | 3 | 12 | 6 | 8 |
| Icosahedron | Triangle | 5 | 30 | 20 | 12 |
| Dodecahedron | Pentagon | 3 | 30 | 12 | 20 |

A carbon allotrope based on the regular dodecahedron (Figure 2), namely $C_{20}$ with all trigonal $sp^2$ carbon atoms, is unfavorable for a number of reasons including the following:

(1) The local curvature of the vertices of a dodecahedron is high enough to introduce significant strain in an $sp^2$ carbon atom with its three hybrid orbitals in the same plane;

(2) Two pentagons sharing an edge form an electronically unfavorable pentalene unit.

The curvature can be reduced significantly and pentalene units eliminated by expanding the regular dodecahedron to the truncated icosahedral $C_{60}$ structure by means of the so-called leapfrog transformation (Figure 3).[5] The leapfrog transformation consists of omni-capping each face followed by dualization and generates a network of $3n$ carbon atoms from any network of $n$ carbon atoms. In the omnicapping stage, each face is capped so that the omnicapped network can be regarded as a "compound" of the original network and its dual. The processes of omnicapping and dualization making up the leapfrog transformation both preserve the symmetry of the original network. Figure 3 illustrates the leapfrog transformation in going from the regular dodecahedron to the $C_{60}$ truncated icosahedron.

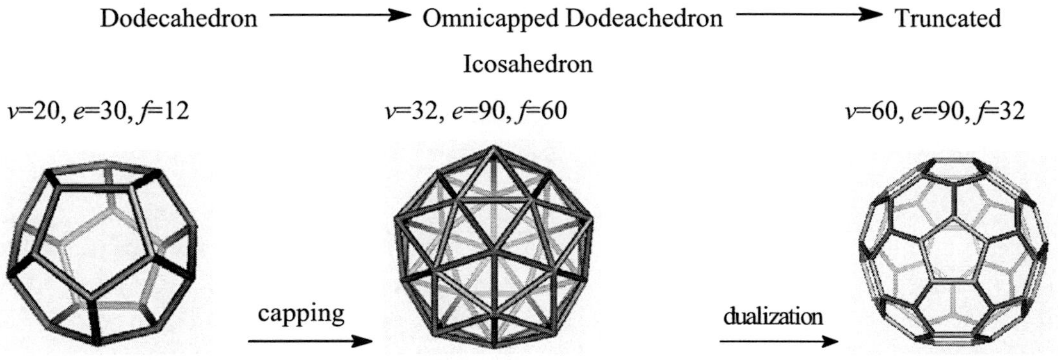


Figure 3: The leapfrog transformation of the regular dodecahedron to the truncated icosahedron of C60.

## 2.2. Riemann Surfaces of Non-zero Genus

The mathematics of Riemann surfaces of non-zero genus is relevant to understanding the zeolite-like carbon and boron-nitride allotropes discussed in this paper. In this connection the concept of a Riemann surface was originally introduced to facilitate the integration of multi-valued functions.[6] The simplest example of such a multi-valued function is the function $w = \sqrt{z}$, which has two values depending on whether the positive or negative square root is taken. In order to illustrate the geometry of Riemann surfaces; consider the more complicated multi-valued function

$$w = \sqrt{4z(1-z)(1-\lambda z)} \tag{8}$$

which arises in the theory of elliptic functions and elliptic curves.[6] In equation (8) there are two values of $w$ for $z \neq 0$, 1, and $1/\lambda$ corresponding to the positive and negative square roots. At $z = 0$, 1, and $1/\lambda$, as well as at $z = \infty$, which are called *critical points*, there is a unique value for $w$, since the two values for $w$ coalesce at these points.

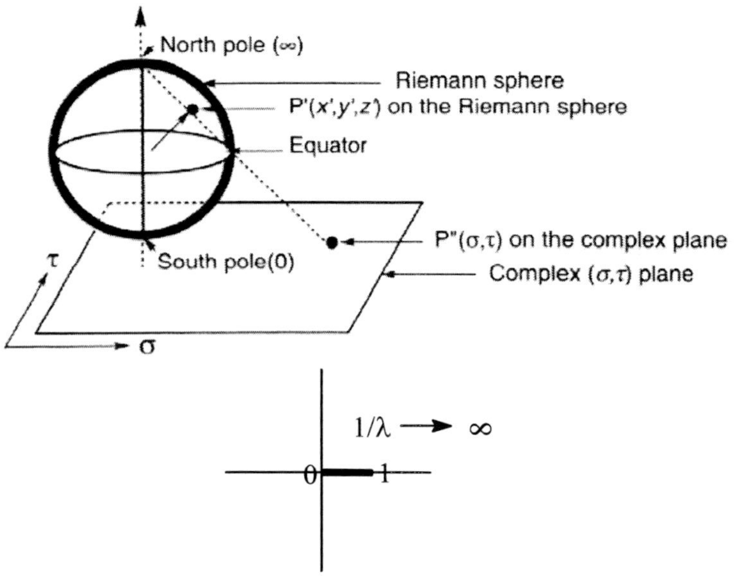


Figure 4: (a) Riemann sphere and the projection of a point $P'(x',y',z')$ on the Riemann sphere to a point $P''(\sigma,\tau)$ in the complex plane; (b) A complex plane sheet of the Riemann surface for the equation $w^2 = 4z(1-z)(1-\lambda z)$ (equation 1). Bold lines indicate the locations of the slits.

A Riemann surface for equation (8) can be constructed by taking the complex (Argand) plane (Figure 4a) and slitting it along the real axis from 0 to 1 and along a half line parallel to the real axis going to the right beginning at $1/\lambda$ and ending at $\infty$ (Figure 4b). An important requirement in slitting the complex plane is that the two slits do not intersect. To represent $w$ defined by equation (8) in its entirety, a second copy of the slit complex plane is taken and attached to the first at the points 0, 1, $1/\lambda$, and $\infty$ and along the slits. The bottom of each slit on either sheet is attached to the top of the corresponding slit on the other sheet, where a *sheet* refers to a single copy of the complex plane. The two sheets can thus be pictured as being parallel and lying over the complex plane so that each point projects down vertically on the point of the complex plane with the same value of $z$. Since each of the two slit sheets may be mapped onto a sphere, the Riemann surface is like a basketball and an internal air bladder joined by the two crossed slits. Such a Riemann surface thus corresponds to a double covering of an underlying Riemann sphere (Figure 4a).

A surface, including a Riemann surface, can be characterized by its genus. Thus the genus of a sphere or a surface topologically homeomorphic[7] to a sphere is zero. In this connection two surfaces $S_1$ and $S_2$ are said to be topologically homeomorphic to each other if $S_1$ can be deformed to $S_2$ without tearing or cutting. Surfaces of non-zero genus are topologically homeomorphic to surfaces generated by drilling holes (or tunnels) through a plastic sphere. The genus, $g$, of such a surface is the number of holes that must be drilled through a plastic sphere to make a surface homeomorphic to the surface in question. Thus the genus of a sphere itself is zero and the genus of a torus (*i.e.*, a typical doughnut with a hole) is one (Figure 5a). Surfaces of genus 3 are particularly significant since genus 3 is the lowest genus that can lead to surfaces isotropic in all three dimensions. This contrasts with genus 1 surfaces, such as the torus, where one dimension must be different from the other two dimensions.

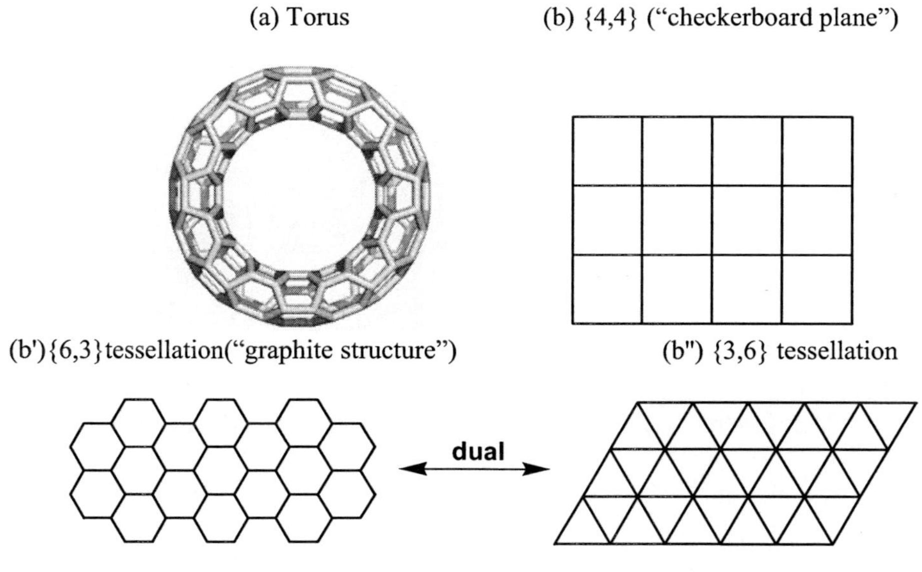


Figure 5: (a) The torus; (b, b', b") The three Platonic planar tessellations. These can be embedded onto the torus by identifying opposite edges.

## 2.3. Platonic Tessellations on Riemann Surfaces of Non-zero Genus

The regular polyhedra (Figure 2) may be regarded as Platonic tessellations of the sphere. In this connection a tessellation of a surface can be described in terms of its *flags*, where a flag is a triple $(V,E,F)$ consisting of a vertex $V$, an edge $E$, and a face $F$, which are mutually incident.[8] A tessellation is called *Platonic* if its symmetry group acts transitively on its flags.[9] This corresponds to the standard definition of regular polyhedra as those in which all faces are identical regular polygons, a condition satisfied by the familiar five Platonic polyhedra (Figure 2). The so-called Schäfli notation $\{p, q\}$ can describe a platonic tessellation consisting of $q$ regular $p$-gons at each vertex. For example, the Schäfli notation is {3,4} for the regular octahedron and {3,5} for the regular icosahedron (Figure 2).

Euler's relationship (equation 1) can be generalized to polygonal networks on a surface of any genus $g$ corresponding to an Euler characteristic $\chi$ by using the following equation:

$$v - e + f = 2(1 - g) = \chi \tag{9}$$

Note that if $g = 0$ so that $\chi = 2$ for polyhedra homeomorphic to a sphere, equation (9) reduces to the familiar Euler's relationship (equation 1).

The concept of duality is important in the study of polyhedra. In this connection a given polyhedron $P$ can be converted into its dual $P^*$ by locating the vertices of $P^*$ above the centers of the faces of $P$ and the centers of the faces of $P^*$ above the vertices of $P$. Two vertices in the dual $P^*$ are connected by an edge when the corresponding faces in $P$ share an edge. The process of dualization has the following properties:

1. The numbers of vertices ($v$ and $v^*$), edges ($e$ and $e^*$), and faces ($f$ and $f^*$) in a pair of dual polyhedra $P$ and $P^*$ satisfy the relationships $v^* = f$, $e = e^*$, $f = v^*$.
2. Dual polyhedra have the same symmetry elements and thus belong to the same symmetry point group.
3. Dualization of the dual of a polyhedron leads to the original polyhedron.
4. The degrees of the vertices of the polyhedron correspond to the number of edges in the corresponding face polygons of its dual. Thus the duals of the deltahedra are trivalent polyhedra, *i.e.*, polyhedra in which all vertices are of degree 3.
5. The dual of a regular polyhedron $\{p,q\}$ in the Schäfli notation is $\{q,p\}$.

In the case of the Platonic (regular) polyhedra dual pairs consist of the cube/octahedron and dodecahedron/icosahedron (Figure 2). The tetrahedron is self-dual, *i.e.*, the dual of a tetrahedron is another tetrahedron. Also the concept of duality can be extended from standard polyhedra embedded in the surface of a sphere to polygonal networks embedded in surfaces of non-zero genus.

The five Platonic polyhedra (Figure 2 and Table 1) correspond to the Platonic tessellations of the sphere. In the case of the simple torus of genus 1 (Figure 5a) there are three planar regular tessellations (Figure 5b) that can be embedded in the torus by identifying opposite edges of the underlying rectangle. These consist of the self-dual $\{4,4\}$ "checkerboard plane" and the dual pair $\{3,6\}$ and $\{6,3\}$.

The Platonic tessellations of surfaces of genus 2 to 5 have been studied in detail by Singerman and Watson[10] based on earlier work by Coxeter,[11] Sherk,[12] and Garbe.[13] Garbe[13] also provides information on Platonic tessellations of surfaces of genus 6. Singerman and Watson[10] investigate the locations of critical points on the underlying Riemann surface called Weierstrass points where the functions described by the Riemann surface have special properties with regard to their singularities. The detailed mathematical definition of a Weierstrass point is too technical for this paper. However, in most cases of interest the Weierstrass points are geometric in that they correspond to the locations of vertices, edge midpoints, and/or face midpoints in the tessellation. The singular points $z = 0$, 1, $1/\lambda$, and $\infty$ in the Riemann surface based on equation (8) (Figure 4b) are examples of the type of singular points that become Weierstrass points in Riemann surfaces of genus 2 or higher.

An important property of non-zero genus Riemann surfaces is their hyperellipticity. In this context a Riemann surface of genus $\geq 2$ is called hyperelliptic if it admits an automorphism of order two (*i.e.*, an involution) such that the quotient of the surface by the involution is of genus zero, *i.e.*, a surface topologically homeomorphic[7] to the sphere. All Platonic tessellations of Riemann surfaces of genus 2 are necessarily hyperelliptic. Thus the lowest genus Platonic tessellations that are non-hyperelliptic have genus 3. Such tessellations are also of particular interest since they can be isotropic in all three dimensions.

Table 2 shows all of the possible Platonic tessellations of genus 2 and 3 in order of increasing size of their automorphism groups. Most of the entries in Table 2 represent dual pairs analogous to those depicted in Figure 2 for the regular polyhedra. Only one member of a dual pair is shown, generally the member in which the number of faces is not larger than the number of vertices. The Platonic tessellations are described in terms of their Schäfli notations as used by Coxeter[11] and Singerman and Watson.[10] In Table 2 $f$, $v$, and $e$ refer to the numbers of faces, vertices, and edges, respectively; $G$ indicates the automorphism group; $|G|$ indicates the order of the automorphism group, and H indicates hyperelliptic tessellations.

**Table 2: Platonic tessellations of genus 2 and 3**

| | $f$ | $v$ | $e$ | $G$ | $\|G\|$ | |
|---|---|---|---|---|---|---|
| **Genus 2** | | | | | | |
| $\{8,8\}_{1,0}$ | $1^1$ | $1^1$ | $4^1$ | $C_8$ | 8H | |
| $\{10,5\}_2$ | $1^1$ | 2 | $5^1$ | $D_5$ | 10H | |
| $\{6,6\}_2$ | 2 | 2 | $6^1$ | $C_6 \times C_2$ | 12H | |
| $\{4,8\}_{1,1}$ | $2^1$ | $4^1$ | 8 | $C_4 \times D_2$ | 16H | |
| $\{6,4\|2\}$ | 4 | $6^1$ | 12 | $D_6 \times C_2$ | 24H | double hexagon |
| $\{3,4+4\}$ | 16 | $6^1$ | 24 | $O \times C_2$ | 48H | double octahedron |
| **Genus 3** | | | | | | |
| $\{12,12\}_{1,0}$ | $1^3$ | $1^3$ | $6^3$ | $C_{12}$ | 12H | |
| $\{14,7\}_2$ | $1^3$ | 2 | $7^3$ | $D_7$ | 14H | |
| $\{8,8\}_2$ | 2 | 2 | $8^3$ | $C_8 \times C_2$ | 16H | |
| $\{4\cdot 2,4\cdot 2\}$ | $2^2$ | $2^2$ | $8^2$ | $\langle 2,2\|2\rangle$ | 16 | |
| $\{2\cdot 6,2\cdot 2\}$ | $2^3$ | $6^3$ | 12 | $C_4 \times D_3$ | 24H | |
| $\{2\cdot 3,2\cdot 3\}$ | $4^3$ | $4^3$ | 12 | $T \times C_2$ | 24H | |
| $\{8,2\cdot 2\}$ | 4 | $8^3$ | 16 | $D_8 \times C_2$ | 32H | double octagon |
| $\{4,4\cdot 2\}$ | $4^2$ | $8^2$ | 16 | $\langle\langle 2,4\|2\rangle\rangle$ | 32 | |
| $\{3,4\cdot 3\}$ | $4^2$ | $16^1$ | 24 | $T \times C_4$ | 48 | Quadruple tetrahedron |
| $\{4,2\cdot 3\}$ | 12 | $8^3$ | 24 | $O \times C_2$ | 48H | double cube |
| **$\{8,3\}_6$** | **$12^2$** | **32** | **48** | **$^4O$** | **96** | **Dyck tessellation (Figure 6a)** |
| **$\{7,3\}_6$** | **$24^1$** | **56** | **84** | **$^7O$** | **168** | **Klein tessellation (Figure 6b)** |

Of particular interest in Table 2 (in bold face) is the listing of the Dyck and Klein tessellations (Figure 6) as the non-hyperelliptic tessellations of genus 3 with the largest automorphism groups.

The Dyck tessellation[14] (Figure 6a), after a leapfrog transformation analogous to that in Figure 3 for the regular dodecahedron, becomes relevant to the understanding of possible zeolite-like carbon or boron nitride allotropes containing only hexagons and octagons. Similarly, the Klein tessellation[15],[16] (Figure 6b) is relevant to the study of such allotropes containing only hexagons and heptagons. The Platonic symmetry of the Dyck and Klein tessellations makes them analogues of regular polyhedra on surfaces of genus 3.

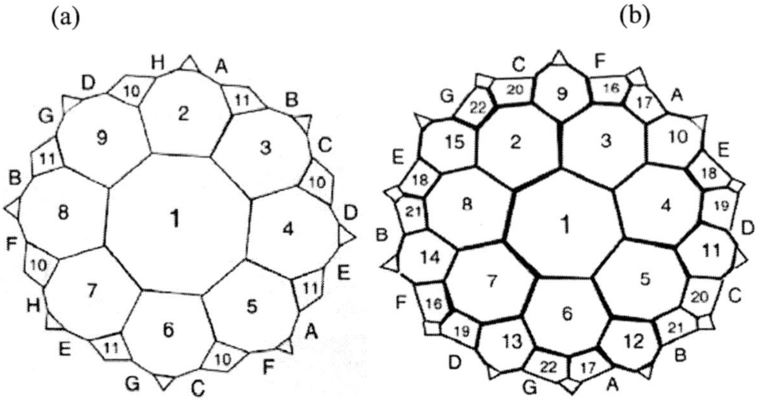


Figure 6: (a) The Dyck tessellation {8,3}; (b) The Klein tessellation {7,3}. In both cases the edges with identical letters (A→H or G) are identified to embed the tessellation in a genus 3 surface.

# 3. Topology of Surfaces of Negative Curvature: Infinite Periodic Minimal Surfaces

## 3.1. The Curvature of Surfaces

Consider a surface consisting of a tessellation of trigonal ($sp^3$) carbon atoms. In such a tessellation the hexagons can be regarded as flat areas of zero curvatures, any squares or pentagons can be regarded as areas of positive curvature, and any heptagons or octagons as areas of negative curvature. Concepts of differential geometry can be used to make such concepts more precise with the objective of describing more fully the effects of the negative curvature from heptagons and/or octagons on the structures and properties of carbon and boron nitride allotropes containing these structural features.

The methods of differential geometry[17] can be used to describe surfaces and their properties. Such methods define a given surface by the following vector equation with the two parameters $u$ and $v$:

$$\mathbf{r} = \mathbf{r}(u,v) \tag{10}$$

The basic tangent vectors are then given by

$$\mathbf{r}_1 = d\mathbf{r} / du \text{ and } \mathbf{r}_2 = d\mathbf{r} / du \tag{11}$$

so that

$$d\mathbf{r} = \mathbf{r}_1 du + \mathbf{r}_2 dv \tag{12}$$

The tangent plane to the surface at the point $(u,v)$ is $\mathbf{r}_1\mathbf{r}_2$. The so-called *metric tensor* has components defined by

$$g_{11} = \mathbf{r}_1 \cdot \mathbf{r}_1;\; g_{12} = g_{21} = \mathbf{r}_1 \cdot \mathbf{r}_2;\, g_2 = \mathbf{r}_2 \cdot \mathbf{r}_2 \tag{13}$$

The properties of the surface are then determined from its metric tensor by the use of two fundamental forms. The first fundamental form for the element of arc d$s$ of the surface is given by

$$ds^2 = d\mathbf{r} \cdot d\mathbf{r} = g_{11}\, du^2 + 2g_{12}\, du\, dv + g_{22}\, dv^2 \tag{14}$$

Now let **n** be the normal vector to the surface $\mathbf{r} = \mathbf{r}(u,v)$, let $b_{ij} = \mathbf{r}_{ij} \cdot \mathbf{n}$, and let $\kappa$ be the normal curvature, different in different directions, of a plane curve formed by a section of the normal plane. The second fundamental form for the element of arc of the surface is given by

$$\kappa\, ds^2 = b_{11}\, du^2 + 2b_{12}\, du\, dv + b_{22}\, dv^2 \tag{15}$$

The first fundamental form (equation 14), which is derived from the tangent vectors, is a measure of the metric of the surface, whereas the second fundamental form (equation 15), which is derived from the normal vectors, is a measure of the flatness of the surface.

When the normal plane rotates, maximum values of $\kappa$ are obtained, namely $\kappa_1$ and $\kappa_2$; these are called *principal curvatures*, and the corresponding curves of intersection are called principal lines of curvature. The principal curvatures $\kappa_1$ and $\kappa_2$ have a product and an arithmetic mean known as the *Gaussian curvature K* and the *mean curvature H*, namely

$$\kappa_1 \kappa_2 = K \tag{16a}$$

$$\kappa_1 + \kappa_2 = 2H \tag{16b}$$

A spherical or ellipsoidal shell has positive Gaussian curvature (*i.e.*, it is "convex"), a hyperbolic sheet has negative Gaussian curvature (*i.e.*, it is "concave"), and a cylinder, cone, or flat surface has zero Gaussian curvature.

## 3.2. Infinite Periodic Minimal Surfaces IPMSs

Consider surfaces with *constant* curvatures that can be embedded in three-dimensional space. Such a surface with constant *positive* curvature is well known to be a sphere. However, such surfaces with constant *negative* curvature do not include any shapes as simple as a sphere. In fact, there is no way of embedding the *entire* hyperbolic plane in three-dimensional space as a surface of constant negative curvature. However, there are certain surfaces that will serve for a portion of the hyperbolic plane of finite area. The simplest example of such a surface is the so-called *pseudosphere*, which is the horn-shaped surface (Figure 7a) described by the parametric equations where $u \geq 0$:

$$x = \text{sech}\, u \cos v \tag{17a}$$

$$y = \text{sech}\, u \sin v \tag{17b}$$

$$z = u - \tanh u \tag{17c}$$

However, because of the singularity (cuspidal edge) of the pseudosphere where $u = 0$, the pseudosphere, unlike its positive curvature counterpart, namely the sphere, is of no use as a surface in which to embed networks of atoms to describe chemical structures or other chemical phenomena. In particular, the singularity in the pseudosphere is incompatible with any use to describe periodic chemical structures.

Surfaces of negative curvature that, unlike the pseudosphere, are useful for describing certain chemical structures are known as infinite periodic minimal surfaces (IPMS's).[18] In this context minimal surfaces are surfaces where the mean curvature $H$ at each point is zero so that $\kappa_1 = -\kappa_2$ by equation (16b) and $K \leq 0$ by equation (16a). Thus all minimal surfaces are negative curvature surfaces constructed from hyperbolic planes. Minimal surfaces are saddle-shaped everywhere except at certain "flat points" which are higher order saddles.

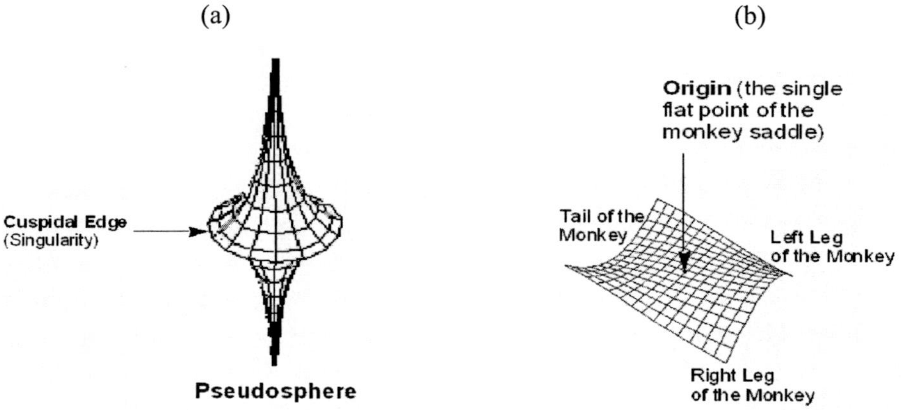


Figure 7: (a) The pseudosphere showing its cuspidal edge; (b) The monkey saddle showing the single flat point in the center.

The simplest example of a (non-periodic) minimal surface excluding the trivial case of the plane is defined by the following cubic equation:

$$F(x, y) = z = x(x^2 - 3y^2) \tag{18}$$

This surface (Figure 7b) is called the *monkey saddle*,[19] since it has three depressions, namely two for the monkey's legs and one for his tail. The *average* curvature of the monkey saddle vanishes at every point so that at every point its "concavity" is equal to its "convexity." However, the only flat point of the monkey saddle is the single point at its origin (Figure 7b).

It is not possible to construct an infinite surface with a *constant* negative Gaussian curvature as noted above. However, H. A. Schwarz found before 1865 that patches of *varying* negative curvature and constant zero mean curvature could be smoothly joined to give an infinite surface with zero mean curvature which is periodic in all three directions. Such surfaces are called infinite periodic minimal surfaces (IPMSs). About five different types of IPMS's were known by 1880[20] and the number of known distinct IPMSs is now more than 50.[21-25] The original five IPMS's have been given the designations P, D, T, H, and CLP (Table 3).[18] These IPMS's can be constructed using monkey saddles (Figure 7b) as basic building blocks so that the single flat point of each monkey saddle is located at a vertex of the polyhedral unit cell.

**Table 3: Polyhedral Unit Cells for the Five Classical Minimal Surfaces**

| Surface | Genus | Polyhedron | Symmetry | Number of Edges |
|---|---|---|---|---|
| P | 3 | Octahedron | $O_h$ | 6 out of 12 |
| D | 3 | Tetrahedron | $T_d$ | 4 out of 6 |
| T | 3 | Cube | $O_h$ | 8 out of 12 |
| CLP | 3 | Trigonal prism | $D_{3h}$ | |
| H | 3 | Triangle | $D_{3h}$ | 3 out of 3 |

The IPMS's described above cannot be defined by analytic functions in the usual Cartesian space of three dimensions but instead require elliptic or hyperelliptic integrals, which must be solved numerically. In this connection Klinowski and coworkers have developed methods for the exact numerical computations on the triply periodic P,[26] T,[27] and CLP[28] surfaces.

The finite surface element building block which is repeated periodically throughout space in an IPMS plays a role analogous to the unit cell in a crystal structure. Hyde and collaborators have exploited this analogy to describe crystal structures using IPMS's.[29],[30]

The description of IPMS's using elliptic or hyperelliptic integrals is facilitated by the fact that minimal surfaces are the only surfaces other than the sphere whose spherical representation is conformal. This means that every complex analytic function can be used to describe a minimal surface. In order to generate such a description, each point P($x,y,z$) of the surface is mapped onto a point P'($x',y',z'$) on the Riemann sphere (Figure 4a) given by

$$P'(x',y',z') = \frac{\left(\frac{\mathrm{d}F}{\mathrm{d}x},\frac{\mathrm{d}F}{\mathrm{d}y},\frac{\mathrm{d}F}{\mathrm{d}z}\right)}{\sqrt{\left(\frac{\mathrm{d}F}{\mathrm{d}x}\right)^2+\left(\frac{\mathrm{d}F}{\mathrm{d}y}\right)^2+\left(\frac{\mathrm{d}F}{\mathrm{d}z}\right)^2}} \qquad (19)$$

Weierstrass showed in the 19$^{\text{th}}$ century that the Cartesian coordinates of a point P($x,y,z$) on an IPMS are related to the coordinates P"($\sigma,\tau$) on the complex plane by the integrals

$$x = \mathrm{Re} \int (1-\omega^2) R(\omega)\, \mathrm{d}\,\omega \tag{20a}$$

$$y = -\mathrm{Im} \int (1+\omega^2) R(\omega)\, \mathrm{d}\,\omega \tag{20b}$$

$$z = \mathrm{Re} \int 2\omega R(\omega)\, \mathrm{d}\,\omega \tag{20c}$$

in which Re and Im denote the real and imaginary parts of the complex integral and $\omega = \sigma + i\tau$ as above.

The Weierstrass function $R(\omega)$ determines the intrinsic parameters of the surface. For example the Gaussian curvature is related to $R(\omega)$ by the following equation:

$$K = \frac{-4}{\left(1+|\omega|^2\right)^4 |R(\omega)|^2} \tag{21}$$

The general form of the Weierstrass function can be expressed as

$$R(\omega) = \frac{1}{(F(\omega))^{1/b}} = \frac{1}{\prod_{i=1}^{k} (\omega - \omega_i)^{1/b}} \tag{22}$$

in which $k$ = 8 and $b$ = 2 for the IPMS's of greatest interest. The integrals in equations (20) are hyperelliptic integrals if $b$ = 2 and the degree of the polynomial $F(\omega)$ is greater than four. The values of the roots $\omega_i$ of the polynomial $F(\omega)$ correspond to the locations of the singular flat points of the monkey saddles (Figure 7b) making up the IPMS.

Table 3 lists the five classical IPMS's. Among these five IPMS's, two of them, namely the D and P surfaces, have the same Weierstrass function namely

$$R(\omega) = \frac{1}{\sqrt{1-14\omega^4+\omega^8}} \tag{23}$$

The integrals (equations 20) determining the coordinates of these surfaces are hyperelliptic integrals since the degree of the polynomial in $\omega$ under the radical sign is greater than four. Furthermore, the locations of the eight roots of the polynomial in the denominator of equation (23) correspond to the vertices of a cube inscribed in the Riemann sphere in accord with the cubic symmetry of the unit cell of the D and P surfaces. Polynomials of this type, which vanish at the vertices of a regular polyhedron inscribed in the Riemann sphere (Figure 4a), are called *polyhedral polynomials* and have been studied for the regular polyhedra, particularly in connection with the solution of the general quintic equation using elliptic functions.[31-33]

The transformation defined by the coordinates

$$x = \mathrm{Re} \int e^{i\theta}(1-\omega^2)R(\omega)\,\mathrm{d}\,\omega \tag{24a}$$

$$y = -\mathrm{Im} \int e^{i\theta}(1-\omega^2)R(\omega)\,\mathrm{d}\,\omega \tag{24b}$$

$$z = \mathrm{Re} \int e^{i\theta} 2\omega R(\omega)\,\mathrm{d}\,\omega \tag{24c}$$

is used to relate IPMS's which have the same Weierstrass function $R(\omega)$. This transformation is called a *Bonnet transformation* and the angle $\theta$ is called the *association parameter*. Surfaces related by a Bonnet transformation are called *associate surfaces*. All associate surfaces have the same metrics so that lengths are preserved during any Bonnet transformation. Thus a Bonnet transformation only bends the surface without stretching. The P and D surfaces are special cases of associate surfaces since the Bonnet transformations converting either the P to the D surface or vice versa both have $\theta = \pi/2$. Surfaces related by Bonnet transformations having $\theta = \pi/2$ are called *adjoint* surfaces.

The use of IPMS's to construct possible carbon allotropes by decorating them with networks of trigonal carbon atoms requires consideration of their genus. In this connection the P and D surfaces suggested for schwarzite structures are found to have unit cells with genus 3. The unit cell of the P surface is depicted in two different ways in Figure 8. Thus it can be viewed as an octahedral junction of six pipes or tubes that has been called a "plumber's nightmare" or equivalently as three hyperboloids whose axes meet at right angles. Connecting the open pipes emerging from each of the three pairs of opposite faces of the plumber's nightmare thereby generating a closed surface leads to a sphere with three handles thereby indicating that a unit cell of this surface has genus 3.

The description of the unit cell of a P surface by the plumber's nightmare (Figure 8) suggests a natural division of the unit cell into eight equivalent octants. Each octant corresponds to a face of the underlying octahedron so that the numbers of vertices, edges, and faces in a unit cell of a P or D surface are eight times those of an individual octant. In the case of the P surface all octants must be equivalent and contain the symmetry elements of the $O_h$ point group describing the symmetry of the underlying octahedron or its dual cube.

(a) (b)

Figure 8: Two descriptions of the unit cell of a P surface: (a) The "plumber's nightmare" octahedral junction of six pipes through the faces of a cube; (b) Three hyperboloids whose axes intersect at right angles.

The local symmetry of the octant of the P surface is that of an equilateral triangle, namely $D_{3h}$. Thus in order to be compatible with the symmetry of the P surface the structure within an individual octant must contain a $C_3$ axis passing through the center of the octant and also must contain a reflection plane ($\sigma_v$ in $O_h$) through each altitude of the triangle representing the octant corresponding to $D_{3h}$ local octant symmetry. This symmetry clearly corresponds to that of a simple cubic lattice. The requirement of reflection planes in an octant of the P surface means that only achiral octants are suitable for P surfaces.

The D surface is adjoint to the P surface as noted above and thus must have the same Weierstrass function with singular points at the vertices of a cube. However, the local symmetry of a D surface is only tetrahedral ($T_d$) which means that the eight vertices are partitioned into two equivalent sets of four vertices each corresponding to the vertices of the two subtetrahedra making up the cube. Three-fold symmetry is still required in the individual octants of the D surface but the reflection plane is no longer necessary. The D surface can thus accommodate chiral octants with the two sets of four vertices each corresponding to enantiomeric pairs of octants.

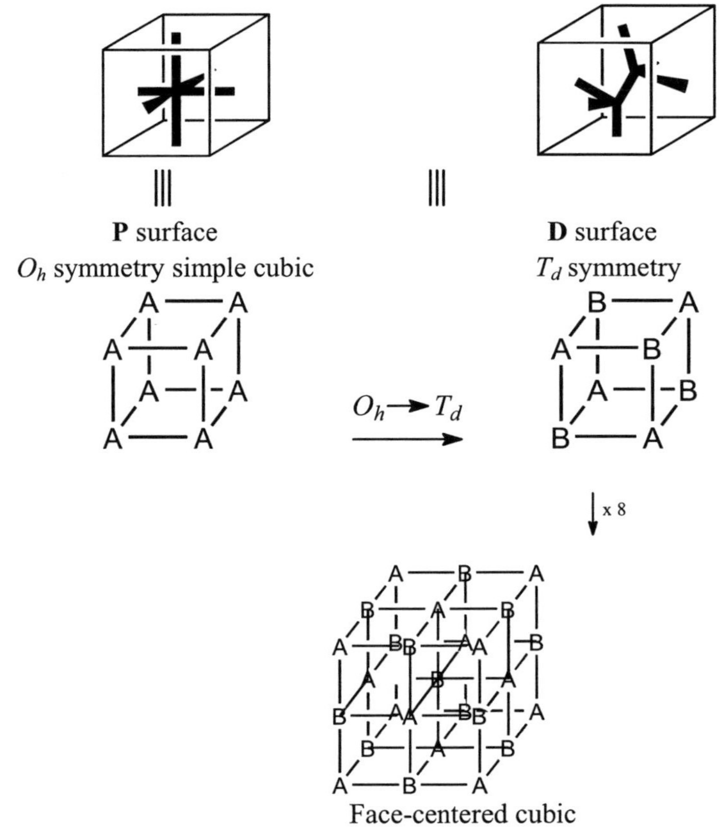


Figure 9: Relationships between the P and D surfaces and cubic structures. The reduction of the $O_h$ symmetry of the unit cell of the P surface to the $T_d$ symmetry of the D surface is depicted schematically either by elongation of the "plumber's nightmare" octahedral junction in the P surface to two tetrahedral junctions in the D surface[34] or by the partitions of eight equivalent octants in the P surface unit cell (represented schematically as A) into two complementary sets of tetrahedrally disposed octants in the D surface unit cell (represented schematically as A and B).

The relationship between the unit cells of P and D surfaces is depicted in Figure 9 as well as the face-centered cubic symmetry of a $2 \times 2 \times 2$ block of eight D surface unit cells. Thus the descent in unit cell symmetry from $O_h$ for the P surface to $T_d$ for the D surface can be depicted in either of the following two ways:

(1) Elongation of the "plumber's nightmare" octahedral junction (Figure 8) in the P surface to two tetrahedral junctions in the D surface[34] (Figure 9);

(2) Partitions of eight equivalent octants in the P surface unit cell (represented schematically as A in Figure 9) into two complementary sets of tetrahedrally disposed octants in the D surface unit cell (represented schematically as A and B in Figure 9).

## 3.3. Platonic Tessellations of IPMSs with Octagons and Heptagons Based on the Dyck and Klein Surfaces and their Leapfrog Transformations

The most symmetrical possible structures for zeolite-like carbon and boron nitride allotropes can be generated from platonic tessellations of IPMSs with octagons or heptagons followed by separation of adjacent octagons or heptagons by the minimum number of hexagons through a leapfrog transformation. The latter is similar to the leapfrog transformation of the regular dodecahedron to the truncated icosahedron of $C_{60}$ depicted in Figure 2. Such processes are most readily visualized if the effect on a single octant of a cubic IPMS (e.g., a D surface) is examined.

### *3.3.1. Networks Containing Only Hexagons and Octagons*

The generalized Euler's theorem (equation 8) indicates that a genus 3 structure of only hexagons and octagons of trigonal carbon atoms has exactly 12 octagons in the unit cell or $1^1/_2$ octagons per octant. The simple configuration of 12 octagons in the unit cell of a genus 3 surface leading to 32 vertices and 48 edges is the Dyck tessellation[14] and is conveniently called the D32 structure. The leapfrog of this configuration (Table 4) has 96 vertices, 144 edges, and 32 hexagons in its unit cell and is conveniently called the D96 structure. Note again that the octants of both the D32 and D96 structures are chiral (Figure 10a) and thus do not satisfy the symmetry requirements for a P surface but require the D surface of lower symmetry (Figure 9).

Mackay and Terrones[1] have proposed negative curvature carbon allotropes containing hexagons and octagons of trigonal carbon atoms with non-adjacent octagons. The simplest such allotrope is P192 in which the cubic unit cell contains the required 12 octagons and 80 hexagons (Figure 10b). Mackay and Terrones[1] also mention another carbon allotrope, conveniently designated as P432, with a large unit cell containing 432 carbon atoms in 12 octagons and 200 hexagons.

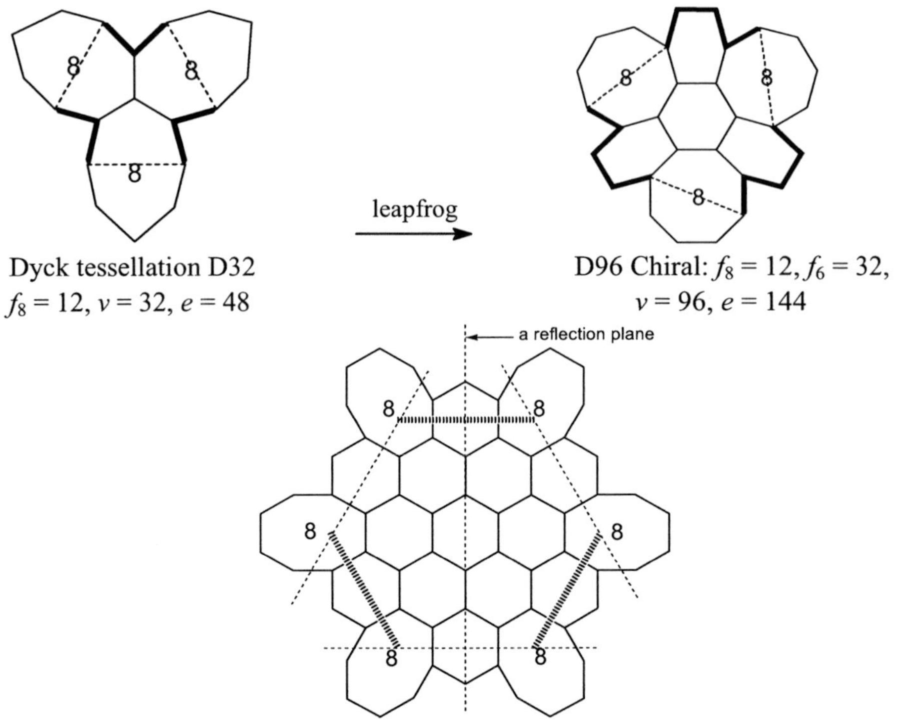


P192 Achiral: $f_8 = 12, f_6 = 80, v = 192, e = 288$

Figure 10: Octants of cubic IPMS's decorated with only hexagons and octagons including the Dyck tessellation, its leapfrog, and the P192 carbon structure proposed by Mackay and Terrones.[1]

### *3.3.2. Networks Containing Only Hexagons and Heptagons*

The generalized Euler's theorem (equation 8) indicates that a genus 3 structure of only hexagons and heptagons of trigonal carbon atoms has exactly 24 heptagons in the unit cell corresponding to 3 heptagons per octant. The simplest version of such a structure with isolated (non-adjacent) heptagons is generated by a leapfrog transformation on the Klein tessellation[15],[16] and can conveniently be called the D56 structure. Table 4 compares this leapfrog transformation with the more familiar leapfrog transformation of a regular dodecahedron to a truncated icosahedron (Figure 2). Figure 11a shows an octant of the original Klein tessellation and an octant of its leapfrog, which has 168 vertices in the unit cell and which was proposed by Vanderbilt and Tersoff[2] as a possible allotrope of carbon designated as D168. Note that the octants of both the Klein tessellation and its leapfrog are chiral (Figure 11) and thus do not satisfy the symmetry requirements for a P surface but require the D surface of lower symmetry. Because of the generation of the D168 structure by a leapfrog transformation from the prototypical network of trigonal carbon heptagons with 24 heptagons in the unit cell, the D168 structure has the minimum number of hexagons required to generate a structure with non-adjacent heptagons.

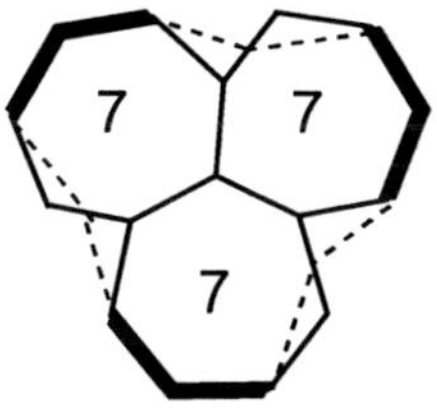


Klein tessellation
D56 Chiral:
$f_7 = 24, v = 56, e = 84$

leapfrog

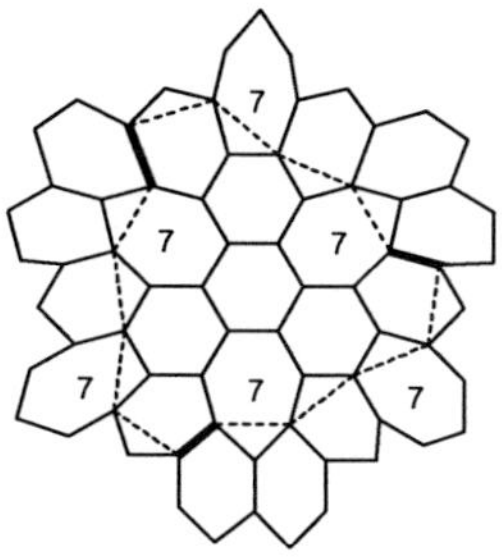


D168 Chiral:
$f_7 = 24, f_6 = 56$
$v = 168, e = 252$

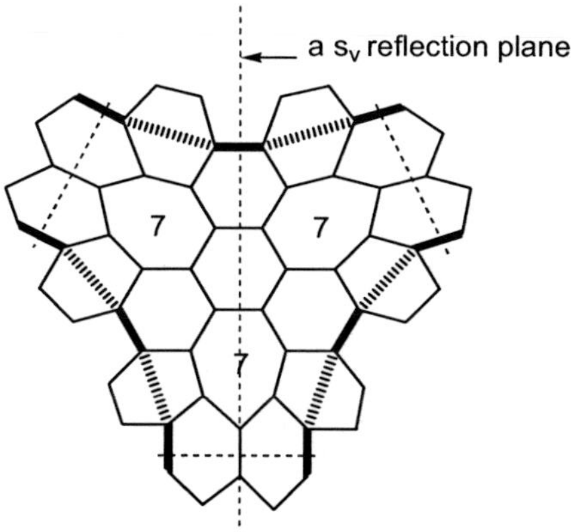


P216 Achiral
$f_7 = 24, f_6 = 80, v = 216, e = 324$

Figure 11: Octants of cubic IPMS's decorated with only hexagons and heptagons including the Klein surface (D56), its leapfrog (D168) and the P216 carbon structure proposed by Lenosky, Gonze, Teter and Elser.[35]

**Table 4: Comparison of Several Leapfrog Transformations**

| **Polygon combination** | **Pentagons + Hexagons** | | **Octagons + Hexagons** | | **Heptagons + Hexagons** | |
|---|---|---|---|---|---|---|
| leapfrog process | $C_{20} \longrightarrow C_{60}$ | | D32 ⟶ D96 | | D56 ⟶ D168 | |
| original figure | $I_h$ Dodecahedron | | Dyck surface | | Klein surface | |
| vertices | 20 | 60 | 32 | 96 | 56 | 168 |
| edges | 30 | 90 | 48 | 144 | 84 | 252 |
| faces | 12 | 32 | 12 | 44 | 24 | 80 |
| $f_{\neq 6}$ | 12 | 12 | 12 | 12 | 24 | 24 |
| $f_6$ | 0 | 20 | 0 | 32 | 0 | 56 |
| $f_{\neq 6}+f_{\neq 6}$ edges | 30 | 0 | 48 | 0 | 84 | 0 |
| $f_{\neq 6}+f_6$ edges | 0 | 60 | 0 | 96 | 0 | 168 |
| $f_6+f_6$ edges | 0 | 30 | 0 | 48 | 0 | 84 |

A different carbon allotrope structure containing only hexagons and heptagons was proposed by Lenosky, Gonze, Teter, and Elser[35] with 216 atoms, 24 heptagons, and 80 hexagons in the unit cell. An octant of this structure (Figure 11b) is achiral and thus meets the symmetry requirements for the P surface as well as the D surface.

# 4. Symmetry and Group Theory

## 4.1. Permutation Groups

The most familiar applications of group theory in chemistry use symmetry point groups, which describe the symmetry of molecules.[36] The elements of symmetry point groups can include only the standard symmetry operations in three-dimensional space, namely the identity ($E$), proper rotations ($C_n$), reflections ($\sigma$), inversion ($i$), and improper rotations ($S_n$). However, the concepts of group theory can also be applied to more abstract sets such as the permutations of a set $X$ of $n$ objects. A set of permutations of $n$ objects (including the identity "permutation") with the structure of a group is called a *permutation group* of *degree n* and the number of permutations in the set is called the *order* of the group.[37] The standard symmetry operations in symmetry point groups (*e.g.*, $E$, $C_n$, $\sigma$, $i$, $S_n$) can be considered to be special types of permutations when applied to discrete sets of points or lines such as the vertices or edges of polyhedra.[38] In such situations, symmetry point groups can be regarded as special cases of permutation groups.

Let $G$ be a permutation group acting on the set $X$ and let $g$ be any operation in $G$ and $x$ be any object in set $X$. The subset of $X$ obtained by the action of all operations in $G$ on $x$ is called the *orbit* of $x$. A *transitive* permutation group has only one orbit containing all objects of the set $X$. Sites permuted by a transitive permutation group are thus equivalent. Transitive permutation groups represent permutation groups of the "highest symmetry" and thus play a special role in permutation group theory.

Let $A$ and $X$ be two elements in a group. Then $X^{1}AX = B$ is equal to some element in the group. The element $B$ is called the *similarity transform* of $A$ by $X$ and $A$ and $B$ are said to be *conjugate*. A complete set of elements of a group which are conjugate to one another is called a *class* (or more specifically a *conjugacy class*) of the group. The number of elements in a conjugacy class is called its *order*; the orders of all conjugacy classes must be integral factors of the order of the group.

A group $G$ in which every element commutes with every other element (*i.e.*, $xy = yx$ for all $x$, $y$ in $G$) is called a *commutative* group or an *Abelian* group. In an Abelian group every element is in a conjugacy class by itself, *i.e.*, all conjugacy classes are of order one. A *normal subgroup N* of $G$, written $N \lhd G$, is a subgroup that consists only of *entire* conjugacy classes of $G$.[39] A *normal chain* of a group $G$ is a sequence of normal subgroups $C_1 \lhd N_{a1} \lhd N_{a2} \lhd N_{a3} \lhd \ldots \lhd N_{as} \lhd G$, in which $s$ is the number of normal subgroups (besides $C_1$ and $G$) in the normal chain (*i.e.*, the length of the chain). A *simple* group is a group having no *normal* subgroups other than the identity group $C_1$. Simple groups correspond to the transitive groups of "highest symmetry" and are particularly important in the theory of finite groups.[40],[41] The only non-trivial simple group found as a symmetry point group is the icosahedral pure rotation group, $I$, of order 60.

### 4.2. The Automorphism Group of the Dyck Tessellation: Relationship to Regular Tripartite Graphs

The permutational symmetry of the Dyck tessellation was already recognized by Dyck[14] to be described by a group of 96 permutations that can be described as follows:

(1) The identity permutation.
(2) Permutations of period 4, each of which leaves four octagons invariant. There are thus three distinct "$C_4$ axes," each of which passes through the midpoints of four octagons. Since "$C_4$," and "$C_4^3$" are in the same conjugacy class, there are a total of $3 \times 2 = 6$ permutations in this class and these may be regarded as analogues of proper rotations $C_4$.
(3) Permutations of period 2, using the same three "$C_4$ axes" as the $C_4$ permutations mentioned above and thus corresponding to $C_4^2$. There are obviously three of these permutations.
(4) Permutations of period 8, each of which leave two octagons invariant. These operations are analogous to an improper rotation $S_8$ rather than a proper rotation $S_8$ since although $S_8$ leaves only two octagons invariant, $S_8^2 = C_4$ leaves four octagons invariant. Furthermore, the number of "$S_8$ axes" is double the number of "$C_4$ axes" since for each of the three "$C_4$ axes" passing through the midpoints of four octagons, there are two ways of choosing the pair of octagons that is permuted and the pair of octagons that remains fixed when an $S_8$ operation is applied. Since "$S_8$," "$S_8^3$," "$S_8^5$," and "$S_8^7$," are in the same conjugacy class, there are a total of $6 \times 4 = 24$ operations in this class.
(5) Permutations of period 3, each of which pass through 2 of the 32 vertices of the D32 structure. Since there are 16 distinct pairs of such vertices and since "$C_3$" and "$C_3^2$" are in the same conjugacy class, there are a total of $16 \times 2 = 32$ operations in this class.
(6) Permutations of period 2 ($C_2$), each of which pass through the midpoints of 4 of the 48 edges so that there are 12 operations in this class.
(7) Permutations of period 4, which are not derived by squaring permutations of period 8. These may be regarded as analogues of improper rotations $S_4$ and there are 18 operations in this class.

These seven classes add up to the 96 operations in the permutation group of the Dyck figure as $E + 24S_8 + 6C_4 + 3C_4^2 + 32C_3 + 12C_2 + 18S_4$. The pure octahedral rotation group, $O$, is a subgroup of index 4 in this group so that this group of order 96 can be described as the tetrakisoctahedral group and designated as $^4O$. However, the octahedral rotation group $O$ is not a *normal* subgroup of the tetrakisoctahedral group since it cannot be constructed from entire conjugacy classes of $^4O$. Nevertheless, the tetrakisoctahedral group is not a simple group since other subgroups of $^4O$, albeit ones unfamiliar in chemistry or as symmetry point groups, can be constructed from entire classes of $^4O$.

The automorphism group of the Dyck tessellation is thus a solvable group of order 96 that does not contain the octahedral group ($O$ or $O_h$) as a normal subgroup. Furthermore $^4O$ is not a normal subgroup of the automorphism group of the four-dimensional analogue of the octahedron. Instead the structure of $^4O$ can be derived from the fact that the dual of the Dyck graph is the symmetrical tripartite graph $K_{4,4,4}$. In this connection a symmetrical tripartite graph, $K_{n,n,n}$, consists of three equivalent sets of $n$ vertices, conveniently designated by the labels {1, 2,...,$n$}, {1', 2',...,$n$'}, and {1", 2",...,$n$"}. Edges connect all possible pairs of vertices in different sets. However, no edges connect any pair of vertices within a single set. Each of the $3n$ vertices of $K_{n,n,n}$ is of degree $2n$ leading a total of $(3n)(2n)/2 = 3n^2$ edges. Furthermore, each vertex of $K_{n,n,n}$ is shared by $2n$ triangular faces (*i.e.*, circuits of length 3) leading to a total of $2n^2$ triangular faces. Note that $K_{2,2,2}$ corresponds to the regular octahedron with the three equivalent pairs of vertices {1,1'}, {2,2'}, and {3,3'} being the pairs of vertices located on the three orthogonal $C_4$ axes. The numbers of vertices, edges, and faces of the regular octahedron are 6, 12, and 8, respectively, corresponding to $3n$, $3n^2$, and $2n^2$, respectively, for $n = 2$. Examples of symmetrical tripartite graphs are depicted in Figure 12.

Each of the vertices of a symmetrical tripartite graph, $K_{n,n,n}$, embedded in a surface of minimum genus, is a part of $2n$ triangular faces. The corresponding dual $K^*_{n,n,n}$ is therefore a trivalent graph in which all of the faces are $2n$-gons. In the simple example of the octahedron as $K_{2,2,2}$ the dual $K^*_{2,2,2}$ is the cube, which has all degree 3 vertices and only square faces (Figure 2). The preservation of the automorphism group upon dualization means that the automorphism groups for these chemically relevant trivalent graphs ($K^*_{n,n,n}$) is the same as that of their duals $K_{n,n,n}$, which follow a simple pattern because of their symmetrical tripartite nature.

Now consider the octahedron (Figure 2) as the symmetrical tripartite graph $K_{2,2,2}$ (Figure 12). Its automorphism group $O$ has the following normal subgroup chain:

$$O \xrightarrow{\div 2} T \xrightarrow{\div 3} D_2 \xrightarrow{\div 2} C_2 \xrightarrow{\div 2} C_1$$

Order: 24 12 4 2 1

This subgroup chain is closely related to the tripartite structure of the octahedron. The process of going from $O$ to its normal subgroup $T$ of index 2 corresponds to removal of the operations of period 2 in the symmetric group $S_3$ permuting the three sets {1,2}, {1',2'}, and {1",2"}. In terms of the underlying octahedron, going from $O$ to its normal subgroup $T$ removes the operations of period 4. The next part of the subgroup chain from $T$ to $D_2$ of index 3 corresponds to removing the remaining operations of this symmetric group $S_3$ (*i.e.*, the operations of period 3). The resulting $D_2$ normal subgroup represents all possible $2^2 = 4$ permutations within the three individual sets {1,2}, {1',2'}, and {1",2"} without any permutations moving members of one of the three sets to another set.

Now consider the Dyck tessellation (Figure 6a), which has 12 octagonal faces and thus 48 edges and 32 vertices. Its dual is the symmetrical tripartite graph $K_{4,4,4}$, which necessarily has 12 vertices, 48 edges, and 32 triangular faces (Figure 12). Its automorphism group $^4O$ has the following normal subgroup chain (using a combination of familiar symmetry point group designations and less familiar designations from Dyck's original paper[14]):

$$^4O \xrightarrow{\div 2} G[3,3,4] \xrightarrow{\div 3} G[4,4,4] \xrightarrow{\div 2} D_{2h} \xrightarrow{\div 2} D_2 \xrightarrow{\div 2} C_2 \xrightarrow{\div 2} C_1$$

Order: 96 48 16 8 4 2 1

$K_{2,2,2}$ Graph = Octahedron

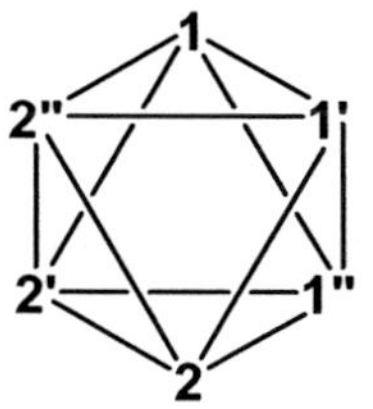


$K_{3,3,3}$ Graph

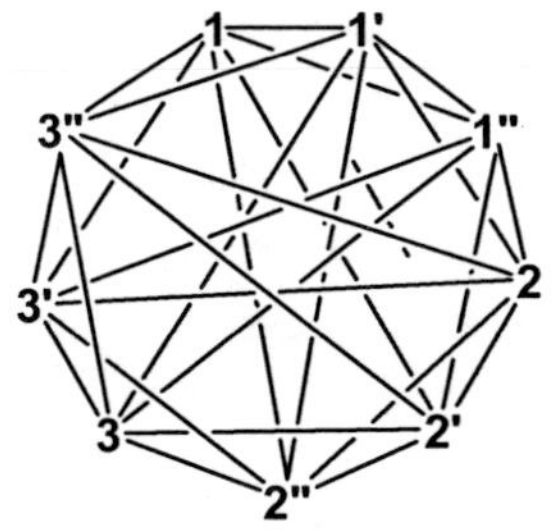


$K_{4,4,4}$ Graph = Dual of Dyck Graph

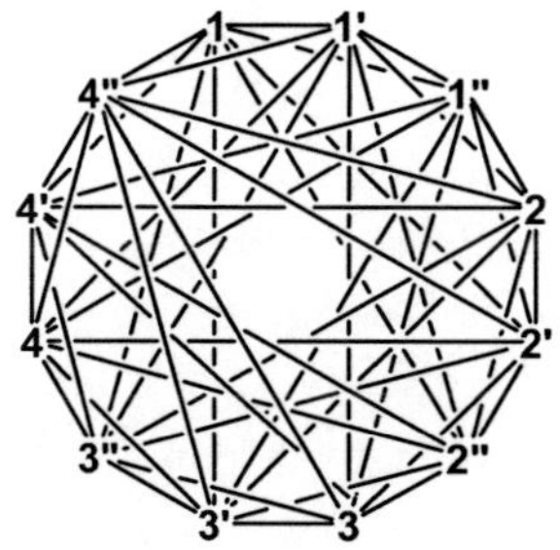


Figure 12: The simplest symmetrical tripartite graphs.

Again this subgroup chain is closely related to the $K_{4,4,4}$ tripartite structure of the dual of the Dyck graph. The process of going from ${}^4O$ to its normal subgroup $G[3,3,4]$ of index 2 corresponds to removal of the operations of period 2 in the symmetric group $S_3$ permuting the three sets {1,2,3,4}, {1',2',3',4'}, and {1",2",3",4"}. In terms of the underlying dual of the Dyck graph, going from ${}^4O$ to its normal subgroup $G[3,3,4]$ removes the operations of periods 8 and 4. The next part of the subgroup chain from $G[3,3,4]$ to $G[4,4,4]$ of index 3 corresponds to removing the remaining operations of $S_3$ (*i.e.*, the operations of period 3). The resulting $G[4,4,4]$ normal subgroup represents all possible $4^2$ = 16 permutations within the three individual sets {1,2,3,4}, {1',2',3',4'}, and {1",2",3",4"} without any permutations moving members of one of the three sets to another set.

### 4.3. The Automorphism Group of the Klein Tessellation: Hidden Symmetry

The structure of the $C_{60}$ fullerene is characterized by unusually high symmetry, namely icosahedral symmetry (*i.e.*, the icosahedral point group $I_h$ or its proper rotation subgroup $I$), which contains $C_5$, $C_3$, and $C_2$ proper rotation axes. A question of interest is whether the D168 schwarzite likewise has special symmetry. The cubic unit cell of D168 has $O_h$ symmetry, which contains $C_4$, $C_3$, and $C_2$ axes. However, the Klein tessellation (Figure 6b) from which the D168 structure is generated also contains a $C_7$ axis, which is lost from the symmetry point group when the open network is folded to a genus 3 surface by joining the outer edge pairs *AA* through *GG*. The open network of 24 heptagons can be considered to be a regular tessellation, *M*, which is folded into a *polyhedral embedding* or a *polyhedral realization*, *P*, in a genus 3 surface.[42] The $C_7$ symmetry that is "lost" when the open map (Figure 6b) is folded into the polyhedral embedding is called the *hidden symmetry* of the resulting polyhedral embedding.[42] This section considers a way that the $C_7$ symmetry element of the open 24-heptagon network can be "remembered" when the network is folded into a genus 3 surface. In order to consider the $C_7$ symmetry elements in the final D168 schwarzite structure, its symmetry must be considered as a permutation group of which its actual symmetry point group is a subgroup.

In order to consider the nature of the simple group relating to the permutational symmetry of the D168 structure, an alternative definition of the icosahedral pure rotation group is required, which can be extended to larger simple permutation groups which do not occur as symmetry point groups.[43] In this connection consider a prime number $p$ and let $F_p$ denote the finite field of $p$ elements which can be represented by the $p$ integers $0,\ldots,p-1$. Larger integers can be converted to an element in this finite field by dividing by $p$ and taking the remainder (*i.e.*, the number is taken "mod $p$"). For example, the finite field $F_5$ contains the five elements represented by the integers 0, 1, 2, 3, and 4 and other integers are converted to one of these five integers by dividing by 5 and taking the remainder, *e.g.*, $7 \rightarrow 2$ in $F_5$ (written frequently as "$7 \equiv 2$ mod 5"). The group SL(2,$p$) is defined to be the group of all $2 \times 2$ matrices with entries in $F_p$ having determinant 1. In addition its subgroup PSL(2,$p$) for odd $p$ is defined to be the quotient group of SL(2,$p$) modulo its center, where the center of a group is the largest normal subgroup that is Abelian. In the case of the groups SL(2,$p$) where $p \geq 5$, the center has only two elements and the quotient group PSL(2,$p$) is a simple group. The group PSL(2,5) contains 60 elements and is isomorphic to the icosahedral pure rotation group *I*.

An important property of the PSL(2,$p$) permutation groups for $p$ = 5, 7, and 11 (Table 5) is that they can function as *transitive* permutation groups on sets of either $p$ or $p$+1 objects. In the case of the group PSL(2,5), these transitive permutation groups on 5 and 6 objects can be visualized as permutations of parts of an icosahedron since PSL(2,5) is isomorphic to the icosahedral pure rotation group. Thus the PSL(2,5) group acts as a transitive permutation group on the six diameters of a regular icosahedron, where a diameter of an icosahedron is defined as a line drawn between a pair of antipodal vertices. In order to obtain in an icosahedron a set of *five* objects that is permuted transitively by the PSL(2,5) group, the 30 edges of an icosahedron are partitioned into five sets of six edges each by the following method:[44]

(1) A straight line is drawn from the midpoint of each edge through the center of the icosahedron to the midpoint of the opposite edge.

(2) The resulting 15 straight lines are divided into five sets of three mutually perpendicular straight lines.

Each of these five sets of three mutually perpendicular straight lines resembles a set of Cartesian coordinates and defines a regular octahedron. The PSL(2,5) permutation group as manifested in its isomorphic $I$ symmetry point group functions as a transitive permutation group on these five sets of three mutually perpendicular straight lines. In fact the PSL(2,5) permutation group is also isomorphic with the so-called *alternating permutation group* on five objects,[45] namely $A_5$, where an alternating permutation group on $n$ objects is the set of all possible *even* permutations and is of order $n!/2$.

**Table 5: Properties of the Pollakispolyhedral Groups Derived from the PSL(2,$p$) Groups ($p$ = 5, 7, 11)**

| Group | Order | Conjugacy Classes | Polyhedral Subgroup |
|---|---|---|---|
| $^5T \approx$ PSL(2,5) | 60 | $E+12C_5+12C_5^2+20C_3+15C_2$ | $T$ |
| $^7O \approx$ PSL(2,7) | 168 | $E+24C_7+24C_7^3+56C_3+21C_2+42C_4$ | $O$ |
| $^{11}I \approx$ PSL(2,11) | 660 | $E+60C_{11}+60C_{11}^2+110C_3+55C_2+132C_5+132C_5^2+110C_4$ | $I$ |

The PSL(2,$p$) ($p$ = 5, 7, 11) groups are simple groups and thus have no non-trivial *normal* subgroups. However, they contain two different sets of $n$ smaller non-normal subgroups corresponding to pure rotation groups of regular polyhedra; these regular polyhedral rotation groups are subgroups of index $p$ of the groups PSL(2,$p$). However, the PSL(2,11) group has been proven to the largest group of the general type PSL(2,$p$) with $p$ a prime which has a subgroup of index $p$.[47] A corollary derived from this theorem is that if $p > 11$, the PSL(2,$p$) group cannot be a transitive permutation group for a set with fewer than $p$+1 elements. This differs from the PSL(2,$p$) ($p$ = 5, 7, 11) groups which can be transitive permutation groups for sets of $p$ elements, namely 5, 7, and 11 respectively (Table 2). Since the PSL(2,$p$) ($p$ = 5, 7, 11) groups contain polyhedral point groups as subgroups they are conveniently designated as pollakispolyhedral groups.[46] Thus the PSL(2,5), PSL(2,7) and PSL(2,11) groups can be called the pentakistetrahedral, heptakisoctahedral, and undecakisicosahedral groups, respectively, and designated as $^5T$, $^7O$, and $^{11}I$, respectively.

The simplest example of the polyhedral subgroups of index $p$ in the pollakispolyhedral groups occurs in the pentakistetrahedral group, $^5T$, which is equivalent to the icosahedral rotation group. Thus, $^5T$ can be decomposed into two different sets of five tetrahedra corresponding to the conjugacy classes $12C_5$ and $12C_5^2$. This is related to the partitioning of the 20 vertices of a regular dodecahedron into five sets of four vertices each corresponding to a regular tetrahedron. The permutations of the group PSL(2,5) act as the icosahedral pure rotation group $I$ on the regular dodecahedron partitioned in this manner and correspondingly as the alternating group $A_5$ on the five subtetrahedra.

The next higher pollakispolyhedral group, namely the heptakisoctahedral group ${}^7O$ of order 168, is highly relevant to understanding the structure and symmetry of the schwarzite D168. This group can be decomposed into two sets of seven octahedral subgroups. This relates to the embedding of the open network (*i.e.*, the Klein tessellation)[15,16,47] of 24 heptagons (Figure 6b) into the unit cell of a D surface of genus 3 having a unit cell of octahedral symmetry such as the "plumber's nightmare" (Figures 8 and 9). The automorphism group of the Klein tessellation (Figure 6b) is ${}^7O$, which thus remains the *automorphism* group of its embedding into the D surface in D168. The symmetry group of the pure rotations of the D168 unit cell is the octahedral rotation group $O$, which, as noted above, is a subgroup of index 7 in ${}^7O$. Thus the D168 schwarzite structure can be seen to have seven-fold ($C_7$) *hidden symmetry.*

Another question of interest is the relationship of the operations of the heptakisoctahedral group to permutations in the Klein tessellation (Figure 6a).[15,16] In this connection the 168 operations of ${}^7O$ can be divided into the following conjugacy classes:

(1) The identity operation $E$.
(2) Permutations of period 7 ($C_7$), each of which leave three heptagons invariant so that the cycle index on the set of 24 hexagons is $x_1^3x_7^3$. There are eight distinct "$C_7$ axes," each of which passes through the midpoints of three heptagons. The resulting 48 operations can be partitioned into two conjugacy classes of 24 operations each, corresponding to $C_7$ and $C_7^3$ rotations.
(3) Permutations of period 3 ($C_3$), each of which leave two vertices invariant so that the 56 vertices of the Klein figure are partitioned into 28 "$C_3$ axes." There are 56 operations in the $C_3$ class.
(4) Permutations of period 2 ($C_2$), each of which leave four edges invariant so that the 84 edges of the Klein figure are partitioned into 21 "$C_2$ axes." There are thus a total of 21 operations in the $C_2$ class. The $C_2$ operations can be generated by combination of a $C_7$ and a $C_3$ operation, *i.e.*, $C_2 = C_7 \times C_3$.
(5) Permutations of period 4 ($C_4$), which partition the 24 heptagons into six groups of 4. The $C_4$ operations are related to the other operations by the relationships $C_4 = C_7^4 \times C_3$ and $(C_4)^2 = C_2$. Because of the latter relationship there are a total of $21 \times 2 = 42$ operations in the $C_4$ class realizing that "$C_4$" and "$C_4^3$" will be in the same conjugacy class. Thus permutations of these five types can be seen to lead to all 168 permutations and the six conjugacy classes of the heptakisoctahedral group listed in Table 5.

The permutational symmetry of the Klein tessellation provides a simple explanation why the D surface rather than the more symmetrical P surface (Figure 9) is required for the Klein tessellation. The high symmetry of the P surface requires the full achiral octahedral group $O_h$, which however cannot be a subgroup of the ${}^7O$ since the order of $O_h$, namely 48, does not divide the order of ${}^7O$, namely 168. For this reason the Klein tessellation cannot be embedded onto the P surface without reducing its symmetry to that of the D surface. Alternatively, it can be seen that an individual octant of the Klein tessellation (Figure 11a) is chiral whereas the P surface requires achiral octants.

## 5. CONCLUSIONS

Possible structures for low-density zeolite-like carbon and boron nitride allotropes can be constructed from networks of trigonal carbon atoms containing hexagons, heptagons, and/or octagons. These structures can be described by tessellations on genus 3 Riemann surfaces corresponding to periodic minimal surfaces of negative curvature. The most symmetrical such tessellations are derived from either the Dyck tessellation of 12 octagons or the Klein tessellation of 24 heptagons using a leapfrog transformation to avoid the presence of adjacent non-hexagons. The automorphism group, $^4O$, of the Dyck tessellation is a solvable group of order 96, the structure of which relates to the dual of the Dyck tessellation as the regular tripartite graph $K_{4,4,4}$. However, the automorphism group, $^7O$, of the Klein tessellation is a simple group of order 168, which can be generated from the prime number 7 in a way analogous to the generation of the icosahedral rotation group $I$ from the prime number 5.

## ACKNOWLEDGMENT

I am indebted to the National Science Foundation for partial support of this project under Grant CHE-0209857.

## REFERENCES

[1] L. Mackay and H. Terrones, Diamond from Graphite, *Nature*, 1991, 352, 762-762.

[2] D. Vanderbilt and J. Tersoff, Negative Curvature Fullerene Analogues of $C_{60}$, *Phys. Rev. Lett.*, 1992, 68, 511-513.

[3] I Silaghi-Dumitrescu, F. Lara-Ochoa, and I. Haiduc, $A_{12}B_{12}$ (A = B, Al; B = N, P) 4/6 Fullerene-Like Cages and their Hydrogenated Forms Stabilized by Exohedral Bonds. An AM1 Molecular Orbital Study, *J. Mol. Struct. (Theochem)*, 1996, 370, 17-23.

[4] G. Seifert, P. W. Fowler, D. Mitchell, D. Porezag, and T. Frauenheim, Boron-Nitrogen Analogues of the Fullerenes: Electronic and Structural Properties, *Chem. Phys. Lett.*, 1997, 268, 352-358.

[5] P. W. Fowler and D. B. Redmond, Symmetry Aspects of Bonding in Carbon Clusters - The Leapfrog Transformation, *Theor. Chim. Acta*, 1992, 83, 367-375.

[6] H. E. Rauch and A. Lebowitz, *Elliptic Functions, Theta Functions, and Riemann Surfaces,* Williams and Wilkins, Baltimore, MD USA, 1973.

[7] M. J. Mansfield, *Introduction to Topology*, D. van Nostrand, Princeton, New Jersey, 1963, p. 40.

[8] B. Grünbaum and G. C. Shephard, *Tilings and Patterns: An Introduction*, Freeman, New York, 1989.

[9] H. Karcher and M. Weber, in *The Eightfold Way: The Beauty of Klein's Quartic Curve*, S. Levy, ed., Cambridge University Press, Cambridge, U. K., 1998, pp. 9-49.

[10] D. Singerman and P. D. Watson, Weierstrass Points on Regular Maps, *Geom. Ded.*, 1997, 66, 69-88.

[11] H. S. M. Coxeter, The Abstract Groups $G^{m,n,p}$, *Trans. Am. Math. Soc.*, 1939, 45, 73-150.

[12] F. A. Sherk, The Regular Maps on a Surface of Genus Three, *Canad. J. Math.*, 1959, 11, 452-480.
[13] D. Garbe, Über die regulären Zerlegungen geschlossener orienterbarer Flächen, *J. reine angew. Math.*, 1969, 237, 39-55.
[14] W. Dyck, Notiz über eine reguläre Riemann'sche Fläche vom Geschlechte Drei und die zugehörige "Normalkurve" vierte Ordnung, *Math. Ann.*, 1880, 17, 510-516.
[15] F. Klein, Über die Transformation siebenter Ordnung der elliptischen Funktionen, *Math. Ann.*, 1879, 14, 428-471.
[16] F. Klein, *Gesammelten Mathematischen Abhandlungen*, Springer Verlag, Berlin, 1923, Vol. 3, pp. 90-136.
[17] A Goetz, *Introduction to Differential Geometry*, Addison-Wesley, Reading, MA, 1970.
[18] S. Andersson, S. T. Hyde, K. Larsson, and S. Lidin, S., Minimal Surfaces and Structures: From Inorganic and Metal Crystals to Cell Membranes and Biopolymers, *Chem. Rev.*, 1988, 88, 221-242.
[19] D. Hilbert and S. Cohn-Vossen, *Geometry and the Imagination*, Chelsea Publishing Co., New York, 1952, pp 183–204.
[20] E. R. Neovius, Bestimmung Zweier Spezieller Periodische Minimalflächen, J. C. Frenkel and Sohn, Helsinki, 1883.
[21] W. Fischer and E. Koch, New Surface Patches of Minimal Balance Surfaces. I. Branched Catenoids, *Acta Cryst.*, 1989, *A*45, 166-169.
[22] W. Fischer and E. Koch, II. Multiple Catenoids, *Acta Cryst.*, 1989, *A*45, 169-174.
[23] W. Fischer and E. Koch, III. Infinite Strips, *Acta Cryst.*, 1989, *A*45, 485-490.
[24] W. Fischer and E. Koch, IV. Catenoids with Spout-Like Attachments, *Acta Cryst.*, 1989, *A*45, 558-563.
[25] W. Fischer and E. Koch, Genera of Minimal Balance Surfaces, *Acta Cryst.*, 1989, *A*45, 726-732.
[26] P. J. F. Gandy and J. Klinowski, Exact Computation of the Triply Periodic Schwarz P Minimal Surface, *Chem. Phys. Lett.*, 2000, 322, 579-586.
[27] D. Cvijović and J. Klinowski, The T and CLP Families of Triply Periodic Minimal Surfaces. Part 2. The Properties and Computation of T Surfaces, *J. Phys I France*, 1992, 2, 2191-2205.
[28] D. Cvijović and J. Klinowski, The T and CLP Families of Triply Periodic Minimal Surfaces. Part 3. The Properties and Computation of CLP Surfaces, *J. Phys I France*, 1992, 2, 2207-2220.
[29] S. Andersson, S. T. Hyde, and H. G. von Schnering, The Intrinsic Curvature of Solids, *Z. Kristallograph.*, 1984, 168, 1-17.
[30] S. T. Hyde, A Topological Characterization of Chemical Structures *Z. Kristallograph.*, 1987, 179, 53-65.
[31] L. E. Dickson, *Modern Algebraic Theories*, Sanborn, Chicago, 1930, Chapter XIII.
[32] R. B. King and E. R. Canfield, Icosahedral Symmetry and the Quintic Equation, *Comp. and Math. with Appl.*, 1992, 24, 13-28.
[33] R. B. King, *Beyond the Quartic Equation*, Birkhäuser, Boston, 1996.
[34] J. F. Sadoc and J. Charvolin, Infinite Periodic Minimal Surfaces and their Crystallography in the Hyperbolic Plane, *Acta Cryst.*, 1989, *A*45, 10-20.
[35] T. Lenosky, X. Gonze, M. Teter, and V. Elser, Energetics of Negatively Curved Graphitic Carbon, *Nature*, 1992, 355, 333-335.

[36] A Cotton, *Chemical Applications of Group Theory,* Third Edition, Wiley, New York, 1990, Chapter 3.

[37] N. L. Biggs, *Finite Groups of Automorphisms*, Cambridge University Press, London, 1971.

[38] J. A. Pople, Classification of Molecular Symmetry by Framework Groups, *J. Am. Chem. Soc.*, 1980, 102, 4615-4622.

[39] J. K. G. Watson, Symmetry Groups of Non-Rigid Molecules, *Mol. Phys.*, 1971, 21, 577-585.

[40] D. Gorenstein, *Finite Groups*, Harper and Row, New York, 1968.

[41] D. Gorenstein, *Finite Simple Groups: An Introduction to their Classification*, Plenum, New York, 1982.

[42] J. Bokowski and J. M. Wills, Regular Polyhedra with Hidden Symmetries, *Math. Intelligencer*, 1988, 10 *(1)*, 27-32.

[43] B. Kostant, The Graph of the Truncated Icosahedron and the Last Letter of Galois, *Notices of the Amer. Math. Soc.*, 1995, 42, 959-968.

[44] F. Klein, *Vorlesungen über das Ikosaeder*, Teubner, Leipzig, 1884: Part I, Chapter II.

[45] R. B. King and D. H. Rouvray, Chemical Applications of Topology and Group Theory. 19. The Even Permutations of the Ligands in Five-Coordinate Complexes Viewed as Proper Rotations of the Regular Icosahedron," *Theor. Chim. Acta*, 1986, 69, 1-10.

[46] A Ceulemans, R. B. King, S. A. Bovin, K. M. Rogers, A. Troisi, and P. W. Fowler, The Heptakisoctahedral Group and its Relevance to Carbon Allotropes with Negative Curvature, *J. Math. Chem.*, 1999, 26, 101-123.

[47] J. Gray, From the History of a Simple Group, *Math. Intelligencer*, 1982, 4, 59-67.

In: Nanostructures: Novel Architecture
Editor: Mircea V. Diudea, pp. 299-309
ISBN: 1-59454-499-9


*Chapter 14*

# THE SYMMETRY OF THE DYCK GRAPH: GROUP STRUCTURE AND MOLECULAR REALIZATION

***Erwin Lijnen and Arnout Ceulemans***[1]
Departement Chemie, K.U.Leuven,
Celestijnenlaan 200F, B-3001 Leuven, Belgium

## ABSTRACT

The full genealogical tree of the group of automorphisms of the Dyck graph is derived. Three different subgroups of index 2 are identified, and their topological relevance is demonstrated. Two of these correspond to orientation-preserving subgroups for regular maps on a triple torus and a torus that form a pair of Petrie duals. Zeolite-type and nanotube-type molecular realizations of the Dyck graph are discussed.

## 1. INTRODUCTION

A graph is a set of nodes, on which pairwise relationships are defined. Molecules have obvious graph-like properties: their atom-bond matrix may be identified as the adjacency matrix of the corresponding molecular graph.[1] This analogy can be used to classify and retrieve molecules in large molecular data sets. Nonetheless one should not overlook that a molecule is also a three-dimensional object, which is characterized by geometrical and topological properties.[2] In principle such information is thrown away in the graph representation, and thus must be added again when graph theory is used to design molecular frames.

One way to introduce topological information is through embedding. An embedding is a representation of a graph on a closed surface such that there are no intersecting lines.[3] Especially relevant are so-called minimal embeddings, where one is looking for the surface with the lowest genus. In previous work [4],[5] we have discussed the embedding and molecular realization of two highly symmetrical graphs with non-trivial minimal topological

[1] E-mail address: arnout.ceulemans@chem.kuleuven.be

embeddings, which were named after their discoverers as the Klein graph and the Dyck graph.[6],[7] Our analysis focused on the symmetry of these graphs, as an unexpected tool to obtain information on the topological properties. In the present contribution we will elaborate this connection further for the case of the Dyck graph. Since the symmetry of this graph may be described as a fourfold extension of the octahedral group, or 'tetrakisoctahedral' group, we will start our treatment by a simple analysis of the octahedral symmetry group itself, and its topological implications. The results of this analysis will then be extended to the high symmetry of the Dyck graph.

## 2. Octahedral Symmetry and the Topology of the Cube

Figure 1 shows the Schlegel diagram of the cube, centered at a face. Its automorphism group is the cubic symmetry group, $O_h$, which has 48 symmetry elements. Recall that the automorphism group of a graph is the group of all vertex permutations that preserve adjacencies. The graph in Figure 1 is a planar graph and it can be embedded on a two-dimensional spherical surface, giving rise to the familiar cubic structure with six faces. Faces are topological characteristics which are added to the graph through the operation of embedding. It is therefore logical that when we talk about the symmetry of the embedding, we should restrict the automorphism group further to only those vertex permutations that preserve edge and face adjacencies. In the case of the spherical embedding of the cube this additional requirement does not introduce further restrictions, and the symmetries of the graph and the embedding coincide. In fact this embedding is regular, which means it realizes the upper limit of an automorphism group of order 4E (with E denoting the number of edges), being the highest possible symmetry for an embedding on a two-dimensional surface.[8] The automorphism group of a graph no doubt offers a powerful structural description of a graph. We should thus not be surprised if one discovers that group-theory throws additional light on the connection between the graph and its embeddings.

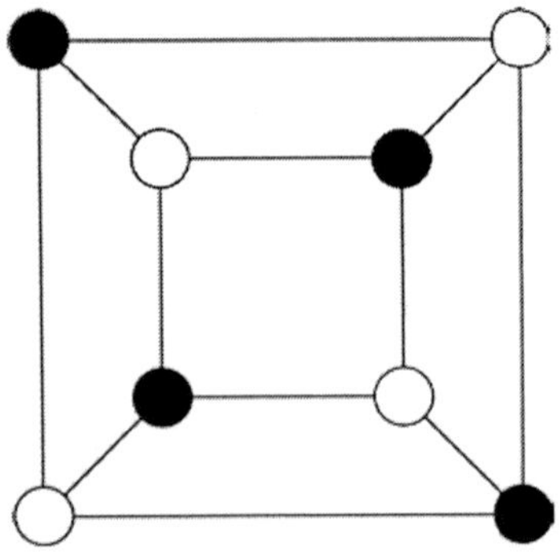

Figure 1: Schlegel diagram for the cube showing its bipartite character.

In this section we will illustrate this aspect for the case of the cubic symmetry group. It is well known that the group $O_h$ has three separate subgroups of index 2, denoted as $T_d$, $O$ and $T_h$. The intriguing feature of these three subgroups is that each of them apparently realizes a natural division of the cubic group into two halves. We ask ourselves if this division is also topologically meaningful. For $T_d$ and also for $O$ this topological meaning is quite straightforward. For $T_h$ it is a bit more intricate, but equally interesting. The existence of the tetrahedral subgroup $T_d$ is a direct group-theoretical expression of the bipartite character of the cubic graph. Indeed, since the nodes of the cubic graph can be subdivided into two sets of nonadjacent nodes, indicated by black and white vertices in the Schlegel diagram (Figure 1), the automorphism group will naturally contain a halving subgroup which covers only those permutations that preserve the color of the vertices. This subgroup thus is already active at the level of the graph itself, as opposed to $O$ and $T_h$, which require the structure of the embedding.

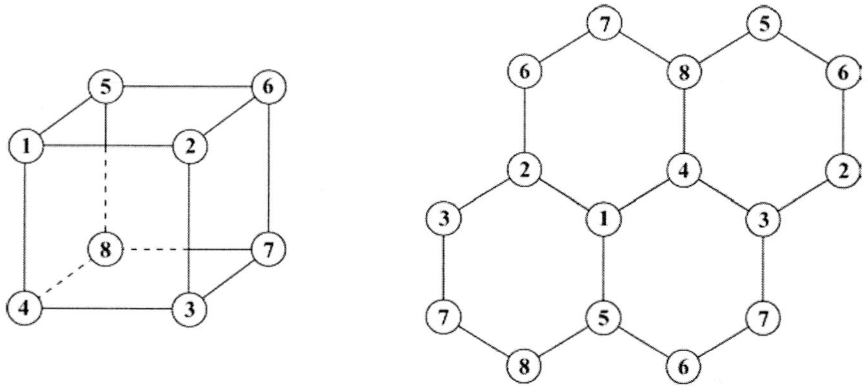


Figure 2: Embedding of the cubic graph on the sphere (left) and torus (right).

The octahedral subgroup, $O$, has a direct and transparent topological meaning. The embedding of the cubic graph on the spherical surface not only creates faces, but also orientability. This concept implies that all face cycles can be given a consistent sense of rotation, such that the vertex pair, $(u,v)$ say, common to two faces occurs as $uv$ in one face cycle and $vu$ in the other. The group $O$ contains all elements that preserve this orientation. The coset of $O$ in $O_h$ then gathers all remaining elements that will invert the orientation. As for $T_d$ such a 'yes' or 'no' criterion is seen to give rise to a halving subgroup. In the usual point group terminology $O$ is called the rotational subgroup, or subgroup of proper rotations. Last but not least we consider the third halving subgroup, $T_h$, which does not seem to have an obvious topological meaning as far as the embedding on the sphere is concerned. Interestingly the topological significance of this group naturally appears when the Petrie dual of the cube is constructed. The Petrie dual preserves the graph, but changes the faces. New faces are composed of all walks on the surface where alternatively left and right directions are taken.[9] Hence no three consecutive edges of an old face form consecutive edges of the new faces, and vice-versa. The Petrie dual of the cube is a toroidal embedding, consisting of four hexagons, as shown in Figure 2. In the case of a regular map, Petrie dualization will of course not change the automorphism of the embedding, hence the toroidal structure will have the same $O_h$ symmetry as the ordinary cube.[10] Since the torus is also orientable, we thus anticipate the existence of a halving subgroup, that will keep the orientation on the toroidal surface. The remaining subgroup $T_h$ precisely corresponds to this 'rotational' subgroup on the torus. As an example, in Figure 3 we show the action of the inversion element of $O_h$ on the torus of four hexagons. It is seen that this vertex permutation preserves the rotation of the hexagonal faces, represented by the circular arrows.

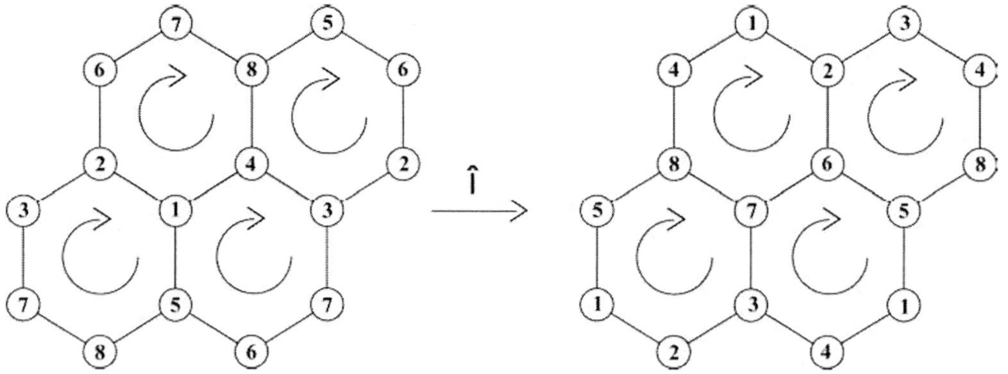


Figure 3: Effect of the inversion element on the toroidal embedding. As one can see, the orientation indicated by the circular arrows is preserved.

The structure of the automorphism group of a regular graph, thus reveals important topological aspects of its embeddings. In particular the existence of halving subgroups could be directly related to the bipartiteness and orientability of the embedding. We will now extend these findings to the group of the Dyck graph, which may be viewed as a fourfold extension of the octahedral group.

## 3. THE SYMMETRY GROUP OF THE DYCK GRAPH

The Dyck graph,[6] first described in Dyck's doctoral thesis of 1879, is a trivalent graph on 32 vertices, characterized by an extremely large symmetry group consisting of 192 automorphisms (Figure 4). It is called the tetrakisoctahedral group [5] and denoted as ${}^4O_h$. In our previous study the structure of this group and the action of its elements on the Dyck graph were intensively studied. The class-structure was derived and shown to divide the 192 permutations into 14 conjugacy-classes:

$$\begin{aligned} {}^4O_h &= E + 24C_8 + 6C_4 + 3C_2 + 32C_3 + 12C_2' + 12C_4' + 6C_4'' + 12\sigma_\perp \\ &\quad + 12\sigma_\parallel + 32S_6 + 4S_6^3 + 24S_8 + 12S_4 \end{aligned} \tag{1}$$

Together with its high symmetry, the Dyck graph also has a very rich subgroup-structure. As in the case of the cube, the full automorphism group, ${}^4O_h$, has three subgroups of index 2, namely ${}^4O$, ${}^4T_h$ and ${}^4T_d$. The discussion of their topological meaning will be deferred to the next section. In earlier work,[5] genealogical trees of all three halving subgroups were derived and discussed. Here we will extend this treatment by providing the genealogical tree of the full group. A generator based computer algorithm was used to extract subgroups efficiently from the group multiplication table. In Figure 5 the genealogical tree of ${}^4O_h$ is shown, where all subgroups are listed down to order 12. Subgroups are numbered from 1 to 36, and can be found in Table 1.

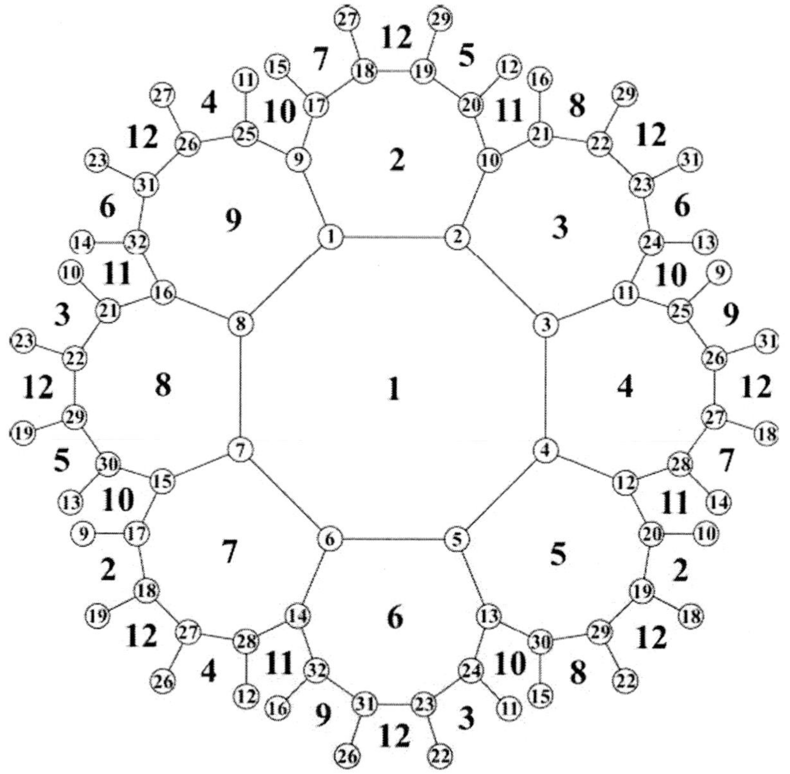

Figure 4: Schlegel-like diagram for the embedding of the Dyck graph on a genus-3 surface viewed from an inner octagonal face.

**Table 1: Listing of the subgroups of $^4O_h$ down to order 12**

| | | | | | |
|---|---|---|---|---|---|
| **1** | $^4O_h$ | **13 (3)** | 32.44 | **25 (3)** | $C_2 \times D_4$ / 16.6 |
| **2** | $^4O$ | **14** | 32.47 | **26** | $M_8$ / 16.11 |
| **3** | $^4T_d$ | **15** | $O$ | **27 (3)** | $C_2 \times D_4$ / 16.6 |
| **4** | $^4T_h$ | **16** | $T_d$ | **28** | $D_8$ / 16.12 |
| **5 (3)** | $Syl2(M_{12})$ / 64.261 | **17** | $T_h$ | **29 (3)** | $C_2 \times D_4$ / 16.6 |
| **6** | $G[3, 3, 4]$ | **18 (3)** | $AES_{16}$ / 16.8 | **30 (3)** | $AES_{16}$ / 16.8 |
| **7** | $O_h$ | **19 (3)** | $C_2 \times D_4$ / 16.6 | **31** | $D_8$ / 16.12 |
| **8** | 32.31 | **20 (3)** | $C_2 \times D_4$ / 16.6 | **32** | $SD_{16}$ / 16.13 |
| **9 (4)** | $E32+$ / 32.42 | **21** | $G[4,4,4]$ / 16.3 | **33 (3)** | $AES_{16}$ / 16.8 |
| **10 (3)** | 32.44 | **22** | $M_8$ / 16.11 | **34** | $SD_{16}$ / 16.13 |
| **11** | 32.31 | **23 (3)** | $AES_{16}$ / 16.8 | **35** | $T$ |
| **12 (3)** | 32.34 | **24 (3)** | $C_2 \times D_4$ / 16.6 | **36** | $D_{3d}$ |

Although some of the subgroups are isomorphic, they all belong to different conjugacy classes under the operations of $^4O_h$. Notice that only a few of the subgroups are identified as point groups, all other subgroups are more general permutation groups. Where possible, groups were given a name. For the point groups we used their well known chemical notation. However, there is no general naming scheme which covers all permutation groups. In our case, because all unnamed permutation group are powers of 2, we followed the numbering scheme of Hall and Senior[11] for groups of order $2^n$. One should note here that the relation between the rank (minimum number of generators) of a group and the rank of its subgroups is not well-behaved.[12] In other words, it can be that a subgroup needs more generators for its complete description than its supergroup does. So, although the full group ${}^4O_h$ can be completely described by means of 2 generators, namely an 8-fold element, $S_8$, through a face center and a 3-fold element, $C_3$, through one of its neighboring vertices, there are some subgroups which need more generators. For these subgroups, the rank is indicated between round brackets in Table 1. Apart from the three halving subgroups there exist two more subgroup classes which are directly positioned under the full group. One of them, subgroup "5", has order 64 and can be identified as the Sylow 2-subgroup of the Mathieu group of order 12: $Syl_2(M_{12})$ or 64.261 in its Hall and Senior notation. The second one, subgroup "7", has order 48 and is nothing else than the octahedral group $O_h$.

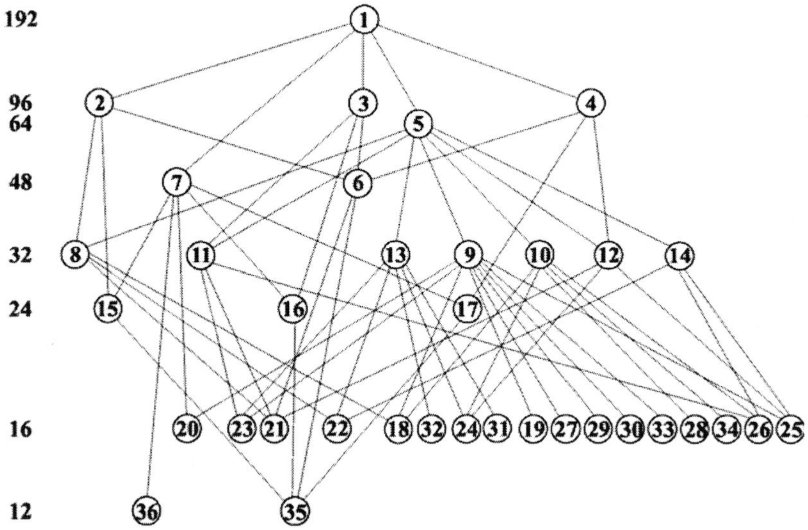


Figure 5: Genealogical tree of the group 4Oh. The numbers refer to subgroups that are identified in Table 1. Subgroup orders are indicated in the left margin.

There is also a further subgroup of order 48, subgroup "6" denoted $G[3, 3, 4]$, which is a maximal subgroup of all three halving subgroups of order 96. This group was already known to Dyck, who derived it by applying consecutive tessellations to the octagons of his genus-3 representation of the Dyck graph.[6] A full characterization of this group can be found in ref. 5. On the next level, the groups of order 32 are divided into seven subgroup classes. Only one of them, subgroup "9", is named in literature, namely $E32+$, representing the extra-special 2-group of order 32 and type + and is the only subgroup which needs four generators. All others are unnamed and could only be labeled by their corresponding Hall-Senior number. For the groups of order 24 one finds, as expected, a total of three subgroup classes corresponding to the three point groups $O$, $T_d$ and $T_h$. The groups of order 16 are divided into an impressive number of 17 subgroup classes: four of type $AES_{16}$ (the almost-extra-special group on 16 elements), six of type $C_2$ x $D_4$ (the direct product of the point groups $C_2$ and $D_4$), two of type $M_8$ (the metacyclic group generated by the direct product of an eightfold and twofold cyclic group), two of type $SD_{16}$ (the semi-dihedral group on 16 elements), two of type $D_8$ (the eightfold dihedral group) and finally one group $G[4, 4, 4]$ (isomorphic to the abelian group formed by the direct product $C_4 \times C_4$). This last group was also retrieved by Dyck in his tessellation process. Finally two subgroup classes of order 12 are found, which are nothing else than the point groups $T$ and $D_{3d}$. Of course there are also subgroups of lower order, but these were omitted from Figure 5 and Table 1 as they would complicate the Figure too much. They will therefore not be discussed here any further.

## 4. EMBEDDINGS OF THE DYCK GRAPH

In the previous section we noted that the Dyck graph, just like the cube, has three index 2 subgroups: $^4O$, $^4T_h$ and $^4T_d$, all of order 96. We shall now see that they can be topologically interpreted in a similar way. Just like the cubic graph, the Dyck graph is bipartite. The meaning of the halving subgroup $^4T_d$ is therefore obvious at once; it consists of all symmetry elements which preserve colors on the vertices. The existence of two other halving subgroups leads one to expect that there exist two regular embeddings of the Dyck graph and that the rotational subgroups of these embeddings will correspond to these index 2 subgroups. One of the regular embeddings was already know to Dyck and consists of 12 octagons for which the Euler equation reads:

$$V - E + F = 2(1-\gamma) \Rightarrow 32 - 48 + 12 = -4 \Rightarrow \gamma = 3 \qquad (2)$$

In this equation $V$, $E$ and $F$ stand for the number of vertices, edges and faces respectively. $\gamma$ denotes the genus (number of handles) of the surface which in this case equals three. This all-octagonal genus-3 embedding is shown in Figure 4. A closer investigation shows that its rotational subgroup corresponds to the halving subgroup $^4O$. The existence of one of the remaining two halving subgroups is thus explained, leaving us with the subgroup, $^4T_h$.

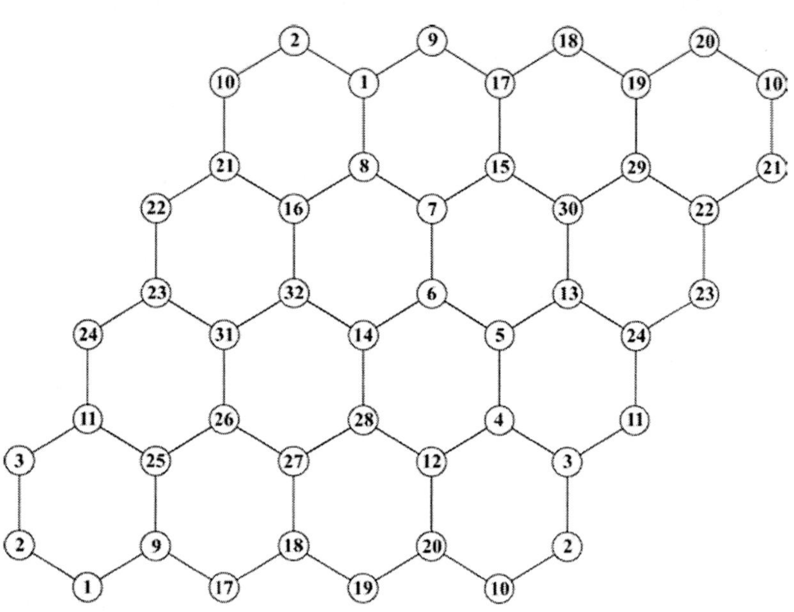


Figure 6: Embedding of the Dyck graph on the torus.

When a single regular embedding is known, a second one can be obtained by Petrie-dualization. When we apply this dualization process to the genus-3 embedding which was just described, all Petrie-paths turn out to be cycles of length six. In the newly found embedding, the original octagonal faces are thus replaced by a set of hexagonal faces. Because every edge belongs to two Petrie paths, the total number of such hexagonal faces will be 16. The existence of an alternative regular map is hereby proven, although the topology of the underlying surface is still unknown. It can however easily be calculated using the Euler equation:

$$V - E + F = 2(1-\gamma) \;\Rightarrow\; 32 - 48 + 16 = 0 \;\Rightarrow\; \gamma = 1 \tag{3}$$

The genus is one, indicating that the newly found embedding, shown in Figure 6, is an all-hexagon network which decorates the torus. Its rotational subgroup exactly matches ${}^4T_h$, being the remaining subgroup which was not yet topological interpreted.

## 5. Molecular Realizations

From a chemical point of view, the Dyck graph is very interesting. It is trivalent and therefore can serve as the blueprint for a new type of $sp^2$ hybridized carbon allotrope. However, in order to be chemically relevant the graph has to be realized in 3-D space. As we shall see, the two regular embeddings of the previous section will be of great help in finding this realization. A first 3-D realization is obviously the all-hexagon embedding on the torus. However, because of its very small dimensions it will incorporate a serious amount of strain and therefore doesn't lead to a chemically plausible structure. In principle the strain of the toroidal closing can be easily removed if one opens up the torus to form a very small cylinder. Connecting several of these cylinders together leads to a nanotube-like structure, which we might characterize as a (4, 0) zigzag tube. Such a tube would have an extremely small diameter and thus probably would still not be realistic.[13]

A better starting point will therefore be the all-octagonal embedding on the triple-torus. The main advantage of such a higher genus surface is that it is characterized by a negative curvature. This means that the network will open up and thereby relieve strain. A first step in finding a 3-D realization consists of finding an embedding of the graph on one of the numerous genus-3 minimal surfaces. The fact that the Dyck group contains the octahedral group $O_h$ as a subgroup, suggests that such an embedding exists on the 'Plumber's nightmare',[14] a genus-3 minimal surface which is characterized by the same octahedral symmetry. It is indeed found that an embedding on the 'Plumber's nightmare' can be realized which has point group symmetry *O*. Hence this embedding is intrinsically chiral. The resulting structure, optimized using MM+ molecular modeling, is shown in Figure 7. Notice that the six 12-rings going around the holes are already part of neighboring unit cells and are only included to visualize the octagonal faces. In fact, the actual unit cell resembles a cube with faces of length 12, in which each face is connected to a neighboring cell by means of four edges. (This cube-like unit cell is indicated in black in the right picture of Figure 7).[15] So although Figure 7 shows 24 octagonal faces, only 12 of them belong to the actual unit cell.

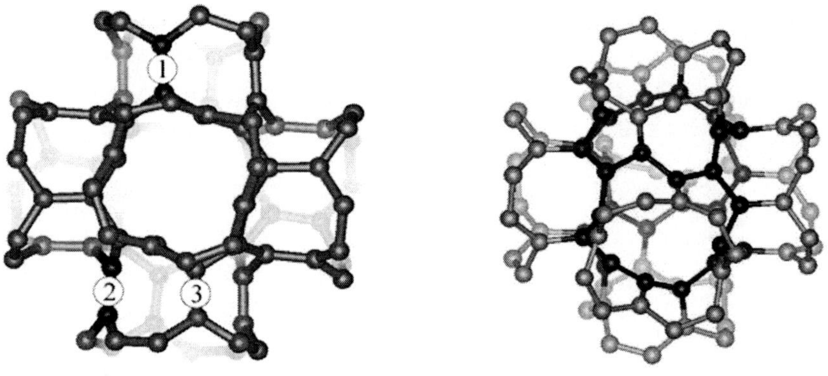


Figure 7: Embedding of the Dyck graph on the "plumber's nightmare" surface viewed from 2 different directions. The Figure shows more than one unit cell as every octagon is equally distributed over two neighboring unit cells.

The shape of the "Plumber's nightmare" easily lends itself to propagation in all three spatial directions, thereby forming a zeolite-like structure. To form such a space-filling model, one only has to indicate in which way subunits should be connected to each other. If one wants to construct a network that completely mimics the original adjacencies of the Dyck graph, one should identify edges 1 and 2 in Figure 7. As one can see from this figure, these edges are not properly located to make this match by simple periodic translation, but instead they are twisted over an angle of 60°. The same mismatch exists when the connections are made in the other two spatial directions. On the left hand side of Figure 8 we show a $2 \times 2 \times 2$ super cell formed by making these connections which preserve the Dyck adjacencies. The structure was optimized by MM+ molecular modeling. As a result of the 60° twists, the shapes of the individual unit cells are quite distorted from the relaxed structure of Figure 7, which means that the structure exhibits considerable strain.

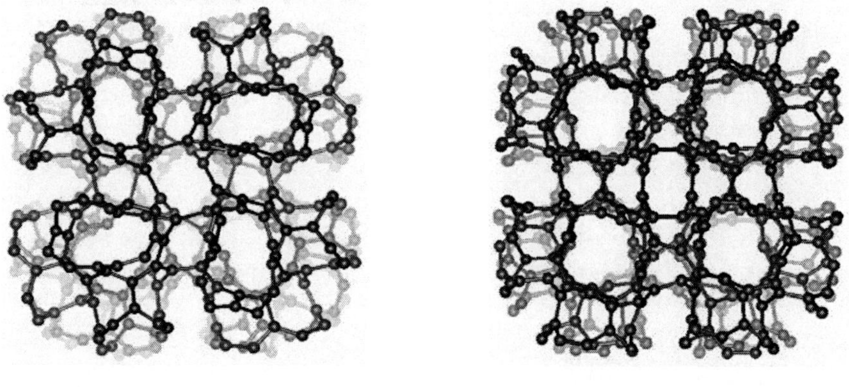

Figure 8: Super cells formed by introducing twists of 60° (left) and 30° (right) between neighboring subunits. The left structure preserves the connectivity of the original Dyck graph.

Aside from this structure, an alternative and less strained allotrope can be found if one identifies edges 1 and 3 instead of edges 1 and 2. The mismatch between the edges is hereby reduced from 60° to 30°. However, one should keep in mind that by making these new identifications, one changes the connectivity of the decorating graph. As in the previous case, a 2×2×2 super cell was constructed and optimized using MM+. The resulting structure is shown on the right hand side of Figure 8. As can be seen from the Figure, the individual 'Plumber' subunits are in this case less distorted from their original shape. The 30° structure will therefore exhibit less strain and make a more plausible chemical candidate than the 60° all-Dyck structure.

## 6. Conclusions

The Dyck graph corresponds to the skeleton of the second highest regular polyhedron on a surface of genus 3. It has recently received attention as a potential target for molecular realization. In the present paper we have derived the full genealogical tree of its symmetry group, and discussed its topological relevance. Our attention was focused on halving subgroups, these are subgroups of index 2. Three of such subgroups are present. They can be described as fourfold extensions of the point groups $T_d$, $O$ and $T_h$. The ${}^4T_d$ subgroup is related to the bipartite character of the graph. The other two have a topological significance as orientation-preserving subgroups on two different embeddings: ${}^4O$ is the rotational subgroup of the regular map on a triple torus, and ${}^4T_h$ is the rotational subgroup of the regular map on a single torus. Since these maps are related by Petrie dualization, we may define ${}^4O$ and ${}^4T_h$ as "Petrie-duals" of each other inside the covering group.

Further interesting extensions of this group-theoretical analogue of the Petrie dualization may be envisaged. As an example, it is possible that the Petrie-path of an orientable structure is non-orientable. This occurs for instance for the case of a tetrahedron, with symmetry group $T_d$. This group has only one halving subgroup, $T$, which corresponds to the rotational subgroup of the tetrahedron, embedded on a two-dimensional sphere. Since this embedding has a non-orientable Petrie dual, we don't expect $T_d$ to have another halving subgroup. And indeed $T$ is the only subgroup of index 2 in $T_d$.

While the topological relevance of these halving subgroups is thus evident, the lower subgroups might have interesting topological ramifications as well. Groups such as subgroup "5", which is the only subgroup of order 64, could point to special lower-symmetry embeddings, containing different types of faces. Further systematic studies are required to analyze these possibilities.

In the final section we also have modeled molecular realizations of the Dyck graph, both on a triple torus and on a torus. The latter corresponds to a (4,0) zigzag-nanotube, but its radius is below the critical value for nanotubes. The triple-torus realization however may be a viable carbon allotrope. It generally resembles a zeolite structure, with cubic cells connected by channels.

## Acknowledgement

E. Lijnen holds a doctoral fellowship from the Belgian Science Foundation FWO.

## References


[1] A. T. Balaban, *Chemical applications of graph theory*, Academic Press, London, 1976.

[2] A. Ceulemans and E. Lijnen, The polyhedral state of molecular matter, *Eur. J. Inorg. Chem.*, 2002, 7, 1571-1581.

[3] J. L. Gross and T. W. Tucker, *Topological graph theory*, Dover Pubs, Dover, 2001.

[4] A. Ceulemans, R. B. King, S. A. Bovin, K. M. Rogers, A. Troisi, and P. W. Fowler, The heptakisoctahedral group and its relevance to carbon allotropes with negative curvature, *J. Math. Chem.*, 1999, 26, 101-123.

[5] A. Ceulemans, E. Lijnen, L. J. Ceulemans, and P. W. Fowler, The tetrakisoctahedral group of the Dyck graph and its molecular realization, *Mol. Phys.* , 2004, 102, 1149-1163.

[6] W. Dyck, Ueber Aufstellung und Untersuchung von Gruppe und Irrationalität regulärer Riemann'sher Flächen, *Mathematische Annalen*, 1880, 17, 473-509.

[7] F. Klein, Ueber die Transformation siebenter Ordnung der elliptischen Funktionen, *Mathematische Annalen*, 1879, 14, 428-471.

[8] H. S. M. Coxeter, Regular polytopes, 3rd Edition, Dover Pubns, Dover, 1973.

[9] T. Pisanski and M. Randić, Bridges between geometry and graph theory. in: Geometry at Work: A Collection of Papers Showing Applications of Geometry (Ed., C. A. Gorini), Math. Assoc. Amer., 2000, 53, 174-194.

[10] P. McMullen and E. Schulte, *Abstract regular polytopes*, Cambridge University Press, Cambridge, 2002.

[11] M. Hall Jr. and J. K. Senior, The groups of order $2^n$ ($n \leq 6$), Macmillan, New York, 1964.

[12] F. Menegazzo, The number of generators of a finite group, *Irish Math. Soc. Bulletin*, 2003, 50, 117-128.

[13] L.-C. Qin, X. Zhao, K. Hirahara, Y. Miyamoto, Y. Ando, and S. Iijima, The smallest carbon nanotube, *Nature*, 2000, 408, 50-51.

[14] R. B. King, Negative curvature surfaces in chemical structures, *J. Chem. Inf. Comput. Sci.*, 1998, 38, 180-188.

[15] A similar drawing can be found in: Cs. L. Nagy and M. V. Diudea, Carbon allotropes with negative curvature, *Studia Universitatis Babeş-Bolyai*, *Chemia*, 2003, 48 (2), 37-46.

In: Nanostructures: Novel Architecture
Editor: Mircea V. Diudea, pp. 311-334
ISBN: 1-59454-499-9


*Chapter 15*

# NANOPOROUS CARBON STRUCTURES

***Csaba L. Nagy and Mircea V. Diudea***[1]
Faculty of Chemistry and Chemical Engineering, Babeş-Bolyai University,
400084 Cluj, Romania

## ABSTRACT

Spongy carbon nano-structures, also called schwarzites, have been recently synthesized. They consist of highly connected $sp^2$ carbon covalent networks, periodic in the three dimensions of the Euclidean space. The intimate structure of schwarzites show the topology of triply periodic minimal surfaces TPMSs. These negatively curved surfaces can be variously tessellated by using some map operations, of crucial importance being two twin septupling-open operations that introduce heptagons within a polyhedral lattice. These new operations were theorized and exemplified. Several schwarzites, with the associate graph embedded in D-, P- and FRD-type surfaces are illustrated and the topology of their dual lattice is rationalized in terms of map operations filiation.

## 1. INTRODUCTION

The traditionally recognized allotropes of carbon are the diamond and graphite. The diamond structure consists of tetrahedral units of $sp^3$ hybridized, covalently bonded, carbon atoms, repeated infinitely in a three dimensional lattice; all atoms and bonds are equivalent. The structure of graphite is represented as a two-dimensional hexagonal tiling with the carbon atoms $sp^2$ hybridized.

Fullerenes and nanotubes, more recently discovered carbon allotropes, can be viewed as assemblies of zero- and one-dimensional $sp^2$-bonded objects held together by van der Waals forces, as in the case of graphite.

A three-dimensional, covalently bonded $sp^2$-carbon network is necessarily conceivable to feel the gap between the above mentioned allotropes and the diamond. In this respect, a number of theoretical predictions and conjectures have been made, in particular, about negatively curved graphite-like structures,[1-7] as described by the infinite triply periodic

[1] E-mail address: diudea@chem.ubbcluj.ro

minimal surfaces TPMSs. These hypothetical structures have been termed schwarzites, in the honor of H. A. Schwarz,[8],[9] who first investigated, in the early nineteen century, the differential geometry of this class of surfaces.

Total energy calculations, carried out with *ab initio* methods[3],[5] as well as with classical[10] and tight-binding[11] simulations, have shown that carbon schwarzites are energetically more favorable than fullerenes of comparable absolute Gaussian curvature.[12] These data suggested that negatively curved carbon networks may appear during the fullerene and nanotube synthesis, at least as random (amorphous) schwarzites, if not in a periodic, crystalline form.

Porous carbon allotropes, synthesized by plasma-enhanced CVD, by Zaakhidov *et al.*,[13] have various structures: diamond or inverse opal, depending on the synthesis conditions. Graphite inverse opals, obtained by surface templating, provide examples of both dielectric and metallic optical crystals. The Raman spectra of the inverse opal materials (tetrahedral and cubic, respectively) indicated the presence of $sp^2$ carbon.

The first synthesis of large amounts of carbon foam (obtained by a pulsed microplasma cluster source, in the presence of a metallorganic catalyst) was reported by Barborini *et al.*[14] and Benedek *et al.*[12]. They brought clear evidences of a nanostructured, fully covalently bonded $sp^2$ carbon, 3D lattice. The simulated HRTEM images proved that the obtained spongy carbons have the topology of random schwarzites.

The extremely high porosity (pores with diameters ranging from 50 to 600 nm)[12] and low density of these nanoporous carbon NPC materials suggested many possible applications: as thermal insulators, in catalysis, in gas and energy storage, in gas and liquid purification, in templating biomaterials and in electrochemistry, as well. Carbon sponges are also useful in the investigation of negatively curved networks.

## 2. Background on Surface Curvature

An embedding is a representation of a graph on a surface $S$ such that no edge-crossing occurs. Two embeddings $\varepsilon_1$ and $\varepsilon_2$ are combinatorially equivalent if there is a one-to-one correspondence between their vertices, edges and faces, such that the incidences between their constitutive substructures are preserved.

*Curvature* is the amount by which a geometric object deviates from being *flat*. It is usually measured either as Gaussian curvature $K$ or as mean curvature $H$. These are defined at each point on the surface as functions of the surface principal curvatures at the point as follows:

$$K = \pm\, k_1 \cdot k_2 \qquad (1)$$

$$H = \frac{1}{2}(k_1 + k_2) \qquad (2)$$

To evaluate these at a given point of the surface, consider the intersection of the surface with a plane containing a fixed normal vector at the point. This intersection is a plane curve, having a curvature $k = 1/r$; if the plane is varied, this curvature will change, and there are two extremal values - the maximal and the minimal curvature, called the *principal curvatures*, $k_1$ and $k_2$. Accordingly, the extremal directions are called principal directions. Usually, a curvature is taken positive if the curve turns in the same direction as the surface's chosen normal, otherwise it is negative.

A surface $S$ is *flat* if $K(p) = 0$ and it is *minimal* if $H(p) = 0$, for every $p \in S$.

The integral of the Gaussian curvature over the whole surface is closely related to the surface's Euler characteristic $\chi(S)$ (*Gauss-Bonnet* theorem):[15]

$$\int_S K dS = 2\pi\chi(S) \tag{3}$$

The above theorem relates the geometric curvature to the topology:

$$\chi(S) = 2(1 - g(S)) \text{ orientable surface} \tag{4}$$

$$\chi(S) = 2 - n(S) \text{ non-orientable surface} \tag{5}$$

A surface is *orientable* when it has two sides and *non-orientable* when it is one sided, like the Möbius strip.

In the above relations, $g$ denotes the *genus*, being the number of handles required to be attached to the sphere to make it homeomorphic to $S$. The parameter $n$ is the number of *cross-caps* needed in the same view, in case of a non-orientable surface.[16]

If $S$ is a topohedral surface in 3D, the count of vertices $V$, edges $E$ and faces $F$, is given by the Euler relation:[17]

$$\chi(S) = V - E + F \tag{6}$$

If $g = 0$, then (4) and (6) lead to the familiar version of Euler's formula for convex polyhedra, *e.g.*, topohedra embedded in the sphere:

$$V - E + F = 2 \tag{7}$$

If a polyhedron has all degree-3 vertices, and each edge is shared by two faces, then:[18]

$$2E = 3V = \sum_s s \cdot f_s \tag{8}$$

where $f_s$ denotes the number of $s$-gonal faces. Substituting this into (6) results in:

$$\sum_s (6 - s) f_s = 12(1 - g) = 6\chi \tag{9}$$

For a surface of a given genus (9) gives the number of $s$-folded polygons. Special cases are the *Platonic tiling*, with a single kind of polygons, and the *Archimedean tiling*, with two different kinds of polygons, one of which being here the hexagon.

If we consider polygons with $n = 3, 4 \ldots 9$ sides, equation (9) reads as:

$$3f_3 + 2f_4 + f_5 - f_7 - 2f_8 - 3f_9 = 12(1 - g) \qquad (10)$$

In (10), $f_3$, $f_4 .. f_9$ represent defects in the graphitic structure, which contribute to the curvature of the polyhedra; $f_6$ is not present since hexagon produces zero Gaussian curvature.

*Descartes*' theorem states that, if $S$ is a topohedral surface in 3D, then the overall angular defects (*i.e.*, disclinations) are proportional to the Euler's characteristic:[19]

$$\sum_{p}^{S} \phi_p = 2\pi\chi(S) \qquad (11)$$

For a Platonic tiling of a surface of genus $g$, with only one kind of polyhedron, $f_s$, is found by setting, in (9, 10), $f_{\neq s} = 0$. Table 1 lists the solutions for surfaces of genus 0, 2 and 3 (the repeat units).

**Table 1: Face counting for Platonic tiling of surfaces of genus 0, 2 and 3**

| | $g = 0$ (sphere) | | $g = 2$ (D surface) | $g = 3$ (P surface) |
|---|---|---|---|---|
| $f_3$ | 4 (tetrahedron) | $f_7$ | 12 | 24 |
| $f_4$ | 6 (cube) | $f_8$ | 6 | 12 |
| $f_5$ | 12 (dodecahedron) | $f_9$ | 4 | 8 |

Relations 9 and 10 are independent of the number of hexagons, which is, therefore, arbitrary. Archimedean fullerenes must always contain 12 $f_5$; thus $f_6$ comes out from (8): $5f_5 + 6f_6 = 60 + 6f_6 = 3v$; $f_6 = (v/2) - 10$.

Mean curvature is closely related to the first variation of surface area. Surfaces having mean curvature $H$ everywhere zero are *minimal surfaces* (*i.e.*, have locally minimal area). The shapes taken by soap films are minimal surfaces. A soap bubble has a constant mean curvature. Unlike Gauss curvature, the mean curvature depends on the embedding, for instance, a cylinder and a plane are locally isometric but the mean curvature of a plane is zero while that of a cylinder is nonzero.

## 3. Carbon Structures Embedded on TPMSs

Triply periodic minimal surfaces TPMSs have, by definition, translational symmetries in three independent directions. They are saddle-shaped, everywhere except at certain *flat points*. When a TPMS is considered in the natural sciences, it is understood that it has no self-intersections. A TPMS belongs to one of the crystallographic space groups as its symmetry group and, if it has no self-intersections, it partitions the space into two disjoint labyrinthine regions. Its topology is characterized by two interpenetrating networks, the *labyrinth graphs*.

A carbon lattice, having all $sp^2$ atoms, can be embedded in a TPMS. According to relations 3 and 4, a TPMS graphene will have a negative average Gauss curvature $\chi < 0$, (unlike fullerenes and capped nanotubes, having $\chi > 0$, and graphite, open and toroidal tubulenes, with $\chi = 0$) and zero mean curvature $H = 0$, for every point belonging to the graphene net. It was conjectured that a covalent lattice, of all $sp^2$ carbon, embedded in a minimal periodic surface is energetically more favorable than fullerenes of comparable Gaussian curvature.[12] The generic name adopted for such carbon allotropes is *periodic schwarzites*.[6],[12]

Among the most prominent scientists who contributed to the development of TPMSs theory we enumerate: Schwartz,[8],[9] Neovius,[20] Schoen,[21] Coxeter,[22] Fischer and Koch,[23] etc. At this effort, the scientists of the last decade have added experimental evidences of the molecular realization of these fascinating structures.

Periodic schwarzites are generated by covalent connections of identical repeat units. The unit cell may contain one or more such units/elements. The centers of the elements can be viewed as the nodes of a corresponding dual lattice, *i.e.*, one of the two complementary labyrinths.

The dual lattice of the 4-connected diamond structure defines a class of cubic schwarzites, with two elements per unit cell (*i.e.*, *bielemental* schwarzites). They are tessellations of a D-type Schwarz minimal surface. *Monoelemental* cubic schwarzites with a *sc*, or *bcc* or *fcc* dual lattice are also possible. P-type schwarzites have the structure of a simple monoelemental cubic lattice. They include a special class of structures known as polybenzenes.[24]

Numerical computations of minimal surfaces may be done with several techniques. The classical method assumes knowledge of the Weierstrass integration formulas and does a numerical integration. Another method relies on the finite element theory and needs the calculation of partial differential equations with numerical techniques.[25]

For visual representation, a minimal surface can simply be approximated by a trigonometric expression[9],[26],[27] corresponding to the lowest-order terms of a Fourier expansion. For example, D- and P-type surfaces are well described in the $(x, y, z)$ space by:

D-type:

$$\cos(x)\cos(y)\cos(z) + \sin(x)\sin(y)\sin(z) = 1 \tag{12}$$

P-type:

$$\cos(x) + \cos(y) + \cos(z) = 0 \tag{13}$$

Each Platonic tiling is the zero$^{th}$ element of an infinite series of larger schwarzites, obtained by inserting in each element an arbitrary number ($f_6 > 1$) of hexagons. For example, the D-type schwarzites of Platonic-Archimedean tessellation (with only heptagons and hexagons) form a family of crystals with $2(28 + 2f_6)$ atoms in the unit cell. The smallest member of this family is *fcc*-$(C_{28})_2$, (with 12 heptagons per element and no hexagon - Figure 1; see also Table 1) being a chiral structure (of $T$ point group) that allows for two enantiomeric lattices. The second member is *fcc*-$(C_{36})_2$, with 4 hexagons per element and the third member is *fcc*-$(C_{40})_2$, with 6 hexagons per element. The elements of *fcc*-$(C_{36})_2$ and *fcc*-$(C_{40})_2$ have a full tetrahedral point group symmetry ($T_d$) and are achiral.[11] Insertion of hexagons, in view of relaxing the strain of the resulting lattice, is easily achieved by some operations on maps, known from the tiling theory.[28-30] Figure 1 illustrates the zero members of the D- and P-type series of schwarzites.

The smallest D-type schwarzite with non-abutting heptagons corresponds to the lattice formula *fcc*-$(C_{84})_2$. This analogue of $C_{60}$, in the sense of being the smallest structure with non-abutting odd polygons, was theoretically predicted to be very stable.[10]

A structure $(C_{84})_2$ having the above tiling can be derived from the tetrahedron $T$ by the operation sequence $Le(Di(Op(S_1(T))))$ (where $Di$ stands for dimerizing, *i.e.*, closing operation – see below).

(a) $C_{28} = Op(S_1(T))$ (b) $C_{56} = Op(S_1(C))$

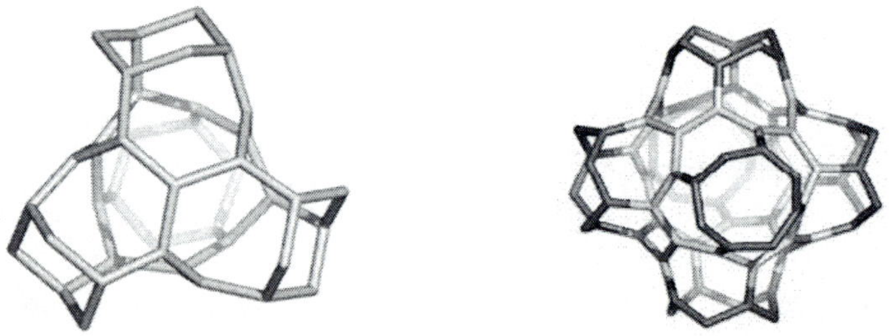

Figure 1: The first three members of the infinite families of D- (a) and P-type (b) schwarzites. The symbols give the number of atoms in the repeat unit.

Isomeric forms of schwarzites are possible, by analogy to the spherical fullerenes. Thus, the elements $C_{28}$ and $C_{56}$ are not unique, as designed by $Op(S_1(M))$. The next sections will provide detailed examples.

Some other surfaces with cubic symmetry have been described (and called following Schoen's notation): IWP, FRD, OCTO, C(D), and G.[25]

Random schwarzites have experimentally been obtained by supersonic cluster beam deposition SCBD of carbon clusters in the presence of very small catalyst nano-particles,[12] and look like a highly porous spongy carbon. The numerical simulation of the TEM images brought evidence that random schwarzites grow in the form of a TPMS.

These new structures, which are fully covalent in 3D, represent the fourth form of $sp^2$ carbon; the first form, fullerite, is the aggregation of 0D objects (fullerenes); the second form, nanotubes bundles, is the aggregation of 1D objects (nanotubes); and the third form, graphite, is the aggregation of 2D objects (the graphite sheets or graphenes).

## 4. Septupling Operations on Maps

Two main operations on maps, leading to Platonic tessellation, are known: the septupling $S_1$ and $S_2$ operations.[31]

The $S_1$ operation was also called [30],[31] Capra $Ca$. It can be written as:

$$S_1(M) = Tr_{P_5}(P_5(M)) \qquad (14)$$

with $Tr_{P5}$ meaning the truncation of each vertex where $P_5$ capping faces of a parent face are incident. It corresponds to the Goldberg[32] parameter $m(2,1) = 7$ (see also Ref. 31).

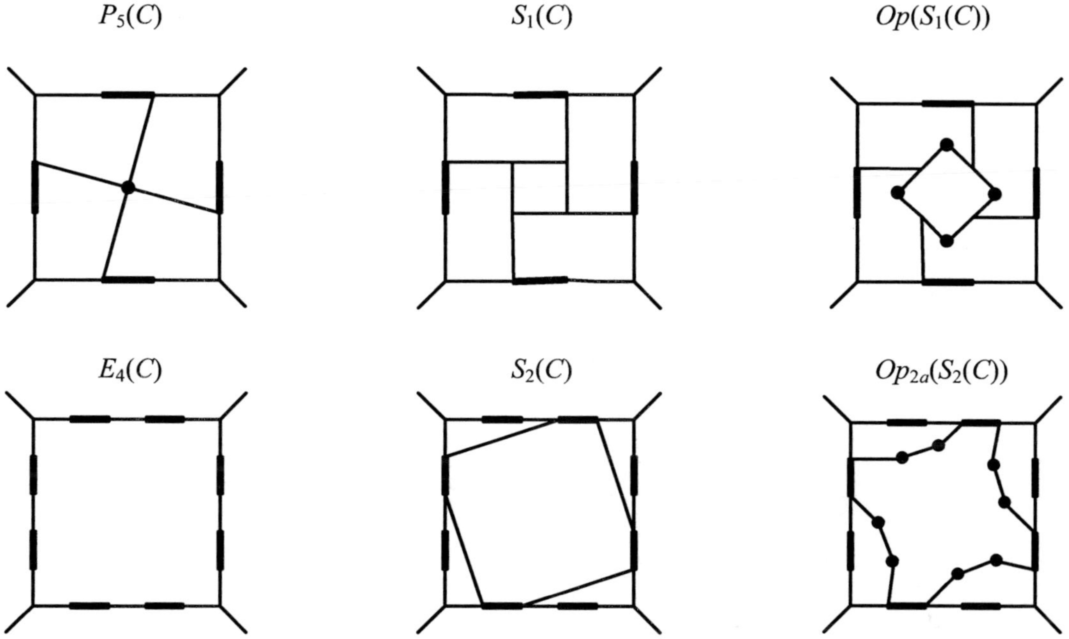


Figure 2: Realization of the septupling $S_1$ and $S_2$ operations on a square face in a trivalent lattice, up to the open structure.

$S_1$ insulates any face of $M$ by its own hexagons, which are not shared with any old face. It is an intrinsic chiral operation and was extensively illustrated in ref. 31. $S_1$ can continue with the "open" operation:

$$Op_n(S_i(M)) \qquad (15)$$

where $n$ represents the number of points added on the boundary of the parent faces, that become the "open" faces. The resulting open objects have all the polygons of the same $(6+n)$ size. The above operation sequence enables the construction of negatively curved networks (see below). Figure 2 gives the steps of $S_1$ realization on a square face in a trivalent lattice, up to the open structure. The transformed net parameters are given in Table 2.

The $S_2$ operation[31] is a simpler one; it can be achieved by putting four vertices on each edge of the parent map $M$ (*i.e.*, the $E_4$ operation) and next join these new vertices in order (-1, +3):

$$S_2 = J_{(-1,+3)}(E_4(M)) \qquad (16)$$

It insulates the parent faces by pentagon triples and the transformed objects are non-chiral ones.

**Table 2: Lattice counting by the Septupling-Open sequence of map operations**

| | Operation | Parameter | |
|---|---|---|---|
| 1 | $S_1(M)$ | $v = v_0 + 2e_0 + s_0 f_0 = (2d_0 + 1)v_0$<br>$e = 7e_0$<br>$f = s_0 f_0 + f_0$ | (17) |
| 2 | $Op(S_1(M))$ | $v_{Op} = v + s_0 f_0 = (3d_0 + 1)v_0$<br>$e_{Op} = 9e_0$<br>$f_{Op} = f - f_0 = s_0 f_0$ | (18) |
| 3 | $S_2(M)$ | $v = (2d_0 + 1)v_0$<br>$e = 7e_0$<br>$f = s_0 f_0 + f_0$ | (19) |
| 4 | $Op_{2a}(S_2(M))$ | $v_{Op} = (4d_0 + 1)v_0$<br>$e_{Op} = 11e_0$<br>$f_{Op} = s_0 f_0$ | (20) |

However, chirality is brought by the *Op* operation, $Op_{2a}$, meaning two points put on alternative edges of a boundary which folding is twice that of the parent faces (in a regular polyhedron - Figure 2). In case of a closed product, the transformed net parameters are identical to those of $S_1$ transforms; difference appears in case of open objects, as shown in Table 2, entries 2 and 4.

The two septupling operations represent twin operations, in that the transformed objects, representing repeat units in D- and P-type surfaces, belong to the complementary labyrinths (see the next section).

## 5. D-TYPE SCHWARZITES

### 5.1. Diamond Lattice

Repeat units of the D-schwarzites can be obtained by applying the septupling operations $S_1$ and $S_2$. Figure 3 illustrates the chiral pair of the Capra-transformed tetrahedron: $Op(S_{1S}(T))$ (3a) and $Op(S_{1R}(T))$ (3b). By identifying the open face between the two structures results in the bi-elemental cell of a diamond lattice (3c). Finally, the $sp^2$ diamond lattice of (7,3) Platonic tessellation, can be easily drawn (4d). For diamond-like structures, the reader can find supplementary information in refs. 33 and 34.

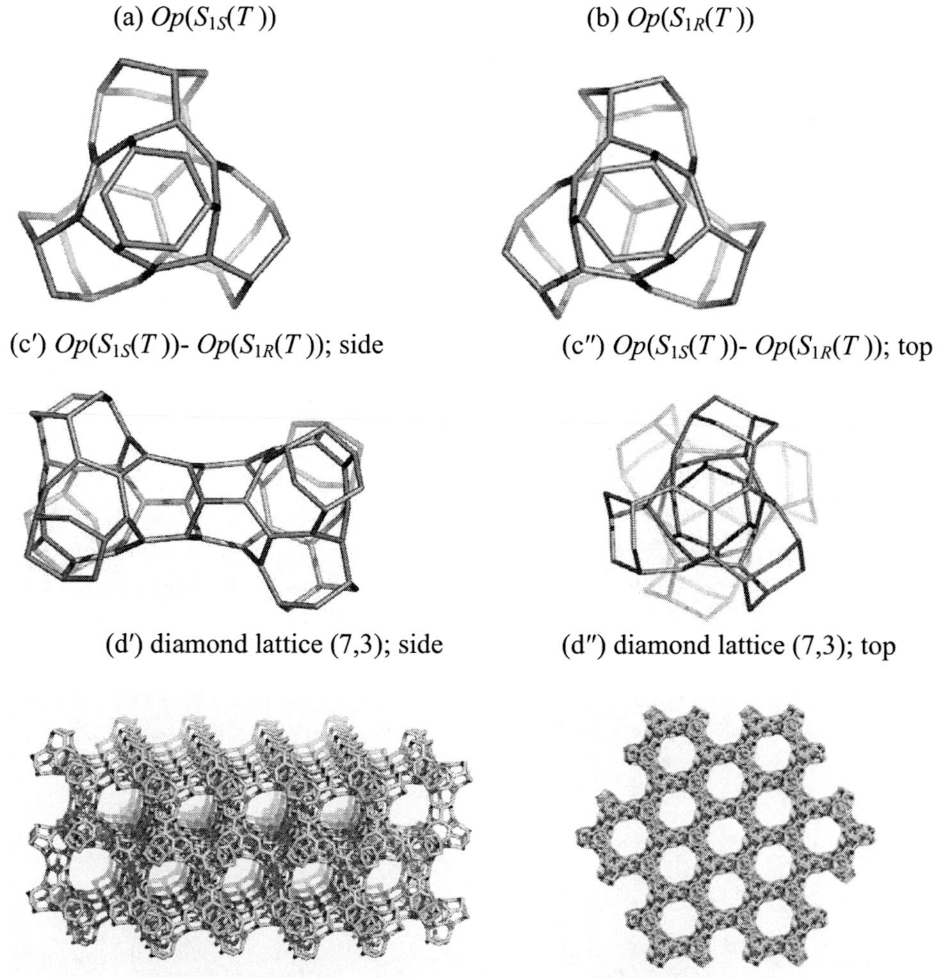


Figure 3: repeat units and the diamond D-type surface of Platonic (7,3) tessellation.

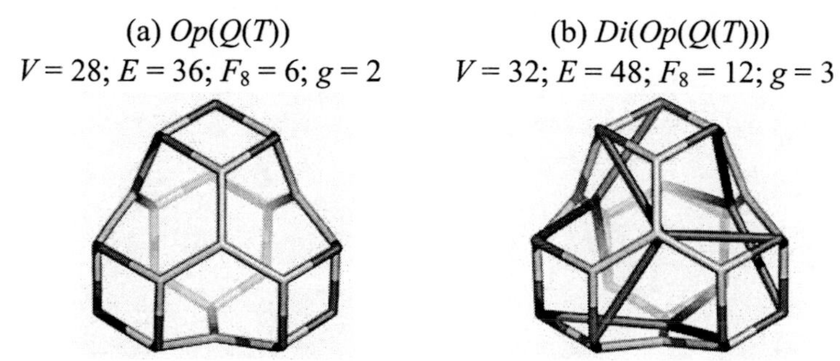


Figure 4: A way to the finite Dyck graph representation.

## 5.2. Klein and Dyck Graphs

The simplest TPMSs of Platonic tessellation are those having the Klein ((7,3) - Figures 1 and 3) and Dyck ((8,3) - Figure 4) tiling.[35-39] The Klein and Dyck graphs are representations of the automorphism of (heptakis and octakis, respectively) octahedral group.[40]

The above examples show repeat units and/or infinite lattices of Klein/Dyck covering of both D- and P-type surfaces. It is possible to built up a finite object, a kind of interlaced dimer $Di(Op(\Omega(M)))$ (Figure 5), by identifying all the opening faces of an $Op(\Omega(M))$ unit, $\Omega$ being a map operation. Among the $6^4 = 1296$ permutations we found[41] the object $Di(Op(Q(T)))_p$ having the all topological and spectral characteristics of the finite, trivalent Dyck graph.[42],[43] A similar procedure, led to $Di(Op(S_1(M)))$ objects,[41] embedded in surfaces of genus 3 or higher, having the Klein tessellation (but not necessarily the all characteristics of the Klein graph).[44],[45] Such an object is presented in Figure 5a. This covering can be modified by Capra (*i.e.*, $T_1$) operation, to give a Platonic ($(Cor_7)3$) tessellation (Figure 5b). This comes from the extension given by Diudea[31],[41] to the Clar benzenoid covering, by taking the corannulenes *Cor* as the units of a supra-organized aromatic tessellation. More important are those structures having a full disjoint PCorS (similar to perfect Clar systems PCS - Figures 5c,d) which show an Archimedean ($((Cor_7,6)3)$) tessellation.

(a) $Di(Op(CaS(T)))= Di(Op(S_{1S}(T)))$ (b) $S_{1R}(Di(Op(S_{1S}(T))))$
Platonic (Klein (7,3)) tessellationPlatonic ($(Cor_7)3$) tessellation ; ($Cor_7 = 24$)
$V = 56$; $E = 84$; $F_7 = 24$; $g = 3$ $V = 392$; $E = 588$; $F = 192$; $g = 3$

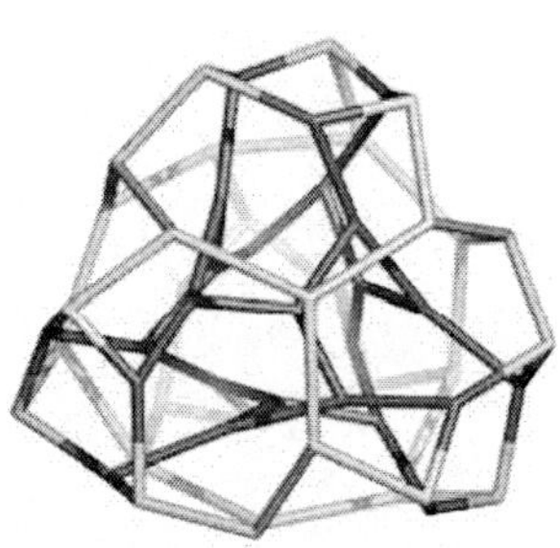

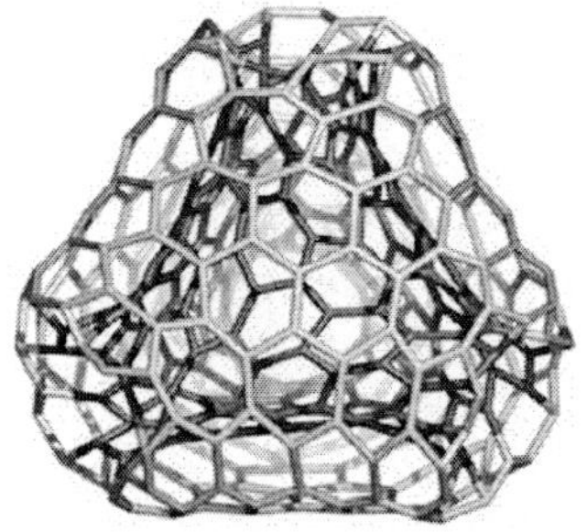

(c) $Le_{22}(Di(Op(S_1(T))))$ (d) $Le_{22}(Di(Op(S_1(T))))$
Archimedean ($((Cor_7,6)3)$) tessellation PCorS
$V = 672$ ; $E = 1008$ ; $F = 332$; $g = 3$

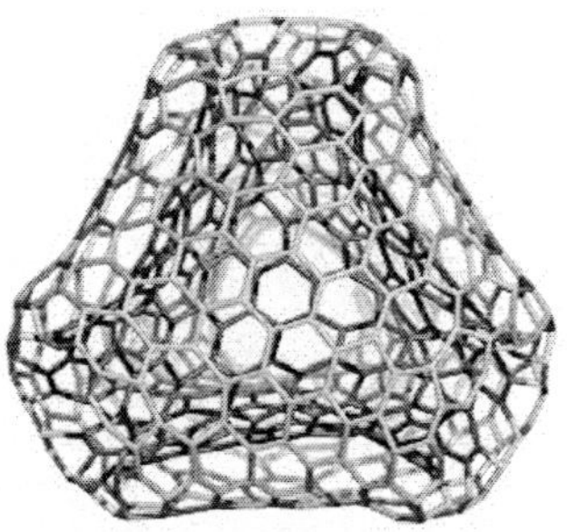

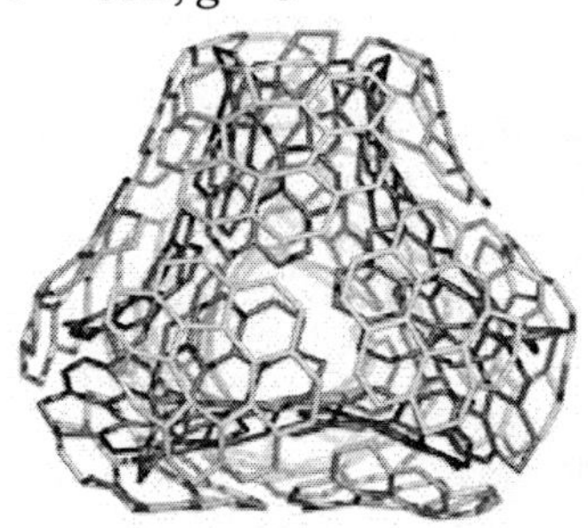

Figure 5: Covering of a dimeric tetrahedral object of genus 3.

## 5.3. Self-assembled Structures of High Genus

*Monoelemental* cubic schwarzites may be obtained by self-assembly of tetrahedral units; they look like a supra-dodecahedral construction (abbreviated hereafter "supra-*D*" or simply *D*, in the operations sequence - see below).

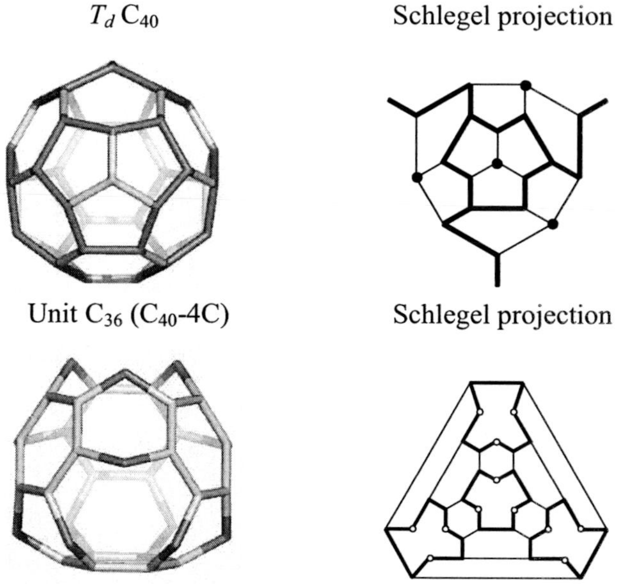


Figure 6: 3D Units of tetrahedral symmetry used in construction of supra-dodecahedral structures of high genus.

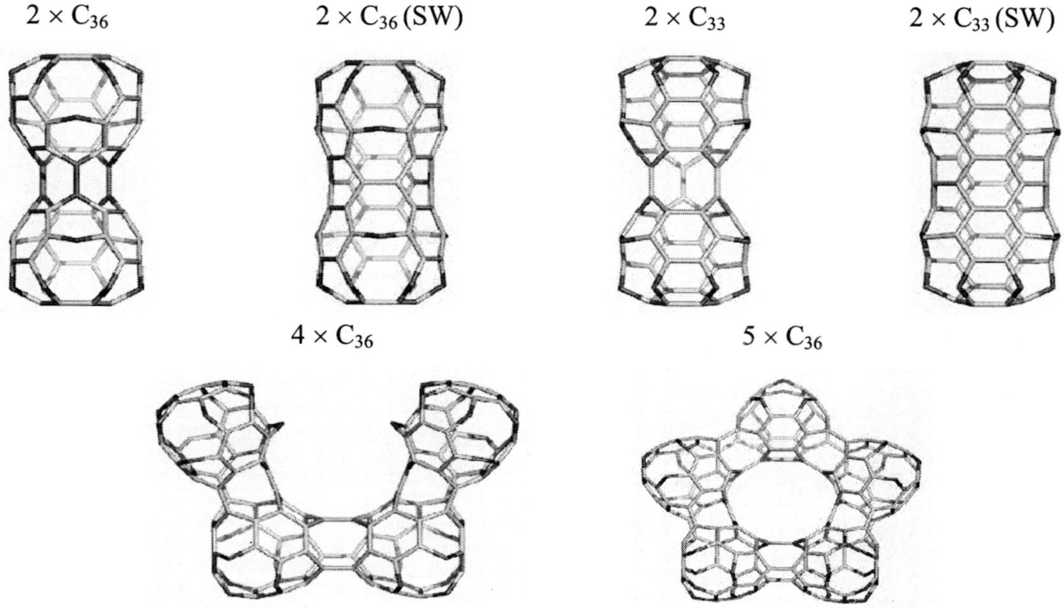


Figure 7: (continued on next page)

$6 \times C_{36}$

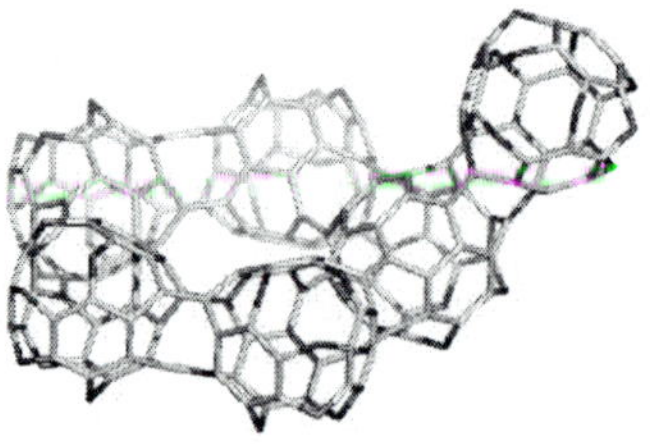

$8\times C_{36}$

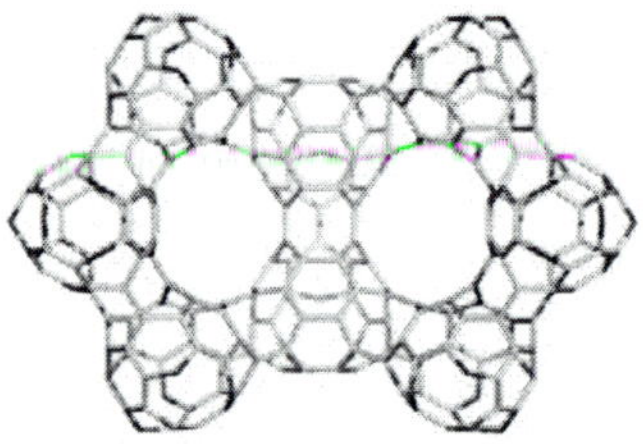

$10 \times C_{36}$
Infinite; $V = 720$; $E = 1050$; $F = 290$; $g = 21$

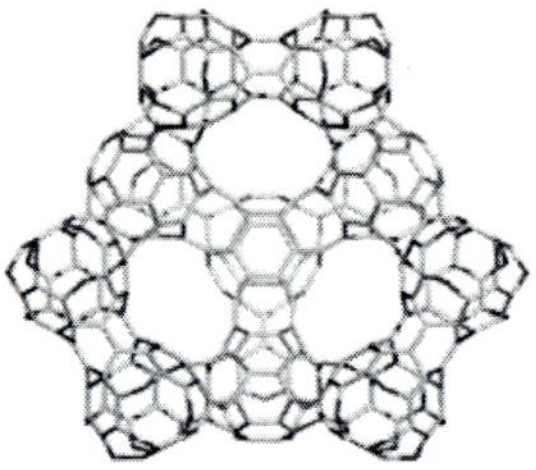

$D(C_{36})$ two-fold axis

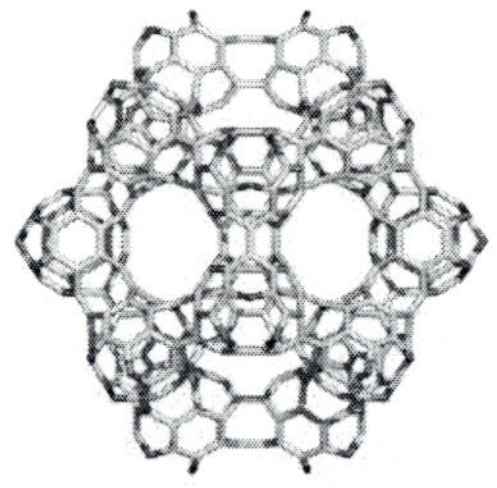

Figure 7: Pathway to the supra-dodecahedral structure of the C36 repeat unit. Dimerizing of 10 × C36 intermediate leads to D(C36).

A possible route for such a process starts by a bi-elemental (eclipsed) unit insertion in more and more connected assemblies, up to the supra-dodecahedron. The exemplified unit is $(C_{36})_2$, derived from the fullerene $T_d$ $C_{40}$. Figures 6 and 7 illustrate these routes, considering several isomeric coverings. Energetics of these units, as closed, open, only carbon and hydrogenated molecules, are given in Tables 3 and 4.

**Table 3: Semiempirical data for fullerenes (possible units) utilizable in construction of spongy carbon structures**

| Structure | *N* | $H_f$ (kcal/mol) | HOMO-LUMO GAP (eV) |
|---|---|---|---|
| $C_{20}$ | 20 | 41.250 | 5.990 |
| $C_{40}$ | 40 | 23.223 | 5.748 |
| $C_{36}$ | 36 | 33.501 | 5.574 |
| $C_{33}$[a] | 33 | 33.467 | 6.601 |
| $C_{56}$ | 56 | 21.2029 | 2.5194 |
| $C_{52}$ | 52 | 25.876 | 5.1575 |
| $C_{49}$[a] | 49 | 24.8261 | 5.0675 |

[a] a hexagonal face is formed by cutting off three atoms of an open face.

Plausible heat of formation have been obtained, if recall that of isolated dodecahedron $C_{20}$ (about 41 kcal/mol, by PM3 Hamiltonian - Table 3).

Also calculated are the POAV1 strain energies[46],[47] *S* of this series of possible intermediates (Table 5). It is known that the curvature elastic[33] energy decreases in the series: sphere > cylinder > saddle > flat surface. Average $S$ values for $C_{60}$ and a nanotube (6,6) TuA [12,$n$] are about 8.3 and 1.6, respectively. In the transformed (by map operations) infinite lattices the strain ranges between 0.31 and 0.85, (see the top of figures) clearly supporting the above elastic energy decreasing sequence. Data in Table 5 show the open/infinite supra-dodecahedral objects more relaxed, in comparison to the closed objects, particularly those having hexagon/ heptagon (SW) covering. The strain is still high, due to the reduced size of the modeled structures.

**Table 4: Semiempirical (MNDO) data for oligomers (hydrogenated species included)**

| Structure | $N$ | $H_f$ (kcal/mol) | HOMO-LUMO Gap (eV) |
|---|---|---|---|
| 2× $C_{36}$ | 72 | 33.201 | 6.545 |
| 2× $C_{36}H$ | 90 | 14.496 | 4.596 |
| 2× $C_{36}$ (SW) | 72 | 26.092 | 5.469 |
| 2× $C_{36}H$ (SW) | 90 | 13.655 | 4.955 |
| 2× $C_{36}C$ | 66 | 30.608 | 6.601 |
| 2× $C_{36}CH$ | 78 | 16.285 | 6.146 |
| 2× $C_{36}C$ (SW) | 66 | 29.098 | 6.051 |
| 2× $C_{36}CH$ (SW) | 78 | 15.256 | 5.316 |
| 3$C_{36}$ | 108 | 30.821 | 6.535 |
| 4× $C_{36}$ | 144 | 29.634 | 6.533 |
| 5× $C_{36}$ | 180 | 26.073 | 6.611 |
| 5× $C_{36}$ (SW) | 180 | | |
| 5× $C_{36}C$ | 165 | 22.922 | 6.562 |
| 5× $C_{36}C$ (SW) | 165 | 20.182 | 5.562 |
| 10× $C_{36}C$ (SW) | 360 | 23.627 | MNDO |

**Table 5: POAV 1 Strain energy $S$ data (in kcal/mol)**

| Unit $C_{36}$ Infinite ((6,8)3) | | | Unit $C_{36}$ (SW) Infinite ((6,7)3) | | | Unit $C_{33}$ Finite ((6,8)3) | | | Unit $C_{33}$ (SW) Finite ((6,7)3) | | |
|---|---|---|---|---|---|---|---|---|---|---|---|
| | $N$ | $S$ | | $N$ | $S$ | | $N$ | $S$ | | $N$ | $S$ |
| 2 | 54 | 5.300 | 2 | 54 | 3.885 | 2 | 54 | 8.490 | 2 | 54 | 6.998 |
| 4 | 114 | 4.965 | 4 | 114 | 3.502 | 4 | 114 | 7.952 | 4 | 114 | 6.313 |
| 6 | 180 | 4.656 | 6 | 180 | 3.158 | 6 | 180 | 7.482 | 6 | 180 | 5.605 |
| 8 | 246 | 4.518 | 8 | 246 | 3.009 | 8 | 246 | 7.266 | 8 | 246 | 5.236 |
| 10 | 312 | 4.437 | 10 | 312 | 2.951 | 10 | 312 | 7.136 | 10 | 312 | 4.987 |
| 20 | 660 | 4.137 | 20 | 660 | 2.584 | 20 | 660 | 6.661 | 20 | 660 | 4.126 |

By Stone-Wales SW edge rotation, it is possible to change the initial covering of the above intermediates and final structure (Figures 7 and 8).

$D(C_{36})$; (SW); two-fold axis; infinite
$V = 720$; $E = 1050$; $F = 290$; $g = 21$; $S = 2.584$

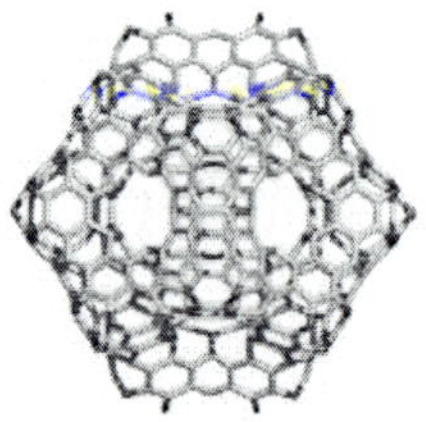

$D(C_{36})$; (SW); five-fold axis

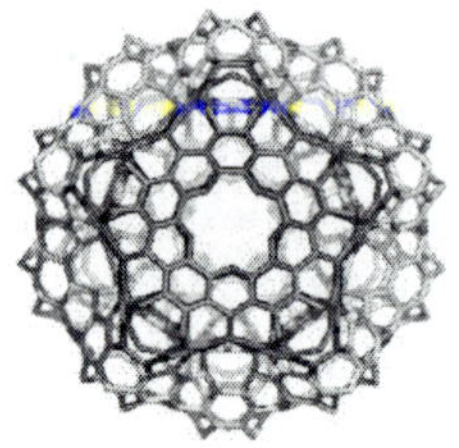

$D(C_{33})$; two-fold axis; finite
$V = 660$; $E = 990$; $F = 310$; $g = 11$; $S = 6.352$

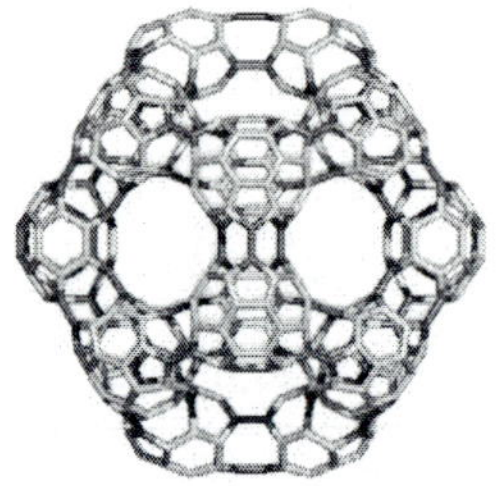

$D(C_{33})$; five-fold axis

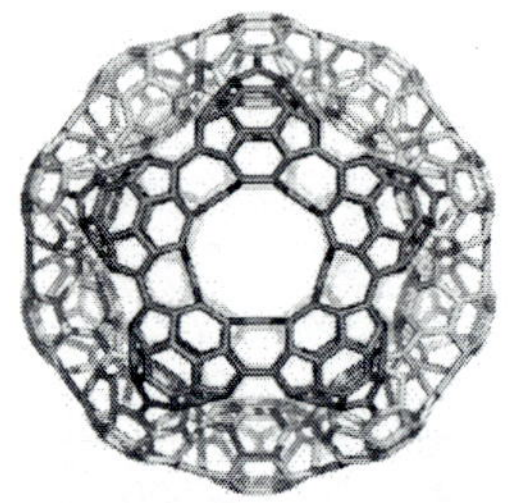

$D(C_{33})$; (SW); two-fold axis; finite
$V = 660$; $E = 990$; $F = 310$; $g = 11$; $S = 3.686$

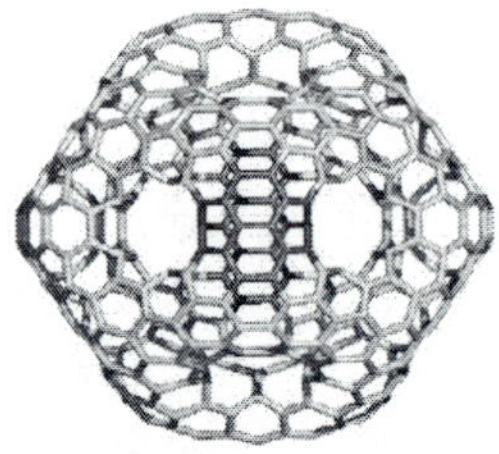

$D(C_{33})$; (SW); five-fold axis

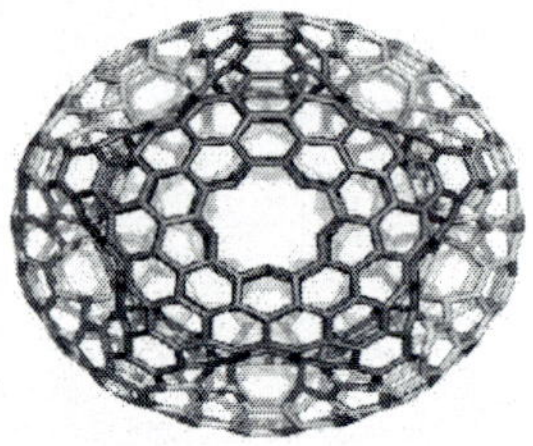

Figure 8: Supra-dodecahedral open (left column) and close (right column) structures derived from the $C_{36}$ (open) and $C_{33}$ (close) units, respectively.

Note that cages of genus 11 have been previously proposed by Terrones and Terrones.[48]

The building unit $C_{24}$, (derived as $Le_{-f}(Le(T))$, with the subscript "$-f$" meaning deletion of all faces of the parent map, at the end of normal *Le* operation, or better as $S_{2-v}(Du(T))$, with $-v$ denoting deletion of all vertices belonging to the dual of the parent map, after $S_2$ - Figure 9a), leads to the supra-dodecahedron $D(S_{2-v}(Du(T)))$ - Figure 9c,d. This has the object $S_{2-v}(Du(D))$ ((Figure 9b) as complement (*i.e.*, core) in the dual labyrinth[49] lattice. The units $S_{2-v}(Du(T))/S_{2-v}(Du(D))$ provide a FRD-type surface[25] (space group *Fd* 3 *m*; unit cell of tetrahedral symmetry). By connecting $S_{2-v}(Du(D))$ with other twelve identical objects results in an FRD-domain, of genus 66 (Figure 9e). It can be closed, by putting back the above deleted pentagons, to obtain the compact array of 13 spherical domains (Figure 9g,h). The above arrays show their core identical to the supra-dodecahedron $D(S_{2-v}(Du(T)))$ (Figure 9c,d). This clearly shows the complementariness of these units (derived by the same operations sequence from tetrahedron and dodecahedron, respectively), interlaced in construction of the FRD-type surface.

(a) $S_{2\text{-}v}(Du(T))$
$V = 24$; $E = 30$; $F = 4$; $g = 2$

(b) Core($D(S_{2\text{-}v}(Du(T)))$) = $S_{2\text{-}v}(Du(D))$
$V = 120$; $E = 150$; $F = 20$; $g = 6$

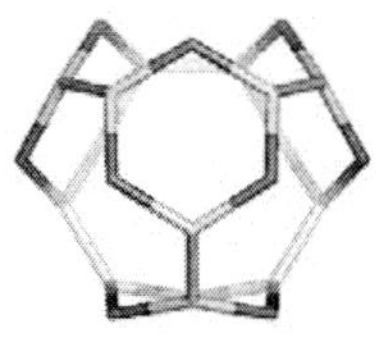

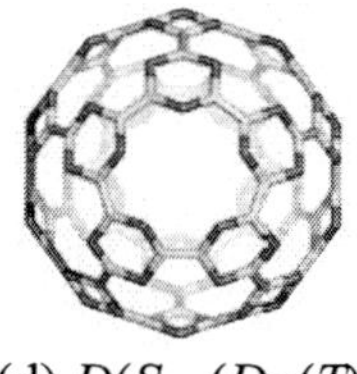

(c) $D(S_{2\text{-}v}(Du(T)))$;
$V = 480$; $E = 690$; $F = 170$; $g = 21$
5-fold axis

(d) $D(S_{2\text{-}v}(Du(T)))$;
$V = 480$; $E = 690$; $F = 170$; $g = 21$
2-fold axis

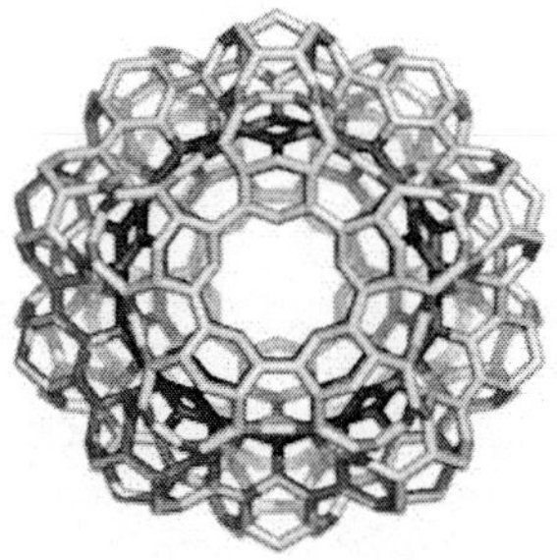

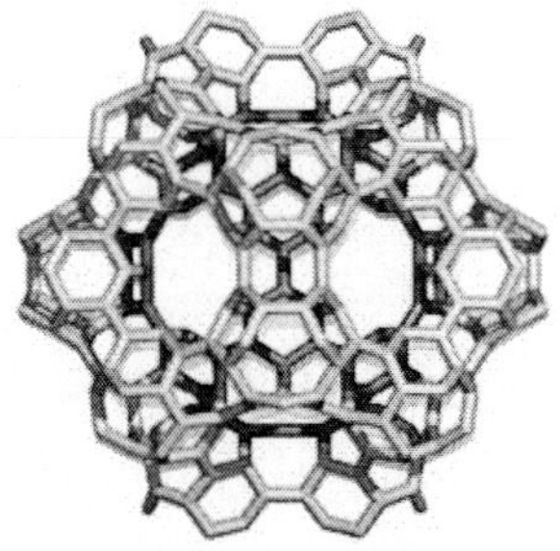

(e) FRD-type surface, $S_{2\text{-}v}(Du(D))$-unit
Infinite; $V = 1560$; $E = 2160$; $F = 470$
$g = 66$; 5-fold axis; $S = 3.72$

(f) Linear lattice of $3 \times D(S_{2\text{-}v}(Du(T)))$
Infinite; $V = 1200$; $E = 1740$; $F = 440$
$g = 51$

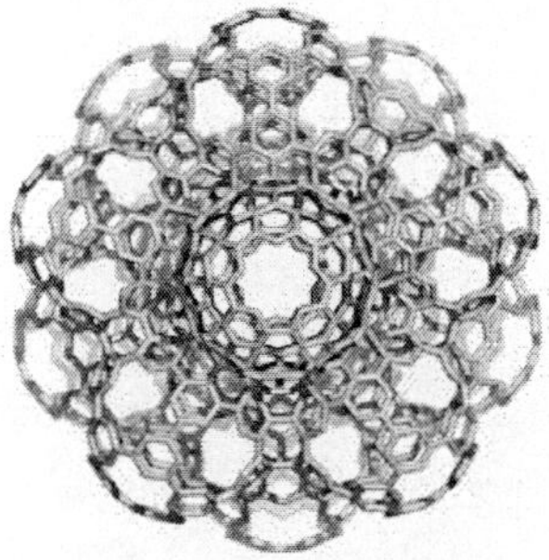

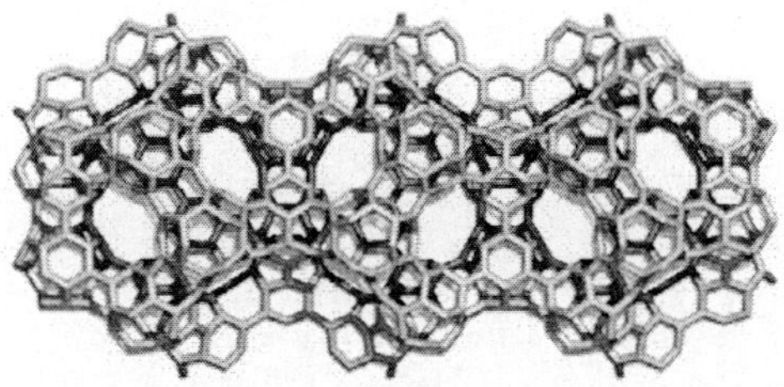

(g) $13 \times (S_{2\text{-}v}(Du(D)))$

(h) $13 \times (S_{2\text{-}v}(Du(D)))$ 3-fold axis

Finite; $V = 1920$; E = 2880; F = 902$g$ = 32 tori - 2 spheres = 30
g = 30; 5-fold axis; $S = 3.73$

Figure 9: A dual labyrinth lattice of the FRD-type surface, built up from either $S_{2\text{-}v}(Du(T))$ or $S_{2\text{-}v}(Du(D))$ units.

The unit $D(S_{2\text{-}v}(Du(T)))$ may also evolve in a linear array, as shown in Figure 9f.

Two nice examples of supra-*D* structure, of Klein (7,3) Platonic tessellation, are given in Figures 10 and 11. The *T*/*D* dual lattice units are as follows: $Op(S_1(T))$/ $Op_{2a}(S_2(D))$ (Figure 10); $Op_{2a}(S_2(T))/Op(S_1(D)$ (Figure 11). The free boundary changes from "zigzag" ($S_1$) to "armchair" ($S_2$) and *vice-versa*. This is just the expected complementarity of the objects derived by the two twin septupling-open operations (see above).

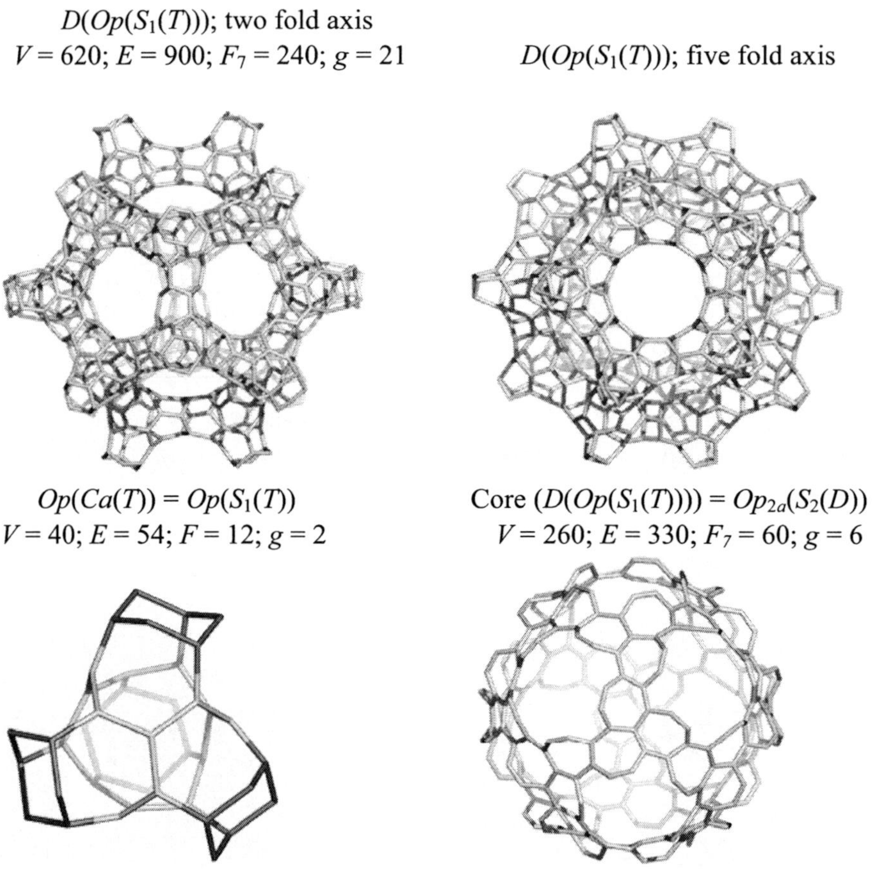


Figure 10: A supra-*D* structure of Platonic (Klein) tessellation and its *T*/*D* dual lattice units; starting process by $S_1$ operation.

The supra-*D* ($S_2$-started) structure (Figure 11) was further transformed in view of obtaining different Platonic or Archimedean tessellations, as shown in Figure 12.

The object transformed by $Le_{22}$, with *T*/*D* dual lattice units $Le_{22}(Op_{2a}(S_2(T)))$/ $Le_{22}(Op(S_1(D)))$, possesses a perfect corannulenic system PCorS (Figures 12a,b). Its covering is an Archimedean ((6,Cor$_7$)3), supplementary information being given in the top of figures.

Next, Capra operation (*i.e.*, $S_1$) provides a Platonic covering, of joint Cor7 supra-covering (Figures 12c,d). The dual lattice units are: $S_{1R}(Op_{2aS}(S_2(T)))$/ $S_{1R}(Op(S_{1S}(D)))$. The same complementarity of the two twin septupling operations is again observed. When the applied operations are *Le* and *Q*, Archimedean ((6,7)3) tessellations could be obtained.

$D(Op_{2a}(S_2(T)))$; two fold axis
$V = 680$; $E = 960$; $F_7 = 240$; $g = 21$

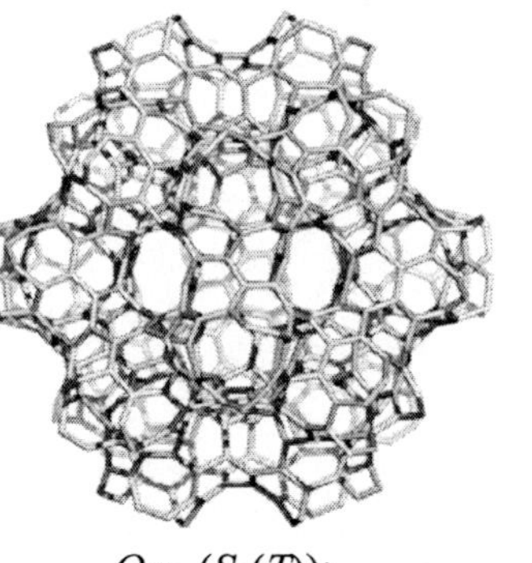

$D(Op_{2a}(S_2(T)))$; five fold axis

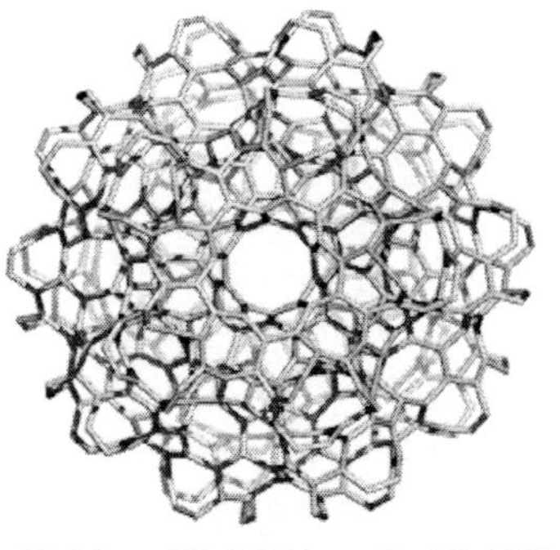

$Op_{2a}(S_2(T))$;
$V = 52$; $E = 66$; $F_7 = 12$; $g = 2$

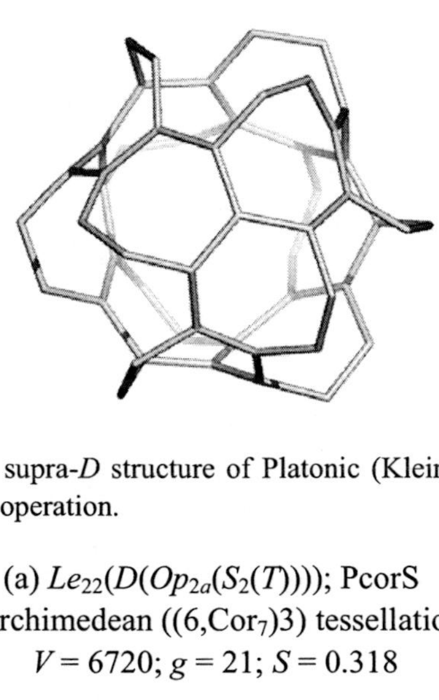

Core$(D(Op_{2a}(S_2(T)))) = Op(S_1(D))$
$V = 200$; $E = 270$; $F_7 = 60$; $g = 6$
$S = 1.574$

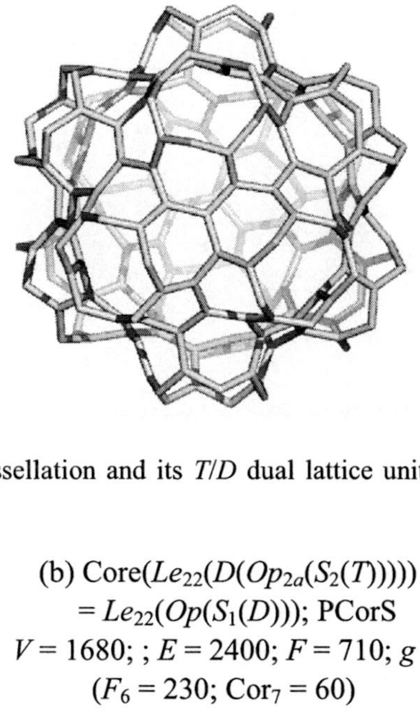

Figure 11: A supra-$D$ structure of Platonic (Klein) tessellation and its $T/D$ dual lattice units; starting process by $S_2$ operation.

(a) $Le_{22}(D(Op_{2a}(S_2(T))))$; PcorS
Archimedean ((6,Cor$_7$)3) tessellation
$V = 6720$; $g = 21$; $S = 0.318$

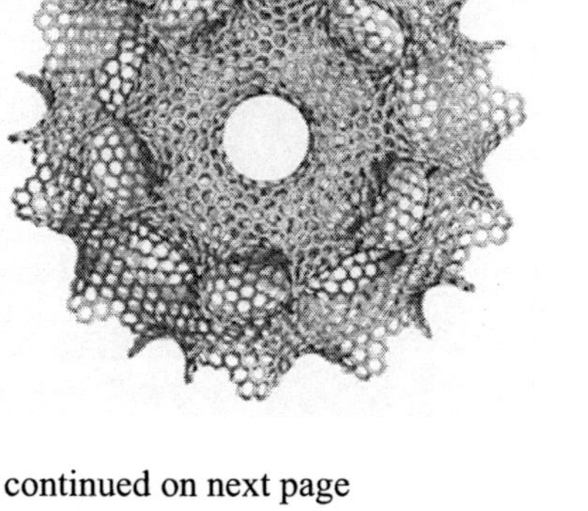

(b) Core$(Le_{22}(D(Op_{2a}(S_2(T)))))$
$= Le_{22}(Op(S_1(D)))$; PCorS
$V = 1680$; ; $E = 2400$; $F = 710$; $g = 6$
($F_6 = 230$; Cor$_7 = 60$)

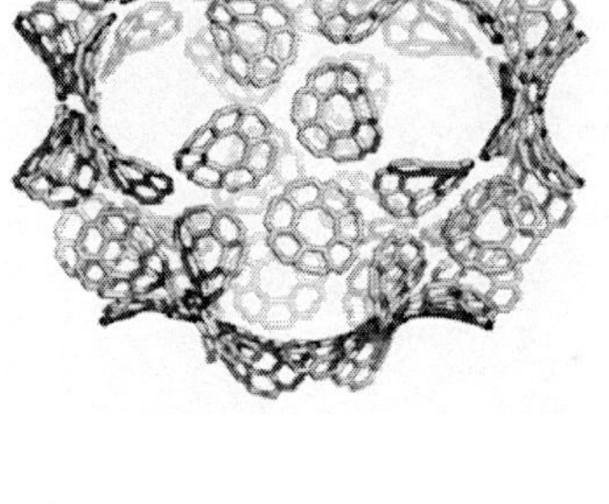

Figure 12 continued on next page

(c) $S_{1R}(D(Op_{2aS}(S_2(T))))$ (d) Core($S_{1R}(D(Op_{2aS}(S_2(T))))$)
Platonic ($Cor_7$,3) tessellation= $S_{1R}(Op(S_{1S}(D)))$
$V = 4280$; $g = 21$ $V = 1160$; $E = 1650$; $F = 480$
($Cor_7 = 60$); $g = 6$; $S = 0.335$

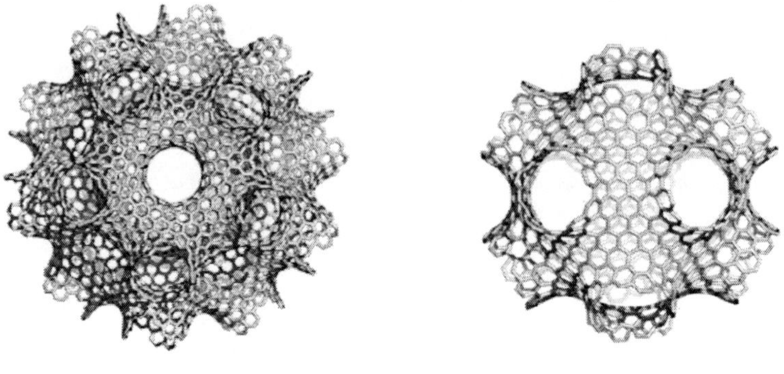

Figure 12: The supra-$D$ ($S_2$-started) structure (Figure 11) transformed by $Le_{22}$ and $S_1$, respectively.

## 6. P-Type Schwarzites

The P-type surface has the space group *Pn* 3 *m*; molecular realization is mainly achieved by the two septupling map operations (Figure 13). Observe the combination of the chiral units, provided by these operations, in obtaining non-twisted (achiral) lattices of high genus. The dual lattice units are easily deduced as: $Op(S_1(C))/Op_{2a}(S_2(C))$, disregarding the chirality. Lattices of this type have been described in refs. [2],[4],[24],[50].

$Op(S_1(C))$ $Op_{2a}(S_2(C))$
$V = 80$; $E = 108$; $F_7 = 24$; $g = 3$ $V = 104$; $E = 132$; $F_7 = 24$; $g = 3$

Lattice of 8 × ($Op(S_{1S}(C))$ *and* $Op(S_{1R}(C))$)Lattice of 8 × ($Op_{2aR}(S_2(C))$ *and* $Op_{2aR}(S_2(C))$)
$V = 544$; $E = 768$; $F = 192$; $g = 17$; $S = 2.127$ $V = 640$; $E = 864$; $F = 192$; $g = 17$;
$S = 0.8567$

Figure 13: Platonic (Klein) (7,3) tessellation of the P-type surface by the two septupling map operations

The sequence $S_i(Op_n(S_i(M)))$ or similar others, may provide variations of the P-surface covering (*e.g.*, those shown in Figure 14). The porosity of such objects is obvious.

$S_{1R}(Op(S_{1S}(C)))$
$V = 320$; $E = 444$; $F = 120$; $g = 3$
$V = 2848$; $E = 4128$; $F = 1248$; $g = 17$

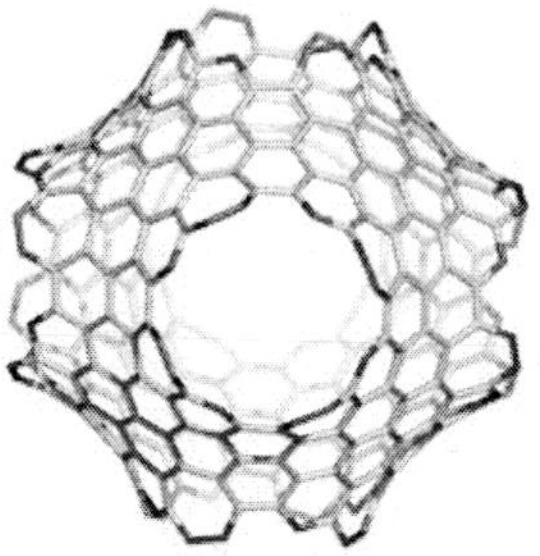

Lattice of $S_{1R}(8\times(Op(S_{1S}(C))))$; top
Platonic ($Cor_7$,3) tessellation

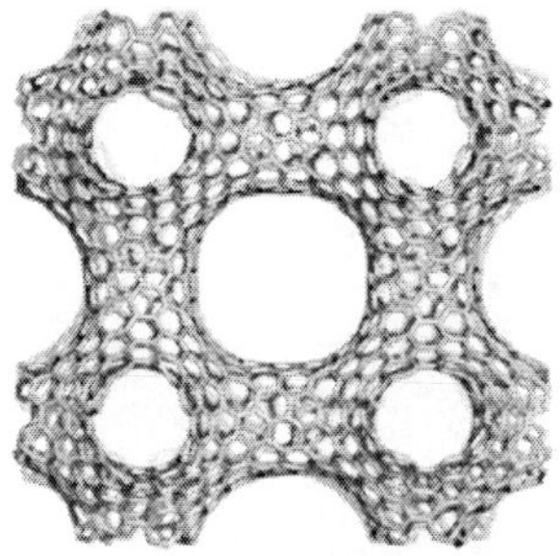

Lattice of $S_{1R}(8\times(Op(S_{1S}(C))))$; side
Platonic ($Cor_7$,3) tessellation
$V= 2848$; $E = 4128$; $F = 1248$; $g = 17$

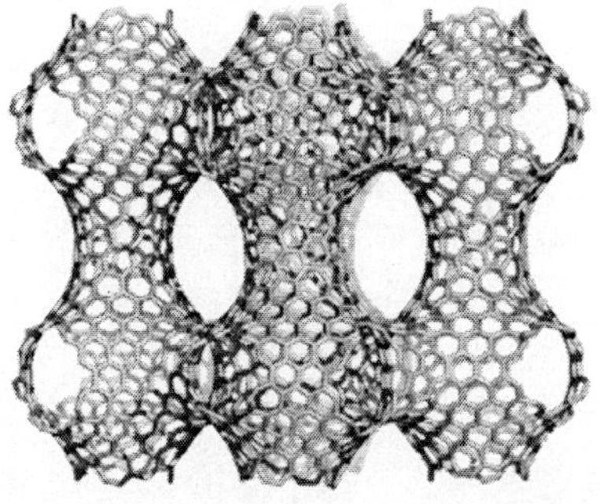

Lattice of $Le_{22}(8\times(Op_{2aR}(S_2(C))))$; *PCorS*
Archimedean ((6,$Cor_7$)3) tessellation
$V = 5376$; $g = 17$

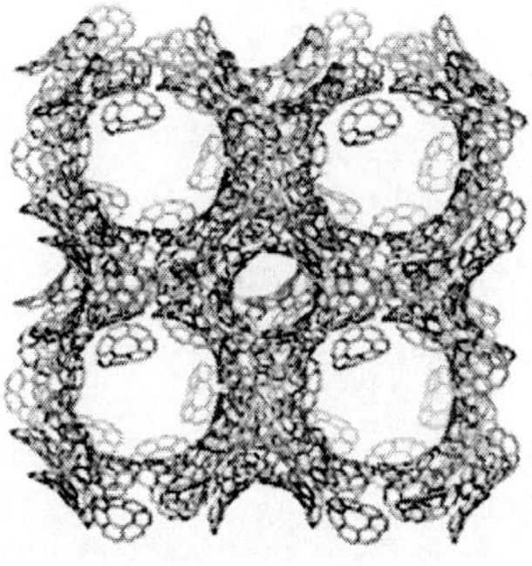

Figure 14: Transformed S1(C) covering of the P-type surface.

Some simpler units, as that in Figure 15a, lead to networks in which the "plumber nightmare" nanotube junctions are evident (Figures 15c,d). The complement/dual unit (Figure 15b) is not always easily found.

(a) $C_{80}$
$V = 80$; $E = 96$; $F =12$; $g = 3$

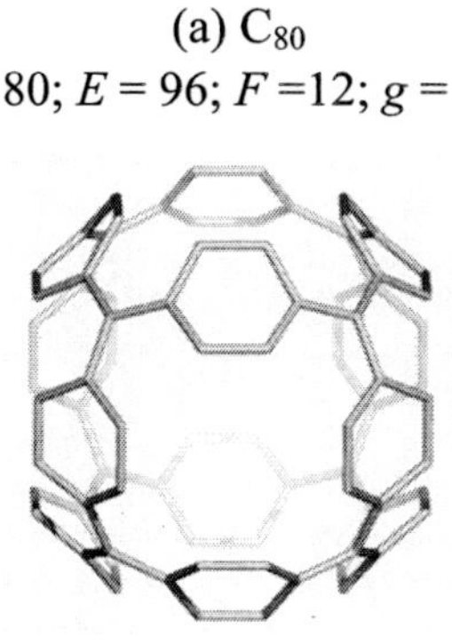

(b) $C_{104}$ (derived from $Op(Q(C))$; (in black))
$V = 104$; $E = 144$; $F = 36$; $g = 3$

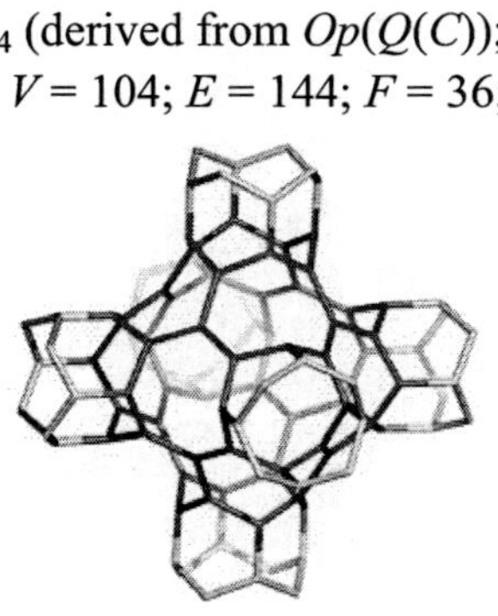

Figure 15 continued on next page

(c) Lattice of 8 × $C_{80}$; top
$V = 544$; $E = 720$; $F = 144$; $g = 17$

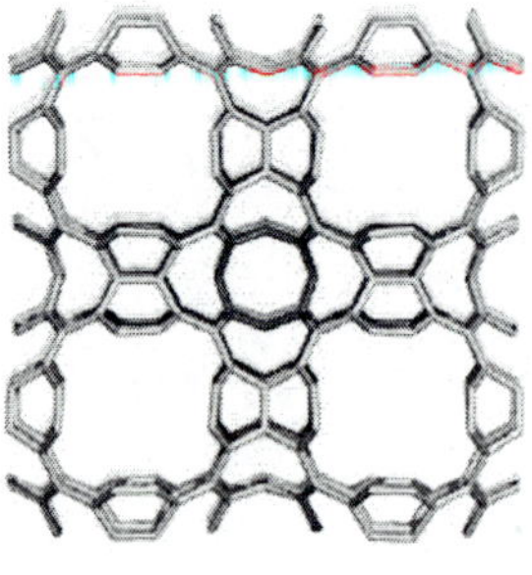

(d) Lattice of 8 × $C_{80}$; side;
$V = 544$; $E = 720$; $F = 144$; $g = 17$

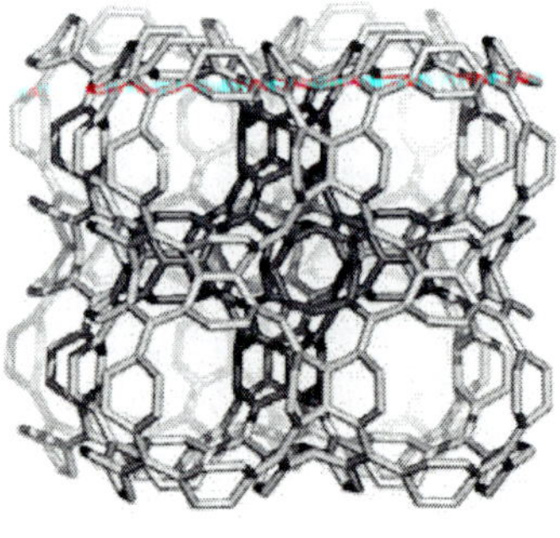

Figure 15: Complementary units and lattice showing a "plumber nightmare" nanotube junction within a P-type surface.

Platonic tessellations are also possible by faces of folding higher than 7: namely 8 (Figure 16) and 9 (Figure 17).

*Op*(*Q*(*C*))
Platonic (Dyck - (8,3)) tessellation
$V = 56$; $E = 72$; $F_8 = 12$; $g = 3$

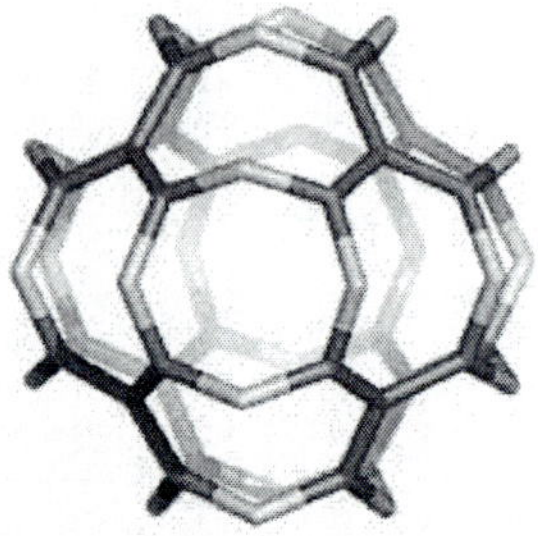

*Op*(*Tr*(*Me*(*C*)))
Core(*Op*(*Q*(*C*)))
$V = 96$; $E = 120$; $F = 14$; $g = 6$

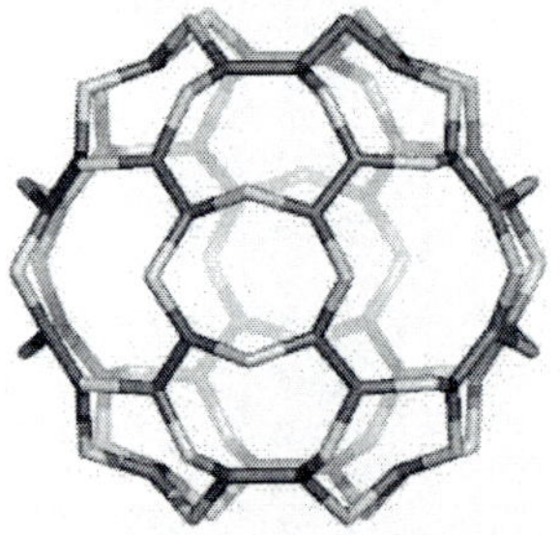

Lattice of 20 × *Op*(*Q*(*C*))
$V = 928$; $E = 1248$; $F_8 = 240$; $g = 41$;
$S = 0.533$

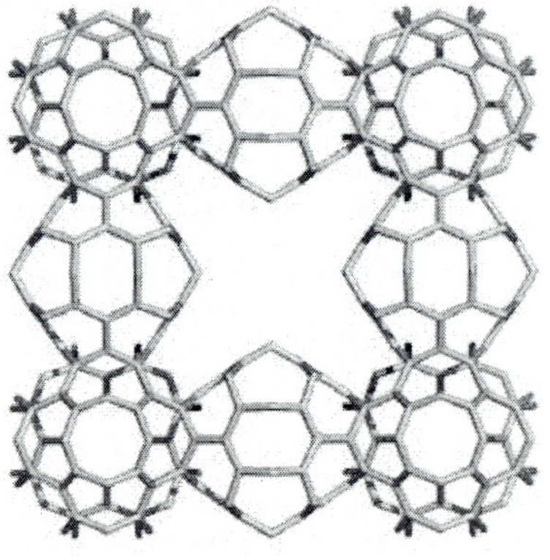

Lattice of (20 × *Op*(*Q*(*C*)) + *Op*(*Tr*(*Me*(*C*))))
$S = 2.392$

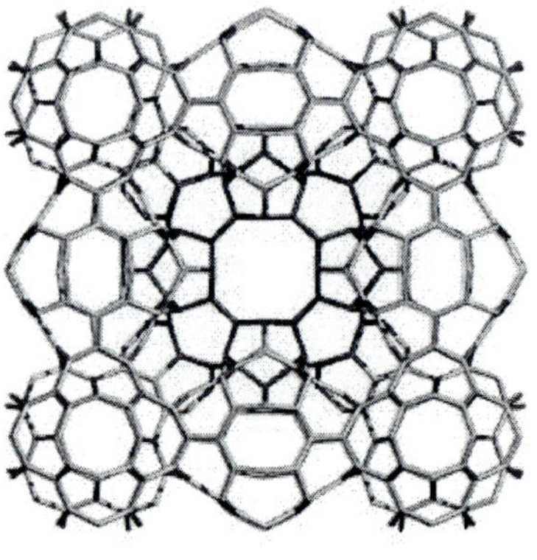

Figure 16: Platonic (Dyck - (8,3)) tessellation of a lattice of the P-type surface units.

| $Op(Le(C))$ | Lattice of $20 \times Op(Le(C))$ |
|---|---|
| $V = 48; E = 60; F_9 = 8; g = 3$ | $V = 768; E = 1008; F_9 = 160; g = 41; S = 1.736$ |

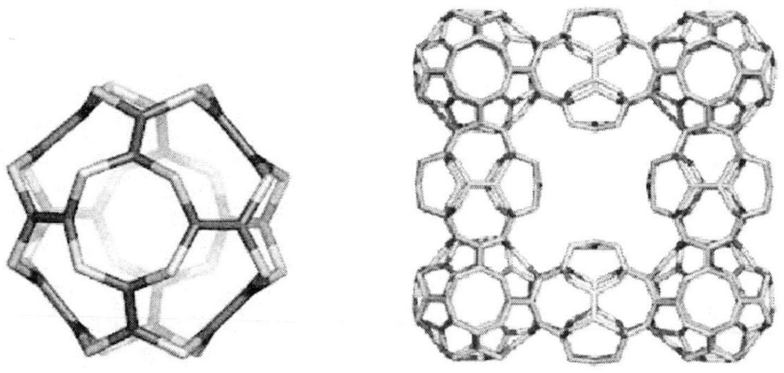

Figure 17: P-type surface decorated with nonagons.

## 7. CONCLUSIONS

The spongy carbon nano-structures, also called schwarzites, consist of highly connected $sp^2$ carbon networks (the cohesion being ensured by covalent bonds), periodic in all the three dimensions of the Euclidean space. The intimate structure of (rather randomly oriented) schwarzites show the topology of triply periodic minimal surfaces TPMSs.

Coordinates of lattices covering the most common D-, P- and FRD-type TPMSs are derivable by means of some composite map operations: leapfrog, quadrupling, septupling, along with the basic dual, medial, truncation, etc. operations. Such networks, of high genus, show pronounced porosity.

POAV1 strain energy calculations for the saddle-shaped structures, particularly for those transformed by some map operations, including thousands atoms, have shown that such structures are very relaxed and approach to the graphite sheet. Data support previous results on curvature elastic energy calculations, indicating a decrease of energy in the series: sphere > cylinder > saddle > flat surface.

Among several proposed synthetic routes of carbon schwarzites, two of them appear essential: one suggests the carbonization of foamed polymers, whereas the other proposes the use of fullerenes as templates. The two routes correspond to the included dual labyrinths of these structures, which could evolve either from a tetrahedral or a spherical (dodecahedral) unit.

## REFERENCES


[1] D. Jones, Ghostly graphite, *Nature*, 1991, 351, 526.

[2] A L. Mackay and H. Terrones, Diamond from graphite, *Nature*, 1991, 352, 762.

[3] T. Lenosky, X. Gonze, M. Teter, and V. Elser, Energetics of negatively curved graphitic carbon, *Nature*, 1992, 355, 333-335.

[4] H. Terrones and A. L. Mackay, From $C_{60}$ to negatively curved graphite, *Prog. Crystal Growth and Charact.*, 1997, 34, 25-36.

[5] S. J. Townsend, T. J. Lenosky, D. A. Muller, C. S. Nichols, V. V. Elser, Negatively curved graphitic sheet model of amorphous carbon, *Phys. Rev. Lett.*, 1992, 69, 921-924.

[6] H. Terrones and A. L. Mackay, Triply periodic minimal surfaces decorated with curved graphite, *Chem. Phys. Lett.*, 1993, 207, 45-50.

[7] L. N. Bourgeois and L. A. Bursill, High-resolution transmission electron microscopic study of a nanoporous carbon consisting of curved single graphitic sheets, *Phil. Mag. A*, 1997, 76, 753-768.

[8] H. A. Schwarz, Über Minimalflächen, *Monatsber. Berlin Akad.*, 1865.

[9] H. A. Schwarz, *Gesammelte Matematische Abhandlungen*, Springer, Berlin, 1890.

[10] D. Vanderbilt and J. Tersoff, Negative-curvature fullerene analog of $C_{60}$, *Phys. Rev. Lett.*, 1992, 68, 511-513.

[11] V. Rosato, M. Celino, G. Benedek, and S. Gaito, Thermodynamic behavior of the carbon schwarzite *fcc* $(C_{36})_2$, *Phys. Rev. B*, 1999, 60, 16928-16933.

[12] G. Benedek, H. Vahedi-Tafreshi, E. Barborini, P. Piseri, P. Milani, C. Ducati, and J. Robertson, The structure of negatively curved spongy carbon, *Diamond Relat. Mater.*, 2003, 12, 768–773.

[13] A A Zaakhidov; R. H. Baughman; Z. Iqbal; Ch. Cui; I. Khayrullin, S. O. Santas, J. Marti, and V. G. Ralchenko, Carbon structures with three-dimensional periodicity at optical wavelengths, *Science*, 1998; 282, 5390.

[14] E. Barborini, P. Piseri, P. Milani, G. Benedek, C. Ducati, and J. Robertson, Negatively curved spongy carbon, *Appl. Phys. Lett.*, 2002, 81, 3359-3361.

[15] Bonnet, *C. R. Acad. Sci. Paris*, 1853, 37, 529-532.

[16] D. J. Klein, Elemental benzenoids, *J. Chem. Inf. Comput. Sci.*, 1994, 34, 453-459.

[17] L. Euler, Elementa doctrinae solidorum, *Novi Comment. Acad. Sci. I. Petropolitanae*, 1758, 4, 109-160.

[18] L. Euler, Solutio Problematis ad Geometriam Situs Pertinentis. *Comment. Acad. Sci. I. Petropolitanae*, 1736, 8, 128-140.

[19] D. J. Klein and H. Zhu, All conjugated carbon species, in: A. T. Balaban (Ed.), *From chemical topology to three-dimensional geometry*, Plenum Press, 1997, 297-341.

[20] E. R. Neovius, Bestimmung zweier speziellen Minimalflächen, *Akad. Abhandlungen*, Helsingfors 1883.

[21] A H. Schoen, Infinite periodic minimal surfaces without self-intersections, NASA Technical Report TN D-5541, Washington DC 1970.

[22] H. S. M. Coxeter, Regular polytopes, 3rd Edition, Dover Pubns, Dover, 1973.

[23] W. Fischer and E. Koch, Spanning minimal surfaces, *Phil. Trans. R. Soc. Lond. A*, 1996, 354, 2105-2142.

[24] M. O'Keeffe, G. B. Adams and O. F. Sankey, Predicted new low energy forms of carbon, *Phys. Rev. Lett.*, 1992, 68, 2325-2328.

[25] H. Karcher and K. Polthier, Construction of triply periodic minimal surfaces, *Phil. Trans. R. Soc. Lond. A*, 1996, 354, 2077-2104.

[26] D. Hoffman, A new turn for Archimedes, *Nature*, 1996, 384, 28-28.

[27] R. Osserman, A survey of minimal surfaces, Dover, New York, 1986.

[28] T. Pisanski and M. Randić, Bridges between geometry and graph theory. in: *Geometry at Work*: A Collection of papers showing applications of geometry (Ed., C. A. Gorini), *Math. Assoc. Amer.*, 2000, 53, 174-194.
[29] Diudea, M. V.; John, P. E.; Graovac,A.; Primorac, M.; Pisanski, T. Leapfrog and related operations on toroidal fullerenes, *Croat. Chem. Acta*, 2003, 76, 153-159.
[30] Diudea, M. V., Capra-a leapfrog related operation on maps, *Studia Univ. "Babes-Bolyai"*, 2003, 48 (2), 3-16.
[31] M. V. Diudea, Covering forms of nanostructures, *Forma*, 2004 (submitted).
[32] M. Goldberg, A class of multi-symmetric polyhedra, *Tôhoku Math. J.*, 1937, 43, 104-108.
[33] F. Valencia, A. H. Romero, E. Hernàndez, M. Terrones and H. Terrones, Theoretical characterization of several models of nanoporous carbon, *New J. Phys.,* 2003, 5, 123.1-123.16.
[34] H. Terrones and M. Terrones, Curved nanostructured materials, *New J. Phys.,* 2003, 5, 126.1–126.37.
[35] R. B. King, Novel highly symmetrical trivalent graphs which lead to negative curvature carbon and boron nitride chemical structures, *Discrete Mathematics*, 2002, 244, 203-210.
[36] R. B. King, Riemann surfaces as descriptors for symmetrical negative curvature carbon and boron nitride structures, *Croat. Chem. Acta*, 2002, 75, 447-473.
[37] R. B. King, Chemical applications of topology and group theory. 29. Low density polymeric carbon allotropes based on negative curvature structures, *J. Phys. Chem.*, 1996, 100, 15096-15104.
[38] R. B. King, Carbon networks on cubic infinite periodic minimal surfaces, *Discrete Mathematics*, 2000, 51, 235-248.
[39] R. B. King, Negative curvature surfaces in chemical structures, *J. Chem. Inf. Comput. Sci.*, 1998, 38, 180-188.
[40] A Ceulemans, R. B. King, S. A. Bovin, K. M. Rogers, A. Troisi, and P. W. Fowler, The heptakisoctahedral group and its relevance to carbon allotropes with negative curvature, *J. Math. Chem.*, 1999, 26, 101-123.
[41] M. V. Diudea, Covering Nanostructures, in: M. V. Diudea, Ed., nanostructures-Novel Architecture, NOVA, New York, 2004 (submitted).
[42] A Ceulemans, E. Lijnen, L. J. Ceulemans, and P. W. Fowler, The tetrakisoctahedral group of the Dyck graph and its molecular realization, *Mol. Phys.*, 2004, 102, 1149.
[43] E. Lijnen and A. Ceulemans, The Symmetry of the Dyck graph: group structure and molecular realization, in: M. V. Diudea, Ed., nanostructures-Novel Architecture, NOVA, New York, 2004 (submitted).
[44] F. Klein, *Vorlesungen über das Ikosaeder*, Teubner, Leipzig, 1884: Part I, Chapter II.
[45] F. Klein, *Gesammelten Mathematischen Abhandlungen*, Springer Verlag, Berlin, 1923, Vol. 3, pp. 90-136.
[46] R. C. Haddon, Rehybridization and π-orbital overlap in nonplanar conjugated organic molecules: π-orbital axis vector (POAV) analysis and three-dimensional Hűckel Molecular Orbital (3D-HMO) Theory, *J. Am. Chem. SOC.* 1987, 109, 1676-1685.

[47] R. C. Haddon, Measure of Nonplanarity in Conjugated Organic Molecules: Which Structurally Characterized Molecule Displays the Highest Degree of Pyramidalization? *J. Am. Chem. Soc.* 1990, 112, 3385- 3389.
[48] H. Terrones and M. Terrones, Fullerenes and nanotubes with non-positive Gaussian curvature, *Carbon*, 1998, 36, 725-730.
[49] S. T. Hyde, Sponges, Foams and Emulsions, in: J. F. Sadoc, N. Rivier (Eds.), *Foams and Emulsions*, NATO ASI Series E: Applied Sciences, vol.354, Kluwer, Dordrecht, 1999, pp. 437–469.
[50] Cs. L. Nagy and M. V. Diudea, Carbon allotropes with negative curvature, *Studia Univ. "Babes-Bolyai"*, 2003, 48 (2), 37-46.

In: Nanostructures: Novel Architecture
Editor: Mircea V. Diudea, pp. 335-344
ISBN: 1-59454-499-9


*Chapter 16*

# CUBANOID AMPHIPHILES AS BUILDING BLOCKS FOR THE DEVELOPMENT OF FUNCTIONAL NANOSTRUCTURED MATERIALS

*Fabio Pichierri*[1]
COE Laboratory, Tohoku University
IMRAM, 2-1-1 Katahira, 980-8577 Sendai, Japan

## ABSTRACT

Amphiphiles made of long alkyl chains attached to a cubanoid core have been designed with the aid of both computer graphics and molecular mechanics calculations. These novel molecules are expected to self-assemble in water solution in a way analogous to that of lipid bilayers. The resulting superstructures display interesting architectures which can be stabilized by further polymerization reactions. The encapsulation of electro-active molecules like, for example, fullerenes paves the way towards the formation of nanostructured functional materials that have the ability to mimic complex biological processes such as the transduction of signals through cell membranes.

## 1. INTRODUCTION

Up to now novel materials have been developed by processing raw matter. Let us think, for example, about alloys such as steel and brass. According to the on-line HyperDictionary,[1] an alloy is defined as "a mixture containing two or more metallic elements or metallic and non-metallic elements usually fused together or dissolving into each other when molten". Steel is made of iron and small amounts of carbon while brass is a mixture of copper and zinc. These binary combinations of elements possess their characteristic phase diagrams whereby composition is plotted versus temperature. By varying these macroscopic parameters, it is possible to produce different materials each characterized by certain mechanical and electrical properties.[2] When observed under the scanning

---

[1] E-mail address: fabio@tagen.tohoku.ac.jp

electron microscope, modern alloys display a microstructure made of grains the size of 30-50 μm. It is clear that in order to reduce the size of these building blocks, it is necessary to adopt a bottom-up approach whereby far smaller material entities (*e.g.* molecules, nanoclusters, etc.) are assembled so as to yield objects 10-100 nm in size. Furthermore, with this bottom-up approach it becomes possible to exert substantial control over the architecture of the material thus produced, a task that is not normally possible with traditional manufacturing methods.

Chemists can contribute greatly to the advancement of nanotechnology since they have so far synthesized a huge number of complex molecules, both organic and inorganic, that can be tailored to meet specific needs. Hence, giant molecules represent the ideal building blocks for the bottom-up assembly of nanostructured materials.[3],[4] The latter are expected to be characterized by novel and unique electronic and mechanical properties which cannot be seen in traditional materials. Colloidal nanocrystals containing either metallic (*e.g.* Au) or semiconducting (*e.g.* CdSe) nanoclusters covered with hydrophobic layers have been realized by several research groups and possible applications in biology have been discussed.[5] As far as non-metallic or organic-based materials (physicists prefer to use the term *soft-matter*) are concerned, carbon nanotubes CNT seem very promising candidates for the creation of novel nanostructured functional materials.[6] As an interesting example of application in nanotechnology it is worth to mention the use of CNT as field-emitters for electronic devices.[7]

A serious problem encountered with the chemical manipulation of CNT concerns their poor solubility in common organic solvents. One possible strategy to overcome such obstacle is by means of their chemical functionalization.[8] Note, however, that unlike organic molecules CNT do not possess hydrogen atoms and, hence, their functionalization is actually an addition of chemical groups. Another possible strategy to follow is that of searching for novel hydrocarbon-based materials which can be functionalized by replacing their hydrogen atoms with different chemical groups. This second approach requires extensive studies of computer-aided molecular design (CAMD)[9] with which is possible to assess whether a new molecule or material is thermodynamically and structurally stable. Modern CAMD approaches integrate computational methods of atomistic simulation based on both classical and quantum mechanics along with methods of multiscale modeling.[10] Classical mechanics methods are known in chemistry as molecular mechanics (MM) and molecular dynamics (MD), both of which are based on force-field parameters and potential energy functions. These methods do not include electrons and they represent the computational equivalent of the well known ball and stick models. On the other hand, quantum mechanics methods solve numerically the Schrödinger equation in order to determine the electronic structure of bulk materials and molecules. By employing such CAMD methods we have recently designed a series of novel carbonaceous molecules dubbed *cubanoids*.[11] These are giant molecules made of cubane units (Figure 1) that are joined through covalent bonds as well as through methylene and sulfur bridges. The possibility of using functionalized cubanoids as building blocks in the synthesis of nanostructured functional materials is the subject of the present contribution.

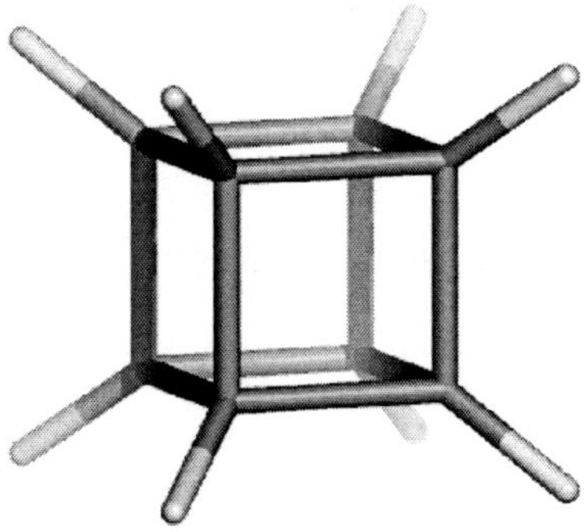

Figure 1: Cubane.

## 2. From Cubane to Cubanoids

The cubane molecule, $C_8H_8$, is well-known amongst chemists owing to its aesthetically pleasing cube-like structure shown in Figure 1. It was synthesized by Eaton and Cole[12] in 1964 and subsequently structurally characterized by Fleisher.[13] Although cubane is a highly strained molecule, Eaton has suggested using it as a precursor for the development of novel materials.[14] So far, several cubane derivatives containing different chemical groups attached to the $C_8$ skeleton have been synthesized and characterized.

Among them, cubyl-cubane is an interesting precursor of one-dimensional cubane polymers. The replacement of all eight hydrogen atoms of cubane with cubyl groups yields the percubyl-cubane (PCC) molecule. PCC has not been synthesized yet but its molecular and electronic structures have been investigated with the aid of molecular orbital calculations.[15] Accordingly, these calculations suggested for PCC a structure of $O$ symmetry with dihedral angles at 15 deg. To the best of our knowledge, no other compact hydrocarbon made of cubane units is yet known. This stimulated us to undertake the design of a series of novel, compact hydrocarbons made of $C_8$ units connected to one another through either covalent bonds or bridging atoms. We dubbed them cubanoids.[11]

The optimized molecular structures of the six cubanoids herein discussed (**1-6**) are shown in Figure 2. Cubanoids **1-3** have a $C_8$ unit at their core whereas cubanoids **4-6** not. Cubanoids **1** and **4** have their outer $C_8$ units joined through single covalent C-C bonds whereas the other four possess either $CH_2$ or S bridging moieties. Semiempirical MO calculations indicate that the strain of the former group is considerably larger than that of the latter group. Furthermore, no imaginary vibrational frequency has been obtained thus indicating that each structure corresponds to a minimum on the corresponding potential energy hypersurface. Each of them possesses a higher molecular symmetry with respect to that of PCC. Further details on these molecules can be found in reference 11.

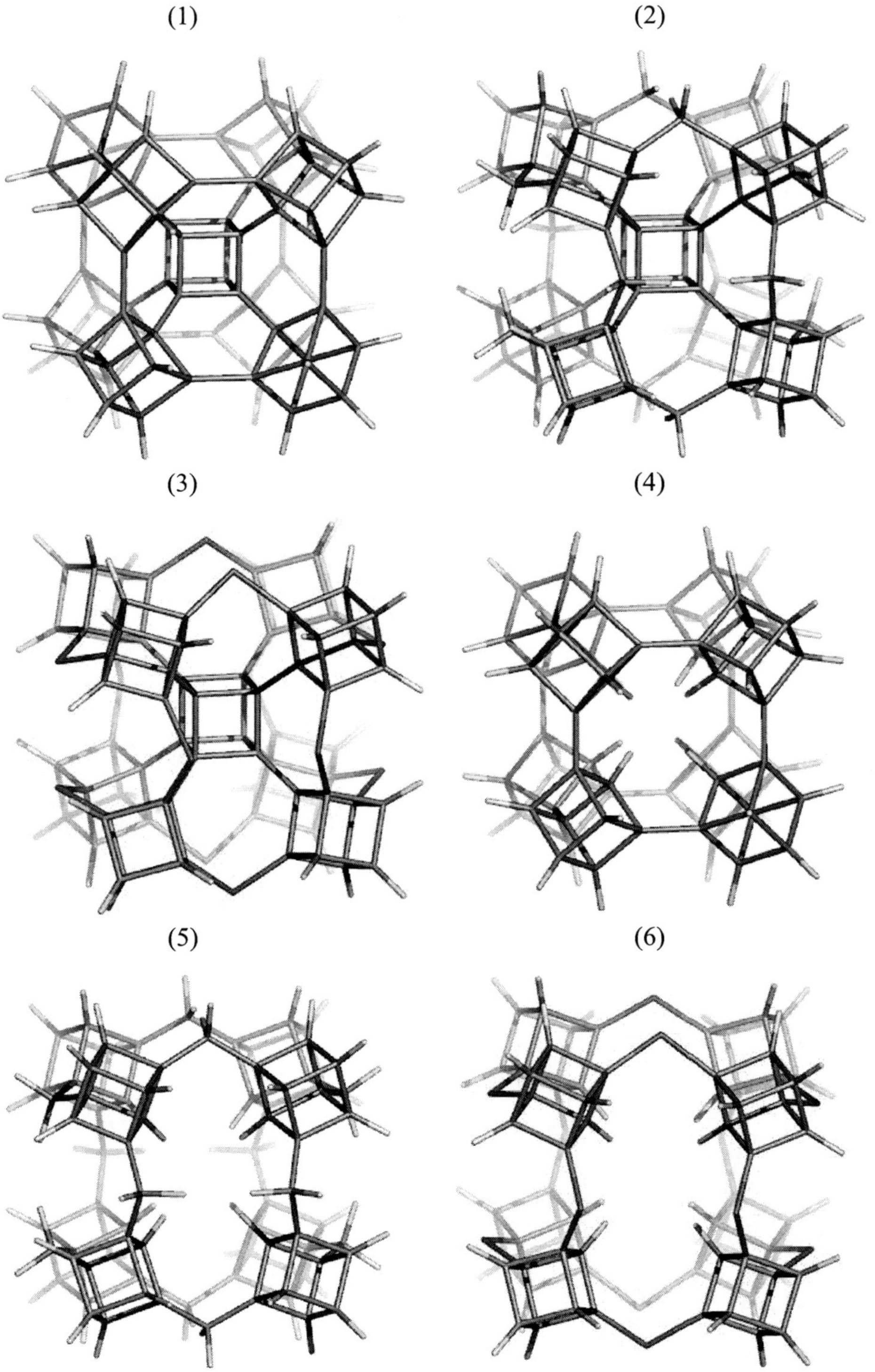


Figure 2: Optimized structures of the six cubanoids.

## 3. From Cubanoids to Functional Nanostructured Materials

A very promising approach for producing nanostructured materials is that based on polymerizable amphiphiles.[16-18] Amphiphiles are molecules that are both lipophilic (or hydrophobic) and hydrophilic (or lipophobic). Typical example of an amphiphile molecule is sodium dodecyl-sulfonate which is commonly used as surfactant for detergents. This molecule is made of a long hydrocarbon chain linked on one end to a negatively charged sulfonate group. The latter group is stabilized in water solution by cations such as $Na^+$ which engage it through ion-ion interactions. Micelles, vescicles, microtubules, Langmuir-Blodgett (LB) layers, and other soft nanostructures can be formed upon self-assembly of amphiphiles in water. These structures, however, can be easily disrupted since their shapes are maintained by non-covalent interactions (*i.e.,* hydrophobic forces, ion-ion and ion-dipole interactions, etc.). However, if the amphiphiles carry reactive chemical groups, polymerization can take place thereby stabilizing the resulting 3D architecture.

On the basis of the concepts highlighted above, we have investigated the hypothetical structures and architectural motifs that might be originated from cubanoid amphiphiles. For this purpose, we have chosen the cubanoid molecule **2** out of the six cubanoids shown in Figure 2. The reason for such a choice is twofold. First, comparison amongst the calculated heats of formation (HOF) of cubanoids **1-6** indicates as the strain of cubanoid **2** is lower with respect to the strains to which the other cubanoids are subject to. Second, cubanoid **2** is highly hydrogenated thus providing a larger number of attachment points for its chemical functionalization with alkyl chains.

Figure 3 shows the molecular structure of cubanoid **2** functionalized with eight alkyl chains each made of ten carbon atoms. The molecular geometry of this molecule has been fully minimized with MM calculations based on the MMX force field implemented within the Hyperchem 6.0 software package.[19] The attachment of the eight alkyl chains is such that the whole shape of the molecule resembles that of the letter X. This giant molecule has linear dimensions of 18.6 × 7.9 × 28.9 Å.

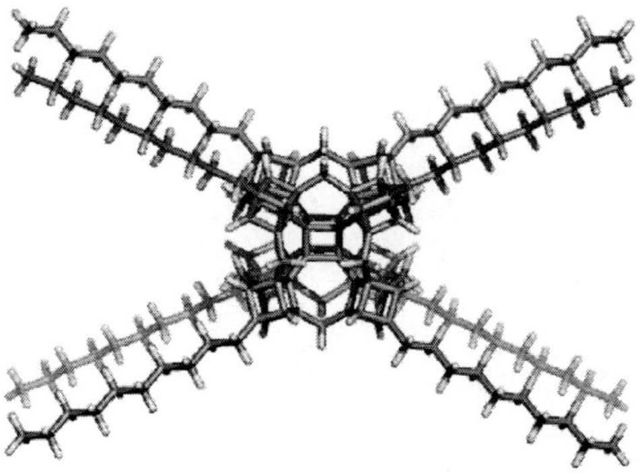

Figure 3: The X-shaped cubanoid amphiphile.

Whenever charged groups, such as sulphates or phosphates, are present at the terminals of each alkyl chain, the resulting molecule could be employed in the self-assembly of micelles and LB layers. The main difference with conventional LB films, however, resides in the rigid cubanoid core which is expected to confer additional structural stability to the resulting nanostructure. Also, the presence of voids among neighboring units would allow the insertion of specific molecules so as to create a functional LB film that can be addressed either photochemically or electrochemically. For example, $C_{60}$ is well-known for its ability to store up to twelve electrons into its lowest-unoccupied molecular orbital (LUMO) levels thereby originating multiply-charged anions whose thermodynamic stability is computed being far greater than that of the corresponding cations.[20] Thence, artificial membranes provided with the necessary room to incorporate it might be charged-up with electrons so as to form arrays of organic quantum dots, as schematically shown in Figure 4.

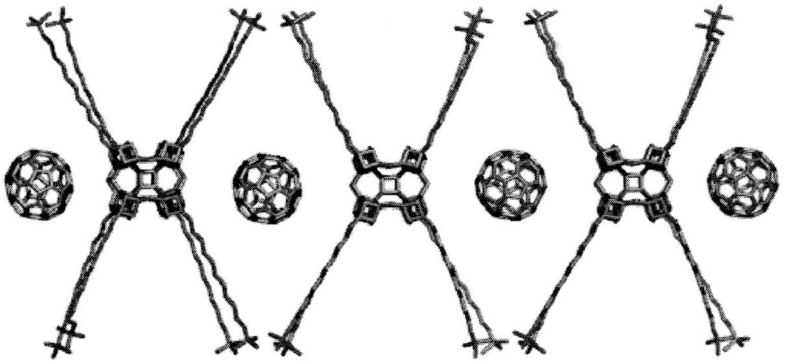

Figure 4: A hypothetical artificial membrane incorporating $C_{60}$ molecules so as to mimic a quantum dot array.

An alternative approach is that of inserting amphiphilic adducts of $C_{60}$ into lipid bilayers, as recently shown by Hirsh and coworkers.[21] As mentioned in the introduction, the presence of reactive chemical groups on the amphiphiles allows polymerization to take place thereby cementing the soft architectures that are based on hydrophobic forces.[16] Without considering in too much detail the chemistry of such polymerization reactions, we have constructed two dimers of cubanoid amphiphile, both linked through oxygen atom (ether) bridges. The MM optimized structures of the two dimers are shown in Figure 5.

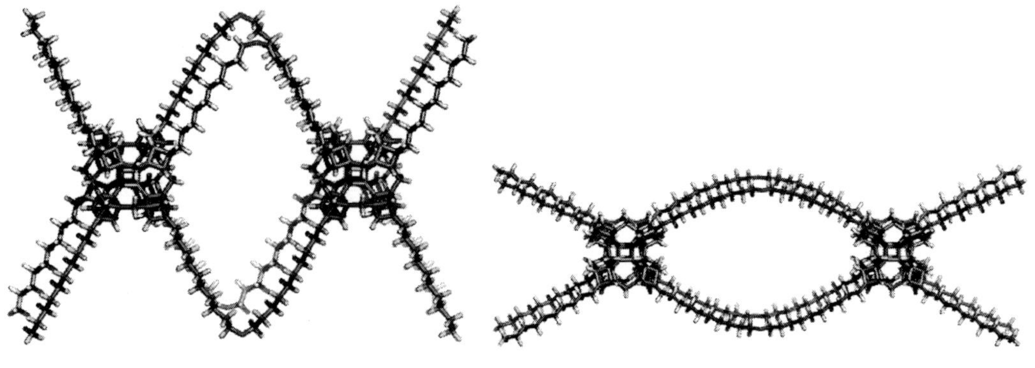

Figure 5: Two O-linked dimers of cubanoid amphiphile.

The separation between cubanoid units is at 14.5 Å in the first dimer and at 24.1 Å in the second one. The second dimer is about 6 nm long. It is clear that these dimers can be further elongated to yield polymers. Their mechanical properties should be quite unique and rather different from those characteristic of, for example, polyethylene or polypropylene. The addition of other four alkyl chains to the X-shaped cubanoid amphiphile (Figure 3) yields the star-shaped cubanoid shown in Figure 6. Interestingly, two pairs of lateral alkyl chains appear to be slightly wrapped to each other as a result of hydrophobic interactions.

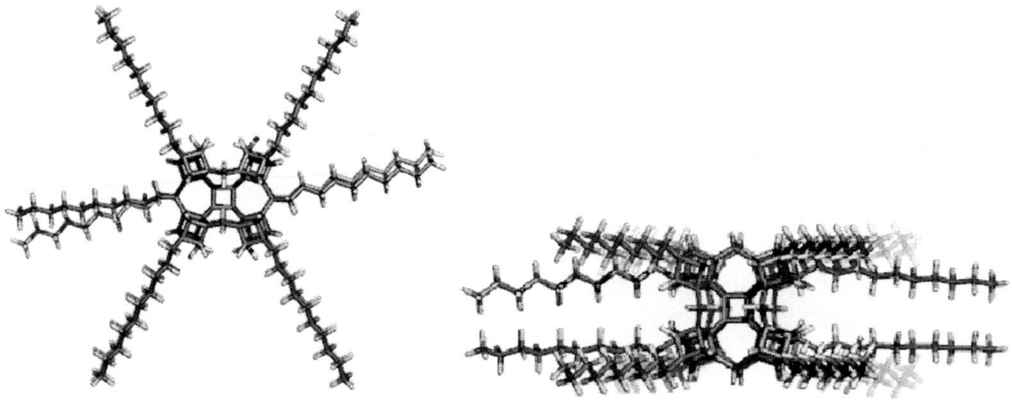

Figure 6: Top and lateral views of the star-shaped cubanoid amphiphile.

Upon self-assembly of star-shaped cubanoid amphiphiles, it should be possible to observe the mosaic-type super-structure shown in Figure 7. Once polymerized, this structure is expected to conserve its constant thickness and, hence, being used, for example, as a nano-platform onto which functional molecules can be attached. Furthermore, it would be interesting to check whether some elements of this 2D superstructure could be employed to build a nano-gear.

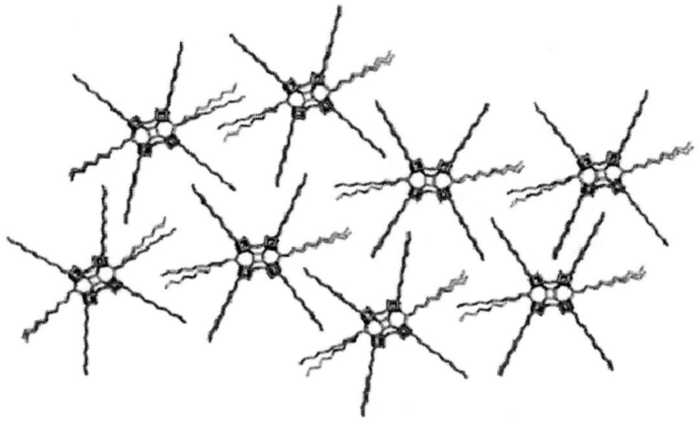

Figure 7: Mosaic-type superstructure resulting from the self-assembly of the star-shaped cubanoid amphiphile.

When other four alkyl chains are added to the star-shaped cubanoid amphiphile (Figure 6), the resulting molecule displays the multi-branched structure shown in Figure 8.

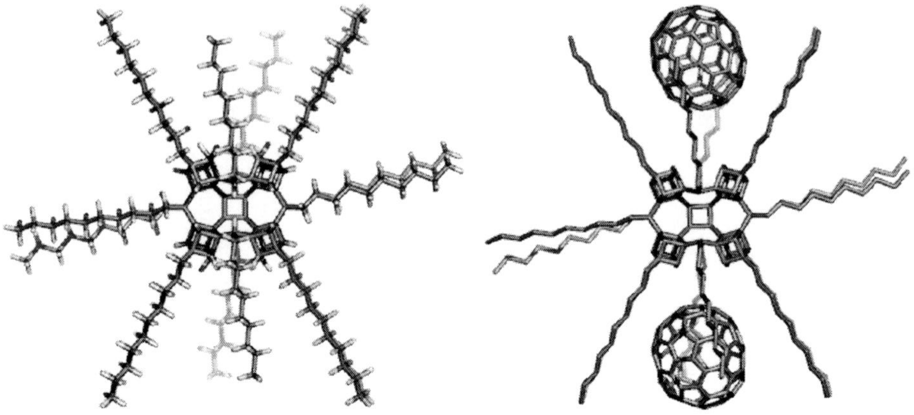

Figure 8: Multi-branched cubanoid amphiphile (left) and the corresponding supramolecular complex with two $C_{70}$ molecules (right).

We notice that twelve alkyl chains, six above and six below the mean molecular plane bisecting the molecule in two halves, do form a sort of chalice-like cavity which might be useful to host cage-like molecules such as $C_{60}$ or $C_{70}$. The supramolecular complex of this multi-branched cubanoid amphiphile hosting a pair of $C_{70}$ molecules is shown on the right side of Figure 6. Whether this complex could ever be formed is yet unknown. However, its synthesis might open up the possibility of realizing an artificial cell membrane whereby molecular recognition on one side gives rise to the transduction of signals to the opposite side, in analogy to the signal transduction events that occur in the cell membranes of living organisms. This artificial membrane could be a precursor for the assembly of artificial cells or being used as a signal-transmission element of nano-sensors.

## 4. Conclusions and Future Prospects

We have shown that the recently designed series of molecules dubbed cubanoids can be functionalized with long alkyl chains to yield amphiphiles. These novel molecules, like the amphiphiles investigated so far, would have the ability to self assemble thereby forming superstructures with interesting nano-architectures. Subsequent polymerization reactions would further stabilize them. The encapsulation of functional molecules such as fullerenes paves the way to the realization of functional nanostructured materials which can mimic complex biological processes such as the transduction of signals through a cell membrane. We wish to conclude with the following words: scientists working in the fields of nanoscience and nanotechnology need stimulating ideas that may drive their work towards fascinating discoveries; both molecular design and computer simulation represent useful means for conveying such ideas.

## ACKNOWLEDGEMENT

The author is thankful to Fujitsu Ltd. (Tokyo) for supplying the MOPAC2002 software package. Financial support for this research was provided by the 21st century Center of Excellence (COE) program "Giant Molecules and Complex Systems" of MEXT activated at Tohoku University.


## REFERENCES


[1] The HyperDictionary is available on-line at http://www.hyperdictionary.com

[2] D.A. Young, *Phase Diagrams of the Elements*, University of California Press, Berkley, 1991.

[3] J. Ying, *Nanostructured Materials*, Academic Press, San Diego, 2001.

[4] H. S. Nalwa, *Nanostructured Materials and Nanotechnology,* Academic Press, San Diego, 2001.

[5] W.J. Parak, D. Gerion, T. Pellegrino, D. Zanchet, C. Micheel, S.C. Williams, R. Boudreau, M.A. Le Gros, C.A. Larabell, A. P. Alivisatos, Biological applications of colloidal nanocrystals, *Nanotechnology*, 2003, 14, R15-R27.

[6] A G. Mamalis, L.O.G. Vogtländer, A. Markopoulos, Nanotechnology and nanostructured materials: trends in carbon nanotubes, *Precision Eng.*, 2004, 28, 16-30.

[7] W. I. Milne, K.B.K. Teo, G.A.J. Amaratunga, P. Legagneux, L. Gangloff, J.-P. Schnell, V. Semet, V. Thien Binh, O. Groening, Carbon nanotubes as field emission sources, *J. Mater. Chem.*, 2004, 14, 933-943.

[8] S.B. Sinnott, Chemical functionalization of carbon nanotubes, *J. Nanosci. Nanotech.*, 2002, 2, 113-123.

[9] J.-P. Doucet, J. Weber, *Computer-Aided Molecular Design*, Academic Press, London, 1996.

[10] N.M. Ghoniem, E.P. Busso, N. Kioussis, H. Huang, Multiscale modeling of nanomechanics and micromechanics: an overview, *Philos. Mag.*, 2003, 83, 3475-3528.

[11] F. Pichierri, Cubanoids: Computer-Aided Molecular Design of Compact Hydrocarbons Constructed by Assembling Cubane Units, *Internet Electron. J. Mol. Des.*, 2004, 3, 134–142.

[12] P. E. Eaton and T. W. Cole Jr., The cubane system, *J. Am. Chem. Soc.*, 1964, 86, 962–964; P.E. Eaton and T.W. Cole Jr., Cubane, *J. Am. Chem. Soc.*, 1964, 86, 3157–3158.

[13] E. B. Fleischer, X–ray structure determination of cubane, *J. Am. Chem. Soc.*, 1964, 86, 3889–3890.

[14] P. E. Eaton, Cubanes: starting materials for the chemistry of the 1990s and the new century, *Angew. Chem. Int. Ed. Engl.*, 1992, 31, 1421–1436.

[15] C. P. Vlahacos and J. O. Jensen, Structure of percubyl–cubane, Relationship to supercubane, S6 and O symmetries and chiral aspects, *J. Mol. Struct.* (*Theochem*), 1996, 362, 225–234.

[16] S.A. Miller, J.H. Ding, D.L. Gin, Nanostructured materials made of polymerizable amphiphiles, *Curr. Op. Colloid Interf. Sci.*, 1999, 4, 338-347.

[17] J. H. Fendler, Self-Assembled Nanostructured Materials, *Chem. Mater.*, 1996, 8, 1616-1624.

[18] I.W. Hamley, Nanotechnology with soft materials, *Angew. Chem. Int. Ed.*, 2003, 42, 1692-1712.

[19] Hyperchem 6.01, Hypercube Inc., Gainesville, US.

[20] F. Pichierri, Proposal for using C60 anions as propellants for ion engines, *Phys. Lett., A* 2003, 318, 425-428.

[21] M. Braun, U. Hartnagel, E. Ravanelli, B. Schade, C. Böttcher, O. Vostrowsky, A. Hirsch, Amphiphilic [5:1]- and [3:3]-Hexakisadducts of $C_{60}$, *Eur. J. Org. Chem.*, 2004, 1983-2001.

In: Nanostructures: Novel Architecture
Editor: Mircea V. Diudea, pp. 345-360
ISBN: 1-59454-499-9


*Chapter 17*

# CONSTRUCTION OF SELF-ASSEMBLED NANOSTRUCTURES BY MEANS OF DIHYDROGEN BONDING

***Mircea Vlassa*[a1]*, Radu Custelcean*[b] *and James E. Jackson*[c]**
[a]"Babeş-Bolyai" University, Faculty of Chemistry and Chemical Engineering,
11 Arany Janos str., 400028, Cluj, Romania
[b]Oak Ridgge Nationaly Laboratory, Oak Ridge, USA
[c]Department of Chemistry, Michigan State University,
East Lansing, USA

## ABSTRACT

The hydridic-to-protonic interaction, or dihydrogen bonding, between negatively charged hydrogen atoms of M-H (M=Al,B,Ga,Ir,Mo,Mn,Os, Re,Ru,W) bonds and conventional proton donors, can influence reactivity and stereoselectivity in solution and in solid state directs crystal packing, thus finding utility in supramolecular synthesis. These bonds were used in construction of novel one-, two-, and three dimensional extended structures with various topolgies. Some of these systems can react topochemically in solid state, releasing hydrogen and exchanging the weak O-H⋯H-M interactions for strong O-M covalent bond to form ordered, self-assembled nanostructured materials.

## 1. INTRODUCTION

Many studies[1] regarding the dynamics of hydrogen-bonded systems in solution have demonstrated that proton transfer from acidic AH partners to the transition metal hydrides MH, along the H⋯H bonds, generally leads to $\eta^2$-$H_2$ non classical complexes, which subsequently eliminate hydrogen upon heating, with the formation of covalent M-A bonds. An analogous process appears to occur in the case of borohydride anion. In aqueous solution, $BH_4^-$ is very likely dihydrogen-bonded to $H_2O$, as suggested by the crystal structure of

[1] E-mail address: mvlasa@chem.ubbcluj.ro

$NaBH_4.2H_2O$[2], as well as theoretical and experimental studies by Eisenstein *et al.*[3] Theoretical work by Elguero *et al.*[4] indicates that $H_2$ generating from dihydrogen-bonded borohydrides can also be induced by the internal forces within a crystal. All these premises, together with the established ability of borohydrides to self-assemble into extended dihydrogen-bonded networks, suggest that A-H···H-B dihydrogen bonds could be employed in topochemical assembly of covalent materials. Such weak H···H interactions, in principle, may be used to organize and hold a structured form while it is more firmly fastened together by A-B bond formation, thus transferring the initial order from the starting crystal to the newly formed covalent frame. This strategy makes dihydrogen bonding a potentially powerful tool for rational assembly of new covalent crystalline materials with tailored nanostructure and desired properties.

## 2. Topochemical Dihydrogen to Covalent Bonding Transformation

In the topochemical transformation outlined above, the initial H-bond arrangement should determine the final covalent structures. However, when going from the A-H···H-B interaction to a covalent A-B bond there is approximately a 1.5-2 Å change in distance between A and B. This is a considerable shrinkage, and if it was cumulative in the reaction direction, it would lead to an increasing gap between the growing covalent network and the unconverted dihydrogen-bonded molecular units as the decomposition advanced, with eventual loss of lattice control over the reaction. Even under these circumstances the transformation may still be topochemical, as demonstrated by the solid state conversion of cyclotrigallazane into nanocrystalline gallium nitride reported by Gladfelter *et al.*[5], which is accompanied by 62.9% shrinkage of the unit cell. The price to pay was the initial loss of crystallinity and the consequent requirement for high annealing temperature to restore it. Although this thermal treatment had no detrimental effect upon the robust GaN product, more delicate structures would not tolerate such high temperatures, limiting the general applicability of this approach.

A low temperature procedure for topochemical dihydrogen to covalent bonding transformation would allow the extension of this strategy into the structurally more diverse domain of organic materials. Like many solid-state processes, this reaction includes two threats to the crystalline order: (a) geometry change upon bond reorganization, and (b) gas generation and diffusion within the lattice. Clearly, careful design of the starting dihydrogen-bonded networks is necessary in order to meet these challenges. Success, however, would mean that the well-developed tools of molecular synthesis could now be applied to the rational construction of crystalline covalent solids with desired structures and functions.

This work involves two strategies to address this problem: (a) design of cations to form closed loops in coordination with hydride-bearing anions, in which case the lattice distortion would not be cumulative, and (b) selection of globular cations large enough that their close packing determines the lattice parameters, with the hydridic anions fitting into the interstitial holes, in which bond formation via flexible arms would induce minimal change in the unit cell (Scheme 1). A dividend of this latter strategy is that the anticipated looseness of the lattice should allow the released hydrogen to diffuse readily through and out of the crystal (Scheme 1).

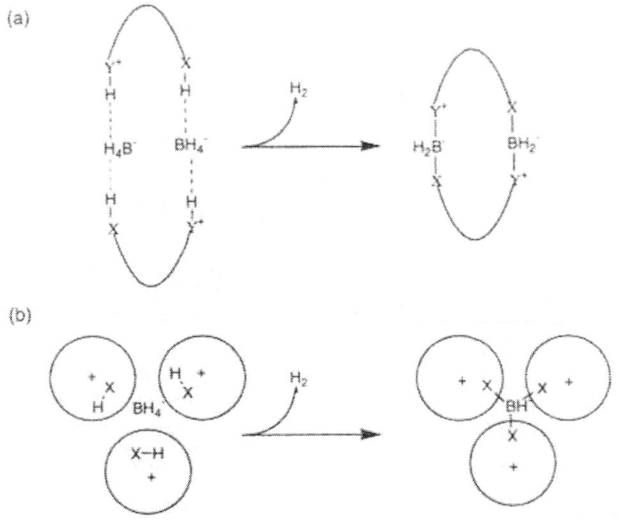

Scheme 1

Crystals of N[2-(6-aminopyridyl]-acetamide cyanoborohydride comprise two independent examples of closed loop self-assembly by dihydrogen bonding (Figure 1).[6] Partial liquefaction during the reaction and decomposition yielded complex mixture precluded topochemical control of decomposition process, accompanied by hydrogen elimination.

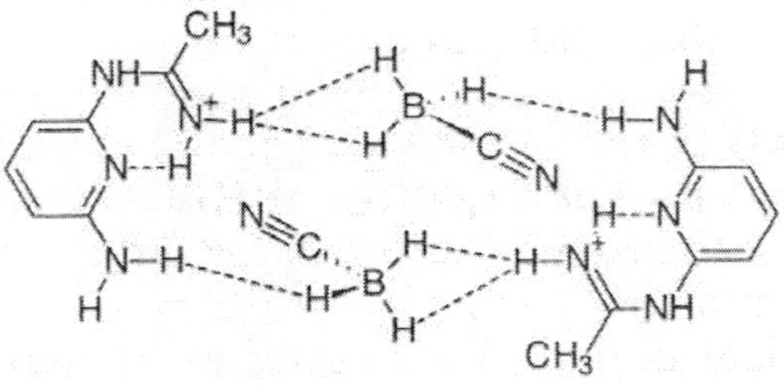

Figure 1: Self-assembly of N-[2-(6-aminopyridyl)]-acetamidine cyanoborohydride in closed-loop dimers.

Other complexes with general formula $MBH_3X$.TEA (M = Na,Li; X = H, CN; TEA=triethanolamine) **1-3** were prepared as candidates for the globular cations strategy.[6],[7] In complex **1**, $Li^+$ is pentacoordinated by the nitrogen and three oxygen atoms from a TEA molecule and by another oxygen atom from a different TEA molecule, forming ($Li^+$ TEA)$_2$ dimers. Multiple hydrogen bonds link the dimers via $BH_4^-$ pairs, giving rise to extended ribbons (Figure 2).The $NaBH_4$.TEA **2** complex (Figure 3) self-assembles in two-dimensional hydrogen-bonded layers in the solid state. Its decomposition leads to a polymeric compound. The decomposition proved to be topochemical, the resulting product being formed from a layered material but its crystallinity was poor. The $NaCNBH_3$.TEA **3** complex (Figure 4), with less basic hydride component, could not be decomposed in the solid state. It seems that as in solution, the relative acidity/basicity of the proton and hydride partners controls the reactivity of the dihydrogen-bonded systems.

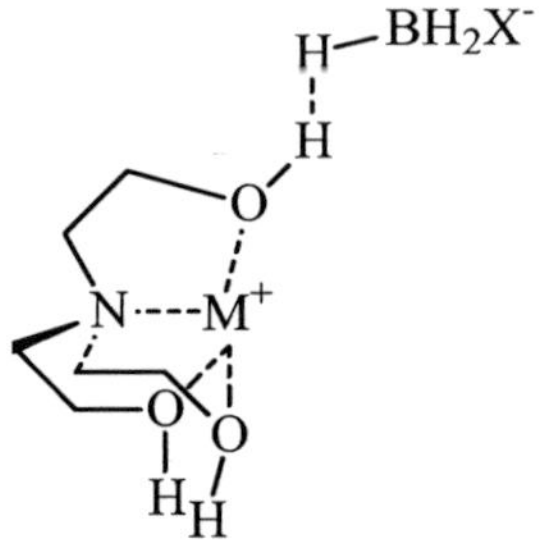


M= Li, X = H (1); M = Na, X = H (2) ; M = Na, X = CN (3);

Because the $Li^+$ cation would complex TEA more strongly, making the OH sites more acidic and hence more reactive, Custelcean and Jackson[7] prepared the $LiBH_4$.complex **1** , which unlike **2**, self-assembles into one-dimensional dihydrogen-bonded ribbons (figure 2).

The decomposition of this compound was again topochemical, leading to a one-dimensional polymeric trialkoxyborohydride structure (**1a**), in direct contrast to decomposition in DMSO solution, which yielded a polymeric borate and unconverted $LiBH_4$, the disproportionation products. Although more crystalline than the Na system, **1a** is still amorphous.

The high reactivity of **1** allowed studying the mechanism of $H_2$ loss and covalent bond formation dihydrogen-bonded complex[8]. Solid-state $^{11}$B-NMR, optical microscopy and H/D isotopic labeling studies showed that this decomposition is heterogeneous, with crystallization of the product phase from the parent crystal. Solid-solid crystal reaction is a mandatory of the supramolecular nature of these nanosized structures.

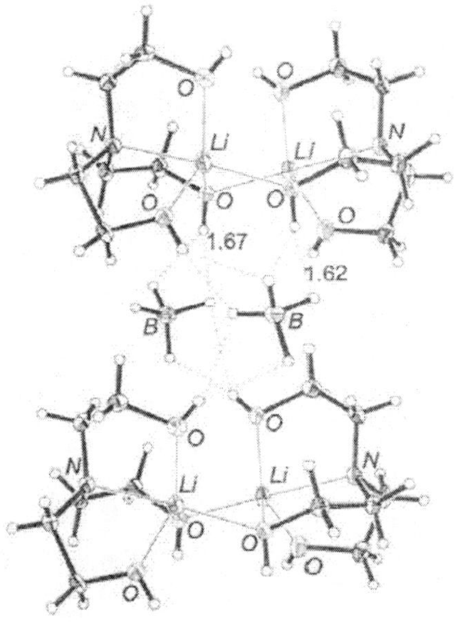


Figure 2: X-ray structure of $LiBH_4$·TEA showing the H···H close contacts in Å.

An example of a dihydrogen-bonded system that combines both the closed loop and the globular cation strategies is the $NaBH_4$.THEC (THEC = N, N', N'', N'''-tetrakis-(2-hydroxyethyl)cyclen) complex (**4**)[9]. This compound presents a crystal structure consisting of $D_2$ symmetrical dimers, held together by four conventional O-H···O hydrogen bonds, complemented by four orthogonal O-H···H-B proton-hydride interaction (Figure 5).

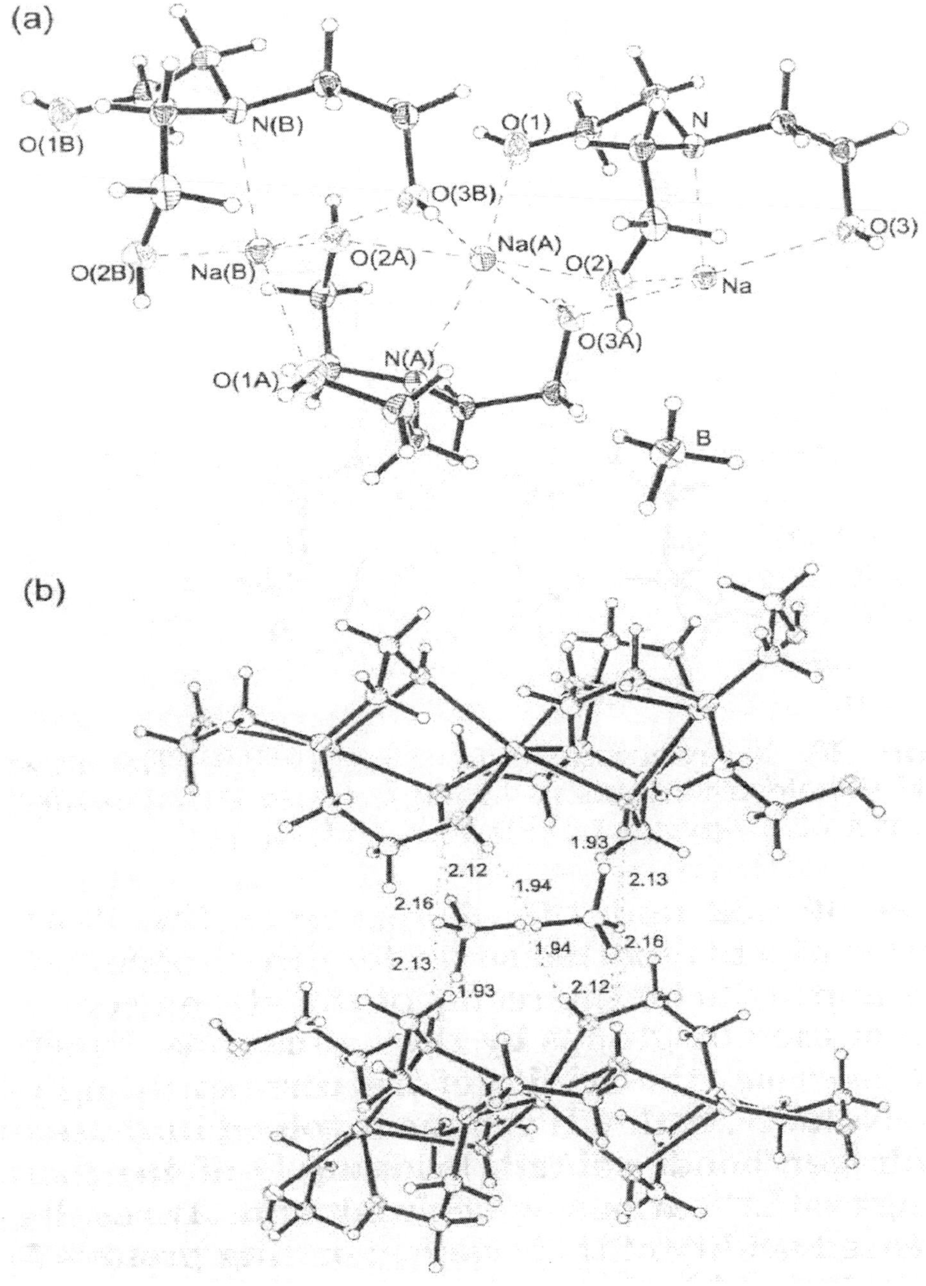


Figure 3: X-ray crystal structure of $NaBH_4$·TEA: (a) coordination of $Na^+$; (b) dihydrogen bonds connecting the chains, with H-H contact distances in Å.

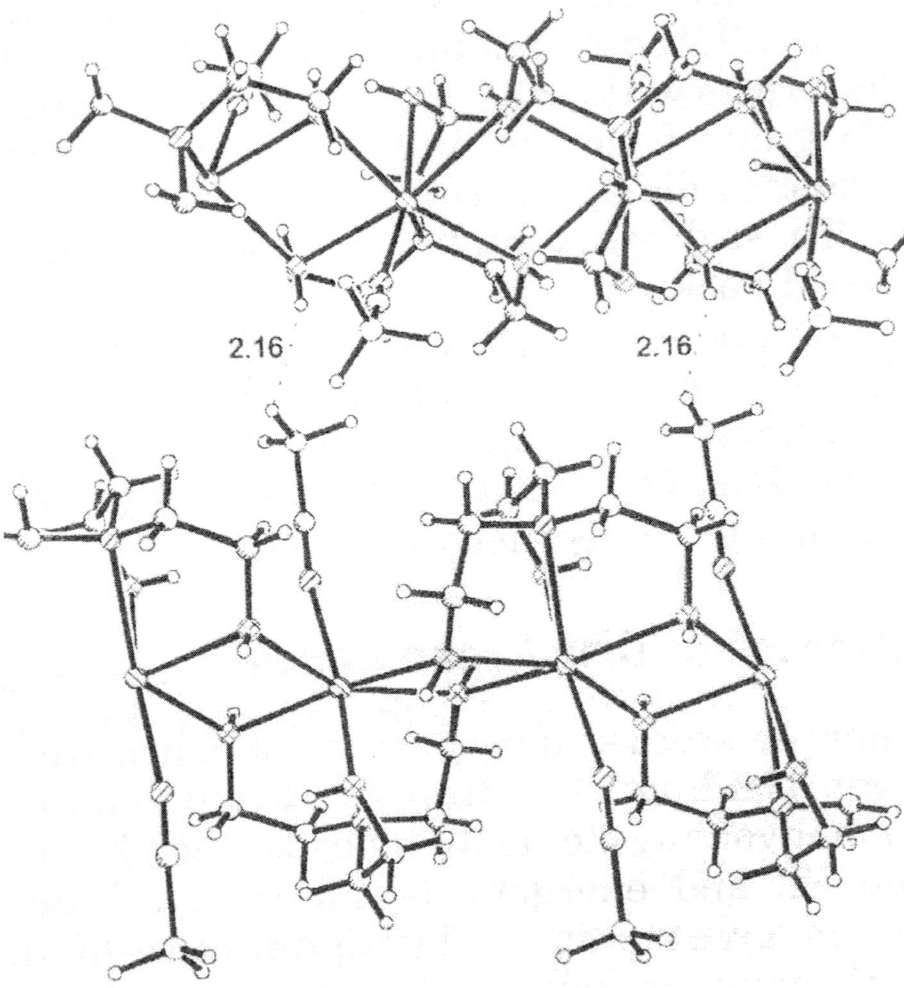


Figure 4: X-ray crystal structure of $NaNCBH_3{\cdot}TEA$ showing the H-H contact distances in Å.

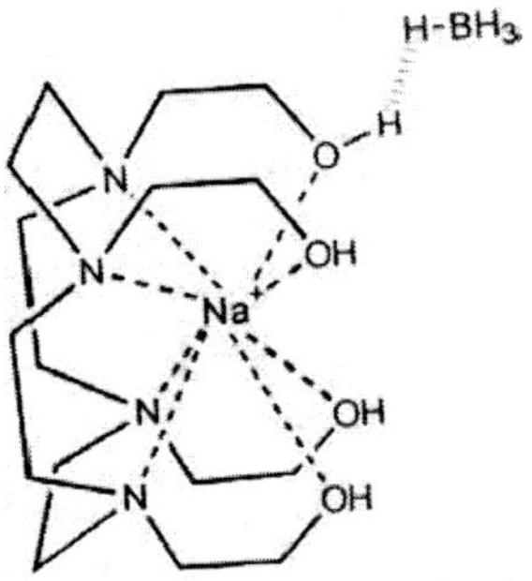


(4)

No hydrogen bonds link the dimers, which are paked into two-dimensional layers (Figure 6). Formation of discrete molecular cages might thus be expected from solid- state decomposition. Heating **4** in the solid state induced complete loss of $H_2$, as indicated by IR and $^{11}$B-NMR spectroscopy as well as TGA, with the formation of highly crystalline product.

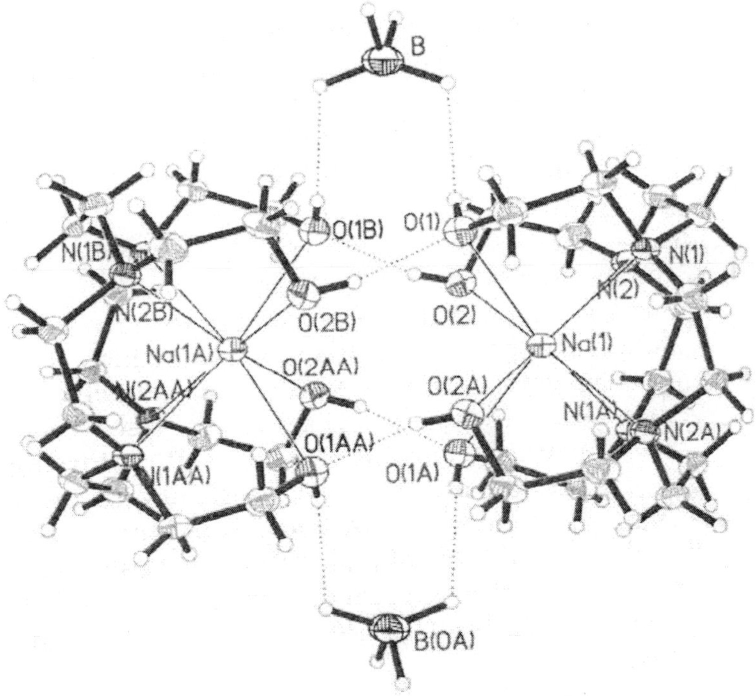


Figure 5: X-ray crystal structures of $NaBH_4$·THEC illustrating the self-assembly in dihydrogen-bonded dimers.

The decomposition product is crystallographically homogeneous, as demonstrated by polarizing microscopy and powder X-ray diffraction, which indicates a 9% shrinkage of the unit cell during the crystal-to-crystal process. The detailed structure of the resulting crystalline product remained elusive to date, due to deterioration of the single crystal, quality and the insolubility in common solvents. Considering the crystal packing of **4**, with no interdimer hydrogen bonding present, formation of discrete molecular cages is expected from the solid-state decomposition. A possible structure for the molecular cages in decomposition product of **4** is presented in Figure 7, optimized with the density functional theory method (DFT), at the B3LYP/6-31 G* level.[10] Selected bond lengths [Å] and angles [°] are : B-O(1) 1.458, B-O(2) 1.482, B-O(3) 1.497, B-O(4) 1.474, N(1)-Na(A) 3.003, N(2)-Na 2.714, N(3)-Na 2.435, N(4)-Na(A) 2.565, O(2)-Na 2.670, O(3)-Na 2.276, O(4)-Na(A) 2.228; O(1)-B-O(2) 112.5, O(1)-B-O(3) 109.1, O(1)-B-O(4) 114.6, B-O(1)-C 123.7, B-O(2)-C 122.3, B-O(3)-C 125.9, B-O(4)-C 125.6 No unusual values for the bond lengths and angles were observed in the resulting $C_2$-symmetrical structure, suggesting the minimal strain in these covalent dimers. When **4** was decomposed in DMSO by heating at 120$^{0}$C for 48 h, a crystalline product precipitated in high yield. X-ray structural analysis of the resulting crystals revealed a centrosymmetric molecular cage very similar to that expected from the solid-state reaction (Figure 8). Selected bond lengths [Å] and angles [°] of this crystal are: B-O(1)

1.502(2), B-O(2) 1.511(2), B-O(3) 1.4502(2), B-O(5) 1.457(2), N(1)-Na 2.519(1, N(2)-Na 2.5401(1), N(3)-Na 2.565(1), N(4)-Na 2.682(1), O(1)-Na 2.371(1), O(2)-Na 2.796(1), O(3)-Na 2.391(1); O(1)-B-O(5) 111.6(1), O(2)-B-O(5) 109.8(1), O(3)-B-O(5) 111.6(1), B-O(1)-C 121.7(1), B-O(2)-C 113.2(1), B-O(5)-C 122.8(1).

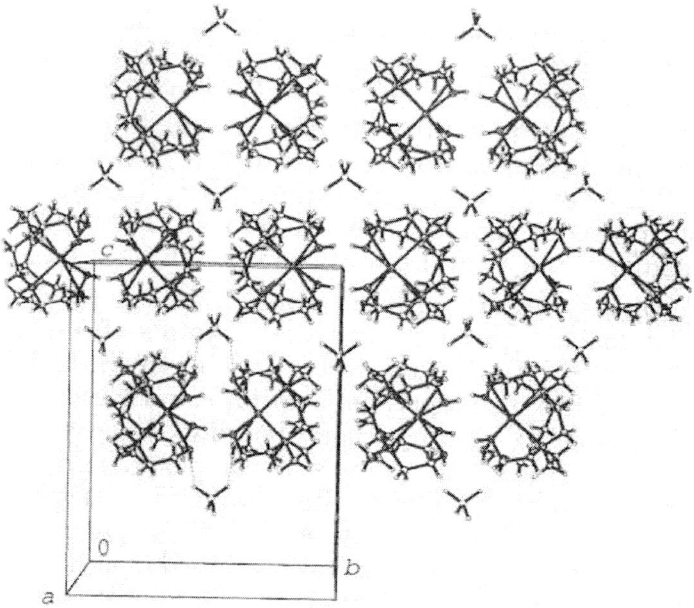


Figure 6: Crystal paking in $NaBH_4$·THEC.

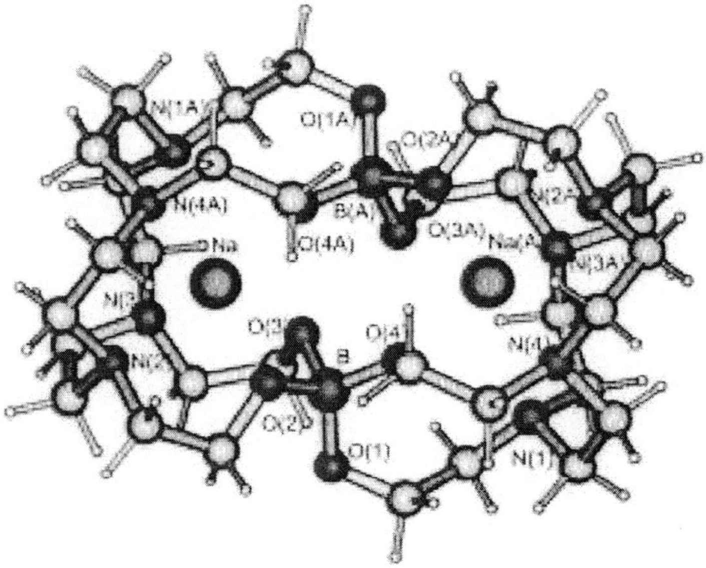


Figure 7: Calculated molecular model of the solid-state decomposition product of 4.

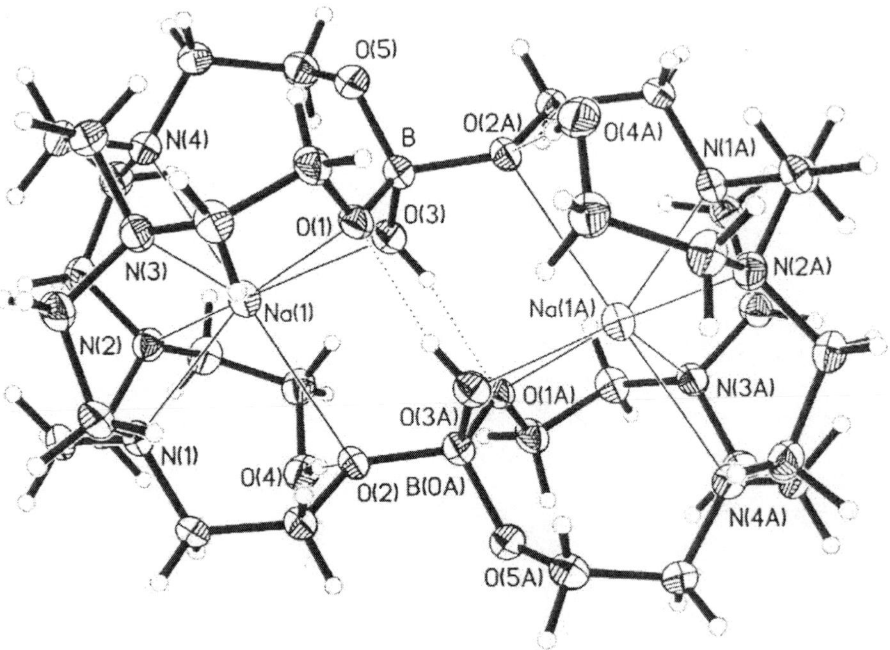


Figure 8. X-ray crystal structures of the major conformation found in the covalent dimer obtained by decomposition of 4 in DMSO.

Thus, judicious engineering of dihydrogen-bonded crystals permits transfer of crystallinity to the covalent products resulting from their solid-state decomposition.

## Supramolecular Synthesis through Dihydrogen Bonds

The high flexibility of THEC does not recommend it as a reliable building block for the construction of supramolecular networks with predictable architectures. However, the coordination of alkali metal cations locks the ligand intro a conformation with all four hydroxyethyl arms projecting to the same side of azaacrown ring, leading to an almost cubic disposition of the nitrogen and oxygen atoms. This geometry favors the arrangement of $NaBH_4$.THEC **4** into dihydrogen-bonded closed loop dimers, which to date appears to be the most promising arrangement for the preservation of crystalinity through solid-state dihydrogen-to-covalent bonding conversions.

To explore the role of the OH···HB dihydrogen bonds in the dimers formation we synthesized the $NaCNBH_3$.THEC complex **5**. With a lower basicity than $BH_4^-$, the $CNBH_3^-$ ion was expected to form weaker dihydrogen bonds. In fact, the crystal structure of this complex reveals the absence of any OH···HB close contacts, the cyanoborohydride ions participating in conventional hydrogen bonding through the N-atoms. Instead of forming dimers in this case the Na-THEC units self-assemble into extended helices through bifurcated OH···N···HO hydrogen bonds involving CN groups (Figure 9).

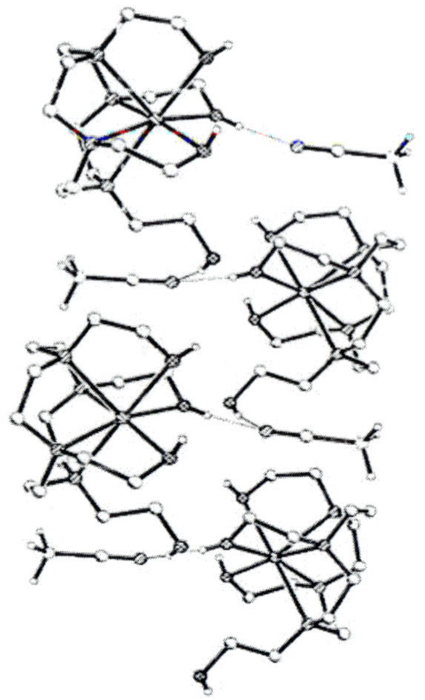

Figure 9. Helix formation by $NaCNBH_3$ complex (5) in the solid state.

It appears that the OH···HB dihydrogen bonds play a central role in the formation of the dimers, as their absence resulted in a completely different supramolecular organization in cyanoborohydride complex.[11]

Further, our intention was to explore the effect of molecular modification in the ligand upon dimer formation in solid state. With this end in mind we prepared $NaBH_4$·THCP (THCP=1,4,7,10-tetrrakis(S)-2-hydroxypropyl)-1,4,7,10-tetraazacyclododecane **6** . The most notable feature of its crystal structure consists in the formation of hydrogen-bonded dimers, which display an S-shaped architecture because of the absence of the steric repulsion between the methyl groups, repulsion which would arise in an O shaped dimer of **6** (see Figure 10). Dihydrogen bonding parameters of this complex are [Å, °]: O(4)···HB(1) (lower) 2.106, O(4)H···HB(1) (upper) 2.266, O(3)H···HB(1) 2.005, O(7)···HB(2) 1.972, O(8)H···HB(2) 1.676, O(2)H···HB(2) 1.904, O(4)H-H-B(1) (lower) 96.1, O(4)H-H-B(1) (upper) 88.7, O(3)H-H-B(1) 113.8, O(7)H-H-B(2) 110.7, O(8)H-H-B(2) 120.2, O(2)H-H-B(2) 133.3, O(4)-H-HB(1) (lower) 162.5, O(4)-H-HB(1) (upper) 143.9, O(3)-H-HB(1) 153.4, O(7)-H-HB(2) 155.4, O(8)-H-HB(2) 153.0, O(2)-H-HB(2) 169.3

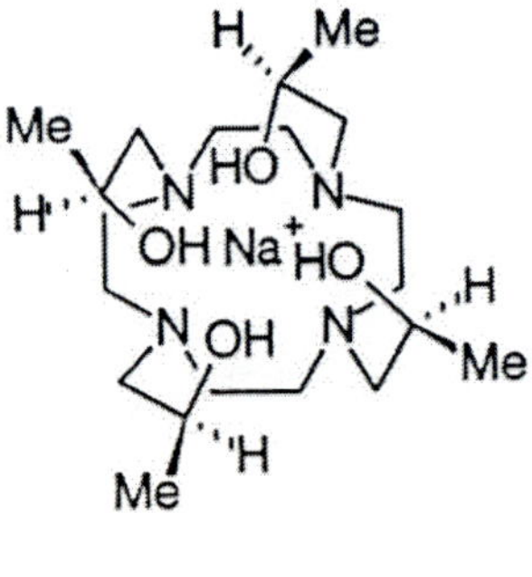


**(6)**

More profound structural variations in the basic THEC supramolecular building block were brought about by replacing one of the hydrohyethyl arms with a 9-methylantracyl group (see Scheme 2). This was done in order to see the ability of the resulting functionalized ligand to maintain the association in dimers. The 9-[1',4',7'-tris-2-hydroxyethyl-cyclen)]methylanthracene ligand **7** was synthesized from 9-(1',4',7',10'-tetraazacyclododecyl)methylanthracene[12] and ethyl oxide. Its X-ray structure is illustrated in Figure 11. Upon mixing with $NaBH_4$ in 2-propanol, complex $NaBH_4$·**7**·2-propanol (**8**·2 propanol) was formed which precipitated as a white crystalline solid with following normalized dihydrogen parameters [ Å, °]: O (1) ··· HB 1.692, O(2)H ···HB 1.661, O(3)H··· HB 1.730, O(1)H···H-B 114.5, O(2)H···H-B 117.8, O(3)H···H-B 116.1, O(1)H ··· H-B 166.8, O(2)-H···HB 155.1, O(3)-H···HB 154.6. Once again, self assembly in diydrogen-bonded dimers was revealed by X-ray crystallography. In contrast to the previous structures, no O-H···O interaction is present in this case; instead, the $BH_4^-$ ions are intercalated between the two ligands, forming three dihydrogen bonds with the OH groups.

These examples demonstrate that various poly-HEC ligands persistently self-assemble into dihydrogen-bonded dimers in the presence of $NaBH_4$, in spite of the fact that the exact connectivity of the dimers varies from one case to another in response to structural modification in the ligand. The consistent recurrence of the general dimeric motif suggested that these dihydrogen-bonded units might be sufficiently reliable building blocks for the controlled assembly of extended networks.

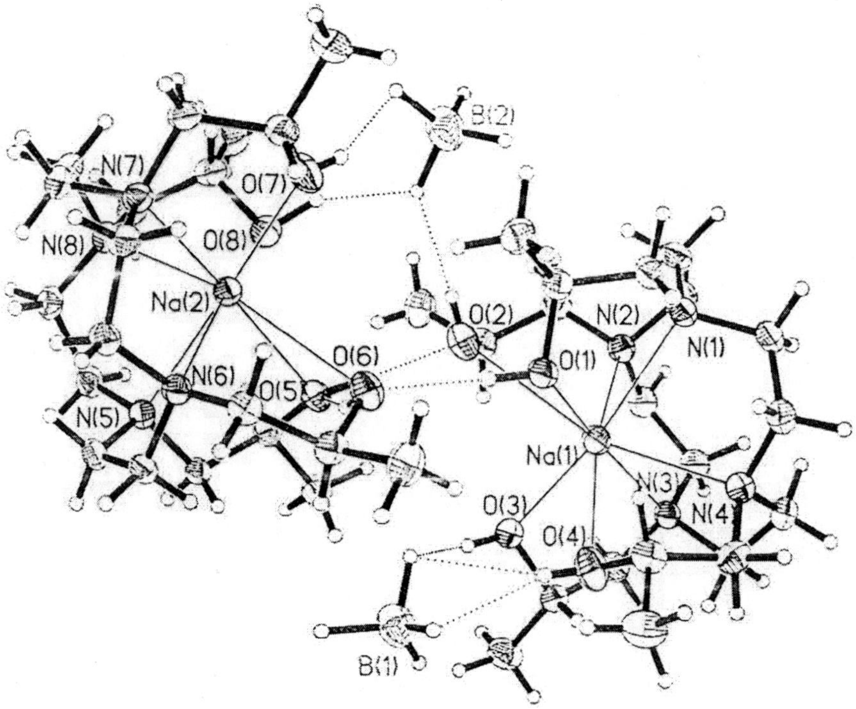


Figure 10. X-ray crystal structure of $NaBH_4$· THPC (6).

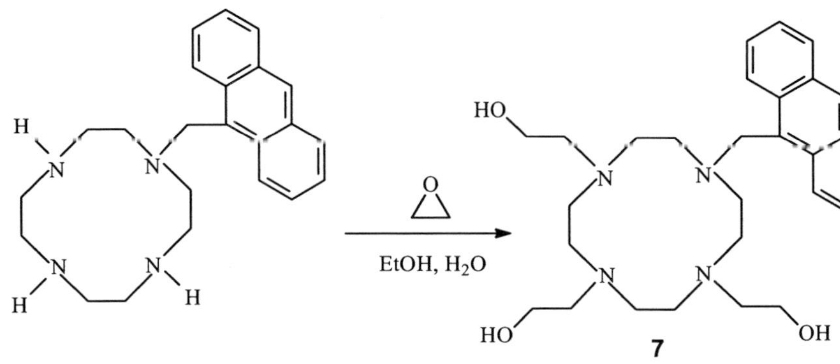


Scheme 2

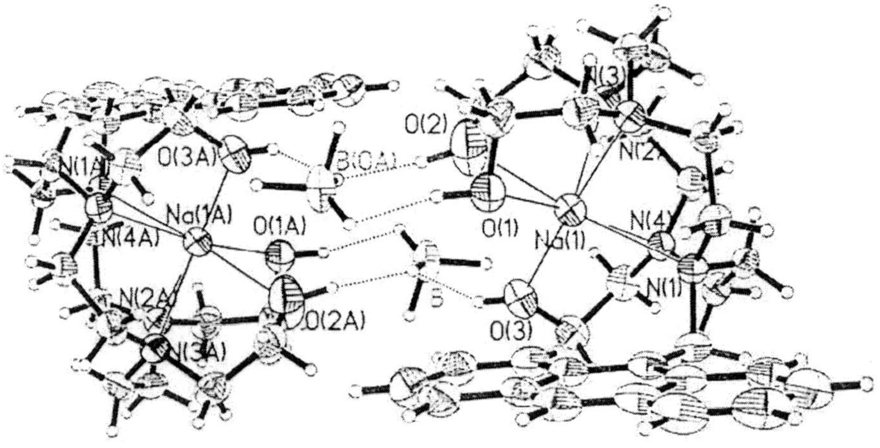


Figure 11: X-ray structure of monocrystal 8·$Et_2O$ (obtained by recrisytallization from acetonitrile/diethyl ether).

It can be anticipated that preparation of ligand **9** with two tris-HEC units connected by a rigid p-xylylene spacer (see Scheme 3) would lead to self-assembly into one-dimensional chains in the presence of $NaBH_4$. Upon mixing **9** with sodium borohydride in a 1:2 ratio in pyridine, complex 2 $NaBH_4$.**9**·pyridine (**10**) was formed (see Scheme 4). X-ray crystal structure of **10** (figure 12) show the two independent dihydrogen-bonded chaines with following dihydrogen bonding parameters: [Å, °]:

O(5)···HB(4) (front) 2.245, O(5)H···HB(4) (back) 2.198, O(1)H···HB(4) (lower) 2.243, O(1)H···HB(4) (upper) 2.076, O(2)H···HB(3) 2.093, O(3)H···HB(3) 2.294, O(6)H···HB(3) 2.269, 2303, O(11)H···HB(2) (front) 1.996, O(11)H···HB(2) (back) 2.216, O(7)H···HB(2) (lower) 2.095, O(7)H···HB(2) (upper) 2.092, O(8)H···HB(1) 2.218, O(9)H···HB(1) 2.174, O(10)H···HB(1) 2.373, O(5)H···H-B(4) (front)82.88, O(5)H···H-B(4) (back) 89.99, O(1)H···H-B(4) (lower) 92.94, O(1)H··H-B(4) (upper) 91.52, O(2)H···H-B(3) 110.14, O(3)H···H-B(3) 100.53, O(6)H···H-B(3) 73.81, 73.83, O(11)H···H-B(2) (front) 94.16, O(11)H···H-B(2) (back) 83.09, O(7)H···H-B(2) (lower) 94.63, O(7)H···H-B(2) (upper) 92.55, O(8)H···H-B(1) 111.71, O(9)H···H-B(1) 107.49, O(10)H···H-B(1) 86.90, O(5)-H···HB(4) (front) 136.70, O(5)-H···HB(4) (back) 170.91, O(1)-H····HB(4) (lower) 163.86, O(1)-

H···HB(4) (upper) 141.95, O(2)-H···HB(3) 160.28, O(3)-H···HB(3) 159.45, O(6)-H···HB(3) 144.00, 139.60, O(11)-H···HB(2) (front) 123.92, O(11)-H···HB(2) (back) 135.75, O(7)-H···HB(2) (lower) 166.56, O(7)-H···HB(2) (upper) 142.26, O(8)-H···HB(1) 156.37, O(9)-H···HB(1) 163.39, O(10)-H···HB(1) 140.19.

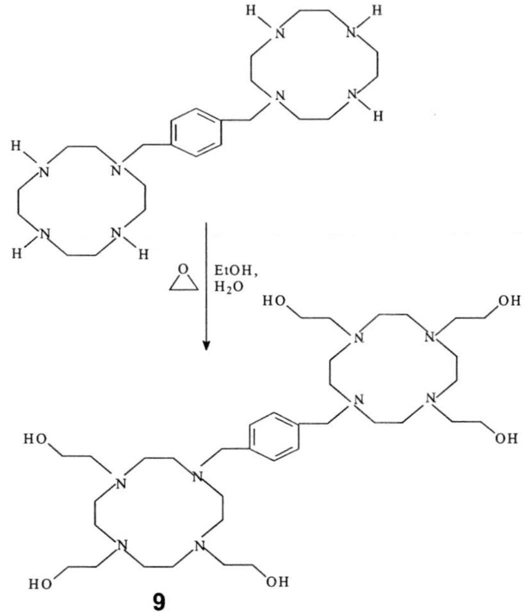


**9**

Scheme 3

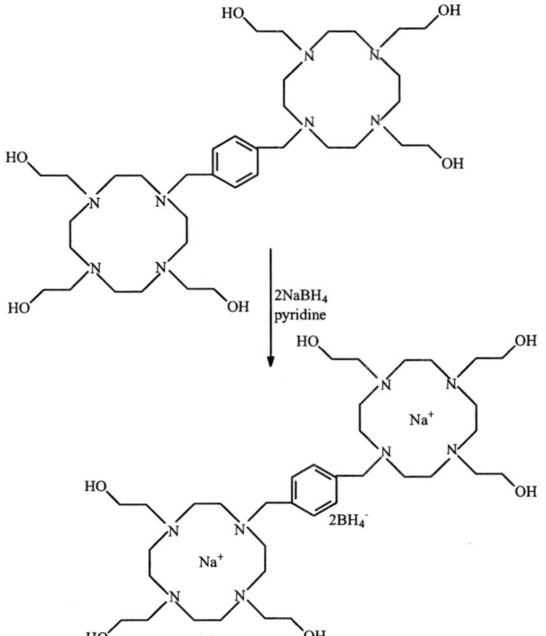


**10**

Scheme 4

As in the previous structures, complexation by $Na^+$ preorganizes the hydroxyethyl arms on the same face of the azacrown ring, thus favoring the self-association of the tris-HEC moieties in dimers through the borohydride anions. The p-xylylene spacers interconnect the resulting dimeric units into extended chains. There are two crystallographically unique molecules of **10** in the unit cell, differing in their hydrogen bond parameters. In the first case, one $BH_4$ connects neighboring tris-HEC units by dihydrogen bonding to O (1)H and O(5)H, generating one-dimensional chains. Additionally, conventional O(4)···O(1) hydrogen bonds also provide intrachain connections. The second $BH_4^-$ crosslinks adjacent chains into two-dimensional layers by dihydrogen bonding to the O(2)H, O(3)H, and O(6)H groups. A similar self-assembly into extended chains, further associated in layers, is exhibited by the second unique molecule of **10**. In this case, O(7)H···HB(2) and O(11)...HB(2) dihydrogen bonds, as well as O(12)···O(7) interactions interconnect the tris-HEC units, while dihydrogen bonds between $B(1)H_4^-$ and O(8), O(9)H, and O(10)H link the chains into layers. The shape of **10** , consisting of a long linear axis capped with bulky tris-HEC groups, precludes the close packing of the p-xylylene spacers within each layer, which leads to the formation of one-dimensional channels along the crystal, which are occupied by pyridine solvent (Figure 13).

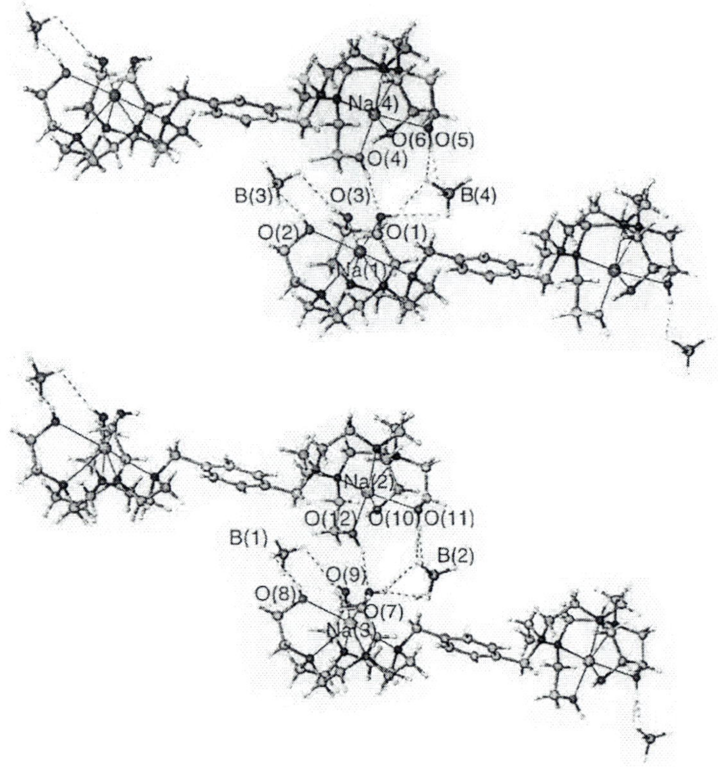


Figure 12: X-ray crystal structure of 9·pyridine.

These dihydrogen-bonded inclusion complexes are particularly attractive, as they might lead to porous two-dimensional covalent networks upon topochemical loss of $H_2$. Such structures are expected to show enhanced stability, possibly comparable with that found in inorganic zeolites. More importantly, these materials could easily be functionalized by taking advantage of the residual $(RO)_xBH_{4-x}$ groups resulting from the solid-state decomposition.

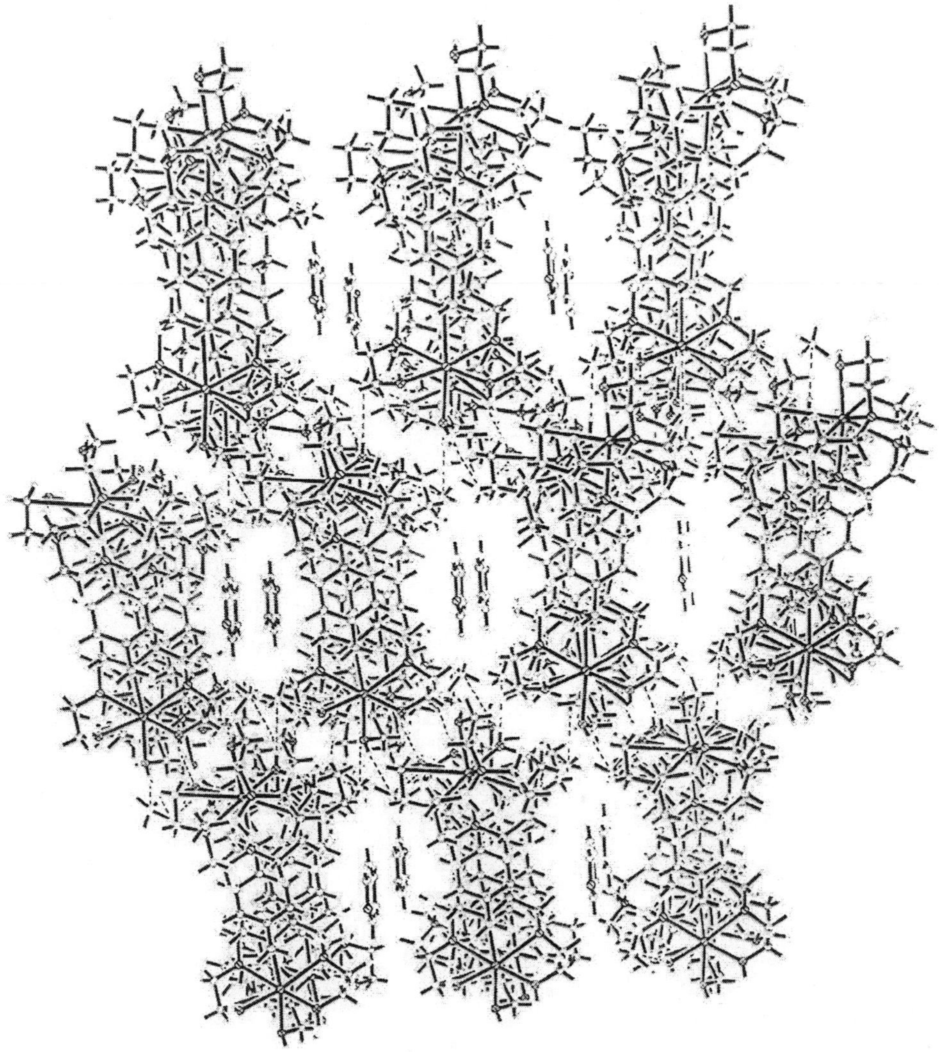

Figure 13: Crystal packing of 10·pyridine showing the formation of extended channels occupied by the pyridine guest.

## 3. Conclusions

We presented a strategy for building dihydrogen-bonded networks with architectures that can be controlled to a significant extent. These systems can react topochemically in the solid state, releasing hydrogen and exchanging the weak O-H···H-B interactions for strong B-O covalent bonds, to form ordered, extended covalent materials. By choosing a correct engineering of this transformation is possible to transfer crystallinity from dihydrogen-bonded crystals to the covalent products obtained by their solid-state decompositions. Solid-state reaction, particulary self-organisation to a new crystalline state, can be considered as mandatory for the supramolecular nanosized structures. Further studies will be focused on preparation on dihydrogen-bonded building blocks that will be able to undergo decomposition to low-temperature, thus permitting a rational construction of extended crystalline covalent solids. These materials are very important from practical point of view due to their nanostructures, finding thus applications in catalysis, hydrogen stocking or in preparation of truly functional organic zeolites.

## REFERENCES

[1] J. C. Lee, E. Peris, A. L. Rheingold, R. H. Crabterr, *J. Am. Chem. Soc.*, 1994, 116, 11014; J. A. Ayllon, C. Gervaux, S. Sabo-Etienne, B. Chaudret, *Organometallics*, 1997, 16, 2000; G. Orlova, S. Scheiner, T. Kar, *J. Phys. Chem. A*, 1999, 103, 514; E. S. Shubine, N. V. Belkove, E. V. Bakhmutova, E. V. Vorontso, V. I. Ionidis, C.Binachini, L. Marvelli, M. Peruzzini, L. M. Epstein, *Inorg. Chim. Acta*, 1998, 280, 302; L. M. Epstien, E. S. Shubina, *Ber. Bunsen-Ges. Phys. Chem.*, 1998, 102, 359; G. Orlova, S. Scheiner, *J. Phys. Chem. A*, 1998, 102, 4813; H. S. Chu, Z. Xu, S. M. Ng, C. P. Lau, Z. Lin, *Eur. J. Inorg. Chem.*, 2000, 993; J. A. Ayllon, S. Sabo-Etienne, B. Chaudret, S. Ulrich, H.-H Limbach, *Inorg. Chim. Acta*, 1997, 259, 1; S. Gründemann, S. Ulrich, H.-H. Limbach, N. S. Golubev, G. S. Denisov, L. M. Stein, S. Sabo-Etienne, B. Chaudret, *Inorg. Chem.*, 1999, 38, 2550; H. S. Chu. C. P. Lau, K. Y. Wong, W. T. Wong, *Organometallics*, 1998, 17, 2768; Y. Musashi, S. Sakaki, *J. Am. Chem. Soc.*, 2000, 122, 3867; J. A. Ayllon, S. F. Sayers, Sabo-Etienne, B. Donnadieu, B. Chaudret, *Organometallics*, 1999, 18, 3981; A. Caballero, F. A. Jalon, B. R. Manzano, *Chem. Comm.*, 1998, 1879.

[2] J. E. Jackson, R. Huang, J. Eckert, M. Sharma, D. Argyriou, R. Sheldon, Unpublished results.

[3] E. Peris, J. C. Lee, J. R. Rambo, O. Eisenstein, R. H. Crabtree, *J. Am. Chem. Soc.*, 1995, 117, 3485.

[4] I Rozas, I. Alkorta, J. Elguero, *Chem. Phys. Lett.*, 1997, 275, 423.

[5] J.-W, Hwang, J. P. Campbell, J. Kozubowski, S. A. Hanson, J. F. Evans, W. L. Gladfelter, *Chem. Mater.*, 1995, 7, 517.

[6] R. Custelcean, J. E. Jackson, *J. Am. Chem. Soc.*, 1998, 120, 12935.

[7] R. Custelcean, J. E. Jackson, *Angew. Chem., Int. Ed. Engl.*, 1999, 38, 1661.

[8] R. Custelcean, J. E. Jackson, *J. Am. Chem. Soc.*, 2000, 122, 5251.

[9] R. Custelcean, M. Vlassa, J. E. Jackson, *Angew. Chem., Int. Ed. Engl.*, 2000, 39, 3299.

[10] A D. Becke, *J. Chem. Phys.* 1993, 98, 5648.

[11] R. Custelcean, M. Vlassa, J. E. Jackson, *Chem. Eur. J.*, 2002, 8, 302.

[12] E. U. Akkaya, M. E. Huston, A. W. Czarnik, *J. Am. Chem. Soc.* 1990, 112, 3590.

In: Nanostructures: Novel Architecture
Editor: Mircea V. Diudea, pp. 361-397

ISBN: 1-59454-499-9


*Chapter 18*

# NANOSIZED INORGANIC METAL-OXYGEN CLUSTERS 1. BASIC STRUCTURES

***Adrian Pătruț***[1]
Faculty of Chemistry, "Babeş-Bolyai" University
11 Arany Janos, RO-400028, Cluj Napoca, Romania

## ABSTRACT

Metal-oxygen clusters are the most important representatives of nanosized inorganic molecular clusters. They are also called polyoxometalate/metal oxide-based clusters. Over the past ten years, a new generation of very large polyoxometalate/metal oxide-based clusters with supramolecular properties has emerged and exerted a significant impact on our overall perception on nanosized systems. The first part of this review is dedicated to the basic structures of the new polyoxometalate clusters. Three basic types and their subtypes are discussed and exemplified by representative clusters: (1) the giant ellipsoid-type (with 63 metal centers); (2) the giant sphere-type, which includes complete giant sphere (with 132 metal centers) and down-sized giant sphere (with 102 metal centers) and (3) the giant wheel-type, which includes common giant wheel or the Bielefeld structure (with 154 metal centers) and enlarged giant wheel (with 176 metal centers). Certain illustrious molecules, such as the $[Mo_{132}]$ giant sphere and the $[Mo_{154}]$ giant wheel, can be considered as true prototypes of the inorganic metal-oxygen clusters. The close relationship between giant spheres and giant wheels, with topological implications, is also discussed.

## 1. INTRODUCTION

Nanosized clusters or nanoclusters is a general term to designate particles whose size is greater than that of typical/common molecules, but is too small to exhibit bulk properties. Examples of such particles/molecules include: carbon-based fullerenes, weakly bound van der Waals clusters in gaseous phase, metal carbonyl clusters, boron carbydes and hydrides, polyoxometalates, metal clusters surrounded by protecting and stabilizing ligands, metal ions connected by bridging ligands etc.[1]

[1] E-mail address: apatrut@chem.ubbcluj.ro

Polyoxometalates are probably the most important representatives of the inorganic molecular nanoclusters. Over the past ten years, polyoxometalates have undergone a spectacular development surpassing even the most optimistic expectations of the relatively few research groups involved in investigations dedicated to this class of substances. The continuous diversification of polyoxometalates through the synthesis of an impressive number of new molecules having interesting electronic structures, high symmetry, unexpected topologies and remarkable magnetic properties has broadened the frontiers of chemistry towards a meso-physical world or nano-world.[2],[3]

According to a widely used definition, polyoxometalates (POMs) are the polyoxoanions of the early transition elements, especially Mo, W and V, made up by linked $MO_n$ units.[4] The metal M centers, which may belong to one or more atomic species, are named addenda.

Under the common definition, POMs are exclusively anionic molecules (or molecular anions), while chemical compounds are electroneutral substances. In the case of POMs, the corresponding electroneutral compounds are in fact salts, typically hydrated, of these anions.

Recently, the situation has become more complicated, especially through the reduction of a number of metal centers and/or the introduction of other metal centers having lower oxidation numbers than those usually found in POMs. Thus, electroneutral molecules having identical or similar structures to well-known anionic POMs were also obtained. POM clusters with giant sphere-type structure and 132 metal centers are well-known especially through their prototype, the $[Mo^{VI}_{72}Mo^{V}_{60}O_{372}(CH_3COO)_{30}(H_2O)_{72}]^{42-}$ anionic cluster,[5] abridged as $[Mo_{132}]$ or $[Mo^{VI}_{72}Mo^{V}_{60}]$. In addition, several electroneutral molecules with similar giant sphere-type structure, containing only 102 metal centers, were also obtained, such as the $[Mo^{VI}_{72}Fe^{III}_{30}O_{252}(CH_3COO)_{12}\{Mo_2O_7(H_2O)\}_2\ \{H_2Mo_2O_8(H_2O)\}(H_2O)_{91}]$ cluster,[6] which contains the giant sphere abridged as $[Mo_{72}Fe_{30}]$ or $[Mo^{VI}_{72}Fe^{III}_{30}]$, respectively the $[Mo^{VI}_{72}Mo^{V}_{30}O_{282}(CH_3COO)_{12}(H_2O)_{78}]$ cluster [7], abridged as $[Mo_{102}]$ or $[Mo^{VI}_{72}Mo^{V}_{30}]$.

It is obvious that for such neutral molecules the term POM with the ending *-ate*, suitable only for anions, can no longer be used. They can be named simply as metal oxide-based clusters, according to the term molybdenum-oxide based clusters,[6],[7] already used for the molecules mentioned above.

The term metal oxide-based (MOB) clusters is of a more general nature and can be applied to both electroneutral and electrically charged molecules (be they a anionic or, why not, even cationic). On the other hand, the traditional term POM clusters, which is valid only for anionic molecules, is well-known and widely used. Therefore, we consider that a hybrid term, such as polyoxometalate/metal oxide-based clusters, POM/MOB clusters for short,[3] would be appropriate for covering the entire range of these molecules.

The POB/MOB molecules have as structural skeleton a closed cage, composed of metal centers linked by bridging oxygen atoms. Thus, they can also be called metal-oxygen clusters.

This review is dedicated to the amazing structures and topologies of the main representatives of a new generation of nanosized POM/MOB clusters.

## 2. A New Generation of Polyoxometalate Clusters

As mentioned above, the chemistry of POM/MOB clusters brought recently about spectacular changes, due to the synthesis and characterization of giant molecules, with new structures, having high symmetry and remarkable topological and electronic properties.

Consequently, there are two distinct generations of POM/MOB clusters: one of classical type and another with supramolecular properties.[3]

The classical POMs, which are exclusively anionic molecules, have the characteristics and properties of traditional coordination compounds, more precisely of the molecules that form the so-called complex. They are obtained by simple intermolecular condensation of certain oxoanions that generates polyoxoanionic buildings. Depending on the nature of the oxoanion(s) involved in the condensation process and the pH value, a large number of POMs with various structures and properties have been obtained. In an exaggeratedly simplified classification, polyoxometalates (POMs) were divided in isopolyoxometalates (isoPOMs) and heteropolyoxometalates (heteroPOMs), while in the condensation process are involved oxoanions from one or several atomic species. Rigorously speaking, for isoPOMs there are exclusively one or more atomic species with addenda role, while in the case of heteroPOMs one or more atomic species present play the role of heteroatom. The positions of the addenda and heteroatom(s) are well defined for every structure.

Two or three decades ago, the chemistry of isoPOMs became a closed area, with very limited perspectives of future development. The maximal number of metal centers seemed to be restrained to 10-12 atoms. On the other hand, the heteroPOM chemistry underwent a constant development, even if not a spectacular one. The presence of other atomic species with heteroatom role, which can be selected from about 70 elements of the periodic table, gives to the heteroPOMs not only more stability, a larger variety or the possibility of increasing the number of metal centers, but also new properties which allow for new applications. Until recent times, POM chemistry was focused almost exclusively on the synthesis and investigation of new heteroPOMs (frequently with Keggin or Keggin-derived structure) and on the search for classical applications.

The new generation of POM/MOB clusters emerged only ca. ten years ago, but its representatives became instantly well-known worldwide and their reputation has gone beyond the realm of chemistry.

The very large POM/MOB molecules are obtained by a novel synthesis strategy. Starting in the 90's, the development of supramolecular chemistry and especially the activity of Achim Müller and his co-workers at Bielefeld University (Germany) opened a new chapter in the synthesis of exceptionally large inorganic molecules.

Currently, chemists construct more complex and large molecules in a sequential manner via a time-consuming and hard-work sequential synthesis, isolation and purification of each intermediate, according to the logic of retrosynthesis. Recent developments in supramolecular chemistry lead to a very different approach for the synthesis of giant molecules, characterized by "intelligence" and efficiency rather than hard-work.

The so-called bottom-up method uses a type of unit construction under a multicomponent one-pot synthesis. Based on a number of simple combinatory linkable building blocks/units at disposition, the novel synthetic strategy allows successive and rapid processes of self-assembly, self-organization and molecular growth for a huge variety of very large molecules. This is the starting point of a new type of chemistry, governed by geometrical and topological rules.

The basic chemical principles for the synthesis of the new giant molecular species include,[8-11] besides the existence of a library of linkable building blocks/units, the presence of reducing agents in the system (making possible the reduction of a number of metal centers, introduction of heteroatoms with different oxidation numbers, introduction and/or exchange of various ligands other than oxygen), which allows for tuning the electron density and other molecular properties.

These conditions can only be optimally fulfilled in POM/MOB systems, containing exclusively or inclusively molybdenum metal centers, which possess the relevant variety of molecular and electronic structures/configurations.

Using the same reaction type, *i.e.*, the acidification of aqueous solutions of the tetrahedral $MoO_4^{2-}$ oxoanions at appropriate pH values, in the presence of adequate reducing agents (especially those which can drive multi-electron transfer), a large variety of discrete giant POM/MOB molecules can be obtained (essentially through a process which involves successive condensations).

In one-pot synthesis, the step-wise building process fundamentally entails the existence of adequate linkable building blocks and possibly of different types of linkers/spacers, as well as a control parameter which influences the kind of successive building blocks to be formed, such as the reduction degree (but also the pH value).

For classical POM/MOB clusters, the most interesting structures belong to the heteroPOMs with tungsten as addendum. Somewhat surprisingly, the new generation of POM/MOB clusters developed inside or within the isoPOMs with molybdenum as addendum. This is mainly due to the fact that molybdenum can offer a much larger variety of linkable building blocks and the reduction of certain $Mo^{VI}$ to $Mo^{V}$ metal centers occur more easily than for tungsten. However, for the new generation of POM/MOB clusters the distinction iso-/hetero- is moot. The specific positions of the metal centers are those of addendum and linker/spacer, which are located in the center of building blocks/building units and of linker/spacer units.

One should emphasize that all bonds between the atoms of the new huge POM/MOB clusters are covalent. This is the mandatory characteristic of usual covalent molecules. In the opposite, supramolecules are characterized by the mandatory presence of noncovalent bonds and by certain special properties, such as multicomponent nature, cooperative interaction of the components, and even self-organization.

It is obvious that a supramolecule is not yet a discrete molecule, but rather an assembly of discrete molecules linked by noncovalent bonds. Initially Lehn, one of the fathers of supramolecular chemistry, considered the unusual properties of supramolecules, which open up new avenues in chemistry, are mainly due to the existence of noncovalent bonds. He even defined supramolecular chemistry as "the chemistry of noncovalent bonds".[12]

Perhaps it would be better to define supramolecules exclusively by their characteristic properties that truly bring something new to chemistry, rather than by the presence of certain types of bonds. Such a shift would be recommended since it has been demonstrated in a reliable manner that there are substances, and we mean especially the new POM/MOB clusters, with the properties of supramolecules but which lack noncovalent bonds. Furthermore, the largest synthesized and structurally characterized molecules, which contain up to 368 metal centers, belong to these very clusters.

Therefore, it has been proposed[3] that the new generation of the POM/MOB molecules should be named POM/MOB clusters with supramolecular properties or pseudosupramolecular POM/MOB clusters.

## 3. A VIRTUAL LIBRARY OF BUILDING BLOCKS

The following question should be addressed: what type of structural building blocks should a giant molecule be classified, within a structural and functional hierarchy starting just from these blocks?

Müller[8-11] made the inventory of the virtual library of building blocks, out of which one can construct all the structures of the POM/MOB clusters known in present. These linkable building blocks/units/fragments/motifs/groups, in the form of Platonic and Archimedean solids, allow the chemist to operate conceptually with a large variety of distinct molecular modules derived from these units.

In principle, all known POM/MOB structures with molybdenum metal centers can be ultimately reduced to [$Mo_1$] units (containing one molybdenum atom surrounded by a number of oxygen atoms in different geometries) or to [$Mo_2$] units (formally derived from two [$Mo_1$] units linked in various ways by edges and/or corners). The most interesting are however the larger building blocks, which can be considered pre-organized molecular fragments and act as directing molecular modules, leading to very large molecular systems of high structural variability and versatility. These basic/essential building blocks, out of which huge POM/MOB clusters can be constructed, are the following: [(Mo)$Mo_5$], [$Mo_8$], [$Mo_{11}$] and [$Mo_{17}$].

The pentagonal [(Mo)$Mo_5$] building block, with $C_5$ or, when distorted, $C_s$ symmetry, consists of a central $MoO_7$ pentagonal-bipyramidal unit sharing its equatorial edges with five additional $MoO_6$ octahedra. Pentagons can be utilized as potential building blocks, especially for constructing spherical clusters of icosahedral symmetry ($I_h$). This symmetry can be achieved even if the pentagons do not share edges, but are interconnected by appropriate linkers, for instance in the giant sphere-type clusters.

The structure of the [(Mo)$Mo_5$] building block is displayed in Figure 1.

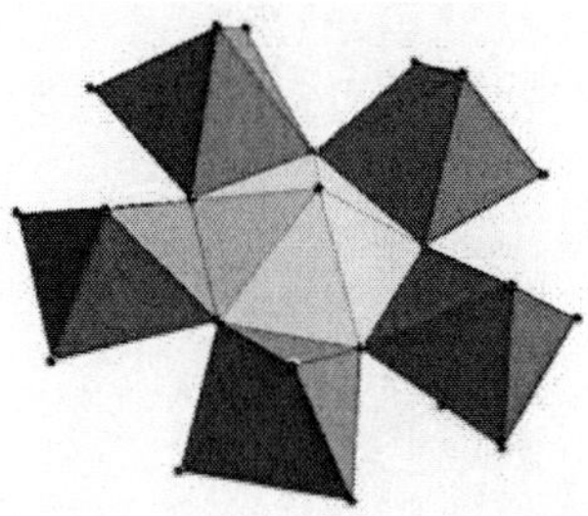

Figure 1: Structure of $[(Mo)Mo_5O_{21}]^{6-} \equiv [(Mo)Mo_5]$ building block in polyhedral representation. The central $MoO_7$ pentagonal bipyramid shares edges with five equatorial $MoO_6$ octahedra. Color code: $MoO_7$ unit, blue-turquoise; $MoO_6$ units, blue; O, small red spheres.

The $[Mo_8] \equiv [(Mo_1)[(Mo)Mo_5](Mo_1)] \equiv [(Mo)^0(Mo_5)^I(Mo_2)^{II}]$ building block, characteristically curved, is built up of the densely packed pentagonal [(Mo)$Mo_5$] unit to which two additional $MoO_6$ octahedra (sharing only corners) are more weakly connected and can more easily be removed (the superscript 0, I, II indices indicate the zero, first and second generation of the corresponding building unit) The [$Mo_8$] unit allows for the synthesis of even larger POM/MOB molecules, such as the giant wheel-type clusters.

The structure of the [$Mo_8$] building block is shown in Figure 2.

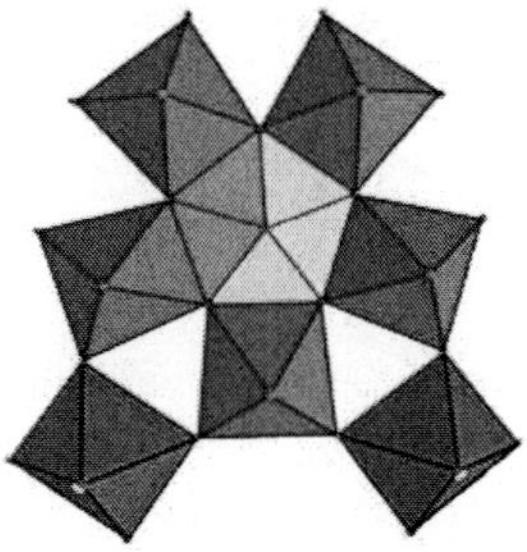

Figure 2: Structure of $[Mo_8] \equiv [(Mo_1)[(Mo)Mo_5](Mo_1)] \equiv [(Mo)^0(Mo_5)^I(Mo_2)^{II}]$ building block in polyhedral representation. The central $MoO_7$ pentagonal bipyramid is connected to five equatorial $MoO_6$ octahedra by sharing edges to which another two $MoO_6$ octahedra are added by sharing only corners. Color code: $(Mo)^0 \equiv MoO_7$ unit blue-turquoise; $(Mo_5)^I \equiv 5\ MoO_6$ units blue; $(Mo_2)^{II} \equiv 2\ MoO_6$ units red; O small red spheres.

The $[Mo_{11}] \equiv [(Mo)^0(Mo_5)^I(Mo_5)^{II}]$ building block has for its central part the pentagonal-bipyramidal $MoO_7$ unit of $(Mo)^0$-type with a directing function. Five $MoO_6$ octahedra of $(Mo_5)^I$-type are condensed to the equatorial plane of the central $MoO_7$ unit sharing edges, while the next five $MoO_6$ octahedra of $(Mo_5)^{II}$-type share corners with the former octahedra (as in the previous case, the superscript 0, I, II indices mention the zero, first or second generation of the building unit). $[Mo_{11}]$ is the basic building block for POM/BOM clusters which can be formulated as $[Mo_{11}]_n$, namely giant spheres and giant wheels.

The structure of the $[Mo_{11}]$ building block is displayed in Figure 3.

Finally, the large $[Mo_{17}] \equiv [(Mo_8)(Mo_1)(Mo_8)] \equiv [(Mo_8)_2(Mo_1)]$ building block can be reduced to two $[Mo_8]$ units fused via eight oxygen atoms and symmetrically connected by an $[Mo_1] = MoO_6$ octahedron which acts as linker. Each $[Mo_8]$ unit also contains the directing group, here the $MoO_6(NO)$ pentagonal bipyramid. The ellipsoidal $[Mo_{57}M_6]$-type clusters are made up of three $[Mo_{17}]$ building blocks.

Figure 3: Structure of the $[Mo_{11}] \equiv [(Mo)^0(Mo_5)^I(Mo_5)^{II}]$ building block in polyhedral representation. The central $MoO_7 \equiv (Mo)^0$ pentagonal bipyramid (zero generation) is connected to five equatorial $MoO_6 \equiv (Mo_5)^I$ octahedra (first generation) by sharing edges to which another five $MoO_6 \equiv (Mo_5)^{II}$ octahedra (second generation) are added by sharing only corners. Color code: $(Mo)^0$ unit blue-turquoise; $(Mo_5)^I$ units blue; $(Mo_5)^{II}$ units green; O small red spheres.

The structure of the $[Mo_{17}]$ building block is shown in Figure 4.

Figure 4: Structure of the $[Mo_{17}] \equiv [(Mo_8)(Mo_1)(Mo_8)]$ building block in a combined representation. The two $[Mo_8] \equiv [Mo_7(MoNO)]$ units represented as polyhedra are connected by a $[Mo_1] = MoO_6$ linker shown in ball-and-stick representation. Color code: $MoO_6(NO)$ units blue-turquoise; $MoO_6$ octahedra of the $[Mo_8]$ units blue; Mo of the $[Mo_1]$ linker large yellow sphere; O bound to Mo small red spheres; O of the NO groups large red spheres; N small green spheres.

## 4. Basic Nanostructures

Units acting as building blocks with intrinsic properties (variable charge and flexible coordination modes) are self-assembling in successive molecular growth processes, generating nanostructures of the new huge POM/MOB clusters with supramolecular properties. In certain cases, also involved in the formation of these structures are units acting as linkers/spacers that connect/separate the different building blocks. (The option of a choice between a linker, which would connect or a spacer, which would separate is only a particular case of the eternal attraction/repulsion dilemma).

As already mentioned, the resulting structures and the corresponding molecules contain exclusively or inclusively (but in majority) molybdenum atoms as metal centers. In addition, each structure contains a large internal cavity and a number of tailored pores/openings on the surface.

Any large POM/MOB cluster molecule can be designated by its complete formula which indicates all the atoms, or, alternatively, it can be written under the form of a short/abridged formula that specifies only the metal centers.

Practically, all the new huge discrete POM/MOB clusters known up to the present fall in three classes of basic structures, to which other special structures can be added.

These classes and their subclasses, listed in increasing order of the number of metal centers, are the following:

1. **Giant ellipsoid**, with 63 metal centers;
2. **Giant sphere**, with the following subclasses:
    2*a*. *Complete giant sphere*, with 132 metal centers;
    2*b*. *Down-sized giant sphere*, with 102 metal centers;
3. **Giant wheel**, with the following subclasses:
    3*a*. *Common giant wheel (Bielefeld structure)*, with 154 metal centers;
    3*b*. *Enlarged giant wheel*, with 176 metal centers.

Until now, the first type of basic structure, with 63 metal centers, has not been named. We propose this structure to be referred as giant ellipsoid, by analogy to the accepted names for the other basic structures, *i.e.*, giant spheres and giant wheels, which make reference to geometrical features.

In fact, if one uses a rigorous mathematical terminology, the first basic structure corresponds to a particular ellipsoid, called spheroid, in which two of the three semi-axes are equal. However, we prefer the general term of giant ellipsoid, as the giant spheroid designation would closely remind that of giant sphere.

The second type of basic structure is a true giant sphere. One should mention that the sphere also represents a particular case of the ellipsoid, in which all three semi-axes are equal.

The third type of basic structure, usually called giant wheel, corresponds to a toroid, namely to the single-holed ring torus.

For certain classes and subclasses, there are also several variants with a slightly different number of metal centers, which does not significantly influence the architecture of the parent structure to justify placing them in a different subclass.

Almost every structure has several representatives, namely different isostructural POM/MOB cluster molecules, with distinct composition of the metal centers. They are differentiated by the nature of the atomic species acting as metal centers and/or the stoichiometric ratio of the different metal centers. Usually, the differences due to these variables are not very important for the structure. Nevertheless, they will be described briefly in each case.

Finally, some representatives with a distinct composition of the metal centers exist in several varieties, which slightly differ by the composition of the ligands. Ordinarily, this exerts even less influence on the structure.

All POM/MOB clusters having one of these basic structures belong to the largest synthesized and structurally characterized molecules; they are only surpassed by other POM/MOB clusters with special structures.

The classes and subclasses of the basic structures and their representatives are described below. In every case, the structure is presented completely for its first representative that exhibits the so-called pattern structure. Only the differences to the pattern structure are mentioned for the other isostructural molecules.

One should stress here that, for certain POM/MOB clusters, the condensation process of the corresponding oxoanions does not necessarily resume to the stage of discrete molecules. It may continue in a solid-state reaction generating extended structures. The resulting infinite clusters are molecular assemblies produced by inflationary multiplication of discrete molecules, with formation of chains or layers. Nevertheless, since the review is dedicated exclusively to discrete POM/MOB clusters, extended structures do not represent a priority for our discussion.

## 4.1. Giant Ellipsoids

The ellipsoidal POM/MOB clusters which belong to the three-fragment giant ellipsoid-type structure, with the $[Mo_{57}M_6]$ general formula, were the first synthesized molecules with over 50 metal centers. Up to now, several representatives, corresponding to M = $V^{IV}$, $Fe^{III}$ and $Cu^{II}$, have been obtained and characterized.

According to a topological principle, which designates the building blocks and the linkers, the giant ellipsoidal clusters can be formulated as {[building block]$_3$ [di-nuclear linker]$_3$ [mononuclear linker]$_6$}.

Also known are several variants of this structure with a slightly increased number of metal centers.

## [$Mo_{57}V_6$]

In 1993, Zhang *et al.*[13] and Müller and coworkers,[14] have reported independently the synthesis of the first giant ellipsoidal POM/MOB cluster of the [$Mo_{57}M_6$] series. The molecular formula [$Mo_{57}V_6(NO)_6O_{183}(H_2O)_{18}]^{6-}$ given by the Chinese team was incorrect with respect to its composition, structure and the oxidation numbers of the metal centers. The molecule was structurally characterized by Müller *et al.* in 1994,[15] who also established the correct molecular formula [$Mo_{57}V_6O_{180}(NO)_6(OH)_3(H_2O)_{18}]^{21-}$ ≡ [$Mo^{VI}_{51}Mo^{V}_{6}V^{IV}_{6}O_{180}(NO)_6(OH)_3(H_2O)_{18}]^{21-}$, abridged as [$Mo_{57}V_6$] or [$Mo^{VI}_{51}Mo^{V}_{6}V^{IV}_{6}$].

The molecule is built up of three very large [$Mo^{VI}_{15}(Mo^{VI}NO)_2O_{58}(H_2O)_2]^{20-}$ building blocks (Figure 4), connected by three dinuclear [$Mo^{V}(H_2O)_2(OH)Mo^{V}]^{9+}$ linkers and by six mononuclear [$V^{IV}O(H_2O)]^{2+}$ linkers. According to the topological principle which indicates the building units and the linkers, namely {[building block]$_3$ [dinuclear linker]$_3$ [mononuclear linker]$_6$}, the [$Mo^{VI}_{51}Mo^{V}_{6}V^{IV}_{6}$] cluster can be formulated as {[$Mo^{VI}_{17}$]$_3$ [$Mo^{V}_{2}$]$_3$ [$V^{IV}_{1}$]$_6$}.

The structure of the [$Mo_{57}V_6$] POM/MOB cluster is displayed in Figure 5.

The giant ellipsoidal-type structure and the corresponding clusters are frequently called three-fragment structure/cluster/molecule, as they comprise three very large [$Mo_{17}$] building blocks.

The [$Mo_{57}V_6$] POM/MOB cluster, with 63 metal centers and 276 non-hydrogen atoms, is characterized by a high $D_{3h}$ symmetry. The doughnut-shaped molecule corresponds to an ellipsoid, more precisely to a spheroid, with the dimensions of 2.2 × 2.2 × 1.0 nm.

Figure 5: Structure of the [$Mo^{VI}_{51}Mo^{V}_{6}V^{IV}_{6}O_{180}(NO)_6(OH)_3(H_2O)_{18}]^{21-}$ ≡ [$Mo_{57}V_6$] ≡ {[$Mo^{VI}_{17}$]$_3$ [$Mo^{V}_{2}$]$_3$ [$V^{IV}_{1}$]$_6$} POM/MOB cluster in polyhedral representation parallel to the $C_3$ axis. Color code: [$Mo^{VI}_{17}$] units blue; [$Mo^{V}_{2}$] linkers red, [$V^{IV}_{1}$] linkers green.

The relatively large cavity inside the [$Mo_{57}V_6$] molecule is formally accessible through two openings in the cluster shell. The cavity (with the diameter perpendicular to the $S_3$ axis of

$\approx$ 0.9 nm and parallel to it of $\approx$ 0.5 nm) is delimited by a central $[O_{33}]$ polyhedron. The two openings are delimited by two alternating Mo-O-V-O 12-membered rings.

## $[Mo_{57}Fe_6]$

In 1995, Müller and his team[16] reported an analogous cluster with M = $Fe^{III}$ with the formula of $[Mo_{57}Fe_6O_{174}(NO)_6(OH)_3(H_2O)_{24}]^{15-}$ $\equiv$ $[Mo^{VI}_{51}Mo^{V}_{6}Fe^{III}_{6}O_{174}(NO)_6$-$(OH)_3(H_2O)_{24}]^{15-}$, abridged as $[Mo_{57}Fe_6]$ or $[Mo^{VI}_{51}Mo^{V}_{6}Fe^{III}_{6}]$. Its structure is similar to that of $[Mo_{57}V_6]$, with the difference that six $Fe^{III}$ centers substitute for the six $V^{IV}$ centers. More precisely, the new molecule contains six mononuclear $[Fe^{III}(H_2O)_2]^{3+}$ linkers instead of six $[V^{IV}O(H_2O)]^{2+}$ linkers (only the terminal ligands are noticed). Considering the remaining four oxygen atoms of the octahedra, shared with the $[Mo_{17}]$ units which they connect, the mononuclear linkers can be completely formulated as $[Fe^{III}O_4(H_2O)_2]^{5-}$ and $[V^{IV}O_5(H_2O)]^{6-}$.

The structure of the $[Mo_{57}Fe_6]$ POM/MOB cluster is displayed in Figure 6.

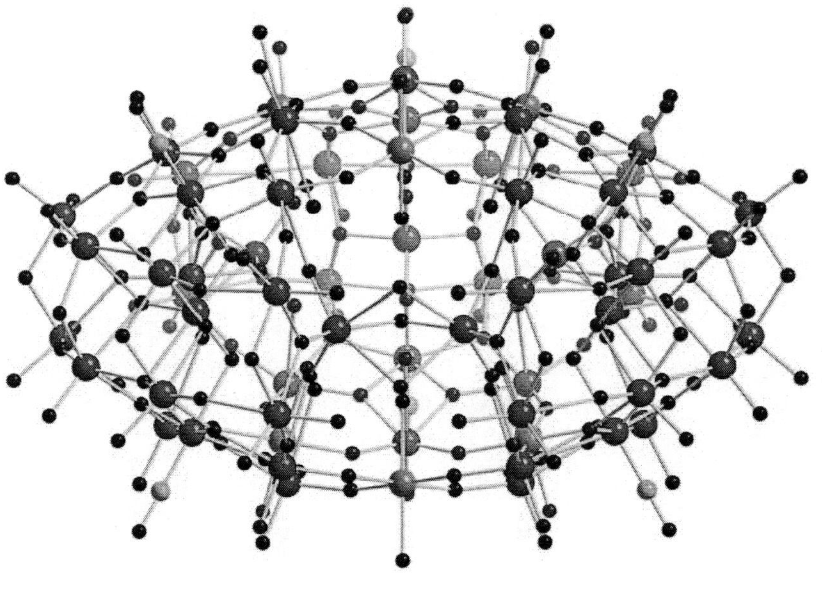

Figure 6: Structure of the ellipsoidal $[Mo^{VI}_{51}Mo^{V}_{6}Fe^{III}_{6}O_{174}(NO)_6(OH)_3(H_2O)_{24}]^{15-}$ $\equiv$ $[Mo_{57}Fe_6]$ $\equiv$ $\{[Mo^{VI}_{17}]_3\,[Mo^{V}_{2}]_3\,[Fe^{III}_{1}]_6\}$ POM/MOB cluster in ball-and-stick representation, along one of the $2(C_2)$ axes. Color code: Mo blue; Fe orange; O red; N green.

According to the three-fragment topological principle, the $[Mo_{57}Fe_6]$ cluster can be formulated as $\{[Mo^{VI}_{17}]_3\,[Mo^{V}_{2}]_3\,[Fe^{III}_{1}]_6\}$.

## $[Mo_{57}Cu_6]$

Since 1999, several attempts have been made in Bielefeld to synthesize homologous clusters with M = $Cu^{II}$. One of them, with the formula $[Mo_{57}Cu_6O_{174}(NO)_6(OH)_3(H_2O)_{21}]^{21-}$ $\equiv$

$[Mo^{VI}_{51}Mo^{V}_{6}Cu^{II}_{6}O_{174}(NO)_6(OH)_3(H_2O)_{21}]^{21-}$, abridged as $[Mo_{57}Cu_6]$ or $[Mo^{VI}_{51}Mo^{V}_{6}Cu^{II}_{6}]$, has been accurately characterized.[17]

In this case, six $Cu^{II}$ centers form the six mononuclear $[Cu^{II}\{H_2O\}]^{2+}$ linkers. Unlike the homologous $[Mo^{VI}_{51}Mo^{V}_{6}V^{IV}_{6}]$ and $[Mo^{VI}_{51}Mo^{V}_{6}Fe^{III}_{6}]$ clusters, in which the $V^{IV}$ and $Fe^{III}$ centers have a number of coordination NC = 6 and an (approximately) octahedral geometry, in the $[Mo^{VI}_{51}Mo^{VI}_{6}Cu^{II}_{6}]$ cluster the $Cu^{II}$ centers have only a number of coordination NC = 5 and a square pyramidal geometry. The four oxygen atoms common to the $[Mo_{17}]$ units, which they connect, are coplanar, forming the basis of the square pyramid, while an $H_2O$ molecule is situated at the apex of the pyramid, being projected into the internal cavity of the $[Mo_{57}Cu_6]$ cluster. The mononuclear $[Cu^{II}O_4(H_2O)]^{6-}$ linker is usually noticed as $[Cu^{II}(H_2O)]^{2+}$ when the oxygen atoms common to the $[Mo^{VI}_{17}]$ units are ignored.

The structure of the $[Mo_{57}Cu_6]$ POM/MOB cluster is shown in Figure 7.

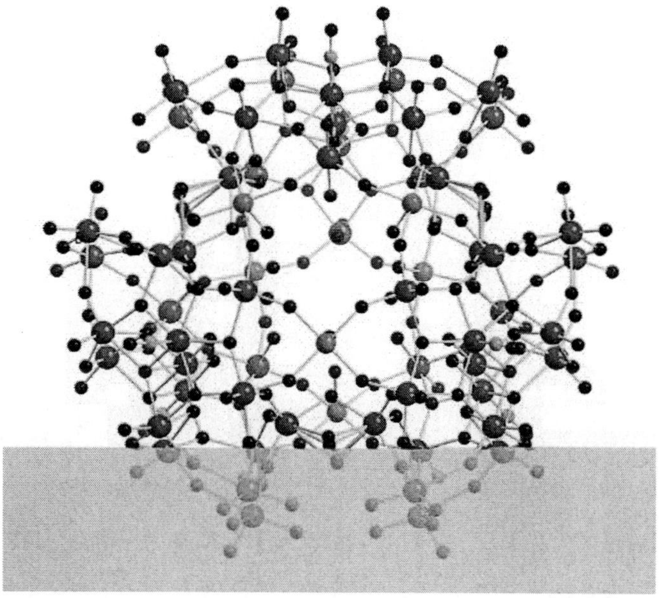

Figure 7: Structure of the ellipsoidal $[Mo^{VI}_{51}Mo^{V}_{6}Cu^{II}_{6}O_{174}(NO)_6(OH)_3(H_2O)_{21}]^{21-} \equiv [Mo_{57}Cu_6] \equiv \{[Mo^{VI}_{17}]_3\,[Mo^{V}_{2}]_3\,[Cu^{III}_{1}]_6\}$ POM/MOB cluster in ball-and-stick representation. Color code: Mo blue; Cu violet; O red; N green.

According to the three-fragment topological principle, the $[Mo_{57}Cu_6]$ cluster can be formulated as $\{[Mo^{VI}_{17}]_3\,[Mo^{V}_{2}]_3\,[Cu^{III}_{1}]_6\}$.

## $[Mo_{57+x}V_6]$

The giant ellipsoid-type structure, with 63 metal centers, also contains three rather large cavities on the outer sphere, between the $[Mo_{17}]$ units. These cavities can be filled with electrophilic metal-oxygen fragments, *i.e.*, positively charged $[MoO]^{4+}$ units, under reducing conditions, which increases the nucleophilicity. As a consequence, in aqueous solutions under strong reducing conditions and in excess of molybdate(VI) a step-by-step molecular growth may proceed, which enables in principle all [$Mo_{57+x}V_6$] ($x$ = 1–6) species to be formed. Each [$MoO]^{4+}$ unit binds to three oxygen atoms (two terminal) of the [$Mo_{57}M_6$] core, yielding a tetrahedral coordination of the inserted additional Mo atoms. The degree of occupation of the cavities (the $x$-value) can be correlated with the reduction degree of the [$Mo_{57}M_6$] cluster. The reduction increases the nucleophilicity of the core-cluster and initiates a growth process, in which up to as many six electrophilic $[MoO]^{4+}$ units can be incorporated. This appears to be the basic principle of the successive molecular growth processes. Aggregation or growth is due to the interaction of nucleophilic growing species with electrophilic units.

Starting from the [$Mo_{57}V_6$] ≡ [$Mo^{VI}{}_{51}Mo^{V}{}_{6}V^{IV}{}_{6}$] parent cluster, two different molecular growth products have been isolated and characterized. They correspond to the insertion of three and six $[MoO]^{4+}$ units and to a 6-electron and 12-electron reduction, respectively. The clusters, comprising 66 and 69 metal centers,[18] have the following molecular formulae: $[Mo^{VI}{}_{54}Mo^{V}{}_{6}V^{IV}{}_{6}O_{186}(NO)_6(OH)_3(H_2O)_{15}]^{21-}$and $[Mo^{VI}{}_{57}Mo^{V}{}_{6}V^{IV}{}_{6}O_{192}(NO)_6(OH)_3(H_2O)_{12}]^{18-}$ abridged as [$Mo_{60}V_6$] ≡ [$Mo^{VI}{}_{54}Mo^{V}{}_{6}V^{IV}{}_{6}$] and [$Mo_{63}V_6$] ≡ [$Mo^{VI}{}_{57}Mo^{V}{}_{6}V^{IV}{}_{6}$].

The two consecutive reduction/molecular growth processes have been monitorized by voltammetry and Resonance Raman spectroscopy [19].

The molecular growth was reported to be quasi-reversible, namely the uptake of $[MoO]^{4+}$ units during reduction can be reversed upon oxidation. The quasi-reversible process, based on switchable sites, represents a model for uptake and release of metal centers in metal storage proteins under redox-active conditions [18, 9].

The quasi-reversible molecular growth is shown in Figure 8.

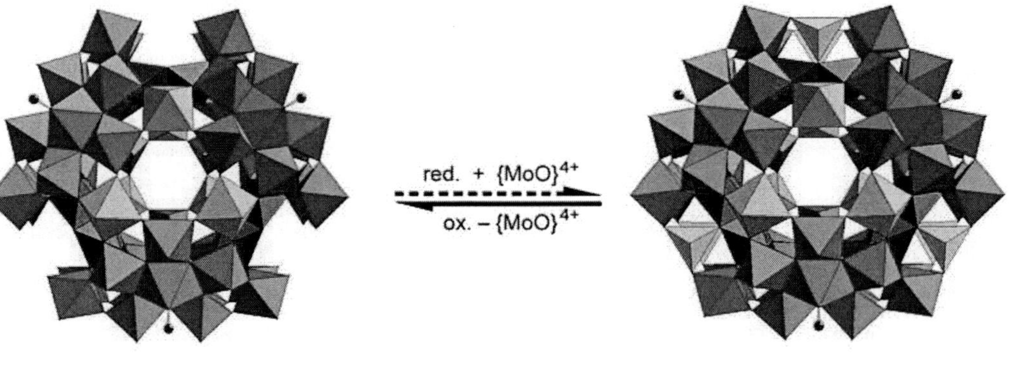


Figure 8: Quasi-reversible molecular growth within the [$Mo_{57}V_6$] (a) / [$Mo_{63}V_6$] (b) POM/MOB cluster system based on a type of switchable sites. Six $[MoO]^{4+}$ units can be taken up under reducing conditions to form the [$Mo_{63}V_6$] cluster and can be expelled upon oxidation. Color code: [$Mo^{VI}{}_{17}$] units blue; [$Mo^{V}{}_{2}$] linkers red; [$V^{IV}{}_{1}$] linker green; additional $[Mo^{VI}O]^{4+}$ units yellow.

## [$Mo_{57}V_{12}$]

Nevertheless, the external cavities of the [$Mo_{57}V_6$] cluster can be filled with metal centers other than Mo. Thus, in 2000, Yang *et al.* [20] synthesized the cluster with six additional $V^V$ metal centers, *i.e.*, $[Mo_{57}(V^{V}O)_6V^{IV}{}_{6}O_{186}(NO)_6(OH)_3(H_2O)_{12}]^{15-}$, abridged as [$Mo_{57}V_{12}$] or [$Mo^{VI}{}_{51}Mo^{V}{}_{6}V^{V}{}_{6}V^{IV}{}_{6}$].

## 4.2. Giant Sphere-Type. Super-Fullerenes and Keplerates

The spherical clusters, belonging to the so-called giant sphere-type structure, show the highest possible (icosahedral $I_h$) symmetry and play a special role in the realm of the new huge POM/MOB clusters.

According to the topological principle, which designates the building blocks and the linkers, these spherical clusters can be formulated as {[pentagonal building block]$_{12}$ [linker]$_{30}$}.

The giant spheres, built up exclusively out of 12 identical building blocks, are called "dodecameric" clusters, and can be formulated as [building block]$_{12}$.

The giant sphere-class structure has two subclasses, namely:

(1) *complete giant sphere*, with dinuclear linkers and 132 metal centers;
(2) *down-sized giant sphere*, with mononuclear linkers and 102 metal centers.

*4.2.1. Complete Giant Sphere*

The POM/MOB clusters, with the [$M_{132}$] general formula, belong to the complete giant sphere-subtype and contain dinuclear linkers; their generic name can be written as {[pentagonal building block]$_{12}$ [dinuclear linker]$_{30}$}.

## [$Mo_{132}$]

In 1998, Müller and his coworkers[5] synthesized the prototype of the spherical POM/MOB clusters, namely the very large spherical mixed-valence $[Mo_{132}O_{372}(CH_3COO)_{30}(H_2O)_{72}]^{42-} \equiv [Mo^{VI}{}_{72}Mo^{V}{}_{60}O_{372}(CH_3COO)_{30}(H_2O)_{72}]^{42-}$ cluster, abridged as [$Mo_{132}$], [$Mo^{VI}{}_{72}Mo^{V}{}_{60}$] or [$Mo_{11}$]$_{12}$. The new cluster was named alternatively *giant ball /giant sphere*.

The topological formulation {[pentagonal building block]$_{12}$ [dinuclear linker]$_{30}$} also differentiate the non-reduced metal centers, that form the building blocks, from the reduced metal centers, which form the linkers. By applying this principle, the [$Mo_{132}$] cluster can be written as $\{[(Mo^{VI})Mo^{VI}{}_5O_{21}(H_2O)_6]_{12}\ [Mo^{V}{}_2O_4(CH_3COO)]_{30}\}^{42-}$, abridged as $\{[(Mo^{VI})Mo^{VI}{}_5]_{12}\,[Mo^{V}{}_2]_{30}\}$.

The 12 pentagonal $[(Mo^{VI})Mo^{VI}{}_5O_{21}(H_2O)_6]^{6-} \equiv [(Mo^{VI})Mo^{VI}{}_5]$ building blocks with the central $MoO_7$ pentagonal bipyramid, having a $C_5$ symmetry (Figure 1), are connected by 30 dumbbell $[Mo^{V}{}_2O_4(CH_3COO)]^{+} \equiv [Mo^{V}{}_2]_{30}$ dinuclear linkers.

The structure of the [$Mo_{132}$] POM/MOB cluster is displayed in Figures 9, 10a and 11a.

The central $Mo^{VI}$ atoms of the 12 [$(Mo^{VI})Mo^{VI}{}_5$] pentagons define the 12 corners, while the 30 dinuclear [$Mo^{V}{}_2$] linkers define the 30 edges of an icosahedron. Thus, the [$Mo_{132}$] molecule shows an overall icosahedral symmetry, having $C_5$, $C_3$ and $C_2$ axes.

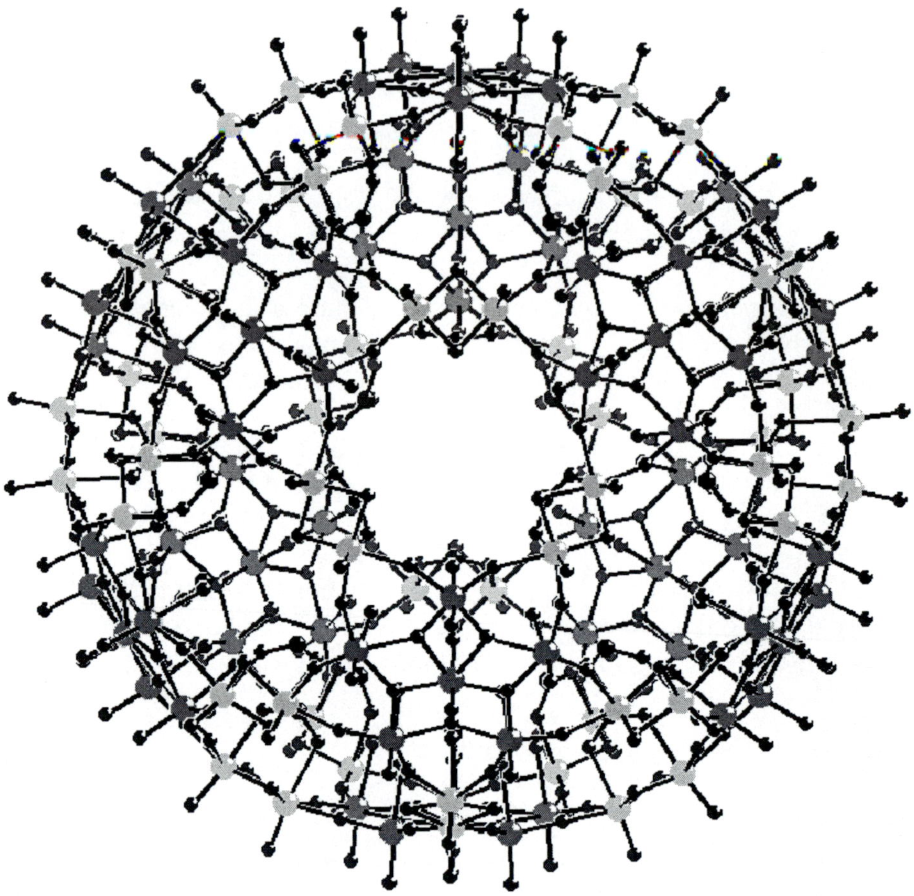

Figure 9: Structure of the $[Mo^{VI}_{72}Mo^{V}_{60}O_{372}(CH_3COO)_{30}(H_2O)_{72}]^{42-} \equiv [Mo_{132}]$ POM/ MOB cluster in ball-and-stick representation. Color code: $Mo^{VI}$ blue; $Mo^{V}$ yellow; C black; O red.

The $[Mo_{132}]$ giant sphere also represents a topological model for spherical viruses (*e.g.*, the $[M_{132}]$ general formula can be associated to the most simple satellite tobacco necrosis virus STNV, coded by only one gene and having 60 identical protein subunits, that correspondingly contain only 12 pentagonal capsomers, each consisting of five protomers). The structural comparison between the $[Mo_{132}]$ giant sphere and the satellite tobacco necrosis virus (STNV) will be shown below ( Figure 12).

The geometry of the fragment $[Mo^{V}_{2}]_{30}$ is also challenging. The 30 dumbbell $[Mo^{V}_{2}]$ linkers span an Archimedean solid of icosahedral symmetry which corresponds to that of the $C_{60}$ fullerene molecule, more precisely a polyhedron with 12 pentagons and 20 hexagons as faces, *i.e.*, the truncated icosahedron. Thus, both systems with 60 atoms, the $C_{60}$ fullerene molecule (d ≈ 0.75 nm) and the much larger $[Mo^{V}_{2}]_{30}$ fragment (d ≈ 2.5 nm) of the $[Mo_{132}]$ giant sphere, correspond to the Archimedean truncated icosahedron which has the same symmetry as the Platonic icosahedron itself. Therefore, Müller[5] also named the $[Mo_{132}]$ molecule as super-fullerene.

a b

Figure 10: Structural comparison between the $[Mo_{132}]$ complete giant sphere (a) and $[Mo_{72}Fe_{30}]$ downsized giant sphere (b) in polyhedral representation along one of the $C_5$ axis. Both clusters include 12 pentagonal $[Mo^{VI}(Mo^{VI}{}_5)]$ building blocks (blue, with the central $MoO_7$ pentagonal bipyramid in blue-turquoise), but the 30 linkers are different, i.e. dinuclear $[Mo^{V}{}_2]$ (red) (a) and mononuclear $[Fe^{III}{}_1]$ (yellow) (b).

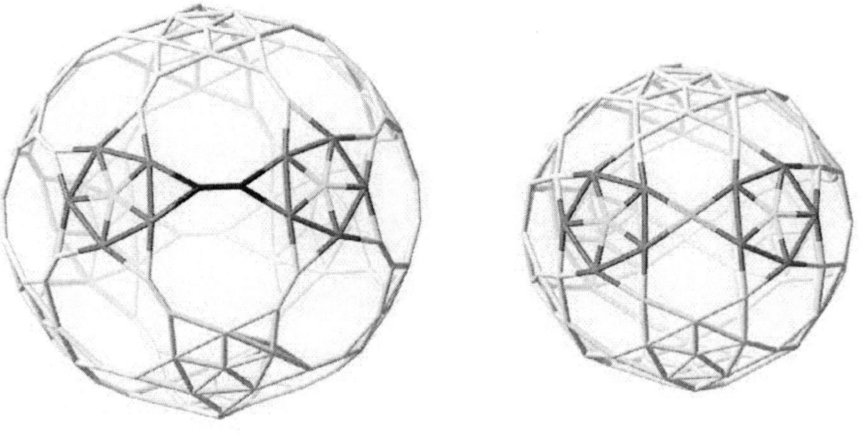

a b

Figure 11: Structural comparison between the $[Mo_{132}]$ complete giant sphere (a) and $[Mo_{72}Fe_{30}]$ downsized giant sphere (b) in wire-frame representation along one of the $C_5$ axis. Highlighted are two of the 12 pentagonal $[(Mo^{VI})Mo^{VI}{}_5]$ building blocks (blue, with the central $MoO_7$ pentagonal bipyramid in blue-turquoise) and one of the 30 linkers, i.e. the dinuclear $[Mo^{V}{}_2]$ (red) (a) and the mononuclear $[Fe^{III}{}_1]$ (yellow) (b).

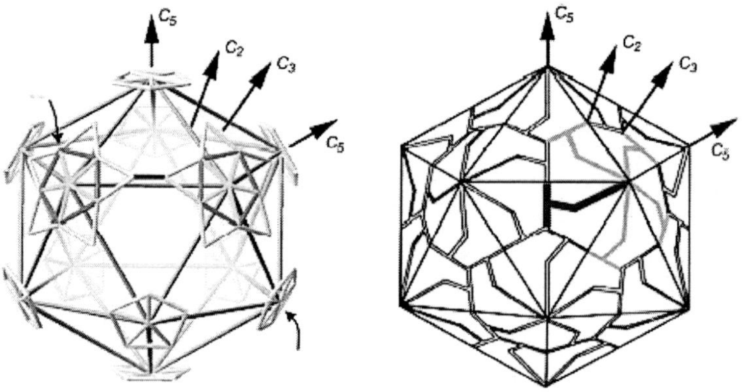


a b

Figure 12: Schematic representation of the icosahedron spanned by the central $Mo^{VI}$ atoms of the 12 $[(Mo^{VI})Mo^{VI}_5]$ pentagons of the $[Mo_{132}]$ giant sphere (a) and of the satellite tobacco necrosis virus (STNV) highlighting five of the 60 protomers (b), showing the $C_5$, $C_3$ and $C_2$ axes.

The structure of the fragment $[Mo^V_2]_{30}$ is displayed in Figure 13a.

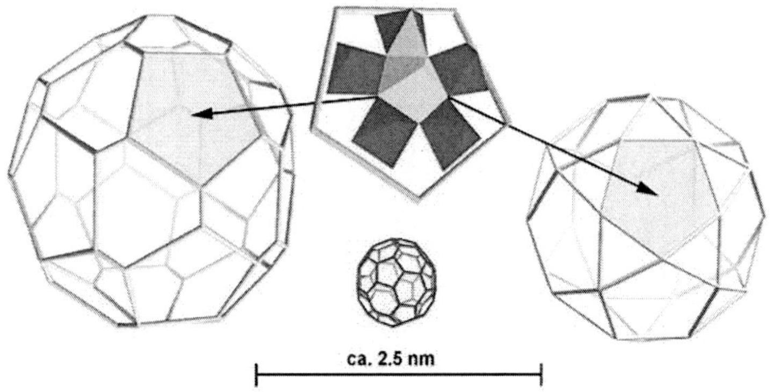


a b c

Figure 13: Structural comparison between the $[Mo^V_2]_{30}$ fragment of the $[Mo_{132}]$ complete giant sphere (a), the $[Fe^{III}_1]_{30}$ fragment of the $[Mo_{72}Fe_{30}]$ down-sized giant sphere (b) and the $C_{60}$ fullerene molecule (c) in wire-frame representation. The three assemblies correspond each to an Archimedean solid, i.e. the truncated icosahedron with 12 pentagonal and 20 hexagonal faces (a and c), respectively the icosidodecahedron with 12 pentagonal and 20 triangular faces (b). A pentagonal $[(Mo)Mo_5]$ building block present in the two clusters is highlighted.

Considering the "dodecameric" topological concept, *i.e.*, [building block]$_{12}$, the $[Mo_{132}]$ molecule can also be formulated as $[(Mo^{VI})^{O}(Mo^{VI}_5)^{I}(Mo^V_5)^{II}]_{12} \equiv [Mo_{11}]_{12}$ (as usual, the superscript 0, I, II indices denote the zero, first and second generation of the corresponding unit). In this case, the term "dodecameric" emphasizes only the fact that giant spheres, as discrete molecules, are considered as formed exclusively of 12 identical $[Mo_{11}]$ building blocks (Figures 3, 24 and 25), that are connected directly, without linkers.

The formulation $[Mo_{11}]_{12}$ takes formally into account that the Mo atoms, in each rather stable $[Mo^V_2]$ dumbbell, belong to two adjacent $[Mo_{11}]$ units. Thus, in this formulation, the

[$Mo_{132}$] giant sphere is built up exclusively of 12 [$Mo_{11}$] building blocks each having a central $MoO_7$ pentagonal bipyramid (Figure 3), such that the fivefold axes are retained in the spherical structure, with the overall icosahedral symmetry ($I_h$).

Since, starting 1985, the $C_{60}$ fullerene or buckyball has been considered "the most beautiful molecule",[21] today the much larger super-fullerene [$Mo_{132}$] can be regarded as "the most aesthetic cluster".[22]

Müller[5] has also proposed to call the [$Mo_{132}$] molecule a *keplerate*, by similarity to Kepler's early cosmological model, as described in his magnum opus *Mysterium Cosmographicum*. Kepler believed that the orbits of the planets could be explained if the ratios between successive orbits were considered to be equivalent to the spheres successively circumscribed around and inscribed within the five Platonic solids (the Tetrahedron, Cube, Octahedron, Dodecahedron and Icosahedron). By analogy, the [$Mo_{132}$] cluster shows concentric spherical shells of the 132 terminal oxygen and 132 Mo atoms, while the centers of the 12 [$(Mo^{VI})Mo^{VI}{}_5$] pentagons, *i.e.*, the 12 $Mo^{VI}$ atoms of the central $MoO_7$ pentagonal bipyramids, span an icosahedron.[5],[9],[11]

The external diameter of the [$Mo_{132}$] giant sphere is of ≈ 2.9 nm and the diameter of the large internal cavity is of ≈ 2.0 nm.

Besides the presented [$Mo_{132}$] cluster, with acetate ligands (also called acetate ball), other similar [$Mo_{132}$] clusters in which the 30 acetate ligands are totally replaced with: formate (formate ball),[23] hypophosphite/phosphinate (hypophosphite ball) [24], monochloroacetate (chloroacetate ball),[25] sulfate (sulfate ball)[26] or monohydrogenphosphate (phosphate)[27] ligands, partially replaced with sulfate ligands (acetate-sulfate balls), [26] have been obtained.

These giant sphere clusters, with a different composition of the ligands, have the following complete molecular formulae: $[Mo^{VI}{}_{72}Mo^{V}{}_{60}O_{372}(HCOO)_{30}(H_2O)_{72}]^{42-}$, $[Mo^{VI}{}_{72}Mo^{V}{}_{60}O_{372}(H_2PO_2)_{30}(H_2O)_{72}]^{42-}$, $[Mo^{VI}{}_{72}Mo^{V}{}_{60}O_{372}(ClCH_2COO)_{30}(H_2O)_{72}]^{42-}$, $[Mo^{VI}{}_{72}Mo^{V}{}_{60}O_{372}(SO_4)_{30}(H_2O)_{72}]^{72-}$,$[Mo^{VI}{}_{72}Mo^{V}{}_{60}O_{372}(HPO_4)_{30}(H_2O)_{72}]^{72-}$, $[Mo^{VI}{}_{72}Mo^{V}{}_{60}O_{372}(CH_3COO)_{12}(SO_4)_{18}(H_2O)_{72}]^{60-}$ and $[Mo^{VI}{}_{72}Mo^{V}{}_{60}O_{372}(CH_3COO)_6(SO_4)_{24}(H_2O)_{72}]^{66-}$, respectively . One should emphasize the high negative charge of the acetate-sulfate balls (60- and 66-) and especially of the sulfate and of the phosphate ball (72-).

The structure of the formate [$Mo_{132}$] giant sphere is displayed in Figure 14.

Figure 14: Structure of the $[Mo^{VI}{}_{72}Mo^{V}{}_{60}O_{372}(HCOO)_{30}(H_2O)_{72}]^{42-}$ POM/MOB cluster in space filling representation. Color code: Mo blue; C black; O red.

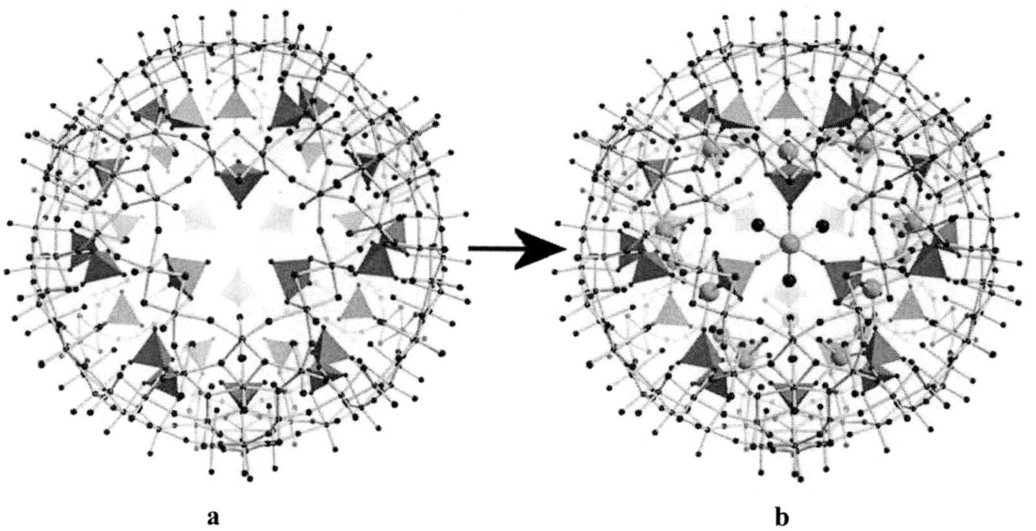

Figure 15: Nucleation process inside the cavity of the $[Mo_{132}O_{372}(HPO_4)_{30}(H_2O)_{72}]^{72-}$ POM/MOB cluster (a). In the presence of molybdate ions in solution and lowering the pH value, 5 new $[MoO_3H]^+$ units are added and the $[H_{15}Mo_{132}(MoO_3H)_5O_{372}(HPO_4)_{30}(H_2O)_{72}]^{52-}$ POM/MOB cluster is formed. The 5 additional $Mo^{VI}$ centers are delocalized over 20 positions (b). Color code: $HPO_4^{2-}$ ligands violet tetrahedra; new $Mo^{VI}$ centers green spheres.

Of special interest is a second phosphate ball, in which the substitution of the acetate ligands occurs at low pH solution in the presence of molybdate ions. In this case, in addition to the replacement of ligands, a nucleation process inside the cavity takes place, such that 5 supplementary $[Mo^{VI}O_3H]^+ \equiv [Mo^{VI}_1]$ units are added to the molecule. The resulting cluster [27] can be abridged as $[Mo_{132}(Mo_5)]$ or $[Mo^{VI}_{72}Mo^{V}_{60}(Mo^{VI}_5)]$. This formula highlights that the 5 inserted $Mo^{VI}$ centers are not of the shell type. The corresponding molecular formula is of $[H_{15}Mo^{VI}_{72}Mo^{V}_{60}(Mo^{VI}O_3H)_5O_{372}(HPO_4)_{30}(H_2O)_{72}]^{52-} \equiv [Mo_{137}O_{387}H_{20}(HPO_4)_{30}(H_2O)_{72}]^{52-}$ or, according to the building block principle, $\{H_{15}[(Mo^{VI})Mo^{VI}_5O_{21}(H_2O)_6]_{12}[Mo^{V}_2O_4(HPO_4)]_{30}[Mo^{VI}O_3H]_5\}^{52-}$.

The formation of the $[Mo_{132}(Mo_5)]$ POM/MOB cluster from the $[Mo_{132}]$ giant sphere via nucleation is shown in Figure 15.

## $[Mo_{102}W_{30}]$

Recently, another complete giant sphere, with a different composition of the metal centers, was synthesized [29]. The new heterometallic giant sphere was formulated as $[Mo_{102}W_{30}O_{372}(CH_3COO)_{30}(H_2O)_{72}]^{42-} \equiv [Mo^{VI}_{42}W^{VI}_{30}Mo^{V}_{60}O_{372}(CH_3COO)_{30}(H_2O)_{72}]^{42-}$, abridged as $[Mo_{102}W_{30}]$ or $[Mo^{VI}_{42}W^{VI}_{30}Mo^{V}_{60}]$.

Due to delocalization of the non-reduced metal centers in the pentagonal building blocks, namely $Mo^{VI}$ and $W^{VI}$, their precise position could not be determined.

Research shows the existence of a series of analogous molybdenum-tungsten mixed metal center giant spheres, with the short formula $[Mo_{132-x}W_x]$, where $x \leq 30$.

*4.2.2. Down-sized Giant Sphere*

The POM/MOB clusters which belong to the down-sized giant sphere-subclass, with the $[M_{102}]$ general formula, contain mononuclear linkers and also can be written as {[pentagonal building block]$_{12}$ [mononuclear linker]$_{30}$}.

## $[Mo_{72}Fe_{30}]$

In 1999, Müller and his team[6] reported the first down-sized giant sphere with the $[Mo_{72}Fe_{30}]$ or $[Mo^{VI}_{72}Fe^{III}_{30}]$ short formula. The $[Mo_{72}Fe_{30}]$ sphere was never isolated as such, but only included in a slightly larger neutral POM/MOB cluster. The complete formula of the larger cluster was presented in several versions,[10],[30],[31] which indicates the difficulties encountered in the attempt of accurate identification in the X-ray diffraction spectra of all atoms of a huge molecule. The definitive formula of the larger neutral POM/MOB cluster was established as $[Mo^{VI}_{72}Fe^{III}_{30}O_{252}(CH_3COO)_{12}\text{-}\{Mo_2O_7(H_2O)\}_2\{H_2Mo_2O_8(H_2O)\}(H_2O)_{91}] \equiv [Mo^{VI}_{78}Fe^{III}_{30}O_{274}H_2(H_2O)_{94}]$, abridged as $[Mo_{78}Fe_{30}]$ or $[Mo_{72}(Mo_6)Fe_{30}]$.

The whole neutral cluster could be considered as a down-sized giant sphere, because only 102 metal centers, *i.e.*, 72 $Mo^{VI}$ and 30 $Fe^{III}$, are located on the shell, while the other 6 $Mo^{VI}$ metal centers are not of the shell type but projected towards the inner cavity. Therefore, the $[Mo_{72}Fe_{30}]$ sphere, also named *iron ball/ iron sphere*, was often presented as being both a discrete molecule and a fragment of a larger cluster.

According to the topological formulation {[pentagonal building block]$_{12}$ [mononuclear linker]$_{30}$}, which differentiates the $Mo^{VI}$ metal centers that form the building blocks from the $Fe^{III}$ metal centers which form the linkers, the $[Mo_{72}Fe_{30}]$ down-sized giant sphere can be written as $\{[(Mo^{VI})Mo^{VI}_5O_{21}(CH_3COO)(H_2O)_n]_{12}\ [Fe^{III}(H_2O)_2]_{30}\}$, abridged as $\{[(Mo^{VI})Mo^{VI}_5]_{12}\,[Fe^{III}_1]_{30}\}$.

The 12 pentagonal $[(Mo^{VI})Mo^{VI}_5O_{21}(CH_3COO)(H_2O)_n]^{7-} \equiv [(Mo^{VI})Mo^{VI}_5]$ building blocks are almost identical, having only slight ligand modifications relative to the $[Mo_{132}]$ complete giant sphere. They are connected however by 30 mononuclear $[Fe^{III}(H_2O)_2]^{3+} \equiv [Fe^{III}_1]$ linkers. By considering the other four equatorial oxygen atoms of the slightly distorted octahedra in addition to the two axial $H_2O$ ligands, the mononuclear linkers can be completely formulated as $[Fe^{III}O_4(H_2O)_2]^{5-}$.

The structure of the $[Mo_{72}Fe_{30}]$ sphere is shown in Figures 10b and 11b.

The central $Mo^{VI}$ atoms of the 12 $[(Mo^{VI})Mo^{VI}_5]$ pentagons define the 12 corners, while the 30 mononuclear $[Fe^{III}_1]$ linkers define the 30 edges of an icosahedron. Thus, the $[Mo_{72}Fe_{30}]$ down-sized giant sphere has an overall icosahedral symmetry.

Similarly to the $[Mo_{132}]$ complete giant sphere, the geometry of the linker fragments, *i.e.*, $[Fe^{III}_1]_{30}$, is very interesting. The 30 $[Fe^{III}_1]$ linkers span an Archimedean solid, in this case an icosidodecahedron, which has 12 triangles and 20 pentagons as faces ($d \approx 2.1$ nm).

The structure of the $[Fe^{III}_1]_{30}$ fragment is shown in Figure 13b. The structure of the $[Mo_{72}Fe_{30}]$ sphere with the two inscribed polyhedra, namely the icosahedron and the icosidodecahedron, is displayed in Figure 16.

Figure 16: Structure of the $[Mo_{72}Fe_{30}]$ down-sized giant sphere in transparent polyhedral representation with the two inscribed polyhedra. The icosahedron (a) is defined by the central $Mo^{VI}$ atoms of the 12 $[(Mo^{VI})Mo^{VI}_5]$ building blocks and the icosidodecahedron (b) by the 30 $[Fe^{III}_1]$ linkers. Color code: $[(Mo^{VI})Mo^{VI}_5]$ units blue, with the $MoO_7$ pentagonal bipyramid blue-turquoise; $[Fe^{III}_1]$ linkers yellow; inscribed polyhedrons red.

By applying the "dodecameric" topological concept for giant spheres, the $[Mo_{72}Fe_{30}]$ sphere can also be formulated as $[(Mo^{VI})^{O}(Mo^{VI}_5)^{I}(Fe^{III}_{5/2})^{II}]_{12} \equiv [Mo^{VI}_6Fe^{III}_{5/2}]_{12}$, analogous to the $[(Mo^{VI})^{O}(Mo^{VI}_5)^{I}(Mo^{V}_5)^{II}]_{12} \equiv [Mo_{11}]_{12}$ formulation for the $[Mo_{132}]$ complete giant sphere. One can observe that by difference to the complete giant spheres, the down-sized giant spheres have only half of the metal centers of second generation.

Subsequently, a second $[Mo_{78}Fe_{30}]$ cluster [32] which contains the $[Mo_{72}Fe_{30}]$ sphere, with the formula $[Mo^{VI}_{72}Fe^{III}_{30}O_{252}(CH_3COO)_{10}\{Mo_2O_7(H_2O)\}$- $\{H_2Mo_2O_8(H_2O)\}_3(H_2O)_{91}]$ was isolated and structurally characterized.

The special interest of the $[Mo_{72}Fe_{30}]$ sphere is due to its magnetic properties, as it contains 30 high-spin $Fe^{3+}$ centers (with s = 5/2), having a total of 150 unpaired electrons (S = 150/2). Thus, the $[Mo_{72}Fe_{30}]$ sphere, more precisely the $[Mo_{78}Fe_{30}]$ POM/MOB cluster, is the discrete molecule with the highest number of paramagnetic centers known up to the present and the largest mesoscopic/nanosized paramagnet. An antiferromagnetic exchange between the near perfect octahedrally coordinated $Fe^{3+}$ centers is mediated by O–Mo–O groups.

The external diameter of the $[Mo_{72}Fe_{30}]$ sphere is of ≈ 2.5 nm and the diameter of its internal cavity is of ≈1.6 nm.

## $[Mo_{102}]$

In 2000, Müller and coworkers[7] reported a new cluster isostructural (but not isoelectronic) to the $[Mo_{72}Fe_{30}]$ sphere, obtained by the direct down-sizing of the $[Mo_{132}]$ complete giant sphere. The new mixed-valence neutral POM/MOB cluster $[Mo^{VI}_{72}Mo^{V}_{30}O_{282}(CH_3COO)_{12}(H_2O)_{78}]$, abridged as $[Mo_{102}]$ or $[Mo^{VI}_{72}Mo^{V}_{30}]$, has rather unusual electronic configuration, with 6 additional electrons delocalized somewhere in the 12 building blocks. Thus, the mean oxidation number of the 72 Mo metal centers of the building blocks slightly decreases from 6+ but $5^{11}/_{12}$+.

According to the topological formulation {[pentagonal building block]$_{12}$ [mononuclear linker]$_{30}$}, the [$Mo_{102}$] cluster can be written as {[($Mo^{VI}$)$Mo^{VI}{}_5O_{21}$ ($CH_3COO$)($H_2O$)$_4$]$_{12}$[$Mo^{V}O(H_2O)$]$_{30}$}, abridged as {[($Mo^{VI}$)$Mo^{VI}{}_5$]$_{12}$ [$Mo^{V}{}_1$]$_{30}$}.

The structure of the [$Mo_{102}$] POM/MOB cluster is shown in Figure 17.

The 30 mononuclear [$Mo^{V}{}_1$] linkers also span an icosidodecahedron, so that the [$Mo_{102}$] sphere is a superfullerene as well.

According to the "dodecameric" topological concept, the [$Mo_{102}$] cluster can be formulated as [$(Mo^{VI})^{O}(Mo^{VI}{}_5)^{I}(Mo^{V}{}_{5/2})^{II}$]$_{12}$ ≡ [$Mo^{VI}{}_6Mo^{V}{}_{5/2}$]$_{12}$.

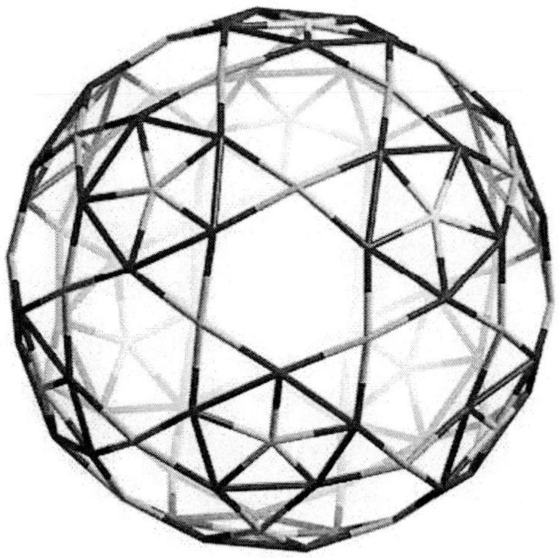

Figure 17: Structure of the neutral [$Mo^{VI}{}_{72}Mo^{V}{}_{30}O_{282}(CH_3COO)_{12}(H_2O)_{78}$] ≡ [$Mo_{102}$] POM/MOB cluster in wire-frame representation along one of the $C_2$ axis. The 12 pentagonal [($Mo^{VI}$)$Mo^{VI}{}_5$]$_{12}$ building blocks (blue, with the central $MoO_7$ pentagonal bipyramid in blue-turquoise) are connected by 30 mononuclear [$Mo^{V}{}_1$] linkers (yellow).

## [$Mo_{78}Fe_{24}$]

In 2002, Müller *et al.*[33] reported a new iron ball, more precisely a (24) iron ball, with the constitutional formula [$Mo_{78}Fe_{24}O_{258}(CH_3COO)_{20}(H_2O)_{84}$]$^{2-}$ ≡ [$Mo^{VI}{}_{72}Fe^{III}{}_{24}Mo^{V}{}_6O_{258}(CH_3COO)_{20}(H_2O)_{84}$]$^{2-}$, abridged as the [$Mo_{78}Fe_{24}$] or [$Mo^{VI}{}_{72}Fe^{III}{}_{24}Mo^{V}{}_6$].

Following the topological formulation {[pentagonal building block]$_{12}$ [mononuclear linker]$_{30}$}, the [$Mo_{78}Fe_{24}$] down-sized giant sphere can be written abridged as {[($Mo^{VI}$)$Mo^{VI}{}_5$]$_{12}$ [$Fe^{III}{}_1$]$_{24}$ [$Mo^{V}{}_1$]$_6$ }. The cluster contains two different mononuclear linkers, *i.e.*, 24 [$Fe^{III}(H_2O)$]$^{3+}$ and 6 [$Mo^{V}O(H_2O)$]$^{3+}$.

Extended Hückel molecular orbital calculations[33] showed that the 6 (4d) electrons of the $Mo^{V}$ linker positions are partially delocalized over the 72 $Mo^{VI}$ positions of the pentagonal [($Mo^{VI}$)$Mo^{VI}{}_5O_{21}(H_2O)_5$]$^{6-}$ building blocks. Consequently, the oxidation numbers of the Mo metal centers are not rigurously charged 6+, respectively 5+ and a short formula of the type [$Mo^{VI/V}{}_{78}Fe^{III}{}_{24}$] is more accurate than [$Mo^{VI}{}_{72}Fe^{III}{}_{24}Mo^{V}{}_6$].

The acetate ligands are located inside the cluster and are coordinated in a bidentate fashion to the metal centers, bridging preferentially Mo and Fe sites.

The structure of the $[Mo_{78}Fe_{24}]$ POM/MOB cluster is displayed in Figure 18.

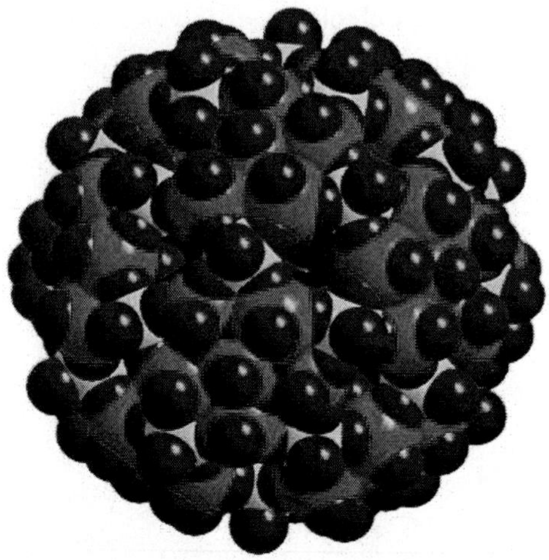

Figure 18: Structure of the $[Mo^{VI}_{72}Fe^{III}_{24}Mo^{V}_{6}O_{258}(CH_3COO)_{20}(H_2O)_{84}]^{2-} \equiv [Mo_{78}Fe_{24}]$ POM/MOB cluster in space filling representation along one of the $C_5$ axis. Color code: $Mo^{VI}$ blue; $Fe^{III}$, $Mo^{V}$ yellow; C black; O red.

The central $Mo^{VI}$ atoms of the pentagonal building blocks define an icosahedron, showing an overall icosahedral symmetry. The 30 mononuclear linkers of the $[Fe^{III}_{24}Mo^{V}_{6}] \equiv \{[Fe^{III}_{1}]_{24}[Mo^{V}_{1}]_{6}\}$ fragment span an icosidodecahedron, which proves that the $[Mo_{78}Fe_{24}]$ down-sized giant sphere is a superfullerene.

Following the "dodecameric" topological formulation, the $[Mo_{78}Fe_{24}]$ sphere can be written as $[(Mo^{VI})^{O}(Mo^{VI}_{5})^{I}((Fe^{III}_{4/5}Mo^{V}_{1/5})_{5/2})^{II}]_{12} \equiv [Mo^{VI}_{6}((Fe^{III}_{4/5}Mo^{V}_{1/5})_{5/2})]_{12}$.

The magnetic properties of the $[Mo_{78}Fe_{24}]$ POM/MOB cluster are also remarkable. The molecule has 24 high-spin $Fe^{3+}$ centers (with s = 5/2) and a total of 120 unpaired electrons (S=150/2). Thus, the $[Mo_{78}Fe_{24}]$ sphere is the second largest mesoscopic/nanosized paramagnet, after the analogous $[Mo_{72}Fe_{30}]$ sphere. An antiferromagnetic exchange between the octahedrally coordinated $Fe^{3+}$ centers is mediated by O–Mo–O groups. Because the 6 (4d) electrons of the $Mo^{5+}$ centers (with s=1/2) are partially delocalized and strongly coupled, their contribution to the overall magnetic moment is negligible.

## $[Mo_{80}V_{22}]$

In 2002, the Bielefeld team[34] obtained another heterometallic down-sized giant sphere, the $[Mo_{80}V_{22}O_{282}(SO_4)_{12}(H_2O)_{66}]^{28-} \equiv [Mo^{VI}_{72}Mo^{V}_{8}V^{IV}_{22}O_{282}(SO_4)_{12}(H_2O)_{66}]^{28-}$ POM/MOB cluster, abridged as $[Mo_{80}V_{22}]$ or $[Mo^{VI}_{72}Mo^{V}_{8}V^{IV}_{22}]$. The $[Mo_{80}V_{22}]$ cluster was also called *vanadium ball/vanadium sphere*.

According to the topological formulation {[pentagonal building block]$_{12}$ [mononuclear linker]$_{30}$}, the $[Mo_{80}V_{22}]$ cluster can be written as $\{[(Mo^{VI})Mo^{VI}_{5}O_{21}(SO_4)(H_2O)_3]_{12}$

$[Mo^{V}O(H_2O)]_8$ $[V^{IV}O(H_2O)]_{22}\}$, abridged as $\{[(Mo^{VI})Mo^{VI}{}_5]_{12}$ $[Mo^{V}{}_1]_8$ $[V^{IV}{}_1]_{22}$ $\}$. The 30 mononuclear linkers are no longer identical, 22 of them having a $V^{IV}$ metal center and the other 8 a $Mo^{V}$ metal center.

As in the case of the $[Mo_{78}Fe_{24}]$ cluster, extended Hückel molecular orbital calculations [34] indicated the 8 supplementary 4d electrons of the $Mo^{V}$ linker positions as being partially delocalized over the 72 $Mo^{VI}$ positions of the pentagonal building blocks, resulting in strong spin pairing and favoring ring configurations, in which the $Mo^{V}$ linkers occupy neighboring sites. Thus, a $[Mo^{VI/V}{}_{80}V^{IV}{}_{22}]$ short formula is more accurate than $[Mo^{VI}{}_{72}Mo^{V}{}_8V^{IV}{}_{22}]$.

The structure of the $[Mo_{80}V_{22}]$ POM/MOB cluster is shown in Figure 19.

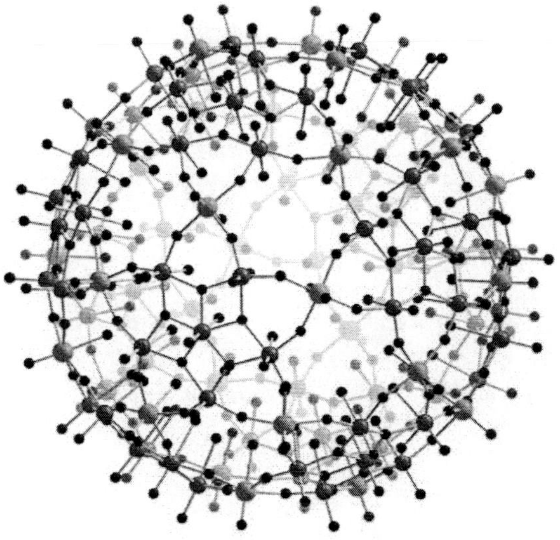

Figure 19: Structure of the $[Mo^{VI}{}_{72}Mo^{V}{}_8V^{IV}{}_{22}O_{282}(SO_4)_{12}(H_2O)_{66}]^{28-} \equiv [Mo_{80}V_{22}]$ POM/MOB cluster in ball-and-stick representation along one of the $C_5$ axis. Color code: $Mo^{VI}$ blue; $Mo^{V}$, $V^{IV}$ green; O red (the distorted $SO_4{}^{2-}$ ligands are omitted for clarity).

The central $Mo^{VI}$ atoms of the pentagonal building blocks and the 30 mononuclear linkers span the usual Platonic solid for giant spheres, *i.e.*, an icosahedron. In turn, the 30 mononuclear linkers of the $[Mo^{V}{}_8V^{IV}{}_{22}] \equiv \{[Mo^{V}{}_1]_8[V^{IV}{}_1]_{22}\}$ fragment span an Archimedean solid, namely a rhombicosidodecahedron, with 20 triangles, 30 squares and 12 pentagons as faces, bearing out that the $[Mo_{80}V_{22}]$ cluster is also a super-fullerene.

In accordance to the "dodecameric" topological concept, the $[Mo_{80}V_{22}]$ down-sized giant sphere can be formulated as $[(Mo^{VI})^{O}(Mo^{VI}{}_5)^{I}((Mo^{V}{}_{4/15}V^{IV}{}_{11/15})_{5/2})^{II}]_{12}$ $\equiv$ $[Mo^{VI}{}_6((Mo^{V}{}_{4/15}V^{IV}{}_{11/15})_{5/2})]_{12}$.

The magneto-chemistry of the $[Mo_{80}V_{22}]$ cluster is also notable. Beside the paramagnetism due to the 22 $V^{4+}$ centers (with s = 1/2), a strong antiferromagnetism, generated by exchange interactions between the 22 $[VO]^{2+}$ linkers and mediated by O–$Mo^{V/VI}$–O groups, can be observed. Because the 8 $Mo^{5+}$ centers (with s = 1/5) are strongly spin-orbit coupled, they have only a negligible contribution to the overall magnetic moment.

As mentioned, all giant spheres with complete or down-sized structure contain an icosahedron and show an overall icosahedral symmetry.

The linker fragments of all giant spheres span an Archimedean solid: the truncated icosahedron, icosidodecahedron or rhombicosidodecahedron, therefore being super-fullerenes.

The complete $[M_{132}]$ and the down-sized $[M_{102}]$ giant spheres obey the mathematical relation N = 12a + 20b + 30c + 60d (where: N = number of metal centers; a, b, c = 0 or 1; d = arbitrary integer) and can be considered as keplerates. [35]

## 5. GIANT WHEELS

The ring-shaped clusters that exhibit the so-called giant wheel-type structure are the main constituents of the well-known molybdenum blue, whose composition remained an enigma for many generations of chemists.

According to the topological principle, giant wheels are considered "tetradecameric" or "hexadecameric" clusters. They are constructed out of 14 or 16 identical building blocks and can be formulated as $[\text{building block}]_{14}$ or $[\text{building block}]_{16}$, more precisely $[Mo_{11}]_{14}$ or $[Mo_{11}]_{16}$.

Consequently, the giant wheel class structure has two subclasses, namely:

*a) common giant wheel* (*Bielefeld structure*), with 154 metal centers, which is "teradecameric";

*b) enlarged giant wheel*, with 176 metal centers, which is "hexadecameric".

### 5.1. Common Giant Wheels

The POM/MOB clusters which belong to the common giant wheel-subtype, with the $[Mo_{154}]$ general formula, exhibit the Bielefeld structure, a term to be discussed later. They are considered "tetradecameric" clusters and can be written as $[\text{building block}]_{14} \equiv [Mo_{11}]_{14}$.

There are several known variants of this structure with deffect(s), *i.e.*, having a slightly reduced number of metal centers.

#### $[Mo_{154}]$

In 1995, Müller and his Bielefeld team[36] reported the first prototype of the ring-shaped POM/MOB clusters, namely the mixed-valence molecule formulated initially as $[Mo_{154}(NO)_{14}O_{420}(OH)_{28}(H_2O)_{70}]^{(25\pm5)-}$ and subsequently[9] as $[Mo_{154}(NO)_{14}O_{448}H_{14}(H_2O)_{70}]^{28-} \equiv [Mo^{VI}_{126}Mo^{V}_{28}(NO)_{14}O_{448}H_{14}(H_2O)_{70}]^{28-}$. The short formula can be written as $[Mo_{154}]$ or $[Mo_{11}]_{14}$ and also as $[Mo_{154}(NO)_{14}]$ or $[Mo_{11}(NO)]_{14}$, to underline the presence of $NO^{3-}$, *i.e.*, nitroso ligands.

The huge cluster was the first synthesized and structurally characterized molecule with over 100 metal centers. Having 154 metal centres (126 $Mo^{VI}$ and 28 $Mo^{V}$), 700 non-hydrogen atoms and a relative molecular mass of 23,638, the new molecule was named giant wheel. This lead to the following metaphorical statement in *New Scientist* : "Big wheels rolls back the molecular frontier" .[37]

We propose the name Bielefeld structure for the common giant wheel-subtype, given that the $[Mo_{154}]$ cluster/structure was first called *Bielefeld giant wheel* (*Bielefelder Riesenrad*), instantly rousing the interest and enthusiasm of the scientific community for the new generation of POM/MOB clusters. Another reason for adopting this name is that it was the Bielefeld University where almost all huge POM/MOB clusters have been synthesized.

The Bielefeld structure of the $[Mo_{154}(NO)_{14}]$ POM/MOB cluster is shown in Figures 20, 21 and 22.

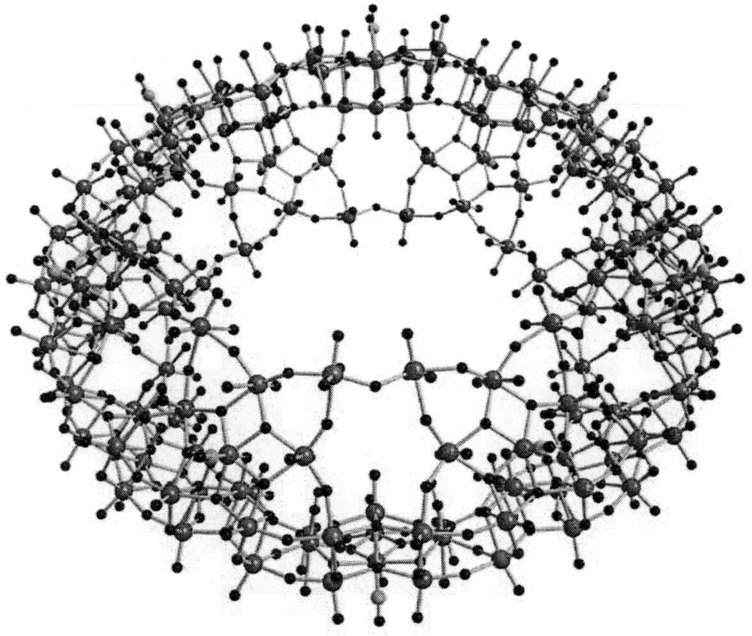

Figure 20: Bielefeld structure of the $[Mo_{154}(NO)_{14}O_{448}H_{14}(H_2O)_{70}]^{28-} \equiv [Mo_{154}(NO)_{14}]$ POM/MOB cluster in ball-and-stick representation. Color code: Mo blue; O red; N green.

Considering the "tetradecameric" topological concept, *i.e.*, [building block]$_{14}$ $[Mo_{154}]$ the common giant wheel is constructed out of 14 identical $[Mo_{11}]$ building blocks (see Figures 27 and 28) and can be formulated as $[Mo_{11}]_{14} \equiv [(Mo^{VI})^{O}(Mo^{VI}{}_{5}))^{I}(Mo^{VI/V}{}_{5})^{II}]_{14}$.

Another topological description is based on the $[Mo_8]$ unit/building block (Figures 2 and 20), considered as a fragment of the larger $[Mo_{11}]$ building block, namely $[Mo_{11}] \equiv \{[Mo_2]$ $[Mo_8]$ $[Mo_1]\}$. Thus, 14 $[Mo_8]$ building blocks are linked by 14 $[Mo_2]$ units, as well as by 14 $[Mo_1]$ units localized at the equatorial plane, to form the complete $\{[Mo_2]$ $[Mo_8]$ $[Mo_1]\}_{14}$ ring of the $[Mo_{154}]$ common giant wheel.

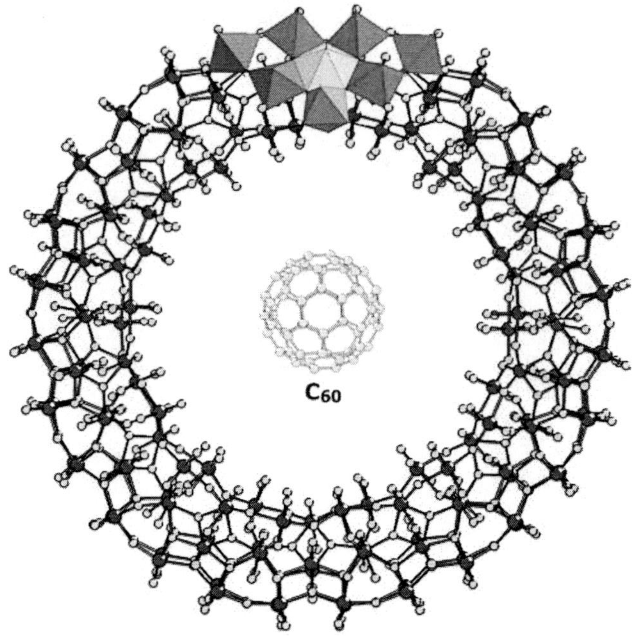


Figure 21: Bielefeld structure of the [Mo154(NO)14O448H14(H2O)70]28- ≡ [Mo154(NO)14] POM/MOB cluster in another ball-and-stick representation, showing one [Mo8] building block with the central MoO7 pentagonal bipyramid in polyhedral representation. For size-comparison a C60 fullerene molecule is also shown. Color code: Mo blue; O red; N green, [Mo8] units blue polyhedra.

In 1999,[38] the pure molybdenum oxide-based homologue (practically an isoPOM) of the first $[Mo_{154}(NO)_{14}]$ giant wheel, which contains oxide instead of nitroso ligands, *i.e.*, 14 $[Mo^{VI}O]^{4+}$ instead of 14 $[Mo^{VI}NO]^{3+}$ groups, with the formula $[Mo_{154}O_{462}H_{14}(H_2O)_{70}]^{14-}$ ≡ $[Mo^{VI}{}_{126}Mo^{V}{}_{28}O_{462}H_{14}(H_2O)_{70}]^{14-}$ was also synthesized. It can be considered as the true "pure" prototype of the giant wheels.

Both $[Mo_{154}(NO)_{14}]$ and the pure $[Mo_{154}]$ giant wheels turned out to comprise better the structure and construction of the common giant wheel. The common giant wheel can be regarded once more as a "tetradecamer" with $D_{7d}$ symmetry if the 14 hydrogen atoms are excluded. It can be formally generated by linking 140 $MoO_6$ octahedra and 14 $MoO_6(NO)$ or $MoO_7$ pentagonal bipyramids, forming the $[Mo_8]$, $[Mo_2]$ and $[Mo_1]$ building blocks/units, which are each present in the structure 14 times. Starting from these units, the complete ring is built up following well-defined rules, as: (i) the two $MoO_6$ octahedra which are not directly connected to the central $MoO_6(NO)$ or $MoO_7$ pentagonal bipyramid are fused to two neighboring $[Mo_8]$ units via corners; (ii) neighboring $[Mo_8]$ units are additionally fused together by the $[Mo_2]$ units, thus completing the inner-ring of the parts of the upper or lower half of the ring structures which are built in the same way and (iii) the complete ring system is constructed when the second half is rotated by around 360/14° relative to the first and fused to it through $[Mo_1]$ units, which are located at the equator of the (complete) ring. [10],[39]

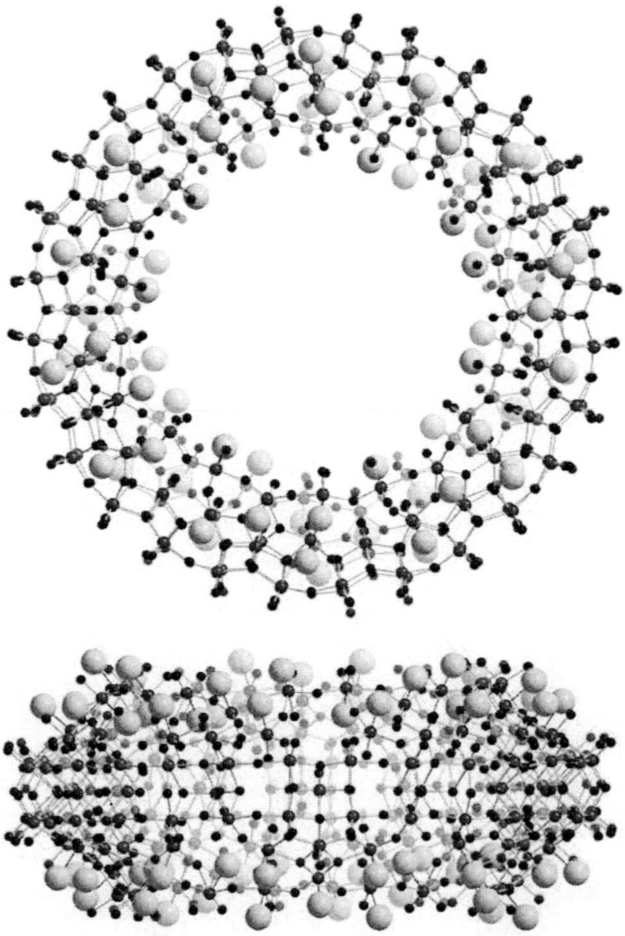

Figure 22: Detailed Bielefeld structure of the $[Mo_{154}(NO)_{14}O_{448}H_{14}(H_2O)_{70}]^{28-}$ ≡ $[Mo_{154}(NO)_{14}]$ POM/MOB cluster in ball-and-stick representation, along the $C_7$ axis (top) and perpendicular to it (bottom), highlighting the 70 $H_2O$ ligands. Color code: Mo blue; O red; N green; $H_2O$ blue-turquoise.

The construction of the Bielefeld structure of the $[Mo_{154}]$ POM/BOM cluster is shown in Figure 23a.

The $[Mo_{154}]$ cluster has an external diameter of ≈ 3.7 nm, a thickness of ≈ 1.5 nm and a large internal cavity with a diameter of ≈ 2.0 nm.

Also synthesized were other $[Mo_{154}]$ clusters, with different ligand composition, such as those with hypophosphite/phosphinate,[40] methanol, [9],[39] α-cyclodextrine[25] and even with various amino acids as ligands, such as α-alanine,[25] valine[25] and diprotonated cystine.[41] The corresponding formulae are as follows: $[Mo_{154}O_{462}H_{14}(H_2PO_2)_7(H_2O)_{54}]^{21-}$, $[Mo_{154}O_{462}H_{14}(CH_3OH)_8(H_2O)_{70}]^{14-}$, $[Mo_{154}O_{462}H_{14}(C_{36}H_{60}O_{30})_2(H_2O)_{70}]^{14-}$, respectively $[Mo_{154}O_{462}H_{14}(CH_3\text{–}CH(NH_3^+)\text{–}COO^-)_{12}(H_2O)_{46}]^{14-}$, $[Mo_{154}O_{462}H_{14}(CH_3\text{–}CH(CH_3)\text{–}CH(NH_3^+)\text{–}COO^-)_{12}(H_2O)_{46}]^{14-}$, and $[Mo_{154}O_{462}H_{14}(COOH\text{–}CH(NH_3^+)\text{–}CH_2\text{–}S\text{–}S\text{–}CH_2\text{–}CH(NH_3^+)\text{–}COO^-)_{11}(H_2O)_{48}]^{3-}$.

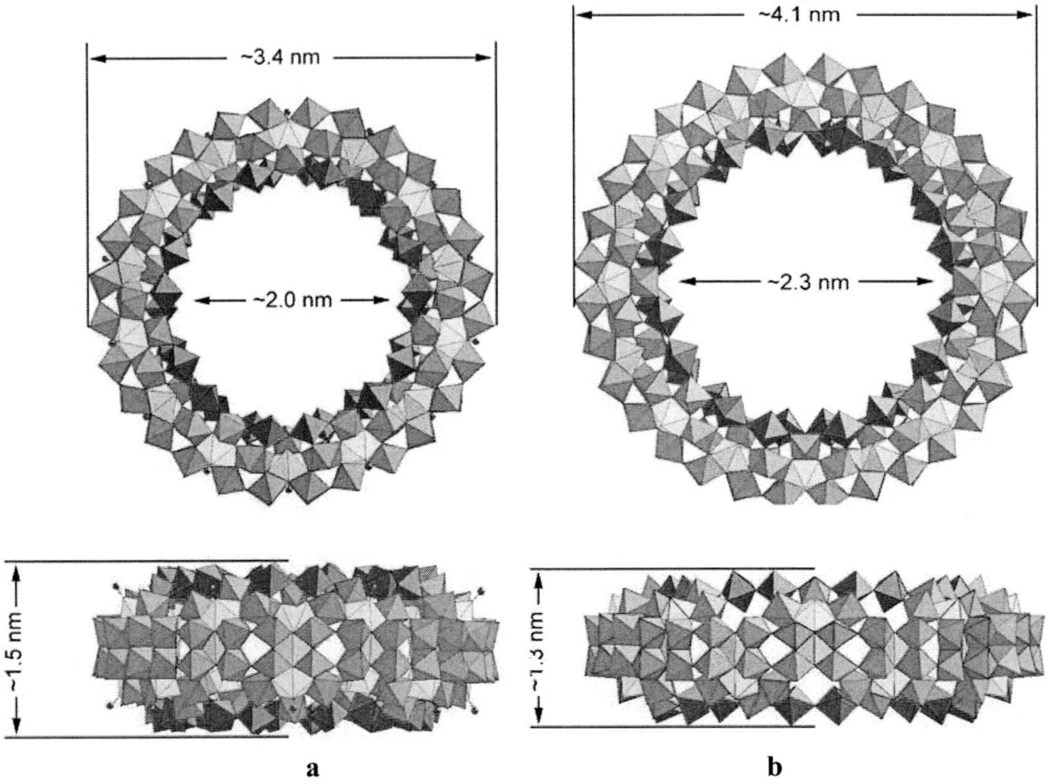


Figure 23: Polyhedral representation of the $[Mo_{154}] \equiv [Mo_{11}]_{14}$ (a) and $[Mo_{176}] \equiv [Mo_{11}]_{16}$ (b) POM/MOB clusters along the $C_7$ axis (a) and $C_8$ axis (b) (top) and perpendicular to it (bottom). Three different units are present: the 14(a)/16(b) $[Mo_8]$ units (blue, with the central $MoO_7$ pentagonal bipyramid in blue-turquoise) which are linked by 14(a)/16(b) $[Mo_2]$ units (red) and by 14(a)/16(b) $[Mo_1]$ units localized at the equatorial plane (yellow).

## $[Mo_{154-2x}]$

Besides the complete ring-system of the $[Mo_{154}]$ common giant wheel, formulated according to the building block principle as $[Mo_{11}]_{14}$ or $\{[Mo_2]_{14}[Mo_8]_{14}[Mo_1]_{14}\}$, a number of species with defect(s) have also been synthesized. In such lacunary species, a number of $[Mo_2]$ units are missing, so that they can be formulated as $\{[Mo_2]_{14-x}\text{-}[Mo_8]_{14}[Mo_1]_{14}\}$, where x = number of missing $[Mo_2]$ units.

Until now, the following lacunary "tetradecameric" species have been reported: $[Mo_{142}]$ (x = 6) [42], $[Mo_{144}]$ (x = 5) [43], $[Mo_{146}]$ (x = 4) [44] and $[Mo_{152}]$ (x = 1) [43]. The complete formulae of these giant wheels with defect(s) are the following: $[Mo_{142}O_{452}H_{14}(H_2O)_{58}]^{26-} \equiv [Mo^{VI}_{114}Mo^{V}_{28}O_{452}H_{14}(H_2O)_{58}]^{26-}$, $[Mo_{144}O_{452}H_{14}(H_2O)_{60}]^{24-} \equiv [Mo^{VI}_{116}Mo^{V}_{28}O_{452}H_{14}(H_2O)_{60}]^{24-}$, $[Mo_{146}O_{442}H_{14}(H_2O)_{58}]^{22-} \equiv [Mo^{VI}_{116}Mo^{V}_{28}O_{442}H_{14}(H_2O)_{58}]^{22-}$ and $[Mo_{152}O_{457}H_{14}(H_2O)_{68}]^{16-} \equiv [Mo^{VI}_{118}Mo^{V}_{28}O_{457}H_{14}(H_2O)_{68}]^{16-}$.

The hexa-lacunary Bielefeld structure of the $[Mo_{142}]$ POM/MOB cluster, a common giant wheel with a high number of defects (x = 6), is displayed in Figure 24.

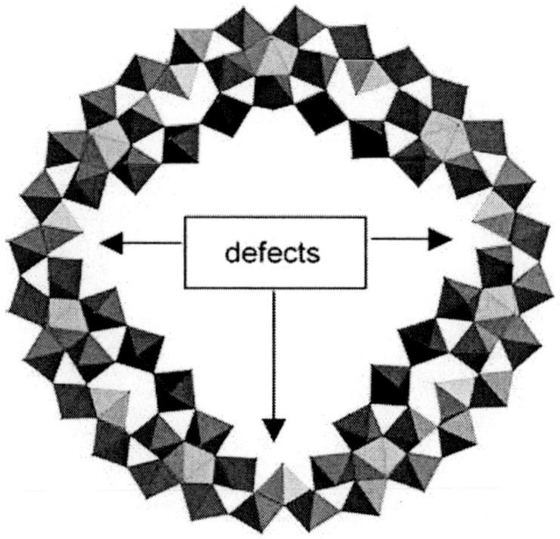


Figure 24: Hexa-lacunary Bielefeld structure of the $[Mo_{142}O_{452}H_{14}(H_2O)_{58}]^{26-}$ ≡ $[Mo_{142}]$ POM/MOB cluster in polyhedral representation. The defects with missing $[Mo_2]$ units are highlighted. Color code: $[Mo_8]$ units blue, with the central $MoO_7$ pentagonal bipyramid in blue-turquoise; $[Mo_2]$ units red; $[Mo_1]$ units yellow.

All reported giant wheels with missing $[Mo_2]$ units are pure molybdenum oxide-based species. Pure "tetradecameric" giant wheels, be they complete or with defect(s), can be generally formulated as $\{[Mo_2]_{14-x}[Mo_8]_{14}[Mo_1]_{14}\}$ ≡ $\{[Mo^{VI}{}_2O_5(H_2O)_2]_{14-x}$ $[Mo^{VI/V}{}_8O_{26}(\mu_3\text{-}O)_2H(H_2O)_3Mo^{VI/V}]_{14}]^{(14+2x)-}$, where $x$ = number of missing $[Mo_2]$ units. The formula shows that all "tetradecameric" giant wheels, with complete ($x = 0$) or lacunary ($x = 1$–$6$) Bielefeld structure, contain 28 $Mo^V$ centers. The difference between the species is made by the number of missing $Mo^{VI}$ centers that build the $[Mo_2]$ units.

## $[Mo_{154\text{-}2x\text{-}y}]$

The situation has became more complicated after the synthesis of a "tetradecameric" giant wheel with defects, in which not only $[Mo_2]$ units are missing, but also $[Mo_1]$ units as well. Such lacunary species can be formulated as $[Mo_{154\text{-}2x\text{-}y}]$ ≡ $\{[Mo_2]_{14-x}$ $[Mo_8]_{14}$ $[Mo_1]_{14-y}\}$, where $x$ = missing $[Mo_2]$ units; $y$ = missing $[Mo_1]$ units.

The reported lacunary common giant wheel, [45] from which 6 $[Mo_2]$ and 4 $[Mo_1]$ units are missing, has the complete formula of $[Mo_{138}O_{416}H_6(CH_3COO)_6(H_2O)_{58}]^{32-}$ ≡ $[Mo^{VI}{}_{110}Mo^V{}_{28}O_{416}H_6(CH_3COO)_6(H_2O)_{58}]^{32-}$, abridged as $[Mo_{138}]$ or $[Mo^{VI}{}_{110}Mo^V{}_{28}]$ ($x = 6$; $y = 4$).

As the absence of several equatorial $[Mo_1]$ linker units could destabilize the structure, these linkers are replaced by acetate ligands, which coordinate metal centers of $[Mo_8]$ and $[Mo_2]$ units. Consequently, lacunary common giant wheels with missing $[Mo_1]$ units are not pure molybdenum oxide-based species and the general formula for all pure "tetradecameric" giant wheels would not describe them accurately.

One should also notice that all missing metal centers are $Mo^{VI}$ which build the $[Mo_2]$ and $[Mo_1]$ units.

## 5.2. Enlarged Giant Wheels

The ring-shaped POM/MOB clusters, which belong to the enlarged giant wheel-subtype, have the [$Mo_{176}$] general formula. They can be considered as "hexadecameric" clusters and written as [building block]$_{16}$ ≡ [$Mo_{11}$]$_{16}$.

### [$Mo_{176}$]

In 1998, Müller and co-workers[46] reported an even larger giant wheel, obtained at least theoretically through enlarging the pure molybdenum oxide-based [$Mo_{154}$] common giant wheel. The huge mixed-valence POM/MOB cluster, with the initial formula of $[Mo_{176}O_{528}H_n(H_2O)_{80}]^{(32-x)-}$, was subsequently formulated[9] as $[Mo_{176}O_{528}H_{16}(H_2O)_{80}]^{16-}$ ≡ $[Mo^{VI}_{144}Mo^{V}_{32}O_{528}H_{16}(H_2O)_{80}]^{16-}$, abridged as [$Mo_{176}$] ≡ [$Mo^{VI}_{144}Mo^{V}_{32}$]. Thus, the [$Mo_{176}$] enlarged giant wheel contains 176 metal centers, *i.e.*, 144 $Mo^{VI}$ and 32 $Mo^{V}$. The [$Mo_{176}$] cluster was also called *large wheel*, as against the [$Mo_{154}$] cluster which was affectionately named *little wheel.*

The structure of the [$Mo_{176}$] POM/MOB cluster is shown in Figures 23b and 25.

According to the "hexadecameric" topological concept, *i.e.*, [building block]$_{16}$, the [$Mo_{176}$] enlarged giant wheel is constructed out of 16 identical [$Mo_{11}$] building blocks (see Fig. 26) and can be formulated as [$Mo_{11}$]$_{16}$ ≡ $[(Mo^{VI})^{O}(Mo^{VI}{}_5))^{I}(Mo^{VI/V}{}_5)^{II}]_{16}$. Thus, the [$Mo_{176}$] enlarged giant wheel contains two additional [$Mo_{11}$] building blocks as compared to the [$Mo_{154}$] common giant wheel.

By taking into account the second topological description which is based on the [$Mo_8$] unit/building block, the [$Mo_{176}$] enlarged giant wheel can be written as {[$Mo_2$] [$Mo_8$] [$Mo_1$]}$_{16}$. The linking of the 160 $MoO_6$ octahedra and 16 $MoO_7$ pentagonal bipyramids, that form the [$Mo_8$], [$Mo_2$] and [$Mo_1$] building blocks/units, which are each present in the structure 16 times, follows the same build up rules to the complete ring as described for the [$Mo_{154}$] common giant wheel (Figure 23b). The enlarged giant wheel can be considered once more as a "hexadecamer" with $D_{8d}$ symmetry if the 16 hydrogen atoms are excluded.

The explicit general formula presented for "pure" common giant wheels can also be extrapolated to the "pure" enlarged giant wheel and once more generalized for all pure molybdenum oxide-based giant wheels, as follows: $\{[Mo_2]_{n-x}[Mo_8]_n[Mo_1]_n\}$ ≡ $\{[Mo^{VI}{}_2O_5(H_2O)_2]_{n-x}\ [Mo^{VI/V}{}_8O_{26}(\mu_3\text{-}O)_2H(H_2O)_3Mo^{VI/V}]_n\}^{(n+2x)-}$, where n = 14 or 16, x = number of missing [$Mo_2$] units. (It should be noticed that lacunary giant wheels with missing [$Mo_1$] units are no longer "pure" species and do not obey this formula).

The [$Mo_{176}$] cluster has an external diameter of ≈ 4.1 nm, a thickness of ≈ 1.3 nm and a very large internal cavity with a diameter of ≈ 2.3 nm. It is somewhat surprising that the thickness of the [$Mo_{176}$] enlarged giant wheel is smaller than that of the [$Mo_{154}$] ordinary giant wheel, *i.e.*, 1.3 *vs*. 1.5 nm. This is due to the fact that the larger structure has a lower curvature and the [$Mo_2$] units are no longer tensioned.

In the same year, Müller's team[47] obtained a second [$Mo_{176}$] giant wheel by exchanging $H_2O$ partially with methanol ligand. The cluster was initially formulated as $[Mo_{176}O_{528}H_n(CH_3OH)_{17}(H_2O)_{63}]^{(32-x)-}$, and subsequently[9] as $[Mo_{176}O_{528}H_{16}(CH_3OH)_{17}(H_2O)_{63}]^{16-}$ ≡ $[Mo^{VI}_{144}Mo^{V}_{32}O_{528}H_{16}(CH_3OH)_{17}(H_2O)_{63}]^{16-}$.

Figure 25: Structure of the $[Mo_{176}O_{528}H_{16}(H_2O)_{80}]^{16-} \equiv [Mo_{176}]$ POM/MOB cluster in space filling representation. Color code: Mo blue; O red.

A few months later, Jiang *et al.* [48] reported independently a neutral enlarged giant wheel, with the formula of $[Mo^{VI}_{148}Mo^{V}_{28}O_{496}(OH)_{36}(H_2O)_{60}]$, which appears to be inaccurate.

## $[Mo_{176-y}]$

The lower curvature of the $[Mo_{176}]$ cluster and the reduced tension of the ring were believed to account for several failures of synthesizing enlarged giant wheels with defect(s).

Nevertheless, recently, [49] a species with missing $[Mo_1]$ units was reported. Such lacunary "hexadecameric" giant wheels can be formulated as $[Mo_{176-y}] \equiv \{[Mo_2]_{16}\ [Mo_8]_{16}\ [Mo_1]_{16-y}\}$, where $y$ = missing $[Mo_1]$ units.

The respective enlarged giant wheel with defects has the complete formula of $[Mo_{173}O_{510}H_{10}(CH_3COO)_3(H_2O)_{83}]^{14-} \equiv [Mo^{VI}_{141}Mo^{V}_{32}O_{510}H_{10}(CH_3COO)_3(H_2O)_{83}]^{14-}$, abridged as $[Mo_{173}]$ or $[Mo^{VI}_{141}Mo^{V}_{32}]$ ($y = 3$).

In the structure of the lacunary $[Mo_{173}]$ enlarged giant wheel the 3 missing $[Mo_1]$ units are replaced by 3 acetate ligands, as previously encountered for the lacunary $[Mo_{138}]$ common giant wheel.

## $[Mo_{176-2x-y}]$

Another enlarged giant wheel with defects was reported, from which not only $[Mo_1]$ units but also $[Mo_2]$ units are missing. Such lacunary species can be formulated as $[Mo_{176-2x-y}] \equiv \{[Mo_2]_{16-x}\ [Mo_8]_{16}\ [Mo_1]_{16-y}\}$, where x = missing $[Mo_2]$ units; y = missing $[Mo_1]$ units.

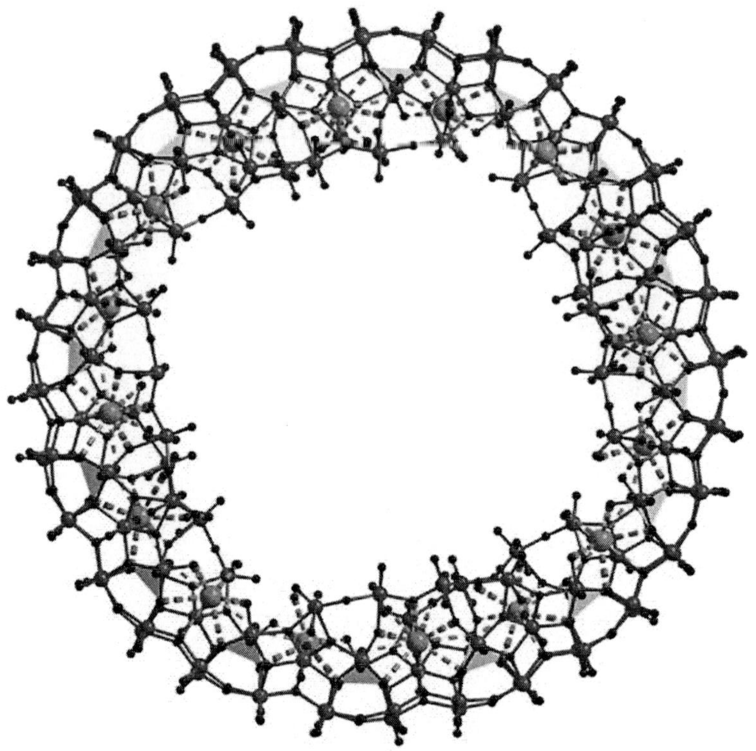

Figure 26: Structure of the $[Mo_{146}]$ POM/MOB cluster showing also the encapsulated $(KSO_4)_{16}$ ring in combined ball-and-stick and polyhedral representation. Color code: Mo small blue spheres; O small red spheres; encapsulated $K^+$ ions large violet spheres; encapsulated $SO_4^{2-}$ ions yellow tetrahedra.

The enlarged lacunary giant wheel from which 7 $[Mo_2]$ and all 16 $[Mo_1]$ units are missing,[50] was formulated as $[Mo_{146}O_{429}(KSO_4)_{16}(H_2O)_{50}]^{30-}$ ≡ $[Mo^{VI}{}_{114}Mo^{V}{}_{32}O_{429}(KSO_4)_{16}(H_2O)_{50}]^{30-}$, abridged as $[Mo_{146}]$ or $[Mo^{VI}{}_{114}Mo^{V}{}_{32}]$. It has an unusual topological formulation for a giant wheel, *i.e.*, without $[Mo_1]$ units, namely $\{[Mo_2]_9\ [Mo_8]_{16}\}$ ($x = 7$; $y = 16$). It also contains a novel inclusion species, which comprises 16 $K^+$ and 16 $SO_4^{2-}$ ions, and is abbreviated as $(KSO_4)_{16}$. Thus, an unusual encapsulated 64-memebered $(KOSO)_{16}$ ring is formed, where the presence of the $K^+$ and $SO_4^{2-}$ ions is important for the stability of the structure. Being encapsulated these ionic species do not belong to the proper discrete cluster molecule. A complete formula of the type $\{16\ K^+,\ 16\ SO_4^{2-} \subset [Mo_{146}O_{429}(H_2O)_{50}]^{14-}\}^{30-}$ seems to be more appropriate. The 16 encapsulated sulfate ions occupy approximately the positions of the 16 missing $[Mo_1]$ units.

The structure of the $[Mo_{146}]$ POM/MOB cluster is displayed in Figure 26.

As in the case of the common giant wheels with defects, all missing metal centers are $Mo^{VI}$ which build the $[Mo_2]$ and $[Mo_1]$ units.

## 5.3. Giant Spheres *vs.* Giant Wheels - A Complex Relationship

The relationship between giant spheres and giant wheels is obvious if one employs the topological principle based on the $[Mo_{11}]$ building block/motif/unit. [24],[10] According to this principle, (complete) giant spheres and giant wheels can be considered as discrete clusters constructed exclusively out of a number of $[Mo_{11}] \equiv [(Mo)^{O}(Mo_5)^{I}(Mo_5)^{II}]$ building blocks and exhibit the $[Mo_{11}]_n \equiv [(Mo)^{O}(Mo_5)^{I}(Mo_5)^{II}]_n$ general formula, where n = 12, 14 or 16.

Shape difference between the spherical and ring-shaped POM/MOB clusters is due to the unlike symmetry of their $[Mo_{11}]$ building blocks.

(Complete) giant spheres are obtained under stronger reducing conditions, so that all five peripheral metal centers of the second generation $(Mo_5)^{II}$ units are reduced to $Mo^{V}$. This ultimately results in a fivefold local symmetry, *i.e.*, $C_5$ of the $[Mo_{11}]$ building blocks, which will generate a $[Mo_{11}]_{12} \equiv [Mo_{132}]$ (complete) giant sphere.

On the other hand, giant wheels are prepared under less reducing conditions, with weaker reducing agents, and only two of the five peripheral metal centers of the $(Mo_5)^{II}$ units are reduced to $Mo^{V}$. This results in a non-equivalence between the five metal centers and a perturbation of the local symmetry. Consequently, the $[Mo_{11}]$ building blocks exhibit a distorted fivefold symmetry, *i.e.*, $C_s$, and will generate a $[Mo_{11}]_{14} \equiv [Mo_{154}]$ common giant wheel or a $[Mo_{11}]_{16} \equiv [Mo_{176}]$ enlarged giant wheel.

Thus, it is the higher or lower symmetry of the $[Mo_{11}]$ building blocks that eventually determines whether a giant sphere or a giant wheel will be formed.

The relationship between the basic structural units that build giant molecular spheres and wheels is displayed in Figures 27 and 28.

Down-sized giant spheres, with the $[Mo_{102}] \equiv [Mo_{72}M'_{60}]$ general formula, where M' = $Mo^{V}$, $Fe^{III}$ or $V^{IV}$, can also be considered as being formed out of 12 homo- or especially heterometallic $[Mo_6M'_{5/2}] \equiv [(Mo)^{O}(Mo_5)^{I}(M'_{5/2})^{II}]$ building blocks, where each such unit has only half of the metal centers of second generation as compared to the $[Mo_{11}]$ building block. Thus, down-sized giant spheres can be topologically formulated as $[Mo_6M'_{5/2}]_{12} \equiv [Mo_{72}M'_{60}]$.

Out of all entirely molybdenum oxide-based giant spheres and giant wheels, the $[Mo_{132}] \equiv [Mo^{VI}_{72}Mo^{V}_{60}]$ complete giant sphere contains the highest percentage of reduced metal centers (45.45%), followed by the $[Mo_{102}] \equiv [Mo^{VI}_{72}Mo^{V}_{30}]$ down-sized giant sphere (29.41%) and by the $[Mo_{154}] \equiv [Mo^{VI}_{126}Mo^{V}_{28}]$ common giant wheel and $[Mo_{176}] \equiv [Mo^{VI}_{144}Mo^{V}_{32}]$ enlarged giant wheel (22.22%).

Consequently, it seems natural that, by oxidation, the "highly reduced" $[Mo_{132}]$ giant sphere can be converted into "less reduced" relative clusters. Thus, by oxidation with (dissolved) atmospheric oxygen the $[Mo_{132}]$ complete giant sphere converts into the $[Mo_{102}]$ down-sized giant sphere, [7] while, by oxidation with $H_2O_2$, respectively with $H_2O_2$ in the presence of $Co^{2+}$ ions, it converts into the $[Mo_{176}]$ enlarged giant wheel,[49] and the $[Mo_{173}]$ enlarged giant wheel with defects,[49] respectively.

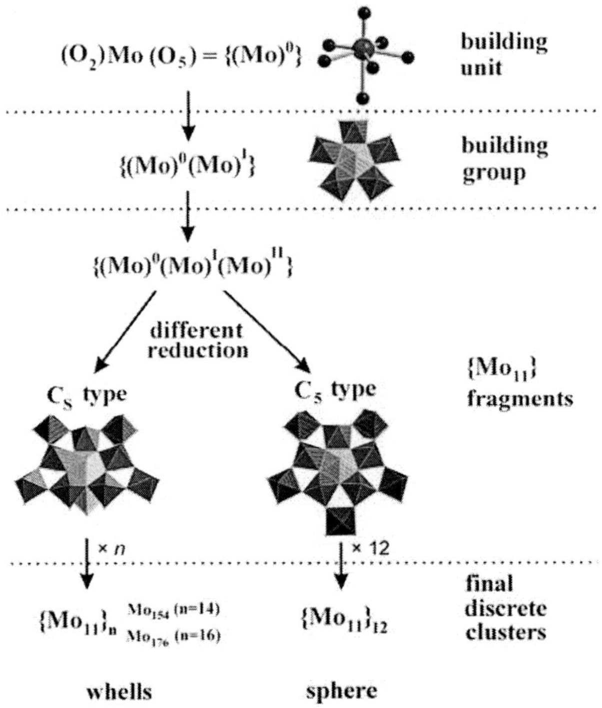


Figure 27: Formal relationship between the basic structural units that construct the $[Mo_{11}]$ building blocks with $C_5$ or $C_S$ symmetry, which further build giant spheres and giant wheels.

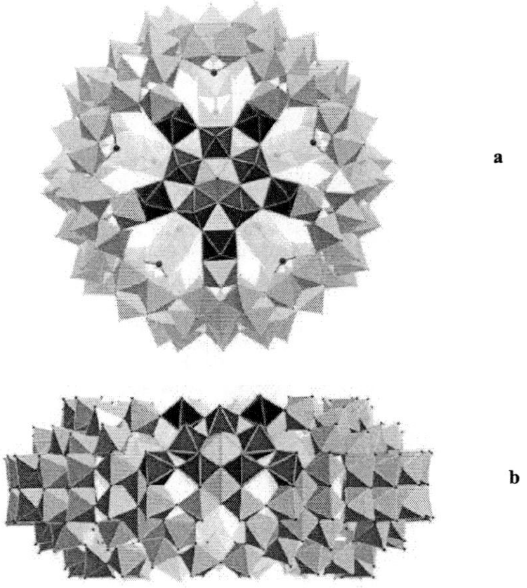


Figure 28: Structural comparison between the $[Mo_{11}]_{12} \equiv [Mo_{132}]$ giant sphere (a) and the $[Mo_{11}]_{14} \equiv [Mo_{154}]$ giant wheel (b), each having one $[Mo_{11}]$ building block highlighted.

Icosahedral giant spheres, also called super-fullerenes or keplerates, as well as ring-shaped giant wheels are probably the most interesting and spectacular nanosized clusters, which are likely to fascinate any chemist.

The upcoming second part of our review will be dedicated to the nanosized inorganic metal-oxygen clusters with special structures, which are related to the basic structures.

## Acknowledgement

The author would like to thank Dr. Hartmut Bögge (Bielefeld University) for providing the figures included in the paper. The author are also grateful to Dr. Daniel A. Lowy (Naval Research Laboratory, Washington, DC) for critically reading the manuscript and useful suggestions.

## REFERENCES

[1] L. Cronin, A. Müller, D. Fenske, in *Encyclopedia of Physical Science and Technology* (ed. R.A. Meyers), 3rd Ed., Vol. 10, Academic Press, New York 2002.
[2] A. Müller, F. Peters, M.T. Pope, D. Gatteschi, *Chem. Rev.*, 1998, 98, 239.
[3] A. Patrut, A. Nicoara, D. Margineanu, *Studia Univ. Babes-Bolyai, Ser. C, Chemia*, 2003, 48, 3.
[4] M.T. Pope, A. Müller, in *Polyoxometalate Chemistry: From Topology via Self-Assembly to Applications* (eds. M.T. Pope and A. Müller), Kluwer, Dordrecht, 2001.
[5] A. Müller, E. Krickemeyer, H. Bögge, M. Schmidtmann, F. Peters, *Angew. Chem. Int. Ed.*, 1998, 37, 3360.
[6] A. Müller, S. Sarkar, S.Q.N. Shah, H. Bögge, M. Schmidtmann, S. Sarkar, P. Kögerler, B. Hauptfleisch, A.X. Trautwein, V. Schünemann, *Angew. Chem. Int. Ed.*, 1999, 38, 3238.
[7] A. Müller, S.Q.N. Shah, H., Bögge, M. Schmidtmann, P. Kögerler, B. Hauptfleisch, S. Leiding, K. Wittler, *Angew. Chem. Int. Ed.*, 2000, 30, 1614.
[8] A. Müller, P. Kögerler, *Coord. Chem. Rev.*, 1999, 182, 3.
[9] A. Müller, P. Kögerler, C. Kuhlmann, *Chem. Commun.*, 1999, 1347.
[10] A. Müller, P. Kögerler, H. Bögge, *Struct. Bond.*, 2000, 96, 203.
[11] A. Müller, P. Kögerler, A.W.M. Dress, *Coord. Chem. Rev.*, 2001, 222, 193.
[12] J.M. Lehn, *Supramolecular Chemistry: Concepts and Perspectives*, VCH, Weinheim, 1995.
[13] S. Zhang, D. Liao, M. Shao, Y. Tang, *J. Chem. Soc. Chem. Commun.*, 1986, 835.
[14] A. Müller, E. Krickemeyer, S. Dillinger, H. Bögge, W. Plass, R. Rohlfing, *Naturwissenschaften*, 1993, 80, 560.
[15] A Müller, E. Krickemeyer, S. Dillinger, H. Bögge, W. Plass, A. Proust, L. Djoczik, C. Menke, J. Meyer, R. Rohlfing, *Z. Anorg. Allg. Chem.*, 1994, 620, 599.
[16] A. Müller, W. Plass, E. Krickemeyer, S. Dillinger, H. Bögge, A. Armatage, A. Proust, C. Beugholt, U. Bergmann, *Angew. Chem. Int. Ed.*, 1994, 33, 849.
[17] P. Kögerler, A. Müller, *J. Appl. Phys.*, 2003, 93, 7101.

[18] A. Müller, J. Meyer, E. Krickemeyer, C. Beugholt, H. Bögge, F. Peters, M. Schmidtmann, P. Kögerler, M.J. Koop, *Chem. Eur. J.*, 1998, 4, 1000.
[19] A. Nicoara, A. Patrut, D. Margineanu, A. Müller, *Electrochem. Commun.*, 2003, 5, 511.
[20] W. Yang, X. Lin, C. Lu, H. Zhuang, J. Huang, *Inorg. Chem.*, 2000, 39, 2706.
[21] H. Aldersey-Williams, *The Most Beautiful Molecule: The Discovery of the Buckyball*, Wiley, New York, 1995.
[22] A. Müller, A. Patrut, *Rev. Chim. (Bucharest)*, 2001, 52, 289.
[23] A. Müller, V.F. Fedin, C. Kuhlmann, H. Bögge, M. Schmidtmann, *Chem. Commun.*, 1999, 927.
[24] A. Müller, S. Polarz, S.K. Das, E. Krickemeyer, H. Bögge, M. Schmidtmann, B. Hauptfleisch, *Angew. Chem. Int. Ed.*, 1999, 38, 3241.
[25] C. Kuhlmann, Ph.D. Thesis, University of Bielefeld, 2001.
[26] A. Müller, E. Krickemeyer, H. Bögge, M. Schmidtmann, B. Botar, M.O. Talismanova, *Angew. Chem.*, 2003, 115, 2131.
[27] A. Müller, S. K. Das, S. Talismanov, E. Beckmann, H. Bögge, M. Schmidtmann, A. Merca, A. Berkle, L. Allouche, Y. Zhou, L. Zhang, *Angew. Chem. Int. Ed.*, 2003, 42, 5039.
[28] A. Müller, S. Roy, *Coord. Chem. Rev.*, 2003, 245, 153.
[29] A. Patrut et al., to be published.
[30] A. Müller, S.K. Das, P. Kögerler, H. Bögge, M. Schmidtmann, A.X. Trautwein, V. Schünemann, E. Krickemeyer, W. Preetz, *Angew. Chem. Int. Ed.*, 2000, 39, 3414.
[31] A. Müller, P. Kögerler, *Coord. Chem. Rev.*, 1999, 182, 3.
[32] A. Müller, S.Q.N. Shah, C. Kuhlmann, M.C. Letzel, *Chem. Commun.*, 2002, 440.
[33] A. Müller, S.K. Das, M.O. Talismanova, P. Kögerler, S.S. Talismanov, M. Luban, E. Krickemeyer, *Angew. Chem.*, 2002, 114, 599.
[34] A. Müller, B. Botar, H. Bögge, P. Kögerler, A. Berkle, *Chem. Commun.*, 2002, 2944.
[35] O. Delgado, A. Dress, A. Müller, in *Polyoxomatalate Chemistry: From Topology via Self-Assembly to Applications* (eds. M.T. Pope and A. Müller), Kluwer, Dordrecht, 2001.
[36] A. Müller, E. Krickemeyer, J. Meyer, H. Bögge, F. Peters, W. Plass, E. Diemann, S. Dillinger, F. Nonnenbruch, M. Randerath, C. Menke, *Angew. Chem. Int. Ed.*, 1995, 34, 2122.
[37] D. Bradley, *New Scientist*, November 1995, 148, 18.
[38] A. Müller, S.K. Das, V.P. Fedin, E. Krickemeyer, C. Beugholt, H. Bögge, M. Schmidtmann, B. Hauptfleisch, *Z. Anorg. Allg. Chem.*, 1999, 625, 1187.
[39] A. Müller, C. Serain, *Acc. Chem. Res.*, 2000, 33, 2.
[40] A. Müller, S.K. Das, H. Bögge, C. Beugholt, M. Schmidtmann, *Chem. Commun.*, 1999, 1035.
[41] A. Müller, S.K. Das, C. Kuhlmann, H. Bögge, M. Schmidtmann, E. Diemann, E. Krickemeyer, J. Hormer, H. Modrow, M. Schindler, *Chem. Commun.*, 2001, 655.
[42] A. Müller, C. Beugholt, M. Koop, S.K. Das, M. Schmidtmann, H. Bögge, *Z. Anorg. Allg. Chem.*, 1999, 625, 1960.
[43] A. Müller, E. Krickemeyer, H. Bögge, M. Schmidtmann, C. Beugholt, S.K. Das, F. Peters, *Chem. Eur. J.*, 1999, 5, 1496.

[44] A. Müller, S.K. Das, V.P. Fedin, E. Krickemeyer, C. Beugholt, H. Bögge, M. Schmidtmann, B. Hauptfleisch, , *Z. Anorg. Allg. Chem.*, 1999, 625, 187.
[45] A. Müller, R. Maiti, M. Schmidtmann, H. Bögge, S.K. Das, W. Zhang, *Chem. Commun.*, 2001, 2126.
[46] A. Müller, E. Krickemeyer, H. Bögge, M. Schmidtmann, C. Beugholt, P. Kögerler, C. Lu, *Angew. Chem. Int. Ed.*, 1998, 37, 1220.
[47] A. Müller, M. Koop, H. Bögge, M. Schmidtmann, C. Beugholt, *Chem. Commun.*, 1998, 1501.
[48] C.C. Jiang, Y.G. Wey, Q. Liu, S.W. Zhang, M.C. Shao, Y.Q. Tang, *Chem. Commun.*, 1998, 1937.
[49] E. Beckmann, Ph.D. Thesis, University of Bielefeld, 2003.
[50] A. Müller, L. Toma, H. Bögge, M. Schmidtmann, P. Kögerler, *Chem. Commun.*, 2003, 2000.

In: Nanostructures: Novel Architecture
Editor: Mircea V. Diudea, pp. 399-413
ISBN: 1-59454-499-9


*Chapter 19*

# Nanosized Inorganic Metal-Oxygen Clusters 2. Special Structures

***Adrian Pătruț[1] and Adrian Nicoară***
Faculty of Chemistry, "Babeş-Bolyai" University
11 Arany János, RO-400028, Cluj Napoca, Romania

## Abstract

Metal-oxygen clusters, also called polyoxometalate/metal oxide-based clusters, are the main representatives of nanosized inorganic molecular clusters. The second part of the review is dedicated to the special structures of the new generation of polyoxometalate/metal oxide-based clusters with supramolecular properties. The special structures are divided into mutant-class and hybrid-class, as they are related to one basic structure or they share elements specific to at least two basic structures. Five special structures, named after the corresponding cluster, are described and explained: $[Mo_{75}V_{20}]$ (with 95 metal centers), $[Mo_{120}Pr_6]$ (with 126 metal centers), $[Mo_{248}]$ (with 248 metal centers), $[Mo_{256}Eu_8]$ (with 264 metal centers) and $[Mo_{368}]$ or the Müller structure (with 368 metal centers). The $[Mo_{368}]$ giant hedgehog cluster, with Müller structure, is the largest synthesized and structurally characterized molecule.

## 1. Introduction

Metal-oxygen clusters are the main representatives of nanosized inorganic clusters.[1] They are also called polyoxometalate/metal oxide-based clusters, POM/MOB clusters for short, a term that encompasses both anionic and neutral molecular species.[2]

The first part of the review was dedicated to the basic structures of the new generation of huge POM/MOB clusters with supramolecular properties, while the second part deals with the special structures.

[1] E-mail address: apatrut@chem.ubbcluj.ro

# 2. Special Structures

In addition to the basic structures and the corresponding POM/MOB clusters, several such clusters have different structures, which we herein call special structures.

These special structures can be further divided into mutant and hybrid structures. Here, the term mutant designates a structure which is closely related to one of the basic structures, while the term hybrid referes to a structure which shares elements specific to at least.two basic structures.

Up to now, five special structures are known, out of which four are mutant-type and one hybrid-type. In present, these structures with no specific name have only one representative. Consequently, each structure can be named after the corresponding cluster.

The five special structures, divided into two classes, according to the increasing number of metal centers, are as follows:

1. **Mutant class**, which includes:
   [$Mo_{75}V_{20}$], with 95 metal centers;
   [$Mo_{120}Pr_6$], with 126 metal centers;
   [$Mo_{248}$], with 248 metal centers;
   [$Mo_{256}Eu_8$], with 264 metal centers.
2. **Hybrid class**, which includes:
   [$Mo_{368}$] (*Müller structure*), with 368 metal centers.

In the construction of special structures out of essential building blocks, molecular growth processes are strongly favored and the resulting clusters can be even larger than those having basic structures. Incidentally, the last three structures correspond to the largest synthesized and structurally characterized molecules known up to the present.

## 2.1. Mutant Structures

The „mutant" structures are closely related to a basic structure, *e.g*, to a giant sphere or giant wheel. Four mutant-type structures are known, one related to the down-sized giant sphere and three to different giant wheels:

[$Mo_{75}V_{20}$], with 95 metal centers;
[$Mo_{120}Pr_6$], with 126 metal centers;
[$Mo_{248}$], with 248 metal centers;
[$Mo_{256}Eu_8$], with 264 metal centers.

### [$Mo_{75}V_{20}$]

In 1999, Müller and coworkers[3] reported an approximately spherical heterometallic POM/MOB cluster, which was formulated as $[Mo^{VI}_{75}V^{IV}_{20}O_{270}(H_2O)_{70}Na_{10}(SO_4)_{10}]^{20-}$, abridged as [$Mo_{75}V_{20}$] or [$Mo^{VI}_{75}V^{IV}_{20}$].

Its structure is related to the down-sized giant sphere-subtype. If down-sized giant spheres[4] are topologically formulated as {[pentagonal building block]$_{12}$ [mononuclear linker]$_{30}$}, the [$Mo_{75}V_{20}$] quasi-sphere corresponds to {[pentagonal building block]$_{10}$ [lacunary pentagonal building block]$_1$ [mononuclear linker]$_{30}$ (encapsulated ring)$_2$}. The corresponding explicit formula is: {[(Mo)$Mo_5O_{21}(H_2O)_3$]$_{10}$ [$Mo_5O_{10}(H_2O)_{10}$] [$MoO_3(H_2O)$]$_{10}$ [$V^{IV}O(H_2O)$]$_{20}$ [$(NaSO_4)_5$]$_2$]}$^{20-}$.

The [$Mo_{75}V_{20}$] cluster contains only 10 pentagonal [$(Mo^{VI})Mo^{VI}{}_5O_{21}(H_2O)_6]^{6-}$ ≡ [$(Mo^{VI})Mo^{VI}{}_5$] building blocks, with the central $MoO_7$ pentagonal bipyramid having $C_5$ symmetry. It also includes a lacunary [$Mo_5O_{10}(H_2O)_{10}$] ≡ [$Mo_5$] pentagonal array, which consists of only 5 $MoO_6$ octahedra, since the central pentagonal bipyramidally coordinated $Mo^{VI}$ center is missing. The lacunary pentagonal building blocks can also be written as two lacunary semipentagonal building blocks, *i.e.*, $[Mo_5O_{10}(H_2O)_{10}]^{10-}$ ≡ $[Mo_{5/2}O_5(H_2O)_5]^{5-}{}_2$, abridged as [$Mo_5$] ≡ [$Mo_{5/2}$]$_2$; thus, a total of 12 building blocks is counted.

The [$Mo_{75}V_{20}$] cluster includes two different mononuclear linkers, namely 10 [$Mo^{VI}O_3(H_2O)$] ≡ [$Mo^{VI}{}_1$] and 20 $[V^{IV}O(H_2O)]^{2+}$ ≡ [$V^{IV}{}_1$] units. This is the only case in which the linker units contain non-reduced $Mo^{VI}$ metal centers.

The two encapsulated $(NaSO_4)^-{}_5$ rings, which complete the structure, are characteristic only for this cluster. The presence of ions that build the ring, *i.e.*, 10 $Na^+$ and 10 $SO_4^{2-}$, is important, maybe mandatory for the stability of the structure. As they are not covalently bound to other atoms of the structure, they do not belong, however, to the proper discrete molecule. Consequently, a complete formula of the type {10 $Na^+$, 10 $SO_4^{2-}$ ⊂ $[Mo_{75}V_{20}O_{270}(H_2O)_{70}]^{10-}$}$^{20-}$, which evidences the encapsulation of the $Na^+$ and $SO_4^{2-}$ ions, is probably more accurate.

The structure of the [$Mo_{75}V_{20}$] POM/MOB cluster is displayed in Figures 1b, 2 and 3.

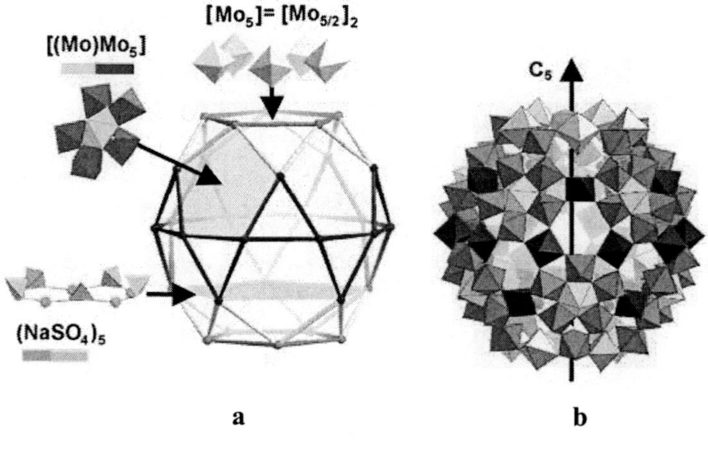


Figure 1: Structure of the [$Mo_{75}V_{20}$] POM/MOB cluster showing also the two encapsulated $(NaSO_4)^-{}_5$ rings: (a) Framework of the [$Mo_{10}V_{20}$] icosidodecahedron with 30 vertices and 32 faces (20 triangles and 12 pentagons) spanned by 10 $Mo^{VI}$ (blue) and 20 $V^{IV}$(red) centers emphasizing the ring-shaped band formed by 10 [$V^{IV}{}_3$] triangles (red). Additional details: (i) one of the 10 [(Mo)$Mo_5$] units which cappe 10 pentagonal faces; (ii) the under-occupied pentagonal [$Mo_5$] array which cappes the other two pentagonal faces (gray) and (iii) one of the two encapsulated $(NaSO_4)^-{}_5$-type rings ($Na^+$ green, $SO_4^{2-}$ yellow tetrahedra). (b) Polyhedral representation of the (approximately) spherical structure of the complete [$Mo_{75}V_{20}$] cluster perpendicular to the $C_5$ axis.

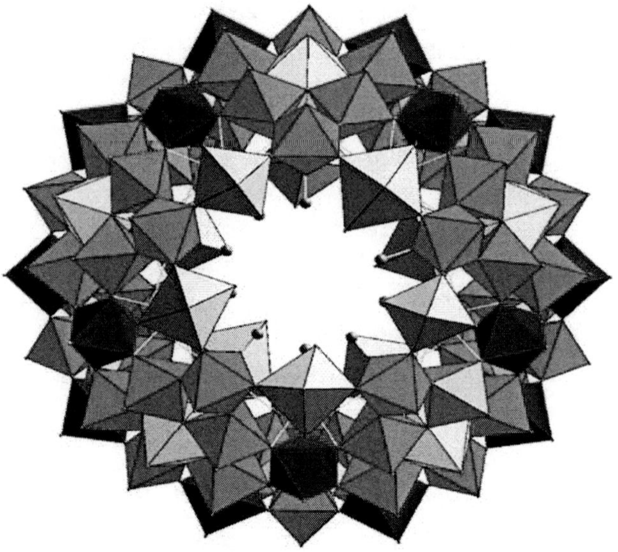

Figure 2: Structure of the $[Mo_{75}V_{20}]$ POM/MOB cluster in another polyhedral representation along the $C_5$ axis, perpendicular to that shown in Figure 1. Color code: $[(Mo)Mo_5]$ units blue, with the central $MoO_7$ pentagonal bipyramid in blue-turquoise; $[V_1]$ units red; $[Mo_5]$ array gray.

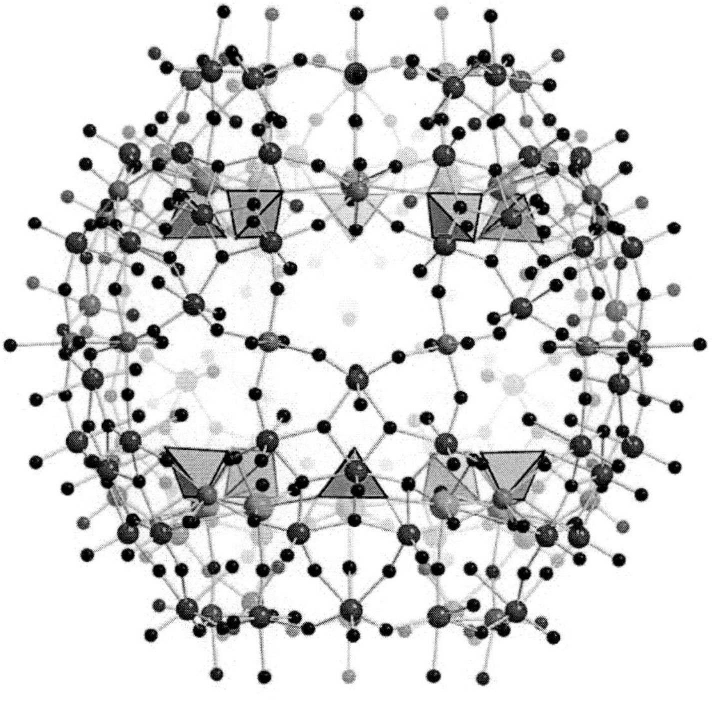

Figure 3: Structure of the $[Mo_{75}V_{20}]$ POM/MOB cluster showing also the two encapsulated $(NaSO_4)^-_5$ rings in combined ball-and-stick and polyhedral representation perpendicular to the $C_5$ axis. Color code: Mo blue spheres; V violet spheres; O red spheres, encapsulated $Na^+$ green spheres; $SO_4^{2-}$ yellow tetrahedra.

As for all giant spheres, the metal centers of the $[Mo^{VI}_{10}V^{IV}_{20}] \equiv \{[Mo^{VI}_1]_{10}[V^{IV}_1]_{20}\}$ linkers fragment span an Archimedean solid, namely an icosidodecahedron (Figure 1a) and can be considered as a super-fullerene. The icosidodecahedron has 12 pentagons and 20 triangles as faces, while the 30 corners are represented by the 30 metal centers of the linkers, *i.e.*, 10 $Mo^{VI}$ and 20 $V^{IV}$ centers. The 20 triangular faces are represented by 10 $[V_3]$ and 10 $[VMo_2]$ triangles. Only 10 of the 12 pentagonal faces, each formed by one $Mo^{VI}$ and four $V^{IV}$ centers, are capped by 10 $[(Mo)Mo_5]$ pentagons. The two remaining pentagonal faces are both spanned by 5 $Mo^{VI}$ centers and are capped by the under-occupied $[Mo_5]$ lacunary pentagonal array.

The second topological formulation of giant spheres, which consider them as "dodecameric" clusters built up exclusively out of 12 identical building blocks without linker units, is not valid for the $[Mo_{75}V_{20}]$ quasi-sphere.

Like the heterometallic $[Mo_{72}Fe_{30}]$ [5], $[Mo_{78}Fe_{24}]$ [6] and $[Mo_{80}V_{22}]$ [7] down-sized giant spheres, the $[Mo_{75}V_{20}]$ molecule has an unusual spin topology with 20 $V^{IV}$ paramagnetic centers (with s = 1/2), exhibiting a total of 20 unpaired electrons (S = 20/2). Remarkably, the 20 $V^{IV}$ centers, which build a ring of 10 linked triangles, form two groups: (i) 10 equatorial $V^{IV}$ centers showing outward-projecting $H_2O$ ligands and trans-terminal oxygen atoms and (ii) 10 remaining $V^{IV}$ centers exhibiting outward-projecting terminal oxygen atoms and trans-$H_2O$ ligands. This results in a new equatorial paramagnetic ring-shaped band with very strong antiferromagnetic exchange interactions.

The $[Mo_{75}V_{20}]$ cluster has a diameter along the $C_5$ axis of ≈ 2.3 nm and perpendicular to it of ≈ 2.5 nm.

Similarly to giant spheres, the $[Mo_{75}V_{20}]$ quasi-sphere is a super-fullerene but, due to topological rules, it is not a keplerate.

## $[Mo_{120}Pr_6]$

In 2002, Müller *et al.*[8] reported a new mixed-valence POM/MOB cluster with the formula of $[Mo_{120}Pr_6O_{366}H_{12}(H_2O)_{78}]^{6-} \equiv [Mo^{VI}_{96}Mo^{V}_{24}Pr^{III}_6O_{366}H_{12}(H_2O)_{78}]^{6-}$, abridged as $[Mo_{120}Pr_6]$ or $[Mo^{VI}_{96}Mo^{V}_{24}Pr^{III}_6]$. The $[Mo_{120}Pr_6]$ cluster contains strong electrophlic open shell metal centers, *i.e.*, lanthanoide cations of large coordination number, incorporated into a reduced highly nucleophilic polyoxomolybdate ring.

The structure is closely related to giant wheels. However, it originates in a hypothetical "dodecameric" down-sized giant wheel and not in the parent "tetradecameric" $[Mo_{154}]$ common giant wheel[9] or "hexadecameric" $[Mo_{176}]$ enlarged giant wheel.[10] This hypothetical down-sized giant wheel with the short formula of $[Mo_{132}]$ or, according to the building block principle, $[Mo_{11}]_{12} \equiv \{[Mo_2]_{12}\ [Mo_8]_{12}\ [Mo_1]_{12}\}$, has not yet been synthesized. Notice that the $[Mo_{132}] \equiv [Mo^{VI}_{108}Mo^{V}_{24}] \equiv [Mo_{11}]_{12}$ down-sized giant wheel, with $D_{6d}$ symmetry, should have the same total number of metal centers as the icosahedral $[Mo_{132}] \equiv [Mo^{VI}_{72}Mo^{V}_{60}] \equiv [Mo_{11}]_{12}$ complete giant sphere.[11]

Thus, the $[Mo_{120}Pr_6]$ molecule can be considered as a new type of "dodecameric" ring-shaped cluster. As compared to the hypothetical "dodecameric" $[Mo_{132}]$ giant wheel, it contains 6 $[Pr_1] \equiv [Pr^{III}(H_2O)_5]^{3+}$ instead of 6 $[Mo_2] \equiv [Mo^{VI}_2O_5(H_2O)_2]^{2+}$ units. Following the building block principle, the $[Mo_{120}Pr_6]$ cluster can be formulated as $\{[Mo_2]_6\ [Pr_1]_6\ [Mo_8]_{12}$

$[Mo_1]_{12}\}$ or, more explicitly, as $\{[Mo^{VI}{}_2O_5(H_2O)_2]_6$ $[Pr^{III}(H_2O)_5]_6$ $[Mo^{VI/V}{}_8O_{26}(\mu_3\text{-}O)_2H(H_2O)_3Mo^{VI/V}]_{12}]^{6-}$. The total number of reduced $Mo^V$ centers in the $[Mo_8]$ units is of 24.

In addition to the 5 $H_2O$ ligands, each $Pr^{III}$ center is bound to the (inner) surface of the cluster by 4 bridging oxygen atoms and exhibits a very high number of coordination, *i.e.*, NC – 9 in the $[Pr^{III}O_4(H_2O)_5]^{5-}$ units. If the disorder of the 5 $H_2O$ molecules of a $[Pr^{III}O_4(H_2O)_5]^{5-}$ unit is ignored, the resulting polyhedron is a tricapped distorted prism.

The structure of the $[Mo_{120}Pr_6]$ POM/MOB cluster is shown in Figure 4.

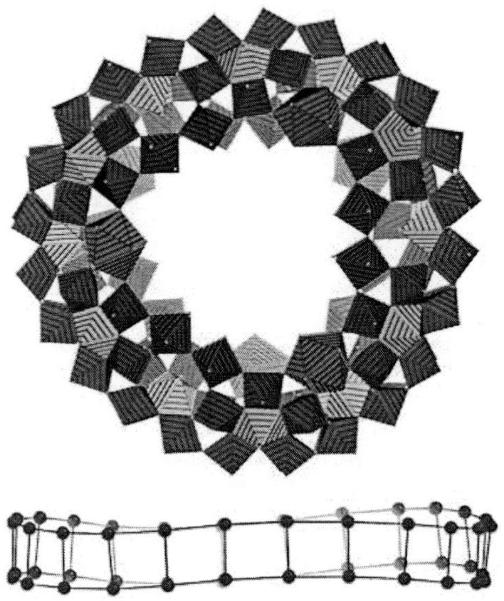

Figure 4: Structure of the $[Mo^{VI}{}_{96}Mo^{V}{}_{24}Pr^{III}{}_6O_{366}H_{12}(H_2O)_{78}]^{6-} \equiv [Mo_{120}Pr_6]$ POM/MOB cluster in polyhedral representation. Color code: $[Mo_8]$ units dark blue, with the central $MoO_7$ pentagonal bipyramid in blue turquoise; $[Mo_2]$ units red; $[Mo_1]$ units yellow; $[Pr_1]$ units green. The marked distances between two equatorial $MoO_7$ pentagonal bipyramids of the $[Mo_8]$ units define the curvature, which is larger if a $[Mo_2]$ unit is replaced by a $[Pr_1]$ unit. The lower part of the figure shows the Mo centers that form the rim of the wheel.

The $Pr^{III}$ centers exert a significant influence on the curvature, which determines the shape and the size of the twisted giant ring. The curvature can be measured by the distance between two equivalent $MoO_7$ pentagonal bipyramids of the $[Mo_8]$ units, separated either by a $[Mo_2]$ unit or by a smaller $[Pr_1]$ unit, *e.g.*, 13 Å *vs.* 11.7 Å.

The $[Mo_{120}Pr_6]$ molecule has a twisted ring-shaped structure with elliptical curvature, by difference to basic giant wheels which are ring-shaped and show a circular curvature. Its external diameter is of ~ 3.2 nm.

## $[Mo_{248}]$

In 1998, Müller and coworkers synthesized a huge POM/MOB cluster which was presented in the early 1999, in *Nature*.[12] The new molecule has the formula: $[Mo_{248}O_{720}H_{16}(H_2O)_{128}]^{16-} \equiv [Mo^{VI}{}_{168}Mo^{V}{}_{80}O_{720}H_{16}(H_2O)_{128}]^{16-}$, abridged as $[Mo_{248}]$ or $[Mo^{VI}{}_{168}Mo^{V}{}_{80}]$.

The structure of the [$Mo_{248}$] molecule is strongly related to the [$Mo_{176}$] enlarged giant wheel.[10] Under special conditions, the [$Mo_{176}$] giant wheel even starts growing, such that two (unstable) [$Mo^{VI}_{12}Mo^{V}_{24}O_{96}(H_2O)_{24}$] ≡ [$Mo_{36}$] neutral fragments cover the cavity of the [$Mo_{176}$] cluster like hub-caps of a wheel, yielding the [$Mo_{248}$] cluster. The two [$Mo_{36}$] hub-caps, which do not exist in separate form, are each connected to 8 terminal oxygen atoms of the 8 [$Mo_2$] units on both the upper and the lower parts of the ring.

The formation of the [$Mo_{248}$] POM/MOB cluster from the [$Mo_{176}$] enlarged giant wheel, is displayed in Figure 5.

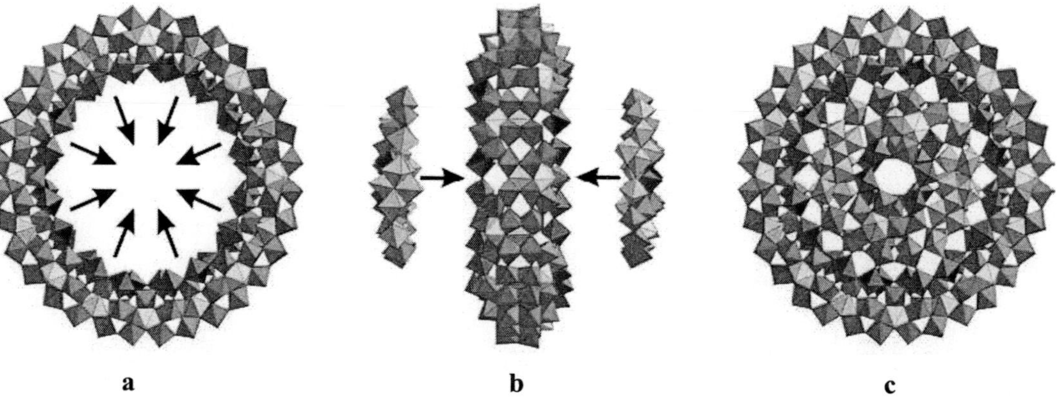

a b c

Figure 5: Schematic view of the relation between the [$Mo_{176}$] and [$Mo_{248}$] POM/MOB clusters in polyhedral representation. Shown are the [$Mo_{176}$] cluster viewed from the top (with arrows indicating the pathway of the growth process) (a) and from the side with the two additional [$Mo_{36}$] fragments acting as hub-caps (b); this ensemble generates the [$Mo_{248}$] cluster (c). Color code: [$Mo_8$] units blue, with the central $MoO_7$ pentagonal bipyramid in blue-turquoise; [$Mo_8$] units only form the [$Mo_{36}$] fragments green; [$Mo_2$] units red; [$Mo_1$] units yellow.

The new type of cavity of the [$Mo_{248}$] POM/MOB cluster is shown in Figure 6.

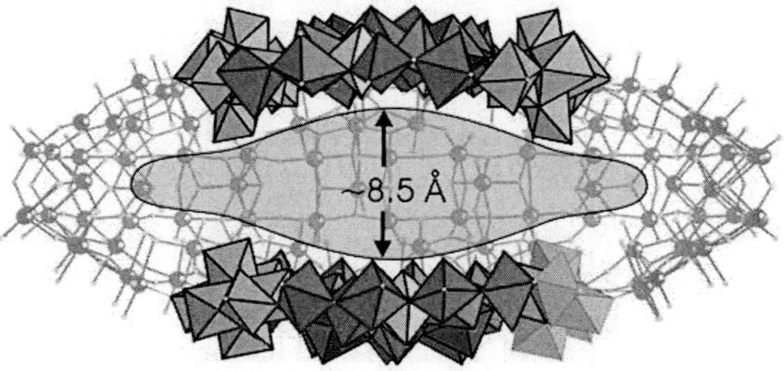


Figure 6: View into the cavity of the [$Mo_{248}$] POM/MOB cluster. The hub-caps are in polyhedral representation and the periphereal ring region is shown in ball-and-stick representation (for simplicity, the front part of the periphereal ring area is omitted). Color code as in Figure 4.

Following the building block principle, the [$Mo_{248}$] POM/MOB cluster can be formulated as {[$Mo_{176}$] [$Mo_{36}$]$_2$} ≡ {[$Mo_{11}$]$_{16}$ [$Mo_{36}$]$_2$} ≡ {[$Mo_2$]$_{16}$ [$Mo_8$]$_{16}$ [$Mo_1$]$_{16}$ [$Mo_{36}$]$_2$]. The corresponding explicit formula is $\{[Mo^{VI}_2O_5(H_2O)_2]_{16}$ $[Mo^{VI/V}_8O_{26}(\mu_3\text{-}O)_2H(H_2O)_3Mo^{VI/V}]_{16}]$ $[Mo^{VI/V}_{36}O_{96}(H_2O)_{24}]_2\}^{16-}$.

## [$Mo_{256}Eu_8$]

In the same 2002 year, the Bielefeld team reported[13] a second mixed-valence POM/MOB cluster containing lanthanoide cations, with the formula $[Mo_{256}Eu_8O_{776}H_{20}(H_2O)_{162}]^{20-}$ $\equiv$ $[Mo^{VI}_{208}Mo^{V}_{48}Eu^{III}_8O_{776}H_{20}(H_2O)_{162}]^{20-}$, abridged as [$Mo_{256}Eu_8$] or [$Mo^{VI}_{208}Mo^{V}_{48}Eu^{III}_8$].

The new cluster contains two elliptical ring-shaped units, linked together covalently by Eu-O-Mo bonds. Each unit can be formulated as $[Mo_{128}Eu_4O_{388}H_{10}(H_2O)_{81}]^{10-}$ $\equiv$ $[Mo^{VI}_{104}Mo^{V}_{24}Eu^{III}_4O_{388}H_{10}(H_2O)_{81}]^{10-}$, abridged as [$Mo_{128}Eu_4$] or [$Mo^{VI}_{104}Mo^{V}_{24}Eu^{III}_4$]. Although a discrete molecule, made up of two very large identical units, the whole cluster was considered a “dimer” and written alternatively as $[[Mo_{128}Eu_4O_{388}H_{10}(H_2O)_{81}]_2]^{20-}$ or [[$Mo_{128}Eu_4$]$_2$]. The interpretation proposed by the authors[13] considers that the “dimeric” cluster is formed by $Eu^{III}$ centers acting as symmetry breakers, which prevent the usual ring closure observed for the parent [$Mo_{154}$] and [$Mo_{176}$] giant wheels.[9],[10]

The structure of the [$Mo_{256}Eu_8$] POM/MOB cluster is displayed in Figure 7.

For reasons of simplicity, the structural and topological discussions will refer only to a single “monomeric” [$Mo_{128}Eu_4$] cluster unit. This large cluster unit includes 128 Mo and 4 Eu centers and can be related, as the previously described [$Mo_{120}Pr_6$] cluster,[8] to the hypothetical “dodecameric” [$Mo_{132}$] $\equiv$ [$Mo^{VI}_{108}Mo^{V}_{24}$] down-sized giant wheel. In this topological consideration, 4 [$Eu_1$] units replace 2 [$Mo_2$] units of the [$Mo_{132}$] down-sized giant wheel. Consequently, the [$Mo_{128}Eu_4$] cluster unit can be formulated, according to the building block principle, as $\{[Mo_2]_{10}$ $[Eu_1]_4$ $[Mo_8]_{12}$ $[Mo_1]_{12}\}$.

Thus, it is demonstrated once again that the presence of strong electrophilic open shell metal centers, such as lanthanoide cations, generates smaller (“dodecameric”) ring-shaped species, as compared to the parent (“tetra-“ and “hexadecameric”) giant wheels. It should be stressed that $Eu^{3+}$ ions are smaller than $Pr^{3+}$ ions and they can replace only half of a [$Mo_2$] unit. In the presence of $Eu^{3+}$ ions, the dramatic increase of the curvature destabilizes the system and ring closure does not take place any more. Thus, two ring-shaped units are linked together to build the stable discrete cluster molecule.

The [$Mo_{128}Eu_4$] unit has a ring-shaped structure with an elliptical curvature. Its external diameter is of $\approx$ 3.8 nm at the most elongated points and $\approx$ 3.0 nm at the most compressed points. The diameter of the large internal cavity is in the range of $\approx$ 1.2 – 2.0 nm.

A more insightful observation of the building blocks forming the [$Mo_{128}Eu_4$] ring reveals several differences relative to parent giant wheels with basic structures. Thus, it comprises not only the “classical” [$Mo_2$], [$Mo_8$] and [$Mo_1$] units, but also several additional units, namely [$Mo_2$]*, [$Mo_7$], [$Mo_9$] and, obviously, [$Eu_1$] units. This new topological consideration starts formally from the parent [$Mo_{154}$] giant wheel, which would generate the [$Mo_{128}Eu_4$] ring by a “cutting process” with a “molecular scissors”.

The construction of the [$Mo_{128}Eu_4$] ring from the parent [$Mo_{154}$] giant wheel by a “cutting” process is shown in Figure 8.

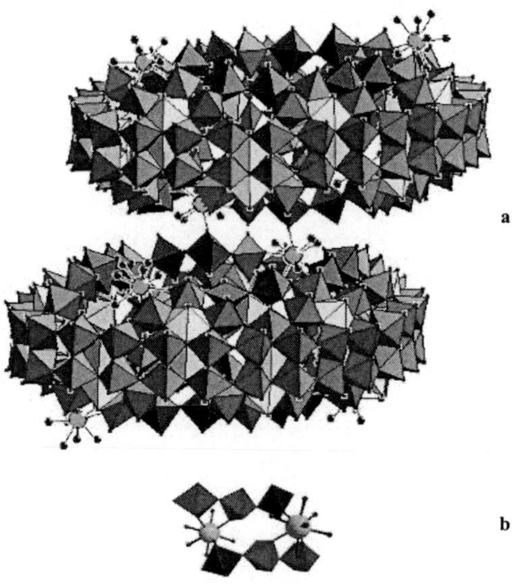


Figure 7: Structure of the $[Mo_{256}Eu_8O_{776}H_{20}(H_2O)_{162}]^{20-} \equiv [Mo_{256}Eu_8]$ POM/MOB cluster in combined representation, with the molybdenum oxide-based units displayed as polyhedra and the $Eu^{III}$ coordination spheres in ball-and-stick representation (a). Expanded view of the Eu-O-Mo groups which link the two rings (b). Color code: $[Mo_1]$ units yellow polyhedra; $[Mo_2]$ units red polyhedra; $[Mo_8]$ units blue polyhedra, with the central $MoO_7$ pentagonal bipyramid in blue-turqoise; $Eu^{III}$ green spheres; O small red spheres; $H_2O$ large red spheres.

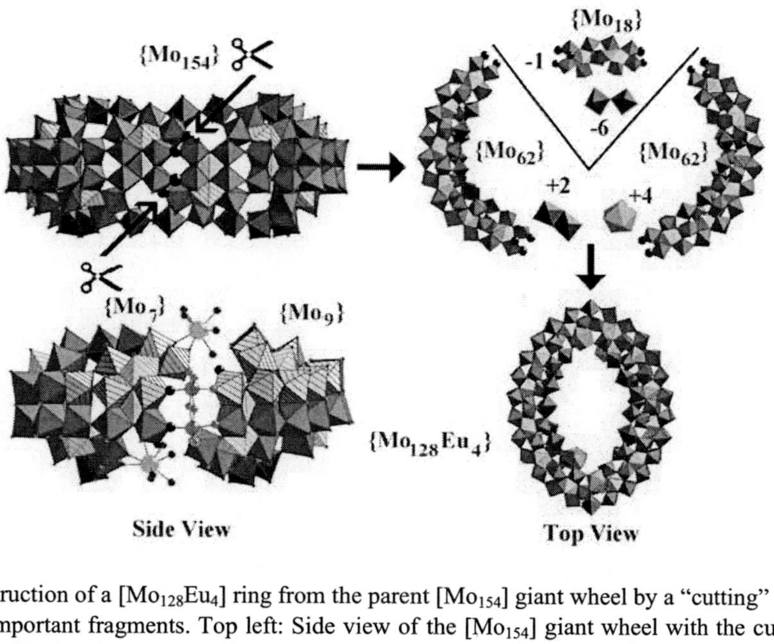


Figure 8: Construction of a $[Mo_{128}Eu_4]$ ring from the parent $[Mo_{154}]$ giant wheel by a "cutting" process giving the two large important fragments. Top left: Side view of the $[Mo_{154}]$ giant wheel with the cutting positions marked as large black spheres; top right: the units which have to be removed from the $[Mo_{154}]$ giant wheel and those which have to be added to the resulting two large fragments (left and right) to generate the $[Mo_{128}Eu_4]$ ring; bottom: the $[Mo_{128}Eu_4]$ ring from a side and top view (color code as in Figure 7) with the the new $[Mo_2]^*$ units in brown which are shown in the side view together with the $EuO_9$ polyhedra in ball-and-stick representation; the $MoO_6$ octahedra of a selected $[Mo_7]$ and a $[Mo_9]$ unit respectively, as hatched violet polyhedra.

The "cutting" process at four positions takes place between the polyhedra with the least number of bonds. The two resulting fragments include the new $[Mo_7]$ and $[Mo_9]$ units, generated formally by "cutting" asymmetrically" two $[Mo_8]$ units.

Each fragment can be formulated as $\{[Mo_1]_6\ [Mo_2]_4\ [Mo_8]_2\ [Mo_7]_2\ [Mo_9]_2\}$. The complete ellipsoidal $[Mo_{128}Eu_4]$ ring is formed when the two fragments are combined with 2 $[Mo_2]^*$ and 4 $[Eu_1]$ units. The explicit formulae of the classical $[Mo_2] \equiv [Mo^{VI}{}_2O_5(H_2O)_2]^{2+}$ and the new mixed-valence $[Mo_2]^* \equiv [Mo^{VI}Mo^{V}O_7(H_2O)]^{3-}$ units are different. The $Eu^{3+}$ centers exhibit a high number of coordination, *i.e.*, NC = 9, in the 6 $[Eu^{III}O_2(H_2O)_7]^-$ and 2 $[Eu^{III}O_3(H_2O)_6]^{3-}$ units. Considering also the new building blocks, the overall topological formulation of the $[Mo_{128}Eu_4]$ ring is $\{[Mo_1]_{12}\ [Mo_2]_8\ [Mo_8]_4\ [Mo_7]_4\ [Mo_9]_4\ [Mo_2]^*{}_2\ [Eu_1]_4\}$.

The $[Mo_{256}Eu_8]$ cluster molecule, which contains two linked $[Mo_{128}Eu_4]$ rings, has twice such units as compared to the ring corresponding to the formula of $\{[Mo_1]_{24}\ [Mo_2]_{16}\ [Mo_8]_8\ [Mo_7]_8\ [Mo_9]_8\ [Mo_2]^*{}_4\ [Eu_1]_8\}$.

## 2.2. Hybrid Structures

We designate as hybrid the structures that include elements specific to at least two basic structures. In present, only one hybrid-type structure is known, namely:

$[Mo_{368}]$ (*Müller structure*), with 368 metal centers, a record-size molecule.

## $[Mo_{368}]$ (Müller Structure)

In November 2001, Müller and his team synthesized and characterized an exceptionally large molecule; this discovery was disclosed to the scientific community a few months later in a paper with the exciting title: *Inorganic Chemistry Goes Protein Size.*[14] The new molecule is a huge POM/MOB cluster, having the size of hemoglobine, with the formula of $[H_xMo^{VI/V}{}_{368}O_{1032}(H_2O)_{240}(SO_4)_{48}]^{48-}$ ($x \approx 16$), abridged as $[Mo_{368}]$. The $[Mo_{368}]$ cluster was also called giant hedgehog or blue lemon.

The giant hedgehog-like molecule, which has approximately $D_4$ symmetry, consists of a central $[Mo_{288}O_{784}(H_2O)_{192}(SO_4)_{32}]$ spherical fragment with $D_{8d}$ symmetry and two $[Mo_{40}O_{124}(H_2O)_{24}(SO_4)_8]$ capping fragments with $C_{4v}$ symmetry. Thus, it can also be formulated as $\{[Mo_{288}]\ [Mo_{40}]_2\}$.

We propose here the name Müller structure to honor Professor Achim Müller's outstanding activity in the field; he is the leader of the Bielefeld research team and the father of the new generation of giant POM/MOB clusters with supramolecular properties.

The Müller structure of the $[Mo_{368}]$ POM/MOB cluster is displayed in Figures 9, 10, 11, 12 and 13.

By defining the formal building blocks/units, the $[Mo_{368}]$ cluster can also be formulated as $\{[Mo(Mo_5)]^*{}_8\ [Mo(Mo_5)]^{**}{}_{32}\ [Mo_2]^*{}_{16}\ [Mo_2]^{**}{}_8\ [Mo_2]^{***}{}_8\ [Mo_1]_{64}\}$, where: $[Mo(Mo_5)]^* \equiv [Mo_6O_{21}(H_2O)_6]$; $[Mo(Mo_5)]^{**} \equiv [Mo_6O_{21}(H_2O)_3(SO_4)]$; $[Mo_2]^* \equiv [Mo_2O_3(H_2O)_2]$; $[Mo_2]^{**} \equiv [Mo_2O_{(t)2}O_{(br)3}(SO_4)]$; $[Mo_2]^{***} \equiv [Mo_2O_{(t)4}O_{(br)}(SO_4)]$; $[Mo_1] \equiv [MoO(H_2O)]$; $O_{(t)}$ = terminal oxygen; $O_{(br)}$ = bridging oxygen).

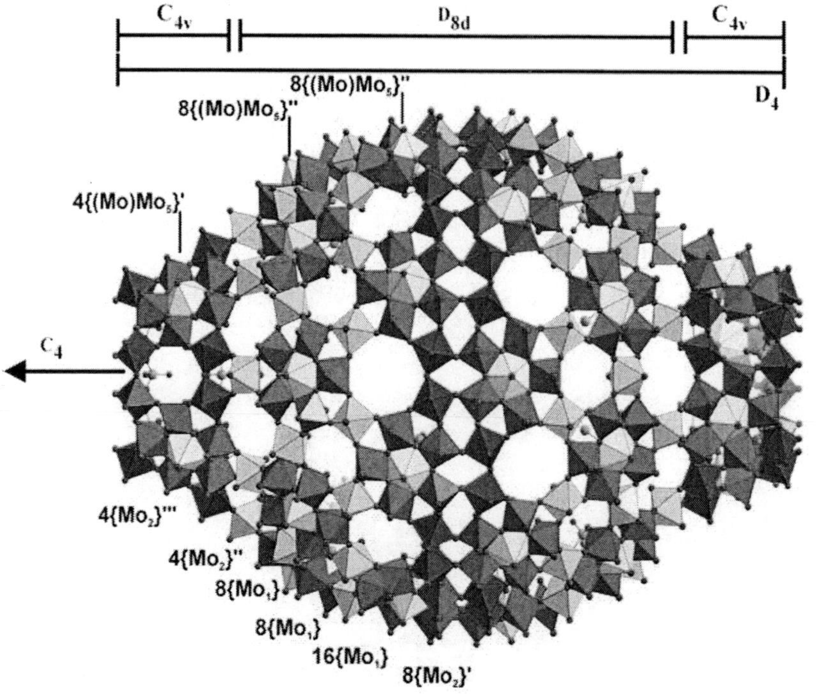


Figure 9: Müller structure of the $[H_xMo_{368}O_{1032}(H_2O)_{240}(SO_4)_{48}]^{48-}$ ≡ $[Mo_{368}]$ POM/MOB cluster in polyhedral representation perpendicular to the $C_4$ axis. Color code: $[Mo_1]$ units yellow, $[Mo_2]$ units red, $[(Mo)Mo_5]$ units blue, with the central $MoO_7$ pentagonal bipyramid in blue-turquoise.

Figure 10: Müller structure of the $[H_xMo_{368}O_{1032}(H_2O)_{240}(SO_4)_{48}]^{48-}$ ≡ $[Mo_{368}]$ POM/MOB cluster in polyhedral representation along the $C_4$ axis. Color code as in Figure 9.

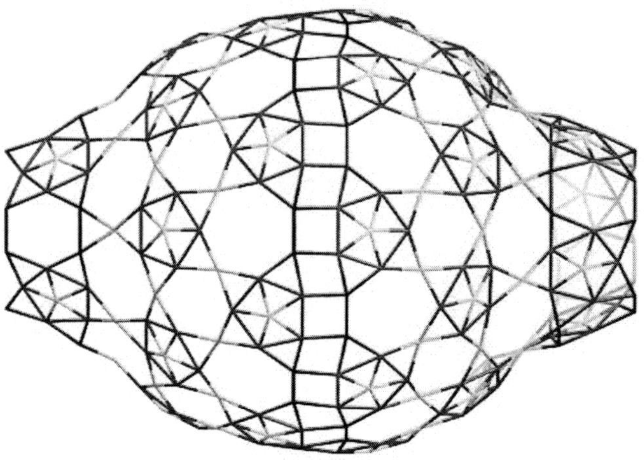

Figure 11: Müller structure of the $[H_xMo_{368}O_{1032}(H_2O)_{240}(SO_4)_{48}]^{48-}$ ≡ $[Mo_{368}]$ POM/MOB cluster in wire-frame representation perpendicular to the $C_4$ axis. Color code as in Figure 9.

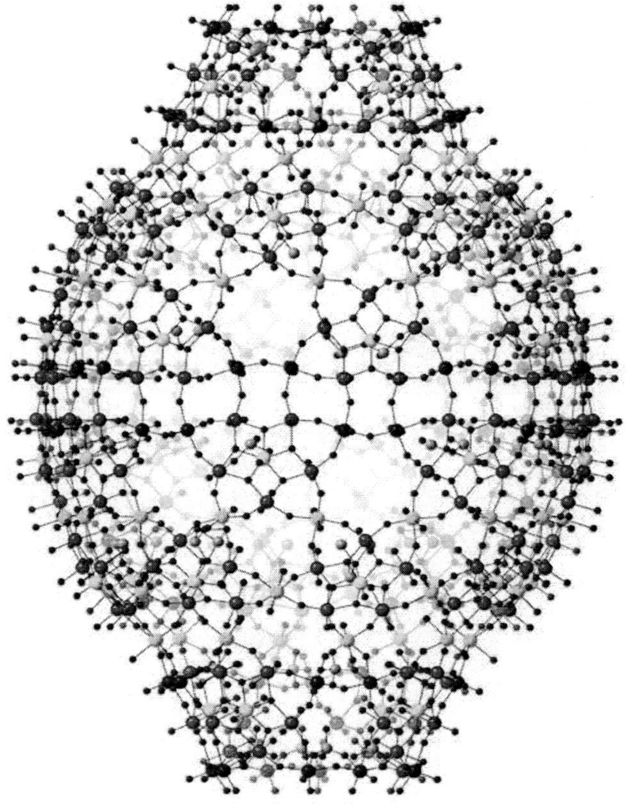

Figure 12: Müller structure of the $[H_xMo_{368}O_{1032}(H_2O)_{240}(SO_4)_{48}]^{48-}$ ≡ $[Mo_{368}]$ POM/MOB cluster in ball-and-stick representation perpendicular to the $C_4$ axis. Color code: Mo of the $[Mo_1]$ units yellow; Mo of the $[Mo_2]$ units red; central Mo of the $[(Mo)Mo_5]$ units blue-turquoise; periphereal Mo of the $[(Mo)Mo_5]$ units blue; S grey; O red.

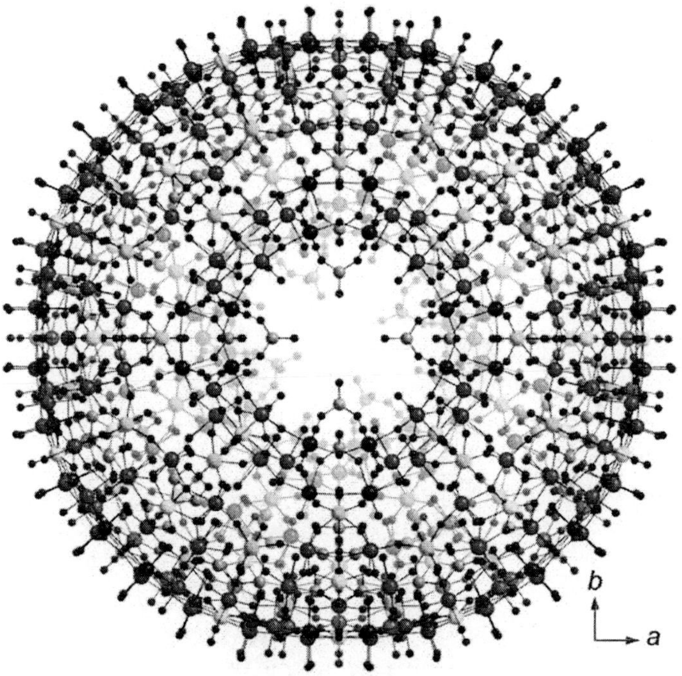


Figure 13: Müller structure of the $[H_xMo_{368}O_{1032}(H_2O)_{240}(SO_4)_{48}]^{48-}$ ≡ $[Mo_{368}]$ POM/MOB cluster in ball-and-stick representation along the $C_4$ axis. Color code as in Figure 12.

The two types of $[Mo(Mo_5)]$ units differ in the presence or absence of the sulfate ligands. On their turn, the three types of $[Mo_2]$ units differ in the number of bidentate sulfate ligands and of terminal and bridging oxygen atoms. The $[Mo_1]$ units, with only $Mo^V$ centers, are classical and contribute to the highly reduced state of the molecule. The other formal $Mo^V$ centers are distributed over several parts of the cluster area, which determines a widespread delocalization of the Mo 4d electrons.

The highly reduced state of this edifice, which contains *ca.* 256 $Mo^{VI}$ and 112 $Mo^V$ centers, demonstrates that strong reducing conditions favor molecular growth.

The $[Mo_{368}]$ POM/MOB cluster contains 24 $C_5$-type $[Mo_{11}]$ units, which are identical to those of the $[Mo_{132}]$ complete giant sphere[5] and 16 $C_s$-type $[Mo_{10}]$ units, which are similar to the $[Mo_{11}]$ units of the $[Mo_{154}]$ [9] and $[Mo_{176}]$ [15] giant wheels, but they contain one octahedron less. It also includes $[Mo_1]$ units, which are identical to the mononuclear linkers found in the $[Mo_{102}]$ down-sized giant sphere.[15] Thus, the Müller structure of the $[Mo_{368}]$ cluster can be regarded as a hybrid between the structures of the $[Mo_{102}]$/$[Mo_{132}]$ giant spheres and of the $[Mo_{154}]$/$[Mo_{176}]$ giant wheels.

The $[Mo_{368}]$ cluster can also be considered a giant container, with a huge cavity (diameter ≈ 2.5 x 4.0 nm at its most extended points), offering room for about 400 $H_2O$ molecules which are encapsulated.

Having 368 metal centers, no less than 1880 non-hydrogen atoms, a relative molecular mass of 60,769 and an external diameter of ≈ 6.0 nm, the $[Mo_{368}]$ POM/MOB cluster is currently the largest synthesized molecule with determined structure.

## 3. Conclusions

We are confident that, this visit to the realm of nanosized metal-oxygen clusters, also called POM/MOB clusters, with basic or special structures, continue to fascinate the chemists. Furthermore, it should be stressed that the relevance of nanosized clusters under discussion goes far beyond the frontiers of chemistry.

All cluster molecules presented in this review have charmingly beautiful structures, hence they may become a source of inspiration for creators of beauty. We believe that the representations of nanosized metal-oygen clusters, especially those of giant sphere- and giant wheel-type molecules with archetypal structures, can be considered as true icons[16] of current natural science. Unlike traditional icons, the objects of worship and adoration for the Christian believer, which embody the Godhead or sacred Biblical characters and whose archetypes are to be found in the subjective reality of consciousness, the new icons, namely the giant spheres, giant wheels and other related marvelous clusters, stand for concrete actual models localized in the here-and-now physical world. More precisely, the molecular patterns (with dimensions of the order of nanometers) belong to a meso-physical world or nano-world and are to be found somewhere in between the microscopic and the macroscopic realm.

All these molecules have for their main constituent flexible pentagonal building blocks (as such or included in larger units), which reproduce the geometrical shape preferred by Pythagoras and also the sign of identification and the secret symbol of the Pythagorean school. Thus, the Pythagorean harmony[17] is reflected in the nano-sized world, dominated by giant metal-oxygen clusters.

It might lack in fairness to restrict our analogy to visual representations stemming from Western spiritual or philosophical traditions, skipping the ancient spiritual traditions of the East. In this respect, the representations of giant spheres, giant wheels and other metal-oxygen clusters could be related to the well-known mandalas, which are images of an abstract geometrical character, who attempt to create an ultimate illusory synthesis between the world of immanent material reality and the transcendental spiritual realm, respectively between the microcosmos represented by the human being and the macrocosmos represented by the holistic universe.

As any mandala, each metal-oxygen cluster molecule appears to possess its own iconography. Mandalas are images and objects for focusing, concentration and meditation for the disciple in search of a so-called spiritual evolution, with the final goal to reintegrate the part in the whole from which it originated. On the other hand, the new icons or mandalas represented by giant spheres, giant wheels and related molecules may become images and subjects of reflection and meditation on the extraordinary abilities of human beings and especially on the human intelligence/mind, whose accomplishments have come to rival the most subtle and exquisite masterpieces of Nature.

## Acknowledgement

The authors would like to thank Dr. Hartmut Bögge (Bielefeld University) for providing the figures included in the paper. The authors are also grateful to Dr. Daniel A. Lowy (Naval Research Laboratory, Washington, DC) for critically reading the manuscript and useful suggestions.

## REFERENCES

[1] L. Cronin, A. Müller, D. Fenske, in *Encyclopedia of Physical Science and Technology* (ed. R.A. Meyers), 3rd Ed., Vol. 10, Academic Press, New York 2002.

[2] A. Patrut, A. Nicoara, D. Margineanu, *Studia Univ. Babes-Bolyai, Ser. C, Chemia*, 2003, 48, 3.

[3] A. Müller, M. Koop, H. Bögge, M. Schmidtmann, F. Peters, P. Kögerler, *Chem. Commun.*, 1999, 1885.

[4] A. Müller, P. Kögerler, H. Bögge, *Struct. Bond.*, 2000, 96, 203.

[5] A. Müller, S. Sarkar, S.Q.N. Shah, H. Bögge, M. Schmidtmann, S. Sarkar, P. Kögerler, B. Hauptfleisch, A.X. Trautwein, V. Schünemann, *Angew. Chem. Int. Ed.*, 1999, 38, 3238.

[6] A. Müller, S.K. Das, M.O. Talismanova, P. Kögerler, S.S. Talismanov, M. Luban, E. Krickemeyer, *Angew. Chem.*, 2002, 114, 599.

[7] A. Müller, B. Botar, H. Bögge, P. Kögerler, A. Berkle, *Chem. Commun.*, 2002, 2944.

[8] A. Müller, C. Beugholt, H. Bögge, M. Schmidtmann, *Inorg. Chem.*, 2000, 39, 3112.

[9] A. Müller, E. Krickemeyer, J. Meyer, H. Bögge, F. Peters, W. Plass, E. Diemann, S. Dillinger, F. Nonnenbruch, M. Randerath, C. Menke, *Angew. Chem. Int. Ed.*, 1995, 34, 2122.

[10] A. Müller, R. Maiti, M. Schmidtmann, H. Bögge, S.K. Das, W. Zhang, *Chem. Commun.*, 2001, 2126.

[11] A. Müller, E. Krickemeyer, H. Bögge, M. Schmidtmann, F. Peters, *Angew. Chem. Int. Ed.*, 1998, 37, 3360.

[12] A. Müller, M. Koop, H. Bögge, M. Schmidtmann, C. Beugholt, *Chem. Commun.*, 1998, 1501.

[13] L. Cronin, C. Beugholt, E. Krickemeyer, M. Schmidtmann, H. Bögge, T.K.K. Luong, A. Müller, *Angew. Chem.*, 2002, 114, 2929.

[14] A. Müller, E. Beckmann, H. Bögge, M. Schmidtmann, A. Dress, *Angew. Chem.*, 2002, 114, 1210.

[15] A. Müller, S.Q.N. Shah, H., Bögge, M. Schmidtmann, P. Kögerler, B. Hauptfleisch, S. Leiding, K. Wittler, *Angew. Chem. Int. Ed.*, 2000, 30, 1614.

[16] A. Patrut, A. Nicoara, D. Margineanu, A. Botar, *Acta Univ. Cibinensis, Ser. F., Chemia*, 2002, 4, 1.

[17] A. Müller, in *Facetten einer Wissenschaft* (eds. A. Müller, H.-J.Quadbeck-Seeger and E. Diemann), Wiley, Weinheim, 2004.

# INDEX

## A



## B



## C

## D



## E



## F

## G



## H



## I



## K



## L



## M

## N



## O



## P



## Q

## R



## S



## T



## U

## V



## W



## X



## Z